GAS TRANSPORT IN POROUS MEDIA

AF768582

Theory and Applications of Transport in Porous Media

Series Editor:
Jacob Bear, *Technion – Israel Institute of Technology, Haifa, Israel*

Volume 20

The titles published in this series are listed at the end of this volume.

Gas Transport in Porous Media

Edited by

Clifford K. Ho
Sandia National Laboratories, P. O. Box 5800, Albuquerque, NM 87185, USA

and

Stephen W. Webb
Sandia National Laboratories, P. O. Box 5800, Albuquerque, NM 87185, USA

Springer

A C.I.P. Catalogue record for this book is available from the Library of Congress.

ISBN-13 978-90-481-7002-9
ISBN-10 1-4020-3962-X (e-book)
ISBN-13 978-1-4020-3962-1 (e-book)

Published by Springer,
P.O. Box 17, 3300 AA Dordrecht, The Netherlands.

www.springer.com

Printed on acid-free paper

© 2010 Springer
All Rights Reserved for Chapter 12
No part of this work may be reproduced, stored in a retrieval system, or transmitted in any form or by any means, electronic, mechanical, photocopying, microfilming, recording or otherwise, without written permission from the Publisher, with the exception of any material supplied specifically for the purpose of being entered and executed on a computer system, for exclusive use by the purchaser of the work.

TABLE OF CONTENTS

CHAPTER 1

INTRODUCTION

CLIFFORD K. HO AND STEPHEN W. WEBB
Sandia National Laboratories, P. O. Box 5800, Albuquerque, NM 87185, USA

Gas and vapor transport in porous media occur in a number of important applications including drying of industrial and food products, oil and gas exploration, environmental remediation of contaminated sites, and carbon sequestration. Understanding the fundamental mechanisms and processes of gas and vapor transport in porous media allows models to be used to evaluate and optimize the performance and design of these systems.

In this book, gas and vapor are distinguished by their available states at standard temperature and pressure (20°C, 101 kPa). If the gas-phase constituent can also exist as a liquid phase at standard temperature and pressure (e.g., water, ethanol, toluene, trichlorothylene), it is considered a vapor. If the gas-phase constituent is non-condensable at standard temperature and pressure (e.g., oxygen, carbon dioxide, helium, hydrogen, propane), it is considered a gas. The distinction is important because different processes affect the transport and behavior of gases and vapors in porous media. For example, mechanisms specific to vapors include vapor-pressure lowering and enhanced vapor diffusion, which are caused by the presence of a gas-phase constituent interacting with its liquid phase in an unsaturated porous media. In addition, the "heat-pipe" exploits isothermal latent heat exchange during evaporation and condensation to effectively transfer heat in designed and natural systems.

The intent of this book is to expose the reader to a variety of important studies of gas and vapor transport in porous and fractured media. The primary focus is on the presentation of fundamental processes, state-of-the-art modeling, experiments, and applications that are relevant to gas and vapor transport in porous and fractured media. The topics in this book span multiple disciplines, ranging from soil science to engineering. This has been done intentionally to integrate the broad audience in this subject area and to provide a compilation of common areas of research. Historically, the treatment of gas and vapor transport processes in porous and fractured media has been segregated according to disciplines or journals. Approaches to understanding these processes have evolved in soil science for many decades, but there has been

C. Ho and S. Webb (eds.), Gas Transport in Porous Media, 1–2.
© 2006 *Springer.*

relatively little cross-fertilization of these studies and findings into other disciplines such as engineering. Some processes such as enhanced vapor diffusion have been investigated by researchers in soil sciences as far back as the 1950s and 1960s, but they are only recently being considered in engineering applications such as drying of porous materials.

Another motivation for this book is to identify applications where gas and vapor transport are important or dominant mechanisms. Often, the importance of gas and vapor transport mechanisms is overlooked or overshadowed by studies of liquid-flow processes. However, advances in numerical modeling and experimental methods have allowed the simulation of coupled gas- and liquid-flow processes in complex media, and we are now able to distinguish the relative importance of mechanisms in various applications. Examples of recent applications in which gas and vapor transport processes are significant include nuclear waste disposal in geologic media and detection of unexploded ordnance in the subsurface.

This book is divided into three parts: Part 1 – Processes and Models; Part 2 – Measurement and Monitoring; and Part 3 – Applications. The first part, Processes and Models, presents fundamental processes associated with gas and vapor transport in porous media. Beginning with gas transport mechanisms, it describes advection and diffusion processes, including the Dusty-Gas Model. Vapor transport processes are then described in a similar manner, illustrating important features of a condensable gas in porous media such as enhanced vapor diffusion and vapor-pressure lowering. Vapor-solid sorption is also discussed because of its importance to recent applications in land-mine detection, and mechanisms involving evaporation and coupled processes are presented. The impact of heterogeneities and scaling on gas and vapor transport processing in porous and fractured media is also discussed. Two-phase processes and characteristics are also described, and various models of gas and vapor transport processes using continuum and Lattice Boltzmann models are presented.

In Part 2, Measurement and Monitoring, various methods are described that have been used to measure gas and vapor transport processes and parameters at the laboratory and field scales. Measurement of the diffusion coefficient, permeability, flow rate, constituent concentration, and mass flux of gas and vapors are described. New technologies and microelectronic sensors that measure gas-phase volatile organic compounds are also introduced.

Finally, in Part 3, Applications of gas and vapor transport in porous and fractured media are presented. Applications include radon transport, landmine detection, environmental remediation, geologic waste disposal, oil and gas exploration, carbon sequestration, and industrial processes.

This book provides a broad and interdisciplinary view of the different processes, models, experimental methods, and applications associated with gas and vapor transport in porous media. We hope that the reader develops an understanding of the many diverse topics and an appreciation for the important applications covered in this book.

PART 1

PROCESSES AND MODELS

CHAPTER 2

GAS TRANSPORT MECHANISMS

STEPHEN W. WEBB

Sandia National Laboratories, P. O. Box 5800, Albuquerque, NM 87185, USA

Gas-phase momentum transport in porous media consists of advective and diffusive components. In this chapter, the individual advective and diffusive components will be presented separately first, followed by a discussion of the combined mechanisms. Gas-only situations will be discussed for all the mechanisms. Two-phase, or unsaturated, flow effects are included in Chapter 5.

The conservation equations presented below are given in a simplified form. For a complete derivation of the various conservation equations, including underlying assumptions, see Whitaker (Chapter 6 of this book).

Energy transport is not discussed in this chapter. Gas-phase energy transport in porous media is treated by Plumb (Chapter 27 of this book). Energy transport is also discussed in Nield and Bejan (1999) and Kaviany (1995).

2.1 GAS-PHASE ADVECTION

2.1.1 Darcy's Law

Gas-phase advection in porous media is generally analyzed using Darcy's law (Darcy, 1856), which simply states that the gas Darcy velocity, u_g, is directly proportional to the gas-phase pressure gradient, ∇P_g, and the gas-phase permeability, k_g. Darcy's law can be written as

$$\overline{u}_g = -\frac{k_g}{\mu_g}\left(\nabla P_g - \rho_g \overline{g}\right)$$

where μ_g is the gas-phase viscosity and $\overline{g}$ is the gravitational constant. In terms of mass flux, the equation is

$$\overline{F}_g = \rho_g \overline{u}_g = -\frac{k_g}{\mu_g}\rho_g\left(\nabla P_g - \rho_g \overline{g}\right)$$

C. Ho and S. Webb (eds.), Gas Transport in Porous Media, 5–26.
© 2006 *Springer.*

Neglecting gravity and rearranging

$$\nabla P_g = -\frac{\mu_g}{k_g}\,\overline{u}_g$$

Note that the Darcy velocity, u_g, is not a physical velocity. Rather, it is a superficial velocity based on the entire cross section of the flow, not just the fluid flow cross-section. The Darcy velocity is related to the pore velocity, V_g, through the porosity, ϕ, or

$$V_g = \frac{u_g}{\phi}$$

The gas-phase permeability, k_g, is a proportionality constant that is usually experimentally determined with units of length^2. The gas-phase permeability may be slightly different than the liquid-phase permeability due to the effects of the fluids. Values of the liquid-phase permeability vary widely, from 10^{-7} to 10^{-9} m^2 for clean gravel down to 10^{-18} to 10^{-20} m^2 for granite (Bear, 1979, pg. 68). The unit Darcy is often used, which is equal to 0.987×10^{-12} m^2.

Darcy's law is applicable to low velocity flow, which is generally the case in porous media flow, and to regions without boundary shear flow, such as away from walls. When wall shear is important, the Brinkman extension can be used as discussed below. For turbulent flow conditions, the Forchheimer equation is appropriate. In some situations (e.g., Vafai and Tien, 1981), the Brinkman and Forchheimer equations are both employed for a more complete momentum equation. For a more detailed discussion of the various flow laws, see Nield and Bejan (1999), Kaviany (1995), or Lage (1998).

2.1.2 Brinkman Extension

The Brinkman extension to Darcy's law equation includes the effect of wall or boundary shear on the flow velocity, or

$$\nabla P_g = -\frac{\mu_g}{k_g}\,\overline{u}_g + \tilde{\mu}\,\nabla^2\overline{u}_g$$

where gravity has been ignored for clarity. The first term on the RHS is immediately recognizable as the Darcy expression, while the second term is a shear stress term such as would be required by a boundary wall no-slip condition. The coefficient $\tilde{\mu}$ is an effective viscosity at the wall, which in general is not equal to the gas viscosity, μ_g, as discussed by Nield and Bejan (1999). For many situations, the use of the boundary shear term is not necessary. The effect is only significant in a region close to the boundary whose thickness is of order of the square root of the gas permeability, $k_g^{1/2}$ (assuming $\tilde{\mu} = \mu_g$), so for most applications, the effect can be ignored.

The Brinkman equation is also often employed at the interface between a porous media and a clear fluid, or a fluid with no porous media, in order to obtain continuity of shear stress. This interfacial condition is discussed in more detail by Nield and Bejan (1999) and Kaviany (1995).

2.1.3 Forchheimer Extension

At low pore velocities, Darcy's law works quite well. However, as the pore velocities increase, the flow becomes turbulent, the flow resistance becomes non-linear, and the Forchheimer equation is more appropriate. From Joseph et al. (1982), the Forchheimer equation is

$$\nabla P_{\mathrm{g}} = -\frac{\mu_{\mathrm{g}}}{k_{\mathrm{g}}}\,\overline{u}_{\mathrm{g}} - c_{\mathrm{F}}\,k_{\mathrm{g}}^{-1/2}\,\rho_{\mathrm{g}}\left|\overline{u}_{\mathrm{g}}\right|\overline{u}_{\mathrm{g}}$$

where c_{F} is a constant and gravity has been ignored. The first term on the RHS is again immediately recognizable as Darcy's law. The second term on the RHS is a non-linear flow resistance term. According to Nield and Bejan (1999), the above equation is based on the work of Dupuit (1863) and Forchheimer (1901) as modified by Ward (1964). The value of c_{F} is approximately 0.55 based on the work of Ward (1964). However, later work indicates that c_{F} is a function of the porous medium and can be as low as 0.1 for foam metal fibers as summarized by Nield and Bejan (1999). In addition, Beavers et al. (1973) showed that bounding walls can change the value of c_{F} significantly.

The above equation can be rearranged in terms of a permeability-based Reynolds number, where the characteristic dimension is the square root of the gas permeability, or

$$\mathrm{Re}_k = \frac{\rho_{\mathrm{g}}\,u_{\mathrm{g}}\,k_{\mathrm{g}}^{1/2}}{\mu_{\mathrm{g}}}$$

The Forchheimer equation can be rearranged in terms of the value of c_{F} and the Reynolds number, or

$$\nabla P_{\mathrm{g}} \propto \left[\frac{1}{\mathrm{Re}_k} + c_{\mathrm{F}}\right]$$

According to Nield and Bejan (1999), the transition from Darcy's law ($c_{\mathrm{F}} = 0.$) to the above Forchheimer equation occurs in the permeability-based Reynolds number range of 1 to 10. Note that this transition is based on liquid flow through an isothermal liquid-saturated porous medium, not an all-gas system. At low Reynolds numbers, Darcy's law is recovered ($c_{\mathrm{F}} \ll 1/\mathrm{Re}_k$). As the Reynolds number increases, the pressure drop increases above that predicted by Darcy's law. For further details, see the discussion in Nield and Bejan (1999).

More recently, porous media approaches have been developed that include a two-equation turbulence model similar to that used in clear fluid computational fluid dynamics codes as exemplified by Masuoka and Takatsu (1996), Antohe and Lage (1997), and Getachew et al. (2000).

2.1.4 Low Permeability Effects

Gas advection through porous media can be idealized as flow through numerous capillary tubes. For large capillary tubes, the gas molecular mean free path is much smaller than the radius, and continuum flow occurs. As the capillary tubes get smaller and smaller, the gas molecular mean free path becomes of the same order, and free-molecule, or Knudsen, diffusion becomes important.

Low permeability effects were experimentally investigated by Knudsen in 1909 for gas flow in capillaries as discussed by Cunningham and Williams (1980, pg. 49). Based on Darcy's law, the mass flux for a given pressure drop should decrease as the average pressure is reduced due to the change in gas density. However, Knudsen found that at low pressures, the mass flux reaches a minimum value and then increases with decreasing pressure, which is due to slip, or the fact that the fluid velocity at the wall is not zero due to free-molecule flow.

Klinkenberg (1941) derived an expression for the effective gas permeability, k_g, of a single gas in the Knudsen diffusion regime, which is a function of the liquid permeability, k_λ, the average pressure, $\overline{P}$, and the Klinkenberg coefficient of gas i, b_i, or

$$k_{g,i} = k_\ell \left(1 + \frac{b_i}{\overline{P}_g}\right)$$

For a large average pressure, the correction factor in parentheses goes to zero, and the gas and liquid permeabilities tend to become equal. As the average pressure decreases, the two permeabilities can deviate significantly from each other. This behavior is confirmed by data presented by Klinkenberg (1941) for glass filters and core samples and by Reda (1987) for tuff. The Klinkenberg parameter for a given porous medium can be derived by plotting the effective gas permeability as a function of the inverse of the average pressure. The slope of the line is related to the Klinkenberg parameter, and the intercept at zero inverse average pressure is the liquid permeability.

The Klinkenberg coefficient, b_i, is a function of the porous medium, the gas, and the temperature. The Klinkenberg coefficient for air can be estimated from the Heid et al. (1950) correlation for air at 25°C as a function of permeability (Thorstenson and Pollock, 1989a, Figure 3), or

$$b_{air} = 0.11\, k_\ell^{-0.39}$$

where b_{air} is in Pa and k_ℓ is the liquid permeability in m^2. The data used in this correlation are from oil-field cores with permeability values between about 10^{-12} and 10^{-17} m^2.

Another expression for the Klinkenberg coefficient is from Jones and Owens (1980), who performed similar measurements for low-permeability gas sands with permeabilities between 10^{-14} and 10^{-19} m^2. They developed the following correlation for air (presumably at 25°C)

$$b_{air} = 0.98\, k_\ell^{-0.33}$$

where the units are the same as for the Heid et al. (1950) correlation. Between 10^{-14} and 10^{-17} m^2, which is where the permeabilities for the data sets overlap, the Klinkenberg factors from both correlations are quite similar.

As discussed later on in Section 2.2.2, the Klinkenberg coefficient for a given porous media is different for each gas and is dependent on the local temperature. The Klinkenberg factor can be corrected for different conditions as follows

$$b_i = b_{\text{ref}} \left(\frac{\mu_i}{\mu_{\text{ref}}\,(T_{\text{ref}})} \right) \left(\frac{m_{\text{ref}}}{m_i} \right)^{1/2} \left(\frac{T_i}{T_{\text{ref}}} \right)^{1/2}$$

where ref refers to the reference gas, which is usually air, and *m* is the molecular weight. The temperature is in absolute units.

As the permeability of the medium gets even lower, the pore dimensions approach those of a single molecule. At this point, the flow mechanisms change, and configurational diffusion (Cunningham and Williams, 1980) becomes important. As discussed in Section 2.2.2, the transition from Knudsen diffusion (Klinkenberg effect) to configurational diffusion is estimated to be at a permeability of approximately 10^{-21} m^2.

2.2 GAS-PHASE DIFFUSION

Diffusion in porous media consists of continuum, or ordinary, diffusion and free-molecule diffusion. Continuum diffusion refers to the relative motion of different gas species. Free-molecule diffusion, or Knudsen diffusion, refers to an individual gas and occurs when the mean-free path of the gas molecules is of the same order as the pore diameter of the porous media. As the pore size decreases further, configurational diffusion is encountered where the gas molecule size is comparable to the pore diameter. Configurational diffusion is briefly discussed in the free-molecule diffusion section.

A number of different models have been used to quantify gas diffusion processes in porous media, some of which will be discussed in the next section. Many of the models are simply models derived for a clear fluid (no porous media) that were simply adapted for a porous media. The clear fluid diffusion models only consider molecular diffusion and do not include Knudsen diffusion. Other models are specifically derived for porous media applications. Molecular diffusion and Knudsen diffusion are included in their formulation.

2.2.1 Ordinary (Continuum) Diffusion

Fick's law is the most popular approach to calculating gas diffusion in clear fluids (no porous media) due to its simplicity. While it is only strictly applicable to clear fluids, it has been extensively applied to porous media situations through introduction of a porous media factor. Another approach often employed is the Stefan-Maxwell equations. This equation set is simply an extension to Fick's law for a multicomponent mixture as discussed by Bird, Stewart, and Lightfoot (1960, pg. 569) (hereafter BSL). While attempts have been made to define effective diffusion parameters to account for the presence of the porous medium, the basic transport equations are not altered.

2.2.1.1 Fick's law

Fick's law is actually two laws. Fick's first law is the relationship of the diffusive flux of a gas component as a function of the concentration gradient under steady-state conditions. Fick's second law relates the unsteady diffusive flux to the concentration gradient. Both laws were originally derived for clear fluids (no porous media).

First Law

Clear Fluids

Fick's first law for a binary system basically states that the mole or mass flux is proportional to a diffusion coefficient times the gradient of the mole or mass concentration. For the mole flux formulation, Fick's first law of diffusion for the mole flux of component A, J_A^M, in one dimension in a clear fluid (no porous medium) is

$$J_A^M = -c\, D_{AB,CF}\, \nabla x_A$$

where c is the concentration of the gas, $D_{AB,CF}$ is the diffusion coefficient in a clear fluid, and x_A is the mole fraction of component A. The above form of Fick's first law is commonly misused. The M superscript on the mole flux denotes that the mole flux is relative to the molar-average velocity, NOT to stationary coordinates (BSL, pg. 502). The mole flux equation relative to stationary coordinates for a binary system is given by

$$N_A^D - x_A \left(N_A^D + N_B^D \right) = -c\, D_{AB,CF}\, \nabla x_A$$

where N is relative to stationary coordinates. The second term on the LHS is the molar-average velocity. The mass flux form relative to stationary coordinates is

$$F_A - \omega_A\, (F_A + F_B) = -\rho_g\, D_{AB,CF}\, \nabla \omega_A$$

where F is the mass flux and ω_A is the mass fraction of component A.

Fick's first law and a number of equivalent forms (mole and mass forms, relative to mole or mass velocities or stationary coordinates) are discussed in great detail by BSL (Chapter 16). The relationships between the various fluxes are also discussed in detail in BSL (Chapter 16). However, many applications that use Fick's first law overlook the coordinate system issues. In particular, many applications use Fick's law for the molar-average velocity and incorrectly apply it to stationary coordinates as discussed later in this chapter.

Porous Media

The above forms of Fick's law are appropriate for clear fluids. For application to porous media, Fick's first law is often modified by the introduction of a porous media factor, β, or

$$D_{AB}^* = \beta\, D_{AB,CF}$$

The term β is defined as

$$\beta = \phi\, S_g\, \tau$$

where D^*_{AB} is the effective diffusion coefficient for the AB gas system in a porous media, $D_{AB,CF}$ is the effective diffusion coefficient of the AB gas system in a clear fluid, ϕ is the porosity, S_g is the gas saturation (equal to 1.0 for all-gas conditions), and τ is the tortuosity. The tortuosity factor is discussed in further detail below.

Similar to the clear fluid equation, the mole flux through a porous media relative to stationary coordinates is given by

$$\begin{aligned} F_A - \omega_A\,(F_A + F_B) &= -\beta\, \rho_g\, D_{AB,CF}\, \nabla\omega_A \\ &= -\rho_g D^*_{AB} \nabla\omega_A \end{aligned}$$

Inclusion of the β term takes into account the effective area for gas flow in the pores ($\phi\, S_g$) and the porous media tortuosity (τ). The diffusion coefficient $D_{AB,CF}$ can be estimated from correlations as discussed by Reid et al. (1987). The diffusion coefficient for gases is inversely proportional to the absolute pressure and directly proportional to the absolute temperature to the 1.75 power as given by the Fuller et al. correlation discussed in Reid et al. (1987).

The tortuosity factor, τ, as defined in this application as the ratio of the length of the "tortuous" path in a porous media divided by a straight line value. The tortuosity factor is evaluated for diffusion, not advection. For clear fluids, the tortuosity is equal to 1.0. Note that sometimes other definitions of the tortuosity factor are used (Dullien, 1992, pg. 311). There are a number of models for the tortuosity factor. The most widely used correlation is that of Millington and Quirk (1961). The tortuosity correlation of Millington and Quirk (1961) is given by

$$\tau = \phi^{1/3} S_g^{7/3}$$

which can be rewritten as

$$\tau = \tau_o\, \tau_{S_g} = \phi^{1/3} S_g^{7/3}$$

where τ_o is the tortuosity due to the structure of the porous medium and τ_{S_g} is the tortuosity due to the partial saturation. For all-gas conditions as discussed in this chapter, the tortuosity factor reduces to the porous medium value, or

$$\tau\,(S_g = 1.0) = \tau_o = \phi^{1/3}$$

Costanza-Robinson and Brusseau (Chapter 7 of this book) discuss porous media tortuosity.

Second Law

Fick's second law of diffusion for clear fluids is concerned with the temporal evolution of the concentration, or (BSL, pg. 558)

$$\frac{\partial c_A}{\partial t} = D_{AB}\, \nabla^2 c_A$$

which is only strictly applicable if the molar-average velocity is zero, or for equimolar counterdiffusion. This equation is similar to the heat conduction equation, so many solutions exist such as in Carslaw and Jaeger (1959).

The rest of the present chapter is concerned with Fick's first law, not Fick's second law. For an excellent discussion of Fick's second law, see Fen and Abriola (2004). Abriola et al. (1992) and Sleep (1998) also evaluate Fick's second law.

2.2.1.2 Stefan-Maxwell Equations

Fick's first law of diffusion presented above is applicable to binary gases. This restriction is due to the fact that the gradients of the two gases are directly related to each other, so only a single gradient needs to be specified. For multicomponent gases, multiple gradients need to be determined. For an ideal mixture, the component mass flux equations can be manipulated resulting in (BSL, pg. 569)

$$\nabla x_i = \sum_{j=1}^{n} \frac{1}{c\, D_{ij}} \left(x_i\, N_j \; - \; x_j\, N_i \right)$$

which are known as the Stefan-Maxwell equations applicable to stationary coordinates in a clear fluid. For a two-component system, the Stefan-Maxwell equations reduce to Fick's first law. For application to a porous medium, the diffusion coefficients need to be modified as discussed above.

2.2.2 Free-Molecule Diffusion

As discussed earlier, when the gas molecular mean free path becomes of the same order as the tube dimensions, free-molecule, or Knudsen, diffusion becomes important. Due to the influence of walls, Knudsen diffusion and configurational diffusion implicitly include the effect of the porous medium. Unlike ordinary (continuum) diffusion, there are no approaches for the free-molecule diffusion regime that use clear fluid approaches modified to include porous media effects.

The molecular flux of gas i due to Knudsen diffusion is given by the general diffusion equation (Mason and Malinauskas, 1983, pg. 16)

$$J_{iK} = -D_{iK}\, \nabla n_i$$

where n_i is the molecular density and D_{iK} is the Knudsen diffusion coefficient. The Knudsen diffusivity of gas i for a capillary of a given radius can be estimated as follows (Cunningham and Williams, 1980, eqns. 2.17 and 2.65) assuming a coefficient of diffuse reflection equal to unity

$$D_{i\mathrm{K,capillary}} = \frac{2}{3}\left(\frac{8RT}{\pi m_i}\right)^{1/2} r_{\mathrm{capillary}}$$

This equation can be written as

$$D_{i\mathrm{K,capillary}} = 97.0\left(\frac{T}{m_i}\right)^{1/2} r_{\mathrm{capillary}}$$

where $D_{i\mathrm{K,capillary}}$ is in m^2/s, T is in K, m is the molecular weight, and $r_{\mathrm{capillary}}$ is in m. While this equation may be appropriate for flow in well-defined capillaries, it is not directly useful for porous media applications. However, there is an alternative way to determine the Knudsen diffusion coefficient that includes the complexity of the porous media.

The equation for the molecular flux from Knudsen diffusion can be rewritten in the same form as Darcy's law. From this expression, and Klinkenberg's formula, the Knudsen diffusion coefficient, $D_{i\mathrm{K}}$, can be related to the Klinkenberg factor as follows (Thorstenson and Pollock, 1989a, eqn 60)

$$D_{i\mathrm{K}} = \frac{k_g\, b_i}{\mu_i}$$

As discussed in the advection part of this chapter, there are a number of correlations for the Klinkenberg coefficient, b_i, as a function of the porous medium, the fluid, and the temperature. The Klinkenberg coefficient can be used in the above equation to evaluate the Knudsen diffusion coefficient for a porous medium. The Klinkenberg coefficient implicitly takes into account the structure of the porous medium as reflected through the permeability. The modifications to the Klinkenberg factor due to the gas (molecular weight and viscosity) and the temperature should be used as discussed earlier in Section 2.1.4.

As the permeability of the medium gets even lower, the pore dimensions approach those of a single molecule. At this point, the flow mechanisms change, and configurational diffusion (Cunningham and Williams, 1980) becomes important. In configurational diffusion, the size of the molecules is comparable to the pore dimensions, and the molecular configuration becomes important (e.g., Xiao and Wei, 1992a, b). Membrane diffusion occurs at even smaller pore sizes where the chemical characteristics of the molecules are important (Cunningham and Williams, 1980).

Cunningham and Williams (1980) suggest that configurational diffusion may be encountered when the pore sizes are less than about 10 Å. Note that the molecular size can be characterized by the Lennard-Jones length constant, σ, which varies from about 2.5 to 7.5 Å as given by BSL (1960, pg. 744). Assuming a porosity of about 10%, and calculating the tortuosity by the Millinton and Quirk relationship given earlier, the effective Knudsen diffusion coefficient will be about 10^{-9} m^2/s. Using the Jones and Owens (1980) correlation for the Klinkenberg coefficient, the permeability is about 10^{-21} m^2. This prediction of the transition from Knudsen diffusion to configuration diffusion is qualitative at best. However, it should indicate the approximate permeability where configurational diffusion should at least be considered to be as the possible diffusion mechanism.

2.3 COMBINED MECHANISMS

The interaction between advection and diffusion in porous media can be significant. Consider two separate volumes connected by a tube containing a light gas and a heavy gas. Diffusion of the light gas is faster than the heavy gas because of the higher molecule velocity. The net flow of molecules is toward the heavy gas volume, so the pressure rises in the heavy gas volume and decreases in the light gas volume. In turn, this pressure difference causes advection from the heavy gas volume to the light gas volume. Thus, diffusion directly leads to advection. Only in the case of equimolar gases will diffusion not result in advection.

As mentioned earlier, this scenario was implicitly included in the diffusion formulation discussed by BSL in that diffusion is relative to the molar-average velocity. However, this effect has generally been ignored. Coupling of the advection and diffusion mechanisms has been formalized with the development of the Dusty Gas Model by Evans, Mason and colleagues (Evans et al., 1961; Evans et al., 1962a; Mason et al., 1963; Mason et al., 1964). The Dusty Gas Model (DGM) takes the gas transport equations a step further by including the effect of the porous media as a "dusty gas" component of the gas mixture. The "dusty gas" is assumed to consist of large molecules fixed in space that is treated as a component of the gas mixture. The kinetic theory of gases is applied to this dusty-gas mixture. One of the key aspects of the DGM is the combination of diffusion (ordinary and Knudsen) and advection. Ordinary and Knudsen diffusion are combined through addition of momentum transfer based on kinetic-theory arguments, and diffusive fluxes (ordinary plus Knudsen) are added to advective fluxes based on Chapman-Enskog kinetic theory.

The DGM, including numerous data-model comparisons, is discussed in detail by Mason and Malinauskas (1983) and by Cunningham and Williams (1980). Other excellent references on application of the Dusty Gas Model for porous media are Thorstenson and Pollock (1989a, 1989b) and Jackson (1977).

The exclusive presentation of the DGM in this chapter does not imply that the DGM is the most comprehensive gas-phase diffusion model available for porous media. There are a number of other models available including Feng and Stewart (1973), who extended the DGM to more complicated pore networks, a mean transport pore model as presented by Arnost and Schneider (1995) (see Šolcová and Schneider (Chapter 14 of this book)), and Shapiro (1993), who developed a model for heterogeneous anisotropic porous media. Altevogt et al. (2003a, b) present an alternate approach for binary gas diffusion. Rather, the DGM is the most widely used model for a mechanistic approach to combine gas advection and diffusion in porous media at the present time.

Ignoring thermal diffusion, which is typically small, the DGM can be written either in terms of the diffusive molar flux, N^{D}, or the total molar flux (diffusive plus advective), N^{T}, which are relative to fixed coordinates (Thorstenson and Pollock,

1989a, eqns. 41 and 48). The two expressions are

$$\sum_{\substack{j=1\\j\neq i}}^{n} \frac{x_i N_j^{\mathrm{D}} - x_j N_i^{\mathrm{D}}}{D_{ij}^*} - \frac{N_i^{\mathrm{D}}}{D_{i\mathrm{K}}} = \frac{\nabla P_i}{RT} = \frac{P\nabla x_i}{RT} + \frac{x_i \nabla P}{RT}$$

$$\sum_{\substack{j=1\\j\neq i}}^{n} \frac{x_i N_j^{\mathrm{T}} - x_j N_i^{\mathrm{T}}}{D_{ij}^*} - \frac{N_i^{\mathrm{T}}}{D_{i\mathrm{K}}} = \frac{P\nabla x_i}{RT} + \left(1 + \frac{k_\ell P}{D_{ik}\mu}\right)\frac{x_i \nabla P}{RT}$$

where the second equation simply includes the advective flux on both sides of the equation. The first term on the LHS considers molecule-molecule interactions and is immediately recognized as being based on the Stefan–Maxwell equations discussed earlier. The second term on the LHS considers molecule-particle (Knudsen diffusion) interactions, while the RHS is the driving force for diffusion and advection, which includes concentration and pressure gradients.

There are many forms of the DGM. One particularly useful form is for the total mass flux of component 1 in an isothermal binary system, or (Thorstenson and Pollock, 1989a, eqn. F4)

$$\begin{aligned} F_1 &= m_1 N_1^{\mathrm{T}} \\ &= -m_1 \frac{\left[D_{1\mathrm{K}} D_{12}^* (P/RT)\nabla x_1 + D_{1\mathrm{K}}(D_{12}^* + D_{2\mathrm{K}}) x_1 (\nabla P/RT)\right]}{(D_{12}^* + x_1 D_{2\mathrm{K}} + x_2 D_{1\mathrm{K}})} \\ &\quad - x_1 m_1 \frac{k_0 P}{\mu} \frac{\nabla P}{RT} \end{aligned}$$

The flux of component 1 has diffusive (first term) and advective (second term) components. The diffusive flux consists of ordinary diffusion (mole fraction gradient) and Knudsen diffusion (pressure gradient) components.

Note that in the special case of isobaric conditions ($\nabla P = 0$), the advective and Knudsen diffusion fluxes are zero. However, this does *not* mean that the Knudsen diffusion coefficients are not important. The ordinary diffusion flux is dependent on *both* diffusion (Knudsen and ordinary) coefficients. The Knudsen diffusion coefficients characterize the impact of the porous media (gas-wall interactions) on ordinary diffusion. This behavior is absent in the clear fluid formulations, such as Fick's law, that are modified for porous media applications.

2.4 COMPARISON TO FUNDAMENTAL RELATIONSHIPS AND EXPERIMENTAL DATA

In the 1800s, Thomas Graham discovered two important relationships for gas diffusion in a porous media that relate the diffusive fluxes of a binary mixture in a porous medium (Mason and Malinauskas, 1983, pg. 3). Graham's law of effusion applies to

Knudsen diffusion as experimentally discovered by Graham (1846), or

$$-\frac{N_1^{\mathrm{K}}}{N_2^{\mathrm{K}}} = \left(\frac{m_2}{m_1}\right)^{1/2}$$

For ordinary diffusion (no advection), Graham's law of diffusion applies as experimentally discovered by Graham (1833) or

$$-\frac{N_1^{\mathrm{D}}}{N_2^{\mathrm{D}}} = \left(\frac{m_2}{m_1}\right)^{1/2}$$

Even though the ratios are identical, each equation applies to a different diffusion regime. In terms of a mass flux ratio, these equations become

$$-\frac{F_1^{\mathrm{K}}}{F_2^{\mathrm{K}}} = \left(\frac{m_1}{m_2}\right)^{1/2}$$

$$-\frac{F_1^{\mathrm{D}}}{F_2^{\mathrm{D}}} = \left(\frac{m_1}{m_2}\right)^{1/2}$$

The above relationships are significant. Graham's laws were "lost" and were independently rediscovered in the 1950s and 1960s (Mason and Malinauskas (1983, pg. 3)). Graham's laws are fundamental relationships for gas diffusion in porous media and were used in the development of the Dusty Gas Model. These relationships will be used in the data-model comparisons that follow.

The data used in these comparisons were obtained by Evans, Watson, and Truitt (1962b, 1963). They performed these experiments to support development of the Dusty Gas Model. The experiments consisted of flow and diffusion of helium and argon across a low-porosity (0.11) and low-permeability (2.13×10^{-18} m^2) graphite. At this low permeability, Knudsen diffusion plays a significant role.

In addition to the data and model predictions, Mason and Malinauskas (1983, pg. 91) provide closed-form solutions for the DGM for special cases, which are also plotted as a continuous line labeled DGM in Figure 2.1 and Figure 2.2. In the data-model comparisons, the DGM is compared to the data as is a model labeled ADM. The ADM Model, which stands for advective-diffusive (or dispersive) model, is simply a linear addition of Darcy's law and ordinary diffusion using Fick's law, which is incorrectly applied to stationary coordinates as discussed above. This simple linear addition is commonly used as exemplified by TOUGH2 (Pruess, 1991) and Abriola and Pinder (1985). Slip effects, or Knudsen diffusion, are included through a Klinkenberg parameter to define an effective permeability for the advective flux as discussed earlier. Porous medium effects for ordinary diffusion are included through a porosity-tortuosity-gas saturation factor applied to the diffusive flux in free space. This simple additive approach, while intuitively appealing, ignores coupling between advective and diffusive mechanisms. The expression for the ADM used in this comparison is

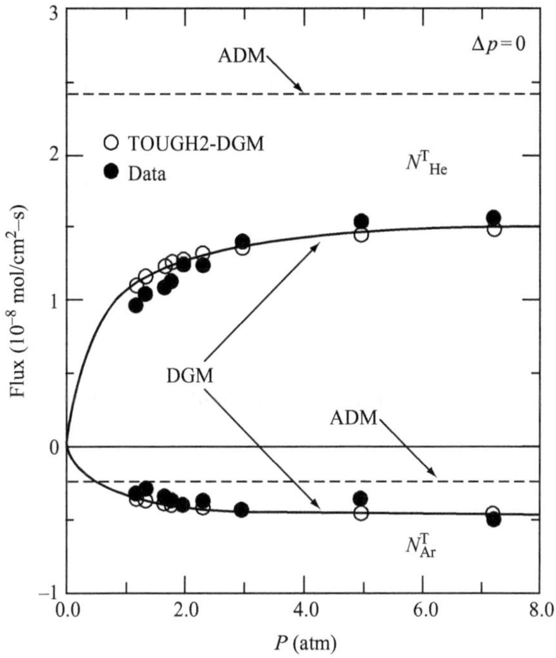

Figure 2.1. Mole flux data-model comparison for zero pressure difference conditions (reprinted from Webb, 1998 with permission with minor modifications)

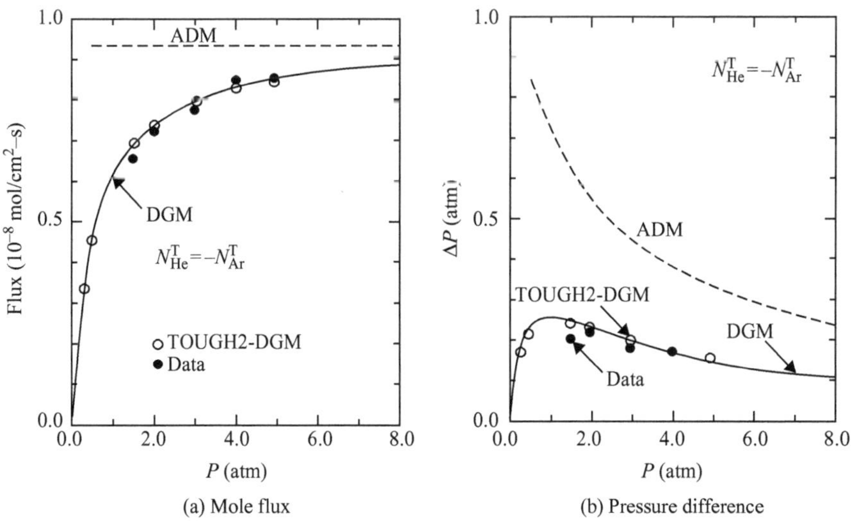

Figure 2.2. Data-model comparison for zero net mole flux conditions (reprinted from Webb, 1998 with permission with minor modifications)

given below

$$F_g = -\frac{k_\ell}{\mu_g}\left(1 + \frac{b_g}{P_g}\right)\rho_g\left(\nabla P_g - \rho_g \overline{g}\right) - \rho_g D_{12}^* \nabla x$$

The DGM and the ADM results presented in this section were calculated by the TOUGH2 code (Pruess, 1991), which was modified to incorporate the DGM by Webb (1998). Pruess (Chapter 12 of this book) summarizes the flux expressions in many commonly used porous media codes including TOUGH2.

2.4.1 Zero Pressure Difference

The first data-model comparison considers zero pressure difference across the porous media. In this case, there is diffusion but no advection. From Graham's laws, the ratio of the mole fluxes (not the magnitude) for Knudsen diffusion and for ordinary diffusion is the same. Graham's laws give a mole flux ratio of helium to argon of 3.2 based on the molecular weights of helium ($m = 4.00$) and argon ($m = 39.944$).

Data-model comparisons are given in Figure 2.1 as a function of the average pressure. Helium mole flux is positive, and argon mole flux is negative. The data for both gases increase with increasing pressure. The DGM data-model comparison is quite good. In addition to the individual values, the ratio of the mole fluxes is consistent with the theoretical value given above. In contrast, the model predictions of the ADM show a constant mole flux value independent of pressure for each component, which is not consistent with the data. The ADM predicts that the mass fluxes of the two components are equal. The mole flux ratio (helium/argon) is simply the inverse of the ratio of the molecular weights, or about 10, which is not consistent with the data. The ADM, which for zero pressure difference reduces to Fick's law, does not match the experimental data very well. Knudsen diffusion is not included because there isn't any advection in the ADM predictions.

2.4.2 Zero Net Mole Flux

The second case is for zero net mole flux. This case simulates what would occur in a closed volume, where the total mole fluxes of the two components are equal. The sum of diffusion and advection of each component are equal to each other resulting in a zero net mole flux. The predicted flux of each component and the pressure difference across the experiment are compared to the experimental data.

Figure 2.2a gives the data-model comparison for the mole flux as a function of the average pressure. The data-model comparison for the DGM is very good including the variation of flux with pressure. For the ADM, the predicted mole flux is constant, unlike the data.

The data-model comparison for pressure difference across the porous media is shown in Figure 2.2b. The pressure difference results in equal and opposite mole fluxes across the graphite. The data-model comparison for the DGM is very good. The ADM data-model comparison is poor, similar to the zero pressure difference case.

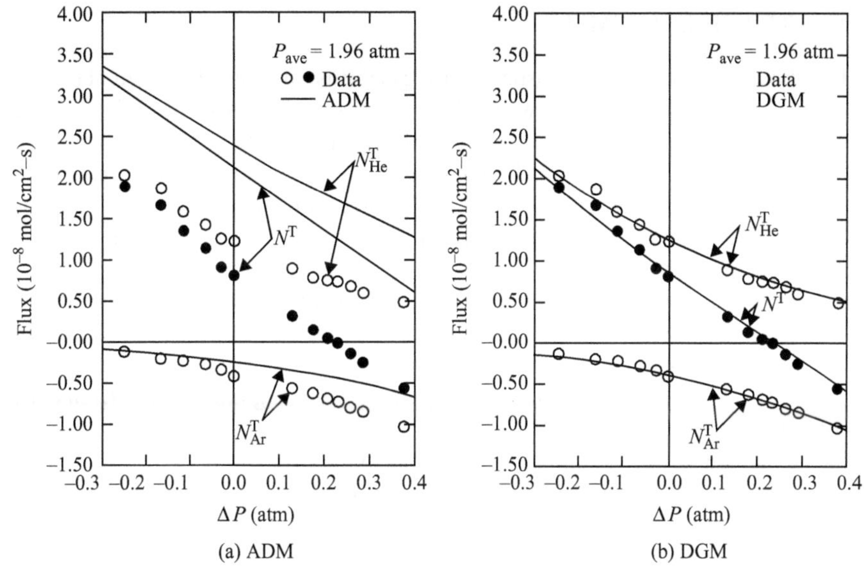

Figure 2.3. Data-model comparison for combined advection and diffusion (reprinted from Webb, 1998 with permission with minor modifications)

2.4.3 Combined Advection and Diffusion

The most general situation of combined advection and diffusion is shown next. The pressure difference across the test section was varied while maintaining the same average pressure. The total individual fluxes of the two gases were measured. Figure 2.3 shows the data-model comparisons. Figure 2.3a shows the ADM comparison, which is generally poor similar to the earlier results. Figure 2.3b shows the DGM results, which show excellent agreement with the experimental data.

2.4.4 Overall Evaluation

In general, the ADM data-model comparisons are poor, while the DGM data-model comparisons are quite good. Overall, the DGM is significantly better than the ADM for the conditions of this experiment. Webb (1998) provides more details on the comparison of these two models.

Many other comparisons of the ADM, or variations of the ADM, and DGM have been performed (Abriola et al., 1992; Fen and Abriola, 2004; Oldenburg et al., 2004). All agree that for higher permeabilities, the ADM is adequate, while for lower permeabilities, the DGM is needed. Just what is "low" and "high" has not been quantified. Some guidance may be developed from the results for trace gas diffusion developed by Webb and Pruess (2003) in that the pressure and permeability were varied for different diffusing species.

2.5 TRACE GAS LIMIT

The above data-model comparison was for gas diffusion where the different gas mass fractions are significant. As shown by Webb and Pruess (2003), in the limit of trace gas diffusion in a binary mixture where one gas has a vanishingly small mass fraction, the DGM and ADM reduce to similar equations. Two "correction" factors are needed to bring the ADM in line with the DGM. The first correction factor is an additional tortuosity term on the diffusion coefficient. The second correction factor is on the Klinkenberg coefficient, b.

The correction factors can also be viewed as ratio of the mass flux predicted by the DGM to that predicted by the ADM. As will be seen, the tortuosity correction factor is always 1 or less, which indicates that ordinary diffusion is always overpredicted by the ADM, in some cases by orders of magnitude. The magnitude of the Klinkenberg correction factor is much smaller and may be less than or greater than 1.0 depending on the molecular weight ratio of the trace gas to the bulk species.

The standard ADM equation incorporating the Klinkenberg coefficient is as follows

$$F_{\mathrm{g}} = -\frac{k_\ell}{\mu_{\mathrm{g}}}\left(1 + \frac{b}{\overline{P}_{\mathrm{g}}}\right)\rho_{\mathrm{g}}\left(\nabla P_{\mathrm{g}} - \rho_{\mathrm{g}}\overline{g}\right) - \rho_{\mathrm{g}} D_{12}^{*} \nabla x$$

Introducing the correction factors gives

$$F_{\mathrm{g}} = -\frac{k_1}{\mu_{\mathrm{g}}}\left(1 + \frac{b_{\mathrm{DGM}} b}{\overline{P}_{\mathrm{g}}}\right)\rho_{\mathrm{g}}\left(\nabla P_{\mathrm{g}} - \rho_{\mathrm{g}}\overline{g}\right) - \tau_{\mathrm{DGM}}\rho_{\mathrm{g}} D_{12}^{*} \nabla x$$

The first term on the RHS is simply the convective flux including the Klinkenberg correction factor. The second term on the RHS is ordinary diffusion with a tortuosity correction factor. For trace gas diffusion, these factors are given by (see Webb and Pruess, 2003)

$$b_{\mathrm{DGM}} = \frac{1 + m_{\mathrm{rat}}^{1/2} D_{\mathrm{rat}}}{1 + D_{\mathrm{rat}}}$$

and

$$\tau_{\mathrm{DGM}} = \frac{1}{1 + D_{\mathrm{rat}}}$$

where D_{rat} is the ratio of the effective ordinary diffusion coefficient to the Knudsen diffusion coefficient, or

$$D_{\mathrm{rat}} = \frac{D_{12}^{*}}{D_{1\mathrm{K}}}$$

and m_{rat} is the ratio of molecule weights

$$m_{\mathrm{rat}} = \frac{m_2}{m_1}$$

where gas 1 is the trace diffusing species and gas 2 is the bulk species.

The above expressions for b_{DGM} and τ_{DGM} have been evaluated by Webb and Pruess (2003) for a bulk species of air at 25°C and a porous media with a porosity of 0.4 and a tortuosity of 0.74. The permeability was varied over a wide range, and the trace gas species include He, water vapor, and TCE.

Figure 2.4 shows the variation in the correction factors as a function of permeability for three different trace gases (He, H_2O, and TCE) at three different gas pressures

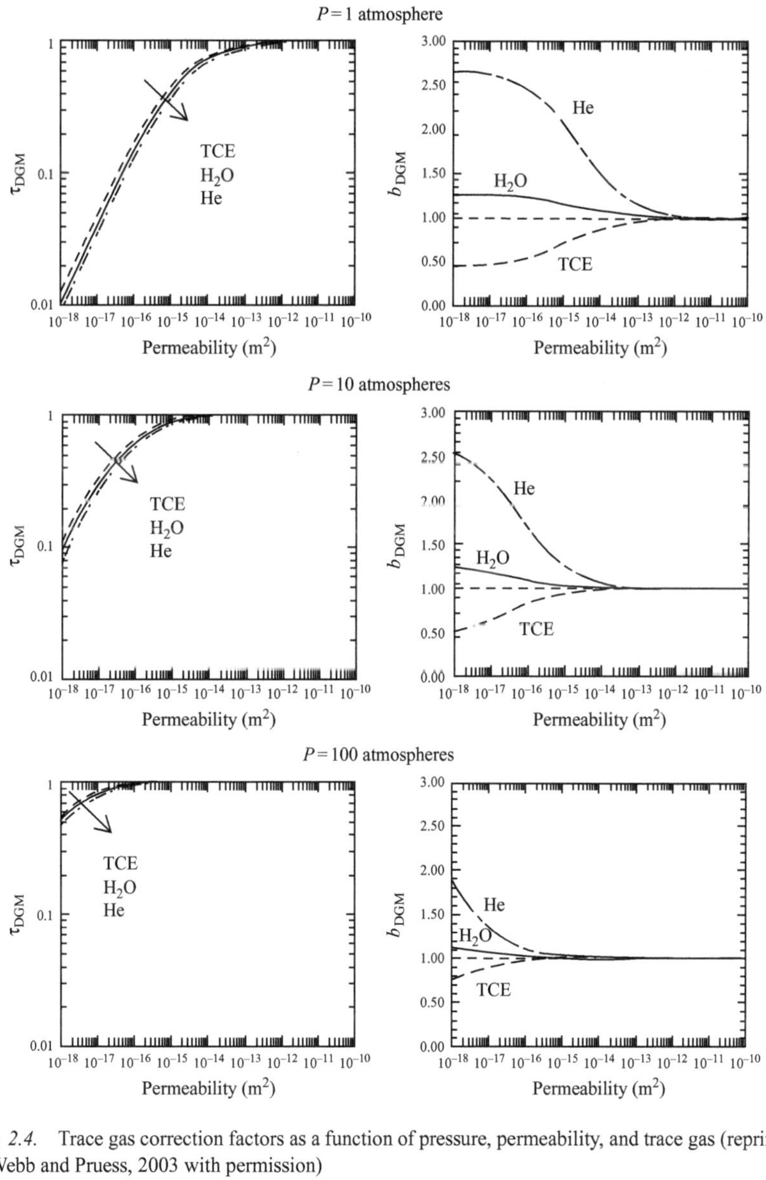

Figure 2.4. Trace gas correction factors as a function of pressure, permeability, and trace gas (reprinted from Webb and Pruess, 2003 with permission)

(1, 10, and 100 atmospheres). The b_{DGM} correction factor depends on the trace gas. If the trace gas is heavier than the bulk species (air), the correction factor is less than 1.0. Conversely, if the trace gas is lighter than the bulk gas, the b_{DGM} correction factor is greater than 1.0. The values have a generally small range, varying between about 0.5 and 2.7 for the trace gases evaluated. The τ_{DGM} correction factor can be viewed as the ratio of ordinary diffusion flux predicted by the DGM divided by the Fick's law value. For example, if the value of τ_{DGM} is 0.01, Fick's law without the correction factor would overpredict the ordinary gas diffusion flux by two orders of magnitude compared to the DGM prediction. The value of τ_{DGM} is about unity for higher permeabilities ($> 10^{-13}$ m^2) at all pressures. For lower permeabilities, Fick's law may dramatically overpredict the gas diffusion flux. The value of τ_{DGM} decreases dramatically to about 0.01 at a permeability of 10^{-18} m^2. The variation of the ratio for different trace gases is small. Therefore, the difference between the DGM and ADM becomes important for permeabilities $< 10^{-13}$ m^2, with larger differences for lower permeabilities and lower pressures.

2.6 APPLICABILITY OF DGM TO REAL POROUS MEDIA

There are questions and concerns about the applicability of the DGM to real porous media and the appropriate parameter values. While a number of authors (Abu-El-Sha'r and Abriola, 1997; Fen and Abriola, 2004; Cunningham and Williams, 1980, pg. 220; Mason and Malinauskas, 1983, pg. 50) address this point, it must be pointed out that the DGM (or the other coupled models as discussed earlier), are the best models to date and are vastly superior to the ADM, which was used for a number of years and is still used today.

While the DGM is widely used, there are modifications to improve its applicability. For example, modifications have been made to the DGM to include things such as mass transfer (e.g., Chen and Rinker, 1979) and chemical reaction (e.g., Veldsink et al., 1995). Note that these references are just representative and do not necessarily represent the latest information on these modifications.

ACKNOWLEDGMENTS

Sandia is a multiprogram laboratory operated by Sandia Corporation, a Lockheed Martin Company, for the United States Department of Energy's National Nuclear Security Administration under Contract DE-AC04–94AL85000.

NOMENCLATURE

b	Klinkenberg coefficient
c	concentration
c_F	constant in Forchheimer equation
D	ordinary diffusion coefficient
D_{iK}	Knudsen diffusion coefficient

F	mass flux
$\bar{g}$	gravity
J^{M}	diffusive mole flux relative to molar-average velocity value
k	permeability
m	molecular weight
N^{D}	diffusive mole flux relative to stationary coordinates
N^{T}	total mole flux relative to stationary coordinates
n	molecular density
P	pressure
$\overline{P}$	average pressure
r	radius
R	gas constant
Re_k	permeability-based Reynolds number
S	saturation
T	temperature
u	Darcy velocity
V	pore velocity
x	mole fraction

Greek

β	porous media factor
τ	tortuosity
ϕ	porosity
ρ	density
μ	viscosity
$\tilde{\mu}$	effective viscosity at wall (Brinkman equation)

Subscript

A,B	components A, B
air	value for air
CF	clear fluid
DGM	Dusty Gas Model
g	gas
K	Knudsen diffusion
ℓ	liquid phase
pore	pore
PM	porous media
rat	ratio
ref	reference
S	saturation
0	all-gas conditions
1,2	component 1, 2

Superscript

D ordinary diffusion
K Knudsen diffusion
T total
* effective porous media value

REFERENCES

Abriola, L.M., C.-S. Fen, and H.W. Reeves, 1992, "Numerical simulation of unsteady organic vapor transport in porous media using the dusty gas model," Subsurface Contamination by Immiscible Fluids, Wyre, ed., Balkema, Rotterdam, pp. 195–202.

Abriola, L.M., and G.F. Pinder, 1985, "A multiphase approach to the modeling of porous media contamination by organic compounds, 1. Equation development," Water Resourc. Res., 21:11–18.

Abu-El-Sha'r, W., and L.M. Abriola, 1997, "Experimental assessment of gas transport mechanisms in natural porous media: Parameter evaluation," Water Resourc. Res., 33:505–516.

Altevogt, A.S., D.E. Rolston, and S. Whitaker, 2003a, "New equations for binary gas transport in porous media, Part 1: equation development," Adv. Water Resources, 26:695–715.

Altevogt, A.S., D.E. Rolston, and S. Whitaker, 2003b, "New equations for binary gas transport in porous media, Part 1: experimental validation," Adv. Water Resources, 26:695–715.

Antohe, B.V., and J.L. Lage, 1997, "A general two-equation macroscopic turbulence model for incompressible flow in porous media," Int. J. Heat Mass Transfer, 40:3013–3024.

Arnost, D. and P. Schneider, 1995, "Dynamic transport of multicomponent mixtures of gases in porous solids," Chem. Eng. J., 55:91–99.

Bear, J., 1979, Hydraulics of Groundwater, McGraw-Hill Book Company, New York.

Beavers, G.S., E.M. Sparrow, and D.E. Rodenz, 1973, "Influence of bed size on the flow characteristics and porosity of randomly packed beds of spheres," J. Appl. Mech., 40:655–660.

Bird, R.B., W.E. Stewart, and E.N. Lightfoot, 1960, Transport Phenomena, John Wiley & Sons, New York.

Carslaw, H.S., and J.C. Jaeger, 1959, Conduction of Heat in Solids, Second Edition, Oxford University Press, London.

Chen, O.T., and R.G. Rinker, 1979, "Modification of the Dusty-Gas Equation to predict Mass Transfer in General Porous Media," Chemical Engineering Science, 34:51–61.

Cunningham, R.E., and R.J.J. Williams, 1980, Diffusion in Gases and Porous Media, Plenum Press, New York.

Darcy, H.P.G., 1856, "Les Fontaines Publiques de la Ville de Dijon," Victor Dalmont, Paris.

Dullien, F.A.L., 1992, Porous Media: Fluid Transport and Pore Structure, Second Edition, Academic Press, Inc., San Diego.

Dupuit, A.J.E.J., 1863, "Etudes Theoriques et Pratiques sur le Mouvement des aux dans les Canaux Decouverts et a Travers les Terrain Permeables," Victor Dalmont, Paris.

Evans, III, R.B., G.M. Watson, and E.A. Mason, 1961, "Gas Diffusion in Porous Media at Uniform Pressure," J. Chem. Physics, 35:2076–2083.

Evans, III, R.B., G.M. Watson, and E.A. Mason, 1962a, "Gas Diffusion in Porous Media. II Effect of Pressure Gradients," J. Chem. Physics, 36:1894–1902.

Evans, III, R.B., G.M. Watson, and J. Truitt, 1962b, "Interdiffusion of gases in a low permeability graphite at uniform pressure," J. Applied Physics, 33:2682.

Evans, III, R.B., G.M. Watson, and J. Truitt, 1963, "Interdiffusion of gases in a low permeability graphite. II. Influence of pressure gradients," J. Applied Physics, 34:2020.

Fen, C.-S., and L.M. Abriola, 2004, "A comparison of mathematical model formulations for organic vapor transport in porous media," Adv. Water Reources, 27:1005–1016.

Feng, C., and W.E. Stewart, 1973, "Practical models for isothermal diffusion and flow of gases in porous solids," Ind. Eng. Chem. Fundam., 12:143–147.

Forchhemer, P., 1901, "Wasserbewegung durch Boden," Z. Vereines Deutscher Ingenieure, 45:1736–1741 and 1781–1788.

Getachew, D., W.J. Minkowycz, and J.L. Lage, 2000, "A modified form of the $\kappa - \varepsilon$ model for turbulent flows of an incompressible fluid in porous media," Int. J. Heat Mass Transfer, 43:2909–2915.

Graham, T., 1833, "On the law of diffusion of gases," Phil. Mag., 2:175, 269, 351; reprinted in Chemical and Physical Researches, pp. 44–70, Edinburgh Univ. Press, Edinburgh, Scotland, UK, 1876.

Graham, T., 1846, "On the motion of gases," Phil. Trans. Roy. Soc., 136:573; reprinted in Chemical and Physical Researches, pp. 88–161, Edinburgh Univ. Press, Edinburgh, Scotland, UK, 1876.

Heid, J.G., J.J. McMahon, R.F. Nielson, and S.T. Yuster, 1950, "Study of the Permeability of Rocks to Homogeneous Fluids," API Drilling and Production Practice, pp. 230–244.

Jackson, R., 1977, Transport in Porous Catalysts, Chem Eng. Monograph 4, Elsevier, New York.

Jones, F.O., and W.W. Owens, 1980, "A Laboratory Study of Low-Permeability Gas Sands," J. Petroleum Techology (September 1980), pp. 1631–1640.

Joseph, D.D., D.A. Nield, and Papanicolaou, 1982, "Nonlinear equation governing flow in a saturated porous media," Water Resources Res., 18:1049–1052 and 19:591.

Kaviany, M., 1995, Principles of Heat Transfer in Porous Media, Second Edition, Springer-Verlag New York, Inc., New York. Check 1995 second edition.

Klinkenberg, L.J., 1941, "The Permeability of Porous Media to Liquids and Gases," API Drilling and Production Practice, pp. 200–213.

Lage, J.L., 1998, "The fundamental theory of flow through permeable media: From Darcy to turbulence," Transport Phenomena in Porous Media, D.B. Ingham and I. Pop, eds., Elsevier, Amsterdam, pp. 1–30.

Mason, E.A., R.B. Evans, III, and G.M. Watson, 1963, "Gaseous diffusion in porous media. III. Thermal transpiration," J. Chem. Phys., 38:1808–1826.

Mason, E.A., and A.P. Malinauskas, 1964, "Gaseous diffusion in porous media. IV. Thermal diffusion," J. Chem. Phys., 41:3815–3819.

Mason, E.A., and A.P. Malinauskas, 1983, Gas Transport in Porous Media: The Dusty-Gas Model, Chem Eng. Monograph 17, Elsevier, New York.

Masuoka, T., and Y. Takatsu, 1996, "Turbulence model for flow through porous media," Int. J. Heat Mass Transfer, 39:2803–2809.

Millington, R.J., and J.M. Quirk, 1961, "Permeability of porous solids," Trans. Faraday Soc., 57:1200–1207.

Nield, D.A., and A. Bejan, 1999, Convection in Porous Media, Second Edition, Springer-Verlag New York, Inc., New York.

Oldenburg, C.M., S.W. Webb, K. Pruess, and G.J. Moridis, 2004, "Mixing of Stably Stratified Gases in Subsurface Reservoirs: A Comparison of Diffusion Models," Transport in Porous Media, 54: 323–334.

Pruess, K., 1991, TOUGH2 A General-Purpose Numerical Simulator for Multiphase Fluid and Heat Flow, LBL-29400, Lawrence Berkeley Laboratory.

Reda, D.C., 1987, "Slip-Flow Experiments in Welded Tuff: The Knudsen Diffusion Problem," Coupled Processes Associated with Nuclear Waste Repositories, C.-F. Tsang, ed., pg. 485–493, Academic Press, Inc., Orlando, 1987.

Reid, R.C., J.M. Prausnitz, and B.E. Poling, 1987, The Properties of Gases & Liquids, Fourth Edition, McGraw-Hill, Inc.

Shapiro, A.A., 1993, "A kinetic theory of the filtration of rarefied gas in an anisotropic porous medium," Theoretical Foundations of Chemical Engineering, 27:140–148.

Sleep, B.E., 1998, "Modeling transient organic vapor transport in porous media with the dusty gas model," Advances in Water Resources, 22:247–256.

Thorstenson, D.C., and D.W. Pollock, 1989a, "Gas Transport in Unsaturated Zones: Multicomponent Systems and the Adequacy of Fick's Laws," Water Resourc. Res., 25:477–507.

Thorstenson, D.C., and D.W. Pollock, 1989b, "Gas Transport in Unsaturated Porous Media: The Adequacy of Fick's Law," Reviews of Geophysics, 27:61–78.

Vafai, K., and C.L. Tien, 1981, "Boundary and inertia effects on flow and heat transfer in porous media," Int. J. Heat Mass Trans., 24:195–203.

Veldsink, J.W., R.M.J. van Damme, G.F. Versteeg, and W.P.M. van Swaaji, 1995, "The use of the dusty-gas model for the description of mass transport with chemical reaction in porous media," The Chemical Engineering Journal, 57:115–125.

Ward, J.C., 1964, "Turbulent flow in porous media," ASCE J. Hydraulic Div., 90 (HY5):1–12.

Webb, S.W., 1998, "Gas Diffusion in Porous Media – Evaluation of an Advective-Dispersive Formulation and the Dusty-Gas Model for Binary Mixtures," J. Porous Media, 1:187–199.

Webb, S.W. and K. Pruess, 2003, "The Use of Fick's Law for Modeling Trace Gas Diffusion in Porous Media," Transport in Porous Media, 51:327–341.

Xiao, J., and J. Wei, 1992a, "Diffusion Mechanism of Hydrocarbons in Zeolites – I. Theory," Chem. Eng. Sci., 47:1123–1141.

Xiao, J., and J. Wei, 1992b, "Diffusion Mechanism of Hydrocarbons in Zeolites – II. Analysis of Experimental Observations," Chem. Eng. Sci., 47:1143–1159.

CHAPTER 3

VAPOR TRANSPORT PROCESSES

CLIFFORD K. HO

Sandia National Laboratories, P. O. Box 5800, Albuquerque, NM 87185, USA

3.1 OVERVIEW

In this chapter, important processes associated with vapor in porous media are investigated. The term *vapor* is used here to identify the gaseous phase of a substance that can be condensed under standard temperatures and pressures. With this definition, compounds such as water, alcohol, and benzene can exist as vapors, but compounds such as nitrogen, helium, and hydrogen exist as gases. The first section describes the partitioning of a vapor with its liquid phase, either as a single component or in a mixture of other components. The second section describes a phenomenon known as vapor-pressure lowering, which reduces the ability of compounds to partition into the vapor phase when strong capillarity exists. The third section describes the rate of evaporation and provides sample calculations. Finally, the fourth section describes a process known as enhanced vapor diffusion, which allows condensable vapors to apparently diffuse through unsaturated porous media at high rates relative to Fickian diffusion.

3.2 PHASE PARTITIONING

Vapor-phase compounds can partition into mobile liquid phases and immobile solid phases in porous media. The mobility of various compounds in porous media depends, in part, on the ability for those compounds to partition into a mobile phase. In addition, the partitioning between vapor and solid phases can cause retardation of transport. This section describes the various mechanisms of vapor-phase partitioning between the liquid and gas phases. A separate chapter is devoted to solid–vapor partitioning.

3.2.1 Single-Component Liquid/Vapor Partitioning

The partitioning of a single-component liquid into its vapor phase is dictated by the vapor pressure of the liquid, which is determined by the temperature of the

C. Ho and S. Webb (eds.), Gas Transport in Porous Media, 27–46.
© 2006 *Springer.*

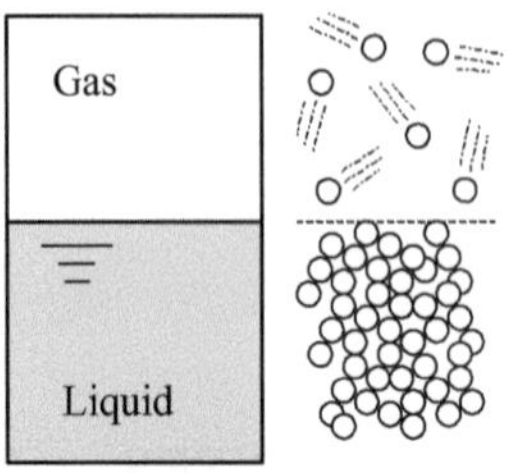

Figure 3.1. Illustration of molecular states of liquid and gas

system. As the temperature increases, the vapor pressure of a liquid increases significantly. At a molecular level, the energy associated with the molecules increases with increasing temperature. At higher temperatures, molecules have greater kinetic energy (they become more "active") and begin to break free of the van der Waals forces that attract the molecules to one another. The process of molecules breaking free from one another is known as evaporation, and the rate of evaporation increases with increasing temperature. The separation distance between vapor-phase molecules is large relative to the separation distance between liquid-phase molecules (Figure 3.1).

The vapor pressure is the pressure that the vapor-phase molecules exert on each other and on the walls of a container. In a closed container partially filled with liquid, evaporation of the liquid will continue until the pressure exerted by the vapor-phase molecules equals the saturated vapor pressure determined by the system temperature for that compound. When this saturated (or equilibrium) vapor pressure is reached, the rate of evaporation is balanced by the rate of condensation at the liquid/vapor interface. Compounds that are more readily evaporated have a higher vapor pressure at a given temperature. Figure 3.2 shows the vapor pressure as a function of temperature for water and trichloroethylene (TCE). For any given temperature, the saturated vapor pressure of TCE is greater than the saturated vapor pressure of water. Therefore, TCE is more volatile than water. Another interesting note is that at 100°C, the saturated vapor pressure of water is 101 kPa, which is equal to the atmospheric pressure at sea level. Thus, the boiling point of water is simply the temperature at which the saturated vapor pressure equals the surrounding pressure. Therefore, according to Figure 3.2, the boiling point of TCE at sea level (101 kPa) is approximately 87°C. At higher elevations, the atmospheric pressure is lower, so the boiling point for liquids is also lower. Expressions relating vapor pressure to temperature for various compounds can be found in Reid et al. (1987).

The concentration of a vapor in equilibrium with its liquid can be calculated using the ideal gas law and the vapor pressure of the compound at the system temperature. For example, if water is poured into a container at 20°C, and then sealed with an amount of head space above the liquid water, the liquid water will evaporate until enough water vapor exists in the head space to exert a pressure equal to the vapor pressure at the system temperature.

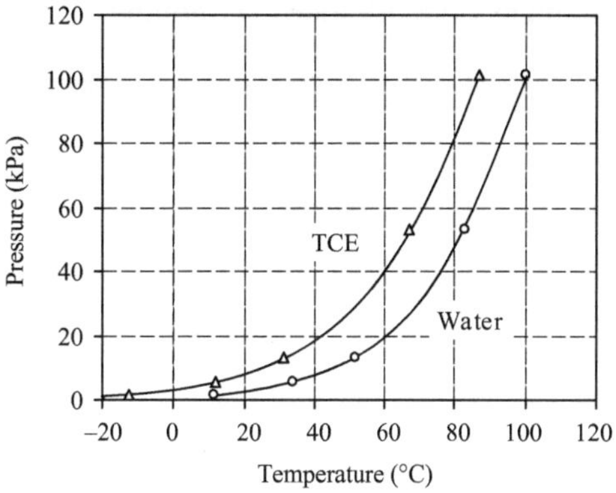

Figure 3.2. Saturated vapor pressures as a function of temperature for trichloroethylene (TCE) and water (data from Lide, 1990)

The vapor concentration can be calculated using the vapor pressure and the ideal gas law:

$$C_i = \frac{P_i}{R_i T} \tag{3.1}$$

where C_i is the vapor concentration of species i [kg/m^3], P_i is the vapor pressure of species i [Pa], R_i is the gas constant for species i [J/kg · K], and T is the absolute system temperature [K]. The concentration of a vapor is also often expressed in units of parts per million by volume (ppmv):

$$\text{ppmv}_i = \frac{P_i}{P} \cdot 1 \times 10^6 \tag{3.2}$$

where P is the total system pressure (Pa). Physically, these units express the number of molecules of i per million molecules of total constituents.

Example:
What is the saturated vapor concentration of water and TCE at 60°C and a surrounding atmospheric pressure of 100 kPa?

According to Figure 3.2, the approximate saturated vapor pressures for water and TCE are 20 kPa and 40 kPa, respsectively. The gas constant for water and TCE can be calculated by dividing the universal gas constant (8314 J/kmol · K) by the molecular weight of each compound. The molecular weight of water is 18 kg/kmol, and the molecular weight of TCE is 131 kg/kmol. Using the absolute system temperature of 60°C + 273 = 333 K, Equation (3.1) yields the vapor concentrations for water and TCE as 0.13 kg/m^3(or 0.13 g/L) and 1.9 kg/m^3 (or 1.9 g/L). The vapor concentrations of water and TCE in parts per million by volume are calculated using Equation (3.2) as 2×10^5 ppmv and 4×10^5 ppmv, respectively.

3.2.2 Multi-Component Liquid/Vapor Partitioning

Equilibrium partitioning between multicomponent liquids and gases is governed by the following equation:

$$P_i = x_i \gamma_i P_i^o \tag{3.3}$$

where P_i is the vapor pressure of species i, x_i is the mole fraction of i in the liquid mixture, γ_i is the activity coefficient of i in the liquid mixture, and P_i^o is the saturated vapor pressure of i. The activity coefficient is a measure of a compound's tendency to be either in the liquid phase or in the gas phase. Values less than one indicate that the compound will tend to remain in the liquid phase, and values greater than one indicate that the compound will tend to be in the gas phase at equilibrium conditions. If the liquid mixture is ideal, then the activity coefficient is one, and the partial vapor pressure of i is given by Raoult's Law:

$$P_i = x_i P_i^o \tag{3.4}$$

Physically, Raoult's Law states that the partial pressure of a compound is equal to its pure-component saturated vapor pressure multiplied by the mole fraction of that compound in the liquid phase. The reduction of a compound's vapor pressure in a mixture is due to the presence of other compounds that dilute the mixture. If two compounds, A and B, exist in an ideal liquid mixture at equal mole fractions (0.5), the molar density of each compound in the mixture is reduced by 50% from its pure-component phase. Therefore, it is reasonable to expect that the equilibrium partial pressure exerted by each compound in the mixture will be reduced by 50% as calculated by Raoult's Law.

In many aqueous solutions containing organic compounds, the equilibrium partial pressure does not follow the ideal behavior expressed by Raoult's Law. The organic compounds that are dissolved in the water have a tendency to escape into the gas phase, and the activity coefficient is greater than one. As a matter of convenience, the product of the activity coefficient and the saturated vapor pressure has been determined for a number of aqueous solutions at prescribed temperatures. The product of the activity coefficient and the saturated vapor pressure is called Henry's constant (H), and the resulting equilibrium expression for the partial pressure of an aqueous compound is known as Henry's Law:

$$P_i = x_i H_i \tag{3.5}$$

It is important to note that the value for Henry's constant can be reported in different units. In the above expression, H_i has units of [Pa $\cdot$ m^3/mol] where x_i is a molar concentration [mol/m^3]. Henry's constant can also be expressed in dimensionless form as follows:

$$C_i = H_{i,\text{dim}} \cdot C_{i,l} \tag{3.6}$$

where

$$H_{i,\text{dim}} = (H_i/RT) \cdot 1000 \tag{3.7}$$

$H_{i,\text{dim}}$ is the dimensionless form of Henry's constant [-], C_i is the vapor concentration of species i [kg/m^3], $C_{i,l}$ is the liquid concentration of species i [kg/m^3], R is the universal gas constant [8310 J/kmol · K], T is temperature [K], and the factor 1000 is a conversion from moles to kilomoles.

For compounds that have low solubility in water (e.g., hydrocarbons), Henry's constant is often approximated by using the following reasoning. The partial pressure of a compound A in water increases as the molar concentration of compound A increases. At the solubility limit, S [mol/m^3], a separate phase will precipitate out of the aqueous solution, and the partial pressure of compound A will equal the saturated vapor pressure of pure compound A. Because compound A has a low solubility in water, it is reasonable to expect that the partial pressure of A increases linearly with the molar concentration of A in water. Using Equation (3.5), the slope of the linear relationship between the partial pressure of A and the molar concentration of A in water is equal to Henry's constant. Using the solubility limit, S, as our data point for the linear regression, Henry's constant is approximated as follows:

$$H_i = P_i^o/S \tag{3.8}$$

3.3 INTERFACIAL PHENOMENA AND VAPOR PRESSURE LOWERING

In the previous section, the equilibrium vapor pressure of a compound was expressed as a function of temperature only. It was assumed that the gas phase existed adjacent to a flat liquid surface. In a porous medium, the curvature of the interface between the liquid and the gas phases may also affect the equilibrium vapor pressure of a compound. Very small pores can produce a very large capillary suction for a wetting fluid. The resulting tension in the liquid phase tends to reduce the equilibrium partitioning of the compound in the gas phase. This is called vapor-pressure lowering, and the equilibrium partial pressure of vapor, P_v [Pa], over a liquid in capillary tension can be expressed by the following expression known as Kelvin's equation:

$$P_\text{v} = P_\text{sat} \exp\left(\frac{-P_\text{c}}{\rho RT}\right) \tag{3.9}$$

where

$$P_\text{c} = \sigma \left(\frac{1}{r_1} + \frac{1}{r_2}\right) \tag{3.10}$$

P_sat is the saturated vapor pressure [Pa], P_c is the capillary pressure [Pa] (defined as the non-wetting phase pressure minus the wetting phase presure), ρ is the liquid density [kg/m^3], R is the gas constant [J/kg · K], T is the absolute temperature [K],

σ is the surface tension of the liquid [N/m], r_1 and r_2 are the radii of curvature [m] of the liquid–vapor interface. Equation (3.9) is derived by equating the chemical potential of the liquid and vapor phases, and Equation (3.10) is derived through either force-balance relations or thermodynamic analysis of the interfacial region (Carey, 1992). It should be noted that the "complete" Kelvin equation includes an additional term inside the exponential in Equation (3.9) that is negligible in most situations except in high-pressure systems (Elliott, 2001). In addition, for capillary tubes, the radii of curvature of the meniscus are equal and related to the radius of the capillary tube, r, and the contact angle, θ, by the following expression:

$$r_1 = r_2 = r/\cos(\theta) \tag{3.11}$$

As the radii of curvature are reduced (i.e., smaller pores), the capillary pressure is increased and the resulting partial pressure in the gas phase is decreased. Note that Equation (3.9) can also be expressed as the ratio of the partial vapor pressure to the saturated vapor pressure (P_v/P_{sat}). For water, this ratio expresses the relative humidity. Figure 3.3 shows a plot of this ratio for TCE and water as a function of radius of curvature (assuming $r_1 = r_2$).

Example:
What is the relative humidity in a porous medium if the concave radius of curvature of the liquid islands in the pores is 0.1 microns at 20°C?

An approximate solution can be obtained from the plot in Figure 3.3, which expresses the humidity of water as a function of the concave radius of curvature at 20°C. For a radius of curvature of 0.1 micron (1×10^{-7} m), the relative humidity of water is just less than one (~0.99). The exact solution using Equations (3.9) and (3.10) with $\sigma = 0.073$ N/m and $R = 462$ J/kg · K is $P_v/P_{sat} = 0.989$.

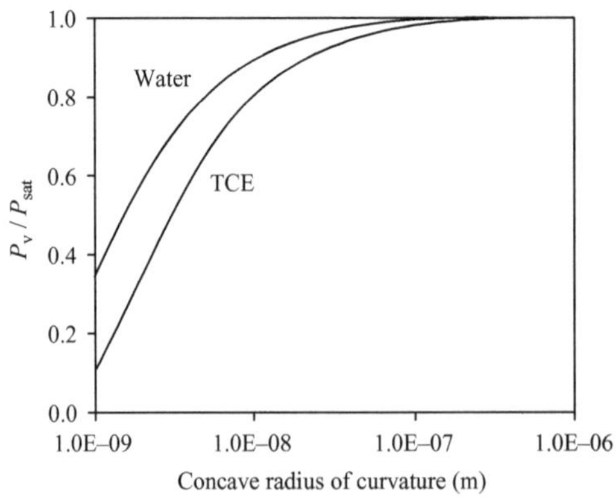

Figure 3.3. Vapor pressure ratio as a function of concave radius of curvature for water and TCE at a temperature of 20°C

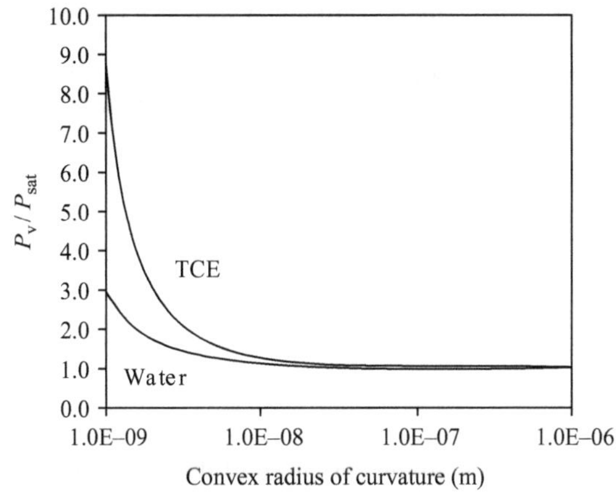

Figure 3.4. Vapor pressure ratio as a function of convex radius of curvature for water and TCE at 20°C

If the liquid curvature is convex instead of concave, a modified form of Kelvin's Equation predicts that the resulting equilibrium vapor pressure can be larger than the equilibrium vapor pressure over a flat surface of liquid. Convex surfaces of liquid can occur when the liquid is non-wetting, or if the liquid is suspended as a pendant droplet. The capillary pressure can be calculated from the Young-Laplace equation, using the liquid as the non-wetting phase in this configuration. Equating the chemical potential of the liquid and vapor phases yields the equilibrium vapor pressure for a convex liquid interface:

$$P_v = P_{sat} \exp\left(\frac{P_c}{\rho RT}\right) \tag{3.12}$$

The only difference between Equation (3.9) and Equation (3.12) is that the term in the exponent is positive in Equation (3.12). Thus, for larger capillary pressures (smaller radii of curvature), the equilibrium vapor pressure can exceed the saturated vapor pressure. In both Equations (3.9) and (3.12), if the radii of curvature is infinite (flat interface), the capillary pressure is zero and the equilibrium vapor pressure equals the saturated vapor pressure. A plot of Equation 3.12 for TCE and water is shown in Figure 3.4.

3.4 RATE OF EVAPORATION

The rate of evaporation of liquids depends on a number of factors including system temperature, liquid composition, ambient vapor concentration, air velocity, extent of stagnation zones, and curvature of the liquid–vapor interface. In this section, we derive analytical solutions that define the rate of evaporation of liquid in porous media under both quiescent (diffusion-limited) and convective conditions.

3.4.1 Steady One-Dimensional Diffusion-Limited Evaporation

In the absence of externally induced flow (e.g., from soil–vapor extraction and barometric pumping in the vicinity of wells), movement of gas in the subsurface can be dominated by diffusive transport. Consider the case of one-dimensional evaporation from the water table to the land surface. We assume that the concentration of water vapor at the surface of the water table, C_s [kg/m^3], is constant and can be derived from the ideal gas law:

$$C_\mathrm{s} = \frac{P_\mathrm{s}}{RT} \tag{3.13}$$

where P_s is the vapor pressure at the liquid surface [Pa], R is the gas constant for the compound of interest [J/kg · K], and T is the system temperature [K]. Note that the vapor pressure, P_s, depends on the curvature of the liquid–gas interface and can be less than or greater than the saturated vapor pressure over a flat liquid surface (see Section 3.3).

If the vapor concentration at the land surface, C_∞, is also assumed constant, the steady-state vapor concentration profile between the water table and the land surface is linear and readily determined from the one-dimensional steady diffusion equation with constant concentration boundary conditions as follows:

$$C(y) = \frac{C_\infty - C_\mathrm{s}}{L} y + C_\mathrm{s} \tag{3.14}$$

where L is the distance between the water table and the land surface. The rate of evaporation, $\dot{m}_\mathrm{e}$ [kg/s], is determined by applying Fick's Law:

$$\dot{m}_\mathrm{e} = -DA\frac{\mathrm{d}C}{\mathrm{d}y} = DA\frac{C_\mathrm{s} - C_\infty}{L} \tag{3.15}$$

where D is the effective vapor diffusion coefficient that accounts for the effects of liquid saturation, porosity, and tortuosity.

3.4.2 One-Dimensional Radial Solution in Spherical Coordinates

If we assume that the shape of the evaporating liquid is spherical, we can derive a steady, one-dimensional solution for the radial concentration profile and evaporation rate. Solutions of this form have been presented by Ho (1997) for evaporation from a pendant water droplet. Analogous conditions can occur if evaporation occurs spherically from a stagnant liquid zone in porous media surrounded by a region with a constant initial vapor concentration. For example, a non-aqueous phase liquid (NAPL) may be trapped in a low-permeability lens and is evaporating slowly to its surroundings. For these scenarios, the steady one-dimensional radial diffusion equation in spherical coordinates can be written as follows:

$$\frac{\mathrm{d}}{\mathrm{d}r}\left(r^2\frac{\mathrm{d}C}{\mathrm{d}r}\right) = 0 \tag{3.16}$$

where

$$C(r = r_o) = C_o \tag{3.17}$$

and

$$C(r \to \infty) = C_\infty \tag{3.18}$$

where r_o is the radius of the evaporating droplet or the radial extent of the evaporating plume. The solution is given as follows:

$$C(r) = (C_o - C_\infty)\frac{r_o}{r} + C_\infty \tag{3.19}$$

The evaporation rate is determined by applying Fick's law at the edge of the evaporating surface:

$$\dot{m}_e = -DA \left.\frac{dC}{dr}\right|_{r=r_o} = 4\pi D r_o (C_o - C_\infty) \tag{3.20}$$

The vapor-phase concentrations can be determined from the vapor pressures using appropriate equations of state, and it should be noted that the vapor concentration at the edge of the evaporating liquid surface may be greater than or less than the saturated vapor concentration above a flat liquid surface depending on the radii of curvature of the evaporating liquid.

3.4.3 Transient One-Dimensional Diffusion-Limited Evaporation

In contrast to the steady solutions provided in the previous sections, transient solutions are presented here for cases where the boundary of the evaporating source is moving or when the distance between the source and the ambient concentration boundary is infinite.

The governing equation for transient one-dimensional gas diffusion in porous media is given as follows:

$$\frac{\partial C}{\partial t} = D\frac{\partial^2 C}{y} \tag{3.21}$$

where C is the gas concentration [kg/m^3], D is the effective diffusivity that accounts for gas saturation and porosity [m^2/s], t is time [s], and y is distance [m]. If $C(y \to \infty, t) = C(y, 0) = C_\infty$ and $C(0, t) = C_s$, the transient concentration profile can be written as follows (adapted from Carslaw and Jaeger, 1959):

$$\frac{C(y,t) - C_s}{C_\infty - C_s} = \text{erf}\left(\frac{y}{2\sqrt{Dt}}\right) \tag{3.22}$$

The evaporation rate is calculated using Fick's law at $y = 0$:

$$\dot{m}_e(t) = \frac{DA\,(C_s - C_\infty)}{\sqrt{\pi D t}} \tag{3.23}$$

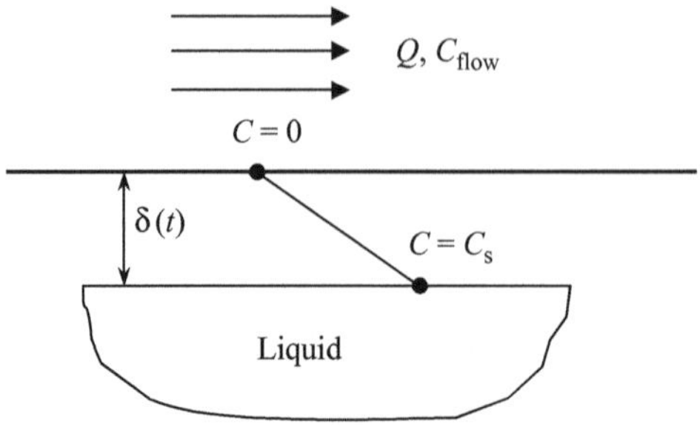

Figure 3.5. Sketch of one-dimensional evaporation from a liquid trapped in a stagnant region with external convection in an adjacent region

Another one-dimensional transient diffusion scenario exists if the evaporating liquid is trapped in a stagnant porous region while external convection exists (say, in an adjacent high-permeability zone). Figure 3.5 illustrates this scenario.

The concentration at the surface of the liquid is a constant, C_s, and the concentration at the boundary of the region with advective flow is approximately zero. The flow rate in the advective region is Q, and the average concentration in the advective region is denoted as C_{flow}. During the initial periods of this external convective drying scenario, mobile liquid will be drawn to the interface between the advective and stagnant regions by capillarity, keeping the evaporating surface stationary at the interface. When the liquid reaches a residual saturation and becomes immobile, the evaporating surface will begin to recede into the stagnant region (Ho and Udell, 1992). The distance between the receding evaporation front and the interface between the advective and stagnant regions is denoted as $\delta(t)$.

Assuming that the transport between the receding evaporation front and the advective zone is governed by diffusion only, and that the recession of the evaporation front is slow (quasi-steady), the evaporation rate at time, t, can be written as follows:

$$\frac{dm}{dt} = -DA\frac{C_s - 0}{\delta(t)} = -QC_{flow} \tag{3.24}$$

where A is the cross-sectional area available for diffusion. Equation (3.24) equates the evaporation rate (i.e., the rate of change in mass of the liquid, m) to the rate of diffusion through the stagnant region and to the rate of mass advected away. The rate of change in mass of the liquid can also be written as follows, assuming that the liquid saturation and porous-media properties are constant:

$$\frac{dm}{dt} = -\phi S_l \rho_l A \frac{d\delta(t)}{dt} \tag{3.25}$$

Equations (3.24) and (3.25) yield the following expression:

$$\frac{DAC_s}{\delta(t)} = \phi S_l \rho_l A \frac{d\delta(t)}{dt} \tag{3.26}$$

Equation (3.26) can be integrated and solved for $\delta(t)$:

$$\delta(t) = \sqrt{\frac{2DC_s t}{\phi S_l \rho_l}} \tag{3.27}$$

Equation (3.24) can then be used to calculate the bulk advective concentration, C_{flow}, and the transient evaporation rate, dm/dt, of a liquid receding into a stagnant region:

$$C_{flow} = \frac{A}{Q}\sqrt{\frac{\phi S_l \rho_l D C_s}{2t}} \tag{3.28}$$

$$\frac{dm}{dt} = A\sqrt{\frac{\phi S_l \rho_l D C_s}{2t}} \tag{3.29}$$

3.4.4 Steady Through-Flow Evaporation

In the preceding sections, the evaporation rate was limited by diffusion. In this section, we consider the case where a gas is flowing through a homogenous, unsaturated region containing a single liquid. Assuming that the flowing gas reaches local equilibrium with the stationary liquid that it passes through, the rate of evaporation of the bulk liquid can be expressed as follows:

$$\frac{dm}{dt} = -QC_{sat} \tag{3.30}$$

where the left-hand side is the time derivative of the mass, m [kg], of liquid in the control volume, Q is the air flow rate [m^3/s], and C_{sat} is the saturated gas concentration coming out of the control volume. Assuming macro-scale equilibrium, the effluent gas concentration, C_{sat}, can be obtained using the ideal gas law:

$$C_{sat} = \frac{P^o M}{RT} \tag{3.31}$$

where P^o is the saturated vapor pressure [Pa] at the system temperature, T [K], M is the molecular weight of the liquid [kg/kmol], and R is the universal gas constant [8300 $J/kmol \cdot K$]. Integration of Equation (3.30) yields a simple expression for the time required to remove the total mass of liquid in the system:

$$t = \frac{m^o}{QC_{sat}} \tag{3.32}$$

Thus, the macro-equilibrium scale model assumes an effluent concentration that is equal to the saturated concentration. This concentration is sustained until all of the liquid is removed at a time given by Equation (3.32).

3.4.5 Multicomponent Evaporation

In many instances, the evaporating liquid is comprised of more than one species. Subsurface contaminants are often mixtures of multiple components. Gasoline, for example, is comprised of numerous hydrocarbons. The compounds in a liquid mixture can have different vapor pressures, and the composition of the mixture can vary. The composition of the mixture impacts the partial vapor pressure in equilibrium with the different compounds in the liquid mixture. Consequently, the rate of evaporation depend on the composition as well as the nature of gas flow in the region of the liquid mixture. Two extreme flow conditions have been considered in multicomponent evaporation of liquid mixtures in porous media: (1) diffusion-limited evaporation and (2) through-flow evaporation.

In diffusion-limited evaporation, the liquid is contained in a stagnant region and the advective flow occurs in an adjacent region (see Figure 3.5). Ho and Udell (1992) observed that for these conditions, the effluent concentration of each species in the advective region corresponds to the bulk concentration in the liquid phase. Experiments were performed in a two-dimensional apparatus with a high-permeability layer overlying a low-permeability layer that contained different liquid mixtures. Ho and Udell (1995) derived an analytical solution for diffusion-limited evaporation of a binary mixture

In through-flow evaporation, advection occurs through the unsaturated liquid mixture. In these cases, the more volatile compounds are evaporated first. The concentration profile in the evaporating liquid propagates as individual waves corresponding to each species through the region (Ho et al., 1994; Ho, 1998). The effluent concentration of each species is dictated by the original composition of the liquid mixture until its wave propagates through the entire liquid-containing region. Then, its effluent concentration drops to zero and the remaining effluent concentrations increase to accommodate the increasing mole fractions in the mixture.

3.5 ENHANCED VAPOR DIFFUSION

Early studies of water-vapor movement in soils under an imposed temperature gradient revealed that vapor-phase diffusion in unsaturated media may be enhanced by several orders of magnitude due to pore-level processes and the presence of liquid "islands" in the porous media (Philip and deVries, 1957; Jury and Letey, 1979). The two factors causing this enhancement were postulated to be an increase in the local temperature gradient in the vapor compared to the bulk porous medium, and vapor transport across liquid islands at pore throats within the porous medium (Figure 3.6). In this section, we review the processes associated with enhanced vapor diffusion caused by vapor transport across liquid islands.

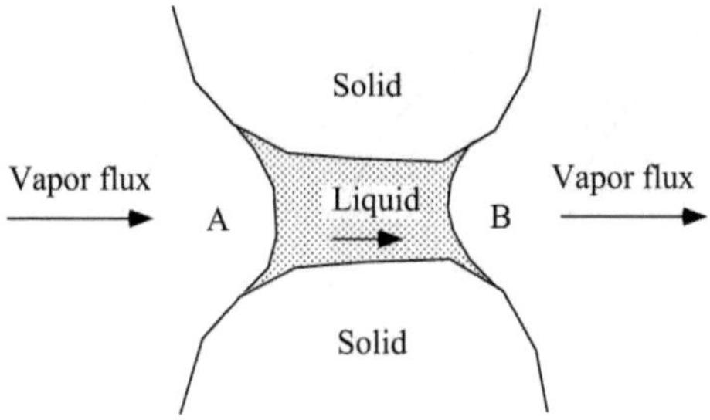

Figure 3.6. Water vapor flux across liquid islands driven by thermal gradients. Condensation occurs on the "hot" side (A) and evaporation occurs on the "cool" side (B) (from Ho and Webb, 1998, reprinted with permission)

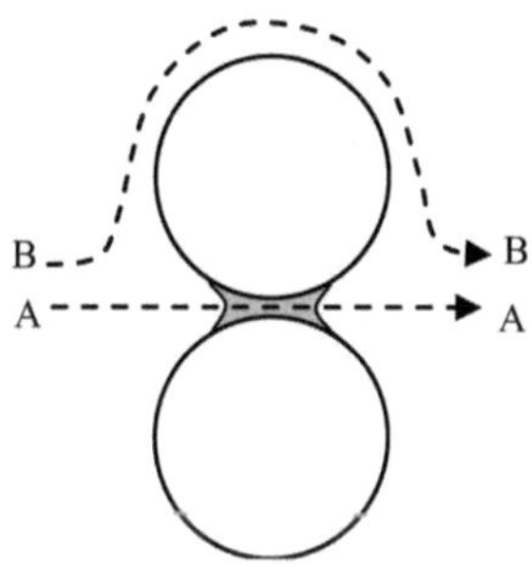

Figure 3.7. Pore-scale transport paths: (A) enhanced vapor-diffusion mechanisms and (B) Fickian diffusion (from Ho and Webb, 1998)

3.5.1 Analytical Model

As detailed in Ho and Webb (1998), a simple pore-scale analysis can be performed to estimate the steady-state mass flow of water across liquid islands due to condensation and evaporation mechanisms (Path A in Figure 3.7). A comparison between the flux due to condensation and evaporation mechanisms and the flux caused by Fickian diffusion (Path B in Figure 3.7) is made to discern the relative importance of enhanced vapor-diffusion mechanisms.

3.5.1.1 Path A

Consider the space between two solid particles that contains entrapped liquid water as shown in Figure 3.8. A thermal gradient in the water-wet porous medium induces vapor diffusion from the hot region (left side) to the colder regions (right side). Condensation on the upstream side of the liquid interface occurs because of vapor pressure lowering. An energy balance is performed on a control volume of the surface of the liquid island exposed to the hotter side (Figure 3.8). The latent heat of condensation

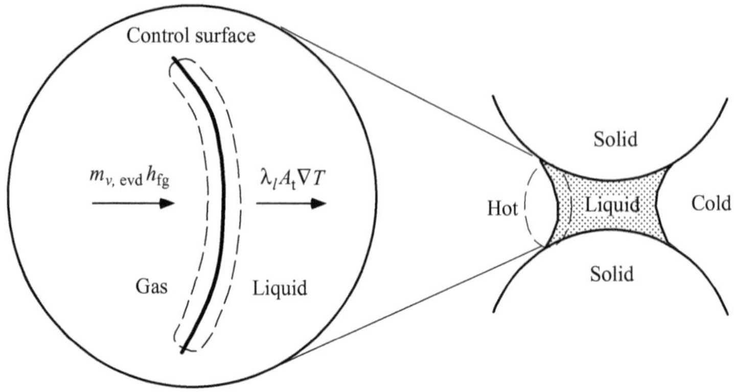

Figure 3.8. Control surface on which the energy balance is performed (from Ho and Webb, 1998)

on the upstream side of the liquid island is balanced by conduction through the liquid island:

$$m_{v,\mathrm{evd}} h_{\mathrm{fg}} = \lambda_l A_{\mathrm{t}} \frac{\Delta T}{L} \tag{3.33}$$

where $m_{v,\mathrm{evd}}$ is the mass flow rate of vapor that condenses on the upstream end of the liquid island [kg/s], h_{fg} is the latent heat of condensation [J/kg], λ_l is the liquid thermal conductivity of water [W/m · K], A_{t} is the cross-sectional area of the surface of the liquid island [m^2], ΔT is the temperature difference across the liquid island [K], and L is the linear distance across the liquid island [m]. Rearranging Equation (3.33) yields an expression for the mass flow of vapor due to the postulated enhanced vapor-diffusion mechanisms of condensation/evaporation:

$$m_{v,\mathrm{evd}} = A_t \frac{\lambda_l}{h_{\mathrm{fg}}} \frac{\Delta T}{L} \tag{3.34}$$

3.5.1.2 Path B

Fick's Law is now used to determine the mass flow of vapor if vapor diffusion has to occur around (rather than through) the liquid island shown in Figure 3.7:

$$m_{v,\mathrm{Fick}} = -DA_{\mathrm{p}} \nabla C = DA_{\mathrm{p}} \frac{\partial C}{\partial T} \frac{\partial T}{\partial x} \tag{3.35}$$

where $m_{v,\mathrm{Fick}}$ is the mass flow rate of vapor around the liquid island due to Fickian diffusion [kg/s], D is the binary diffusion coefficient of air and water vapor in free space [m^2/s], A_{p} is the cross-sectional area of pore space available for diffusion [m^2], C is the concentration (density) of water vapor [kg/m^3], and x is the direction along the actual path length [m]. The ideal gas law is used to express the water vapor concentration, and the temperature gradient is expressed in terms of the temperature difference across the liquid island and the tortuosity, τ (actual path length divided by linear path length, L):

$$m_{v,\text{Fick}} = DA_{\text{p}} \frac{M_{\text{v}}}{RT} \left(\frac{\partial P_{\text{v}}}{\partial T} - \frac{P_{\text{v}}}{T} \right) \frac{\Delta T}{\tau L} \tag{3.36}$$

where M_{v} is the molecular weight of water [18 kg/kmol], R is the ideal gas constant [8300 J/kmol-K], and P_{v} is the saturated vapor pressure of water [Pa].

3.5.1.3 Ratio

Equation (3.34) is divided by equation (3.36) to obtain a ratio of the mass flow rates due to enhanced vapor diffusion mechanisms and Fickian diffusion:

$$\frac{m_{v,\text{evd}}}{m_{v,\text{Fick}}} = \frac{A_{\text{t}}}{A_{\text{p}}} \frac{\lambda_l}{h_{\text{fg}} D} \frac{RT}{M_{\text{v}}} \frac{\tau}{\left(\frac{\partial P_{\text{v}}}{\partial T} - \frac{P_{\text{v}}}{T} \right)} \tag{3.37}$$

A value of this ratio on the order of one or more indicates that enhanced vapor phase diffusion can be significant relative to Fickian diffusion in transporting water vapor through a water-wet porous medium under a thermal gradient. Table 3.1 gives values for several parameters in equation (3.37) at 20°C.

Plugging the values in Table 3.1 into Equation (3.37) yields the following expression:

$$\frac{m_{v,\text{evd}}}{m_{v,\text{Fick}}} = 12 \frac{A_{\text{t}}}{A_{\text{p}}} \iota \tag{3.38}$$

Assuming that the cross-sectional area of the liquid island and the pore space available for vapor diffusion are similar and that the tortuosity is on the order of 1, Equation (3.38) yields the following:

$$\frac{m_{v,\text{evd}}}{m_{v,\text{Fick}}} \approx 10 \tag{3.39}$$

This result indicates that the contribution of water vapor transport by condensation and evaporation mechanisms across liquid islands (i.e., enhanced vapor diffusion)

Table 3.1. Summary of parameters used in analytical model for enhanced vapor diffusion

T[K]	298
λ_l [W/m · K]	0.6
h_{fg} [J/kg]	2.45×10^6
D [m^2/s]	2×10^{-5}
P_{v} [Pa]	2340
$\frac{\partial P_{\text{v}}}{\partial T}$ [Pa/K]	146

may be significant when compared to Fickian diffusion around liquid islands. The following sections present more detailed numerical models and experimental results that confirm this finding.

3.5.2 Numerical Model

Numerical, mechanistic models of the pore-scale processes associated with enhanced vapor diffusion were developed by Webb and Ho (1997) and subsequently applied by Webb (1998, 1999). The Dusty-Gas Model was used to simulate air–vapor advection and diffusion in a pore network including Knudsen and ordinary (Fickian) diffusion. Kelvin's equation was used to estimate vapor-pressure lowering effects at the liquid-island gas/liquid interface, and the Young-Laplace equation was used to evaluate gas–liquid pressure differences at both ends of the liquid island. Concentration gradients were applied to the pore-scale model to calculate the vapor and air flow rates.

The porous media was conceptualized to be a series of randomly-arranged spheres. Heat transfer was simulated to occur between the spheres by particle-to-particle contact, while flow of gas occurred around the spheres and around liquid islands. The liquid saturation was assumed to be sufficiently low such that the liquid was confined to pendular rings, or liquid islands. Bulk flow of liquid was not simulated. The numerical code TOUGH2 (Pruess, 1991) was employed for the simulations. Figure 3.9 shows an example of the simulated water-vapor-diffusion mass flow vectors using the numerical model.

Results of the numerical studies showed that significant enhancement of vapor diffusion in porous media was possible in the presence of liquid islands. The enhancement increased dramatically as the length of the liquid islands increased. In contrast, gas (air) diffusion was found to decrease slightly in the presence of liquid islands. The concentration gradient was found to have a dominant effect on enhancing the vapor diffusion. This contrasted previous modeling results of Philip and deVries (1957) that postulated enhancement would only occur due to temperature gradients. Although the concentration gradient and temperature gradient are often related (e.g., vapor pressure

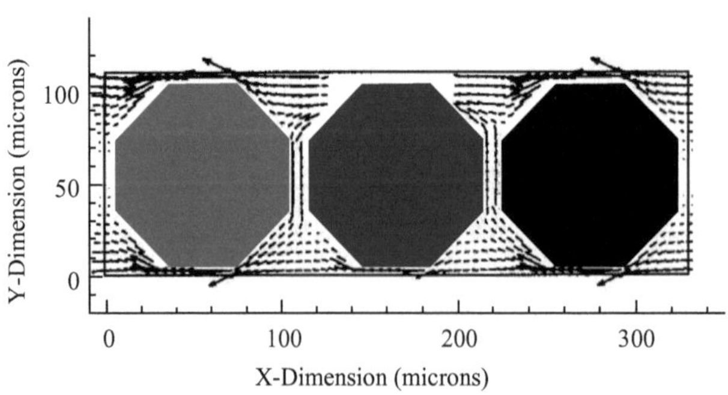

Figure 3.9. Simulated water-vapor-diffusion mass flow vectors in a pore network with a liquid island above the middle particle (from Webb and Ho, 1997, reprinted with permission)

variation with temperature), there are other situations that can cause isothermal vapor concentration gradients (e.g., spatial variability of a liquid species, the existence of brine, etc.). The next section discusses experimental studies that were performed to verify the results of both the analytical and numerical studies.

3.5.3 Experimental Studies

Experiments were performed to confirm the postulated processes of enhanced vapor diffusion. Experiments were conducted at both the pore scale and at a larger porous-media scale (containing a network of many pores).

The pore scale experiments performed by Silverman (1998) consisted of a small fabricated single-pore test cell (Figure 3.10). The column on the left contained pure water, and the column on the right contained a saturated brine solution with lithium chloride to reduce the saturated vapor pressure of water in the right column. The top of each column was covered by septa through which the different liquids could be introduced. Tests were conducted with and without the liquid island present, and the drawdown of liquid in the left column was used to evaluate the mass flow of water through the pore.

Results of the single-pore tests are shown in Figure 3.11. The plot clearly shows that the drawdown (and, hence, mass flow through the pore) was significantly greater when a liquid island was present in the pore. In addition, the drawdown rate increased as the length of the liquid island increased. Because the water vapor condenses on the upstream side and then evaporates on the downstream side of the liquid island, a longer liquid island produces a shorter effective pathway for the transport of water molecules. Additional tests were performed with a two-pore test cell, and results also confirmed that the presence of a liquid island within the pores augmented the transport of water from one column to another. Measured enhancement of vapor diffusion compared to diffusion through an open pore was up to a factor of three for a single pore. Additional enhancement is expected for a larger network of pores.

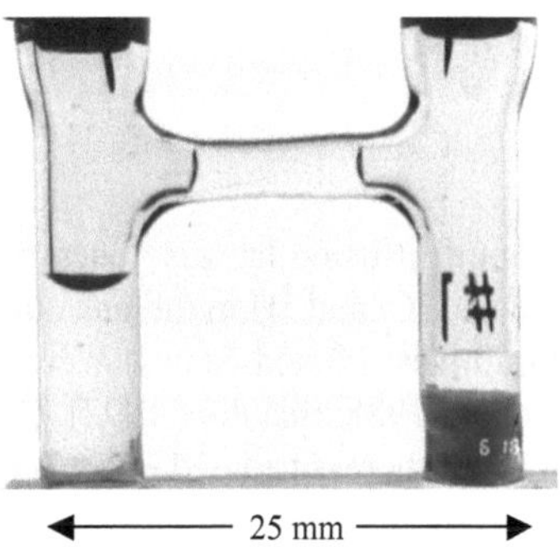

Figure 3.10. Pore-scale test cell with liquid island (Silverman, 1998)

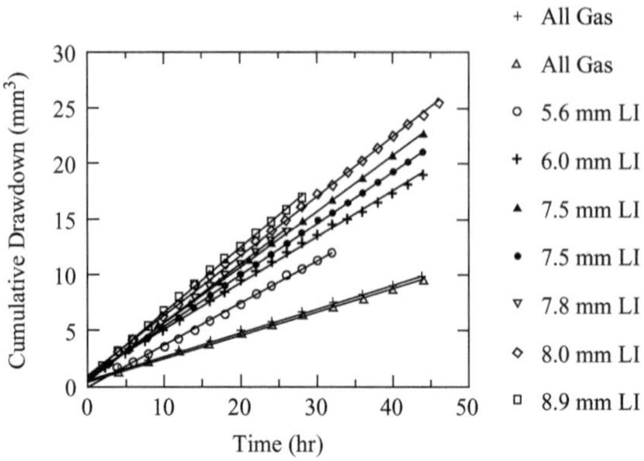

Figure 3.11. Drawdown rates of water for different liquid-island (LI) lengths in the pore (Silverman, 1998)

Additional experiments were performed by Gu et al. (1998) to investigate enhanced vapor diffusion processes in a porous packed bed. The apparatus consisted of an acrylic chamber (51 mm wide × 51 mm deep × 95 mm high). A bed of glass beads were packed in the chamber, which was capped with a water-circulated jacket that was maintained at a constant temperature. Beneath the water jacket was a fibrous wick that was connected to an external reservoir that provided a supply of saturated water at a fixed temperature. An electronic balance was used to monitor the water loss from the reservoir. The bottom of the chamber consisted of a desiccant tray and a water jacket. The tray could be removed for weighing to determine the accumulation of water. The entire apparatus was insulated with foam, and thermocouples were fitted along the chamber. A gamma attenuation system was used to measure the saturation profile of the packed bed during the experiment. During the experiment, water vapor would diffuse from the top toward the bottom desiccant. The rate of diffusion was monitored and compared with conditions when no porous media were present (i.e., diffusion in free space). Results showed that significant enhancement of vapor diffusion occurred under both transient and steady-state conditions when moist porous media was present in the chamber.

3.5.4 Field Studies

Evidence for enhanced vapor diffusion has also been observed in field studies. The distribution of anthropogenic ^{36}Cl and ^{3}H in the unsaturated zone in the Chihuahuan and Sonoran Deserts (Southwest United States) were used to evaluate liquid and vapor transport (Phillips et al., 1988; Scanlon, 1992). Chlorine-36 is nonvolatile and restricted to liquid phase flow whereas tritiated water is volatile and can move in both liquid and vapor phases. In the Chihuahuan Desert, tritium penetrated 1 m deeper than ^{36}Cl although ^{3}H fallout occurred later than that of ^{36}Cl. Deeper penetration of ^{3}H (1.4 m) relative to that of ^{36}Cl (0.5 m) was attributed to enhanced downward movement of ^3H in the vapor phase (Scanlon, 1992). The difference in moisture fluxes between the two tracers suggests a vapor flux of ~6 mm/year. Unsaturated-zone nonisothermal liquid- and vapor-flow modeling was used to simulate these processes, and enhanced vapor diffusion was required in the vapor-transport models to obtain the observed results. Additional details of these studies can be found in Scanlon (1992) and Scanlon and Milly (1994).

REFERENCES

Carey, V.P., 1992, Liquid-Vapor Phase-Change Phenomena, Hemisphere Publishing Corp, Washington.

Carslaw and Jaeger, 1959, Conduction of Heat in Solids, Clarendon Press, Oxford.

Elliott, J.A.W., 2001, On the Complete Kelvin Equation, Chemical Engineering Education, 35(4), 274–278.

Gu, L., C.K. Ho, O.A. Plumb, and S.W. Webb, 1998, Diffusion with Condensation and Evaporation in Porous Media, ASME Proceedings of the 7th AIAA/ASME Thermophysics and Heat Transfer Conference, HTD-Vol. 357-2, pp. 213–220.

Ho, C.K., 1997, Evaporation of Pendant Water Droplets in Fractures, Water Resources Research, Vol. 33, No. 12, pp. 2665–2671.

Ho, C.K., 1998, Analytical Inverse Model for Multicomponent Soil Vapor Extraction, J. Environmental Engineering, v. 124, no. 6, 504–509.

Ho, C.K. and K.S. Udell, 1992, An Experimental Investigation of Air Venting of Volatile Liquid Hydrocarbon Mixtures from Homogeneous and Heterogeneous Porous Media, J. Contam. Hydrol., 11, 291–316.

Ho, C.K., S.W. Liu, and K.S. Udell, 1994, Propagation of Evaporation and Condensation Fronts During Multicomponent Soil Vapor Extraction, J. Contam. Hydrol., 16, pp. 381–401.

Ho, C.K., and K.S. Udell, 1995, Mass Transfer Limited Drying of Porous Media Containing an Immobile Binary Liquid Mixture, Int. J. Heat Mass Transfer, Vol. 38 No. 2, pp. 339–350.

Ho, C.K., and S.W. Webb, 1998, Review of Porous Media Enhanced Vapor-Phase Diffusion Mechanisms, Models, and Data-Does Enhanced Vapor-Phase Diffusion Exist?, Journal of Porous Media, 1(1), pp. 71–92.

Jury, W.A., and J. Letey, Jr., 1979, Water Vapor Movement in Soil: Reconciliation of Theory and Experiment, Soil Sci. Soc. Am. J., Vol. 43, No. 5, pp. 823–827.

Lide, D.R. (ed.), 1990, CRC Handbook of Chemistry and Physics, CRC Press, Inc., Boca Raton.

Philip, J.R., and D.A. deVries, 1957, Moisture Movement in Porous Materials under Temperature Gradients, Trans. Am. Geophys. Union, Vol. 38, No. 2, pp. 222–232, p. 594.

Phillips, F. M., J. L. Mattick, and T. A. Duval. 1988. Chlorine 36 and tritium from nuclear weapons fallout as tracers for long-term liquid movement in desert soils. Water Resources Research 24:1877–1891.

Pruess, K., 1991, TOUGH2 – A General-Purpose Numerical Simulator for Multiphase Fluid and Heat Flow, LBL-29400, Lawrence Berkeley Laboratory, Berkeley, CA.

Reid, R.C., J.M. Prausnitz, and B.E. Poling, 1987, The Properties of Gases & Liquids, 4th Ed., McGraw-Hill Inc., New York.

Scanlon, B. R. 1992. Evaluation of liquid and vapor flow in desert soils based on chlorine-36 and tritium tracers and nonisothermal flow simulations. Water Resources Research 28:285–297.

Scanlon, B. R., and P. C. D. Milly. 1994. Water and heat fluxes in desert soils 2. Numerical simulations. Water Resources Research 30:721–733.

Silverman, T., 1998, A Pore-Scale Experiment to Evaluate Enhanced Vapor Diffusion in Porous Media, Master of Science Thesis submitted to the Department of Hydrology at the New Mexico Institute of Mining and Technology, Socorro, New Mexico.

Webb, S.W., 1998, Pore-Scale Modeling of Transient and Steady-State Vapor Diffusion in Partially-Saturated Porous Media, ASME Proceedings of the 7th AIAA/ASME Thermophysics and Heat Transfer Conference, HTD-Vol. 357-2, pp. 221–231.

Webb, S.W., 1999, Temperature Gradient Effects in Vapor Diffusion in Partially-Saturated Porous Media, in proceedings of the ASME/JSME Thermal Engineering Joint Conference, San Diego, CA.

Webb, S.W., and C.K. Ho, 1997, Pore-Scale Modeling of Enhanced Vapor Diffusion in Porous Media, in Proceedings of the ASME Fluids Engineering Division, FED-Vol. 244, pp. 457–468.

Webb, S.W., and C.K. Ho, 1998, Enhanced Vapor-Phase Diffusion in Porous Media LDRD Final Report, SAND98-2772, Sandia National Laboratories, Albuquerque, NM.

CHAPTER 4

SOLID/GAS PARTITIONING

SAY-KEE ONG

Dept. of Civil, Construction and Environmental Engineering, Iowa State University, Ames, IA, USA

Some of the earliest environmentally related studies on sorption of vapors onto soils involved pesticides, herbicides and fumigants. In these studies, the objective was to assess the extent and effects of adsorption on the performance of the pesticides (Hance, 1965). Over the last two decades, information on adsorption of pesticides onto soils is being complimented with information on adsorption of volatile organic compounds (VOCs) from poor management of waste disposal sites, landfills and leaking underground storage tanks. Soils in the unsaturated zone are generally coated with a film of water as the air pores of the unsaturated zone is saturated with water at a relative humidity greater than 99%. The amount of water retained by soil is a function of soil grain size, pore sizes, pore volume distribution, and the total surface area. Soils with very low or dry surfaces are generally found only on the surface of the ground (1–2 cm from the surface) and soils in arid regions. Moisture content in the unsaturated zone may be assumed to be approximately close to field capacity.

Vapor partitioning may be visualized as follows; under moist conditions, VOCs from the vapor phase distribute into soil matrices by (i) sorption at the gas-water vapor interface, (ii) dissolution into the aqueous phase, (iii) adsorption onto the mineral surface from the aqueous phase, (iv) partitioning into the organic matter from the aqueous phase, and (v) condensation of VOCs into pores. Under very low moisture conditions, typically of surface soils, direct sorption onto the mineral and direct partitioning into organic matter from the gas phase may occur in competition with water molecules or sorption may occur on top of the sorbed water molecules. Figure 4.1 illustrates the various mechanisms.

Sorption of VOCs onto oven-dried soils (typically surface soils) has been found to be greatly affected by the external surface area of the soil (Call, 1957; Jurinak, 1957; Jurinak and Volman, 1957; Chiou and Shoup, 1985; Rhue et al., 1988; Ong and Lion, 1991a; Ong and Lion, 1991b). For low concentrations typically encountered in ambient conditions, partitioning into solid phase may be characterized by a linear partition coefficient. However for high concentrations or vapor pressures that are

C. Ho and S. Webb (eds.), Gas Transport in Porous Media, 47–54.
© 2006 *Springer.*

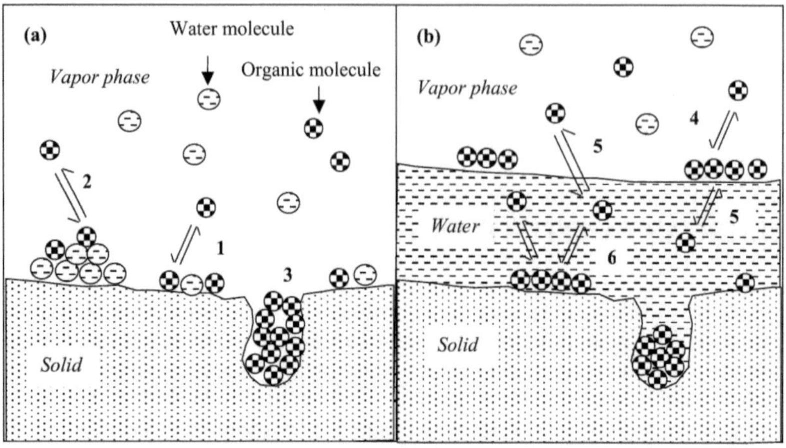

Figure 4.1. Vapor sorption (a) with less than a monolayer of water molecules on the surface, (b) with a film of water on the surface; Mechanisms: 1 – direct solid–gas sorption; 2 – sorption onto water molecules; 3 – condensation; 4 – sorption at the gas–water interface; 5 – dissolution; 6 – solid–water sorption

close to vapor saturation pressures as typically found in the areas with nonaqueous phase liquids, the adsorption isotherms for oven-dried soils are generally curvilinear in nature. The Brunauer-Emmett-Teller (BET) equation presented below is usually used to model the sorption phenomenon.

$$\frac{\Gamma}{\Gamma_m} = \frac{c\left(\frac{P}{P_o}\right)}{\left(1 - \frac{P}{P_o}\right)\left(1 + \frac{P}{P_o}(c-1)\right)} \tag{4.1}$$

where Γ is the amount of sorbate adsorbed per unit surface area (g/m^2), Γ_m is the monolayer of sorbate adsorbed per unit surface area (g/m^2), c is a constant related to the energy of adsorption, P is the vapor pressure of sorbate (mm Hg) and P_o is the saturated vapor pressure of the sorbate (mm Hg). Sorption of organic vapors to soils typically exhibits a type II curve with hysteresis at high relative vapor pressures ($P/P_o > 0.35$). The hysteresis generally indicates the occurrence of multilayer sorption and/or vapor condensation in pores (Call, 1957; Jurinak, 1957; Jurinak and Volman, 1957; Chiou and Shoup, 1985; Rhue et al., 1988; Ong and Lion, 1991b). Some researchers have proposed that sorption of VOCs onto dry soil may be considered as consisting of two fractions, that is, the mineral fraction which behaves as a conventional solid adsorbent and the organic matter fraction which acts as a partition medium (Boyd et al., 1988; Chiou et al., 1988). Under very dry conditions, the mineral surface area available for sorption is usually much larger than the organic carbon content present, making surface adsorption reactions more likely to dominate vapor uptake than partitioning into organic matter (Rhue et al., 1988).

Efforts have been made to correlate the linear gas–solid adsorption coefficients for sorption of various VOCs on different surfaces with various thermodynamic

Table 4.1. Examples of solid–gas sorption coefficients onto dry soils

Soil type	Organic carbon content (%)	Surface area (m^2/g)	Solid–gas partition coefficients, K'_d (cm^3/g)	Solid–water partition coefficient, K_d (cm^3/g)
Rhineback[a]	3	82.5	1,377	1.41
Harford[a]	1.6	44.0	1,669	0.96
Gila[a]	0.4	98.5	1,831	0.3
Montmorillonite[a]	0.02	733	23,300	0.31
Iron Oxides[b]	N/A	273	15,840	N/A
Kaolinite[b]	N/A	10	440	N/A
Aquifer Soil[c]	0.1	18.7	1,464	0.1
Yolo[c]	1.1	80.6	3,401	0.58

[a] Ong and Lion (1991a)
[b] Shimizu et al. (1992)
[c] Peterson et al. (1995)
N/A – not available

properties of the sorbate and sorbent. Most of these equations relate the adsorption coefficients to the enthalpies of vaporization, enthalpies of adsorption, and saturation vapor pressures (e.g., Goss and Schwarzenbach, 1998; Goss and Schwarzenbach, 1999). However, these equations do not consider the effects of moisture or humidity on the sorption of VOCs. Typical solid–vapor partitioning coefficients, K'_d, for trichloroethylene (TCE) on oven-dried soils are presented in Table 4.1. As a comparison, the aqueous solid–water partitioning coefficients, K_d, are also presented.

In a moist environment, the mass of VOC sorbed is less than the mass of VOC sorbed on oven-dried surfaces due to competition for sorption sites between water molecules and VOCs (Call, 1957; Chiou and Shoup, 1985; Rhue et al., 1989; Ong and Lion, 1991b; Thibaud et al., 1992, Goss, 1993; Storey et al., 1995). The charged surfaces of minerals and soils have a higher affinity or preference for water molecules than the hydrophobic vapor compounds. Figure 4.2 shows the impact of different relative humidities on the sorption of TCE onto alumina.

Assuming that the mass of VOC sorbed for each process is additive, the mass of VOC sorbed can be written as:

$$
\begin{aligned}
\text{Total mass sorbed} = {} & \text{Mass sorbed at the solid–water interface} \\
& + \text{Mass dissolved in the liquid phase} \\
& + \text{Mass sorbed at the gas–water interface} \\
& + \text{Mass sorbed at the solid–gas interface} \\
& + \text{Condensation into pores}
\end{aligned}
$$

If sufficient moisture is present on the surface of the soil, the mass sorbed at the solid–gas interface can be neglected. Ignoring condensation, the vapor partitioning

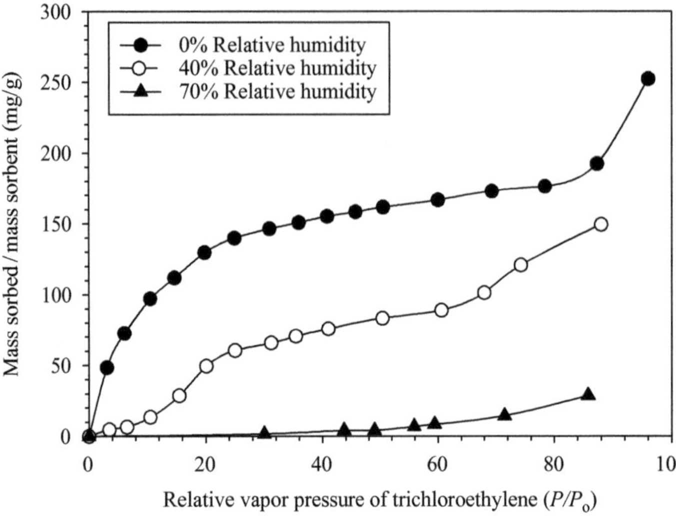

Figure 4.2. Effect of relative humidity on trichloroethylene adsorption on alumina

coefficient, K'_d can be written as:

$$K'_d = \frac{K_d}{K_H} + \frac{\theta}{K_H \gamma_\rho} + \frac{K_{iw}A}{M} \tag{4.2}$$

where K_d is the solid–gas partition coefficient (m^3/g), K_H is the dimensionless Henry's law constant, θ is the moisture content (in fraction), K_{iw} is water–gas partition coefficient (m) for sorption at water–gas interface (g), A is the water–gas interfacial area (m^2), γ is the aqueous activity coefficient, and M is the mass of sorbent (g). Applying the above equation and using the number of monolayers of water coverage instead of the moisture content, Ong and Lion (1991c) were able to generalize the impact of moisture on the solid–gas sorption. Figure 4.3 plots the total mass partitioned onto the soil less the mass sorbed at the solid–water interface (represented by the term, K_d/K_H) versus the number of monolayers of water on the surface. As shown in Figure 4.3, Ong and Lion found that for oven-dried minerals, direct adsorption onto the solid surface of the minerals for TCE was a controlling factor. From oven-dried to one monolayer coverage of water, direct sorption was evident with strong competition from water for adsorption sites. The adsorption coefficient was found to decrease by several orders of magnitude. Between a monolayer and six layer of water molecules coverage, likely interactions include sorption of TCE onto surface bound water and limited TCE dissolution into the sorbed water with some salting out effects caused by the sorbed water structure. Beyond six monolayer of water molecules, TCE dissolution into the condensed water, obeying Henry's Law, and partitioning at the solid–water interface were evident. At this moisture level, water behaves like bulk water on the surface of the soil for the dissolution of VOC vapors (Ong and

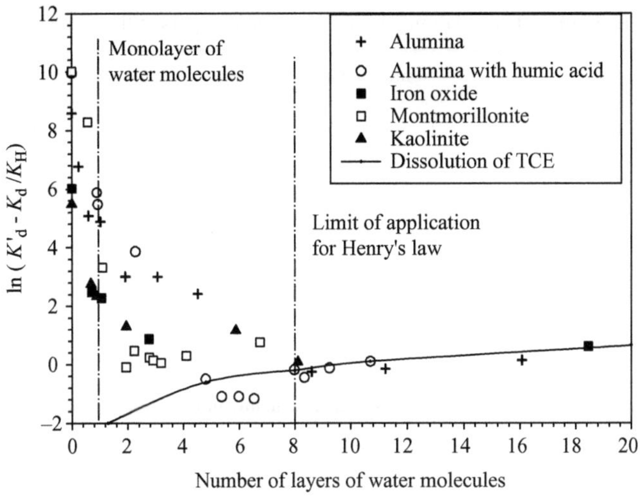

Figure 4.3. Impact of water molecules on mineral surfaces on trichloroethylene vapor sorption and the applicability of Henry's law (adapted from Ong and Lion, 1991c)

Lion, 1991c). The number of monolayers of water on the surface as presented in Figure 4.3 was based on surface area measurement using the ethylene glycol method. If the surface areas of the minerals were determined by the BET nitrogen method, the number of monolayers of water coverage when Henry's Law becomes applicable will be approximately eight monolayers. Similar results were obtained by Petersen et al. (1995) using toluene and TCE as the vapor sorbates.

In the work by Ong and Lion (1991c), they concluded that sorption at the water–gas interface was negligible. However, there are evidence that sorption at the water–gas interface may be significant. Valsaraj and Thibodeaux (1992) showed that model predictions based on the dissolution of VOCs into organic matter underpredicted the mass sorbed from gas phase. Therefore, the assumption that vapor sorption at high relative humidity behaves like aqueous phase sorption and dissolution into the water may be too simplistic and may not take into account other sorption processes such as sorption at the water–gas interface. Mass balance calculations by Thibaud et al. (1992) on the sorption of toluene and chlorobenzene onto an EPA standard soil at a relative humidity of 87%, clearly showed that aqueous phase sorption and dissolution into the water could not account for the mass sorbed from the vapor phase. Pennell et al. (1992) found that sorption of p-xylene at the water–gas interface may account for up to 50% of the total mass adsorbed from the vapor phase. Similarly, Hoff et al. (1993a, b) estimated that the water–gas interface may be responsible for up to 50% of the observed sorption of alkanes in aquifer material. Mass balances conducted by Conklin et al. (1995) on their experimental data showed that up to 60% of the total mass of p-xylene sorbed in their experiments may be at the water–gas interface. A recent review indicated that vapor sorption at the water–gas interface in soils may

Table 4.2. Examples of correlations for the sorption of vapors at the water–gas interface

Equation	Nomenclature	Reference
$\Gamma = -(p/RT)(d\gamma/dp)$	Γ – surface excess (mol cm^{-2}) γ – surface tension p – partial pressure R – universal gas constant T – temperature	Gibbs equation (as presented in Chattoraj and Birdi, 1984)
$\log K_{IW} = -5.53 + 0.2735 I_x$	K_{IW} – water–gas partition coefficient, I_x – first order molecular connectivity index	Valsaraj (1988)
$\log K_{IW} = -8.58 - 0.769 \log C_W^s$	K_{IW} – water–gas partition coefficient (m) C_W^s – saturated solubility of the VOC	Hoff et al. (1993b)
$\ln K_{IW} = A^* - (\Delta H_s/R)(1/T - 1/323.15) - C(100 - RH)(3)$ where $A^* = -0.615 \ln vp + 7.86ß - 5.80$ $C = -0.054\beta - 0.00070\,mR - 0.0041\mu + 0.00061\Delta H_s = 3.20 \ln vp - 50.2\beta - 55.0.$	K_{IW} – water–gas partition coefficient (m), vp – vapor pressure (pascal), β – hydrogen bond acceptor, mR – molar refraction, μ – dipole momentum, ΔH_s – heat of adsorption (kJ mol^{-1}) R – universal gas constant (kJ $mol^{-1}K^{-1}$)	Goss (1994)

be a dominant retention mechanism depending on the moisture content (Costanza and Brusseau, 2000). Obviously as the moisture content approaches saturation the water–gas interface (A) available will be reduced resulting in insignificant sorption at the water–gas interface. It must be pointed out that VOCs with high aqueous solubility would prefer to be in the aqueous phase rather than at the water–gas interface. TCE has an aqueous solubility of 1100 mg/L which is higher than the solubility of p-xylene (198 mg/L). Therefore it is probable that because of the physical characteristic of the TCE used by Ong and Lion (1991c), sorption at the water–gas interface was found to be negligible.

Several researchers have proposed correlations for the estimation of mass sorbed at the water–gas interface (see Table 4.2). Sorption at the water–gas interface can be modeled using the Gibbs equation where the surface excess is determined by the change in the surface tension with varying vapor pressure. The equations proposed by Goss (1994) imply that for a VOC with high saturated solubility and high vapor pressure vapors such as TCE, the air–water partition coefficients would be low and that the VOC that are most likely to be sorbed at water–gas interface are those with a high β, hydrogen bond acceptor, value.

Most vapor sorption studies were conducted using a single VOC in competition with water for sorption sites. There are only a few studies investigating sorption of multiple vapors under unsaturated zone conditions even though many different VOCs may be found at a typical contaminated site. For simplicity sake, it is assumed that each VOC behaves independently and does not affect the other (Baehr and Corapcioglu, 1987; Nye et al., 1994). This assumption which may be true for aqueous phase sorption of hydrophobic VOCs (Chiou et al., 1988) but may not be true for vapor phase sorption under dry conditions. For soils with low moisture conditions, VOCs may compete with each other for adsorption sites while on the other hand, sorption of one class of VOC may be enhanced by another class of VOC. If a condensed VOC phase is formed, this phase, in turn, may act as a medium in which other hydrophobic VOCs may partition into. Lindner (1990) showed that certain types of VOCs would compete for sorption sites onto moist minerals while, in some cases, sorption of some types of VOCs were enhanced. The work done by Lindner was for low concentrations ($P/P_o < 2\%$) and the VOCs used were chlorinated solvents and components of fuel. Nye et al. (1994) investigated the sorption of m-xylene and n-dodecane vapors on air-dried and oven-dried soil. They found that on air-dried soil the addition of a second component increased the sorption of the first for all relative vapor pressures. However, work done by Amali et al. (1994) showed that the ternary system of TCE, toluene and water did not indicate any increase in the sorption of one of the VOCs in the presence of the other compounds.

REFERENCES

Amali S., Peterson L.W., Rolston D.E., Modeling multicomponent volatile organic and water adsorption on soils. J. Environ. Quality, 1994; 36:89–108.

Baehr A.L., Corapcioglu M.Y., A compositional multiphase model for groundwater contaminated by petroleum products: ii numerical solution. Water Resourc. Res., 1987; 23:201–213.

Chattoraj D.K., Birdi K.S., Adsorption and the Gibbs Surface Excess, Plenum Press, NY. 1984.

Boyd S.A., Mortland M.M., Chiou C.T., Sorption characteristics of organic compounds on hexadecyltrimethylammonium-smectite. Soil Sci. Soc. Am. J., 1988; 52:652–657.

Call F., The mechanisms of sorption of ethylene dibromide on moist soils. J. Sci. Food Agri., 1957; 8:137–142.

Chiou C.T ., Shoup T.D., Gas sorption by organic vapors and effects of humidity on sorptive mechanisms and capacity. Environ. Sci. Technol., 1985; 19:1196–1200.

Chiou C.T., Kile, D.E., Malcolm R.L., Sorption of vapors of some organic liquids on humic acid and its relation to partitioning of organic compounds in soil organic matter. Environ. Sci. Technol., 1988; 22:298–303.

Conklin M.H., Corley T.L., Roberts P.A., Davis J.H., van der Water J.G., Nonequilibrium processes affecting forced ventilation of benzene and xylene in a desert soil. Water Resources Res., 1995; 31(5):1355–1365.

Costanza M.S., Brusseay M.L., Contaminant vapor adsorption at the gas–water interface in soils. Environ. Sci. Technol., 2000; 1–11.

Goss K., Effects of temperature and relative humidity on the sorption of organic vapors on clay minerals. Environ. Sci. Technol., 1993; 27(10):2127–2132.

Goss K., Adsorption of organic vapors on polar minerals surfaces and on a bulk water surface: development of an empirical predictive model. Environ. Sci. Technol., 1994; 28(4):640–645.

Goss K.U., Schwarzenbach R.P., Gas/solid and gas/liquid partitioning of organic compounds: critical evaluation of the interpretation of equilibrium constants. Environ. Sci. and Technol., 1998; 32(14): 2025–2032.

Goss K.U., Schwarzenbach, R.P., Empirical prediction of heats of vaporization and heats of adsorption of organic compounds. Environ. Sci. and Technol., 1999; 33(19):3390–3393.

Hance R.J., Observations on the relationship between the adsorption of diuron and the nature of the sorbent. Weed Res., 1965; 5:108–114.

Hoff J.T., Gillham R., Mackay D., Shiu W.Y., Sorption of organic vapors at the air–water interface in a sandy aquifer material. Environ. Sci. Technol., 1993a; 27(13):2789–2794.

Hoff J.T., Mackay D., Gillham R., Shiu W.Y., Partitioning of organic chemicals at the air–water interface in environmental systems. Environ. Sci. Technol., 1993b; 27(13):2174–2180.

Jurinak J.J., Adsorption of 1, 2-dibromo-3-chloropane vapor by soils. J. Agri. Food Chem., 1957; 5:598–601.

Jurinak J.J., Volman D.H., Application of Brunauer, Emmett and Teller equation to ethylene dibromide adsorption by soils. Soil Sci., 1957; 83:487–496.

Lindner S.R., Sorption of organic compounds and their mixtures on synthetic and natural soils. Ph.D. Dissertation, Cornell University, Ithaca, NY, 1990.

Nye P.H., Gerstl Z., Galin T., Prediction of sorption by soils and volatile hydrocarbon mixtures. J. Environ. Quality, 1994; 23:1031–1037.

Ong S.K., Lion L.W., Effects of soil properties and moisture on the sorption of TCE vapor. Water Research, 1991a; 25(1):29–36.

Ong S.K., Lion L.W., trichloroethylene vapor sorption onto soil minerals. Soil Sci. Soc. Am. Jour., 1991b; 55(6):1559–1568.

Ong S.K., Lion L.W., Mechanisms for trichloroethylene vapor sorption onto soil minerals. J. Environ. Quality, 1991c; 20 (1):180–188.

Pennell K.D., Rhue, R.D., Rao, P.S.C., Johnston, C.T., Vapor-phase sorption of p-xylene and water on soils and clay minerals. Environ. Sci. Technol., 1992; 26 (4):756–763.

Petersen L.W., Moldrup P., El-Farhan Y., Jacobsen O.H., Yamaguchi T., Rolston D.E., The effect of moisture and soil texture on the adsorption of organic vapors, J. Environ. Quality, 1995; 24:752–759.

Rhue R.D., Rao P.S.C., Smith R.E., Vapor-phase adsorption of alkylbenzenes and water on soils and clays. Chemosphere, 1988; 17(4) 727–741.

Rhue R.D., Pennell K.D., Rao P.S.C., Reve W.H., Competitive adsorption of alkylbenzene and water vapors on predominantly mineral surfaces. Chemosphere, 1989; 18(9/10):1971–1986.

Shimizu Y., Takei N., Tereshima Y., Roles of solid components on the sorption of trichloroethylene onto solids from vapor phase, Water Sci. Technol., 1992; 30:1–11.

Storey J.M.E., Luo W., Isabelle L.M., Gas/solid partitioning of semivolatile organic compounds to model atmospheric solid surfaces as a function of relative humidity. 1. Clean quartz. Environ. Sci. and Technol., 1995; 29: 2420–2428.

Thibaud, C., Erkey C., Akgerman A., Investigation of the effect of moisture on the sorption and desorption of chlorobenzene and toluene from soil. Environ. Sci. Technol., 1992; 27(11):2373–2380.

Valsaraj, K.T., Binding constants for non-polar hydrophobic organics at the air–water interface:comparison of experimental and predicted values. Chemosphere, 1988; 17:2049–2053.

Valsaraj, K.T., Thibodeaux L.J., *In* Fate of Pesticides and Chemicals in the Environment, J.L. Schnoor (ed.) John Wiley & Sons, Inc., New York, 1992; 155–174.

CHAPTER 5

TWO-PHASE GAS TRANSPORT

STEPHEN W. WEBB
Sandia National Laboratories, P.O. Box 5800, Albuquerque, NM 87185, USA

Chapter 2 was limited to all-gas conditions. This chapter addresses two-phase, or unsaturated, gas transport in the presence of a liquid. While the same general transport mechanisms (advection and diffusion) exist as discussed in Chapter 2, gas transport under two-phase conditions must consider the influence of the liquid phase. Not only does the liquid phase interfere with gas-phase transport in the gas phase, but gas transport may occur in the liquid phase as a dissolved species. The presentation in this chapter complements the discussion given in Chapter 2.

As in Chapter 2, energy transport is not discussed in this chapter. It is treated by Plumb (Chapter 27 of this book), Nield and Bejan (1999), and Kaviany (1995).

5.1 GAS-PHASE ADVECTION

5.1.1 Darcy's Law

Darcy's law as presented in Section 2.1.1 is often extended to two-phase conditions with separate equations for the gas and liquid phases. Relative permeabilities for each phase are introduced into the momentum equations to account for the influence of the other phase. Each phase may be at a different pressure due to interfacial curvature, so a pressure relationship between the two phases (capillary pressure) is also included.

For unsaturated flow conditions, the Darcy equations for mass flux can be written as

$$\overline{F_g} = -\frac{k_g\, k_{r,g}}{\mu_g}\, \rho_g \left(\nabla P_g - \rho_g \overline{g}\right)$$

$$\overline{F_\ell} = -\frac{k_\ell\, k_{r,\ell}}{\mu_\ell}\, \rho_\ell \left(\nabla P_\ell - \rho_\ell \overline{g}\right)$$

The equation for the gas mass flux is the same as in Section 2.1.1 except for the addition of the term $k_{r,g}$, which is the gas-phase relative permeability. The relative

C. Ho and S. Webb (eds.), Gas Transport in Porous Media, 55–70.
© 2006 *Springer.*

permeability, k_r, is the ratio of the permeability of the porous media to that particular phase under unsaturated conditions divided by the permeability of the porous medium for saturated conditions for that particular phase. Under all-gas conditions, the gas-phase relative permeability is equal to 1.0. The liquid mass flux equation is similar to the gas flux equation with liquid-phase parameters.

There are different pressures for the gas and liquid phases due to capillary forces and interfacial curvature (see Ho (Chapter 3 of this book)). The difference between the gas (nonwetting) and liquid (wetting) pressures is referred to as the capillary pressure, or

$$P_c = P_g - P_\ell$$

For porous media where liquid is the wetting phase, the liquid phase pressure is less than the gas phase pressure, and the capillary pressure is a positive quantity. (Note that the liquid phase pressure can be greater than the gas phase as is the case for droplets as discussed in Ho (Chapter 3 of this book)).

In order to close the equation set, relationships for the capillary pressure and relative permeability are needed. These relationships are often referred to as two-phase characteristic curves. In general, the two-phase characteristic curves are a function of the pore structure, phase saturation, surface tension, contact angle, and hysteresis, (Dullien, 1992, pg. 338; Kaviany, 1995, pg. 428, 435). In addition, the phase relative permeabilities may be coupled, which leads to a revised form of the two-phase Darcy equations (Dullien, 1992, pg. 358). In practice, the two-phase characteristic curves are usually posed as a function of the liquid saturation, S_ℓ, in the porous medium, or the fraction of pore space occupied by liquid (deMarsily, 1986).

There are numerous sets of two-phase characteristic curves. Relationships for the capillary pressure as a function of saturation are usually based on experimental data. Expressions for relative permeability are then derived using theoretical relationships derived by Burdine (1953) or Mualem (1976) based on the capillary pressure expression. Two commonly used sets of characteristic curves are van Genuchten (1978, 1980) and Brooks and Corey (1964) as discussed below.

5.1.2 Two-Phase Characteristic Curves

5.1.2.1 van Genuchten

The van Genuchten curves (van Genuchten, 1978, 1980) are the most widely used set of two-phase characteristic curves. The capillary pressure relationship is given by

$$S_e = \left[\frac{1}{1 + (\alpha P_c)^n}\right]^m$$

which can be rearranged as

$$P_c = \frac{1}{\alpha}\left[S_e^{-1/m} - 1\right]^{1/n}$$

S_e is the effective saturation, or

$$S_e = \frac{S_\ell - S_{\ell,r}}{1 - S_{\ell,r}}$$

where α is the capillary pressure parameter, m and n are fitting parameters, and $S_{\ell,r}$ is the residual liquid saturation.

The relative permeability of the liquid phase can be predicted from the capillary pressure curve expression using the integral expressions of Mualem (1976) or Burdine (1953). The resulting liquid phase relative permeability integral can be expressed as a closed form expression for certain relationships between the m and n fitting parameters as detailed by van Genuchten (1978, 1980). For Mualem (1976), the relationship $m = 1 - 1/n$ leads to a closed form solution for the liquid-phase relative permeability, which is given by

$$k_{r,\ell} = S_e^{1/2}\left(1 - \left(1 - S_e^{1/m}\right)^m\right)^2 \quad (m = 1 - 1/n)$$

For Burdine (1953), the relationship $m = 1-2/n$ leads to a closed form expression, or

$$k_{r,\ell} = S_e^2\left(1 - \left(1 - S_e^{1/m}\right)^m\right) \quad (m = 1 - 2/n)$$

In both cases, m must be between 0 and 1.

With these restricted forms, the limits on the parameter n are $n \geq 1$ for Mualem and $n > 2$ for Burdine. As discussed by van Genuchten (1978, 1980) and van Genuchten and Nielsen (1985), both forms give similar results, but the Mualem form is preferred because of it has more general applicability due to the larger range for n, and the Mualem restricted form with $m = 1 - 1/n$ is generally recommended. The values of α, n (or m), and $S_{\ell,r}$ are determined from fitting experimental data for capillary pressure or head. A general program for fitting the experimental capillary pressure data is RETC (van Genuchten et al., 1991), which is freely available on the web.

An important point for the present discussion is that van Genuchten (1978, 1980) did not address gas-phase relative permeability. Parker et al. (1987) extended the van Genuchten–Mualem characteristic curves to include the gas-phase relative permeability, or

$$k_{r,g} = (1 - S_e)^{1/2}\left(1 - S_e^{1/m}\right)^{2m}$$

Similarly, Luckner et al. (1989) present a gas-phase relative permeability expression for the van Genuchten–Mualem characteristic curves.

$$k_{r,g} = (1 - S_e)^{1/3}\left(1 - S_e^{1/m}\right)^{2m}$$

The only difference in the two expressions is the exponent on the first expression; this exponent is referred to as the pore connectivity parameter.

The pore connectivity parameter value was originally proposed by Mualem (1976), who determined that a value of 1/2 minimized the data-model comparison error for the liquid-phase relative permeability. The value of 1/2 has recently been questioned by Schaap and Leij (2000), who found a better fit to the data using a value of -1.0 and a reduced value of the reference permeability. In any event, the pore connectivity value is a fitting parameter that has only been determined for the liquid phase, not the gas phase.

For gas-phase relative permeability, Parker et al. (1987) simply used the pore connectivity parameter based on the Mualem (1976) investigation. The value of 1/3 proposed by Luckner et al. (1989) was not discussed in the reference. Neither extension compared the proposed gas-phase relative permeability expressions to experimental data.

Figure 5.1 shows some generic van Genuchten–Mualem two-phase characteristic curves for some typical parameters as noted on the figures. The van Genuchten capillary pressure curve exhibits unphysical behavior as the liquid saturation is reduced (gas saturation is increased). As the liquid residual value (0.2 in this case) is approached, the value of capillary pressure goes to infinity; this behavior is discussed in more detail below. The liquid relative permeability starts out low and increases dramatically with increasing liquid saturation. The gas-phase relative permeability decreases with increasing liquid saturation and is concave down. The difference between the Parker et al. (1987) and Luckner et al. (1989) expressions is small for these parameter values.

Typical values of the van Genuchten–Mualem parameters for 34 soils are tabulated by Stephens (1996, pg. 186, Table 8), although the third column should be cm^{-1} instead of m^{-1} as can be verified by checking the original tabulations in Stephens et al. (1987) and van Genuchten (1978, 1980). The parameter ranges are α from 0.004 to 0.12 cm^{-1} ($1/\alpha$ from 817 to 24,500 Pa), n from 1.17 to 7.62 (m from 0.15 to 0.87), and $S_{\ell,r}$ from 0 to $0.4/\phi$ (Stephens et al., 1987, used a water content form of the effective saturation equation).

In order to alleviate the problem of an infinite capillary pressure as the liquid residual saturation is approached, or as one approaches all-gas conditions, a number of modifications to the van Genuchten capillary pressure curve have been proposed that extend the curve to zero liquid saturation. Webb (2000) presents a simplified approach to extending the van Genuchten capillary pressure relationship; he also reviews other approaches. The approach simply extrapolates the van Genuchten curve with a logarithmic extension (linear on a semi-log plot) to a capillary pressure value of 10^9 Pa at zero liquid saturation imposing continuity of the capillary pressure derivative. As an example, Figure 5.2 shows the result for Palouse soil (Webb, 2000). The approach is simple to use and fits the limited existing low liquid saturation (high gas saturation) capillary pressure data of Campbell and Shiozawa (1992) for a number of soils.

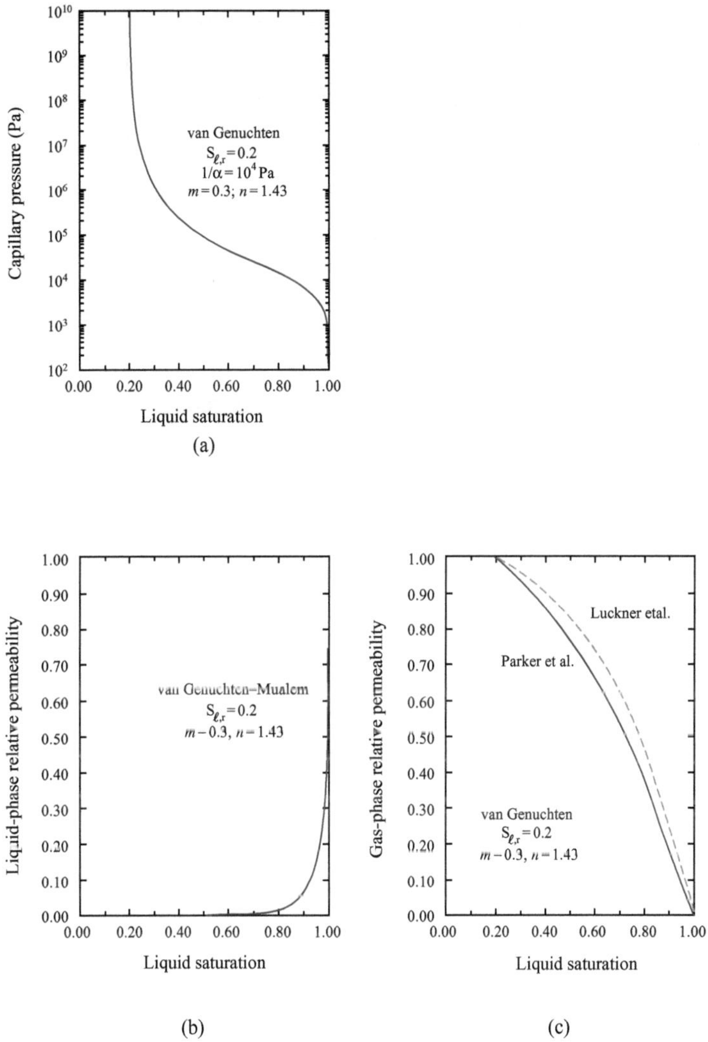

Figure 5.1. Representative van Genuchten–Mualem two-phase characteristic curves (a) Capillary pressure, (b) Liquid relative permeability, (c) Gas relative permeability

5.1.2.2 Brooks and Corey

Brooks and Corey (1964, 1966) developed another popular set of two-phase characteristic curves. Based on experimental observations, the effective saturation is a linear function of capillary pressure on a log-log plot, or

$$S_e = \left(\frac{P_b}{P_c}\right)^{\lambda} \quad \text{for } P_c \geq P_b$$

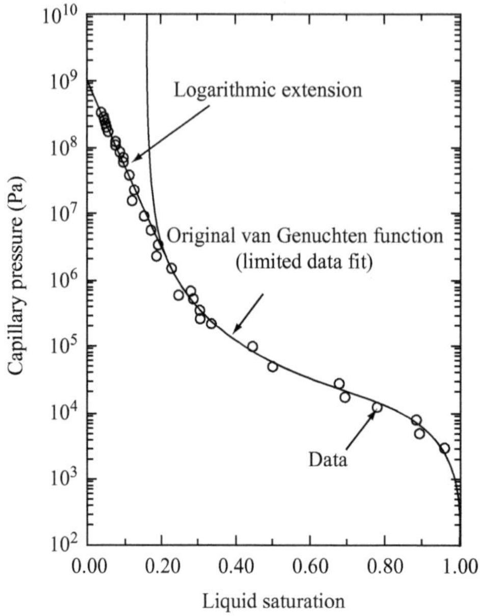

Figure 5.2. Modified van-Genuchten capillary pressure curve, (reprinted from Webb, 2000, with permission)

which can be rearranged as

$$P_c = \frac{P_b}{S_e^{1/\lambda}}$$

where

$$S_e = \frac{S_\ell - S_{\ell,r}}{1 - S_{\ell r}}$$

P_c is the capillary pressure, P_b is the bubbling pressure, S_e is the effective saturation (same as van Genuchten expression), $S_{\ell,r}$ is the residual liquid saturation, and λ is the pore-size index. The bubbling pressure, which is also called the displacement pressure (Corey, 1986, pg 39, 46) is the extrapolated capillary pressure at full liquid saturation. The linear relationship breaks down as full liquid saturation is approached. Values of the bubbling pressure and pore-size index are obtained by fitting the experimental data, which can be done using the RETC program (van Genuchten et al., 1991).

Brooks and Corey used the Burdine (1953) theory to derive expressions for the wetting (liquid) and nonwetting (gas) phases, or

Liquid relative permeability

$$k_{r,\ell} = S_e^{(2+3\lambda)/\lambda}$$

Gas relative permeability

$$k_{r,g} = (1 - S_e)^2 \left(1 - S_e^{(2+\lambda)/\lambda}\right)$$

van Genuchten (1976, 1980) presents Mualem-based expressions for the Brooks and Corey liquid-phase relative permeability. A gas-phase expression could be similarly derived. However, because these expressions have not been compared with experimental data, they will not be presented here.

Figure 5.3 shows some generic Brooks and Corey two-phase characteristic curves. Note that the Brooks and Corey capillary pressure does not go to zero as the liquid saturation goes to 1.0. This behavior is expected because the straight-line fit does not fit the data in this region ($P_c < P_b$). In contrast, the van Genuchten capillary pressure curve in Figure 5.1 does go to zero as the liquid saturation goes to 1.0. This behavior at high liquid saturations is one reason why the van Genuchten characteristic curves are much more popular than the Brooks and Corey formulation. The behavior of the Brooks and Corey capillary pressure curve is similar to van Genuchten as the saturation approaches the residual value. Note that Webb (2000) only discussed extension of the van Genuchten capillary pressure curve, although the same procedure could be applied to Brooks and Corey. The liquid-phase relative permeability increases significantly as liquid saturation is increased, similar to van Genuchten–Mualem shown earlier. The gas-phase relative permeability decreases with increasing liquid saturation and is concave up. The shape of the curve is opposite of the van Genuchten–Mualem curves (Parker et al., 1987; Luckner et al, 1989 extensions) shown earlier. Note that Brooks and Corey (1964, 1966) present data-model comparisons for the capillary pressure as well as the wetting (liquid) and non-wetting (gas) relative permeabilities, which show good results for all relationships including non-wetting relative permeability. The two van Genuchten Mualem relationships have not been compared to data.

The range of the Brooks and Corey parameters for the nine soils investigated by Brooks and Corey (1966) are P_b from 14 to 75 cm (1400 to 7500 Pa), λ from 1.8 to 7.3, and $S_{\ell,r}$ from 0.085 to 0.577.

The Brooks and Corey characteristic curves can be shown to be a limiting case of van Genuchten (van Genuchten (1978, 1980). For large values of capillary pressure ($(\alpha P_c)^n \gg 1$), the following relationships can be derived

$$P_b = \frac{1}{\alpha} \quad (\alpha P_c)^n \gg 1$$

and

$$\lambda = mn \quad (\alpha P_c)^n \gg 1$$

Using the Mualem relationship, $\lambda = m/(1-m)$; for Burdine, $\lambda = 2m/(1-m)$. These relationships were used for the Brooks and Corey capillary pressure curves presented in Figure 5.3 based on the van Genuchten parameters in Figure 5.1.

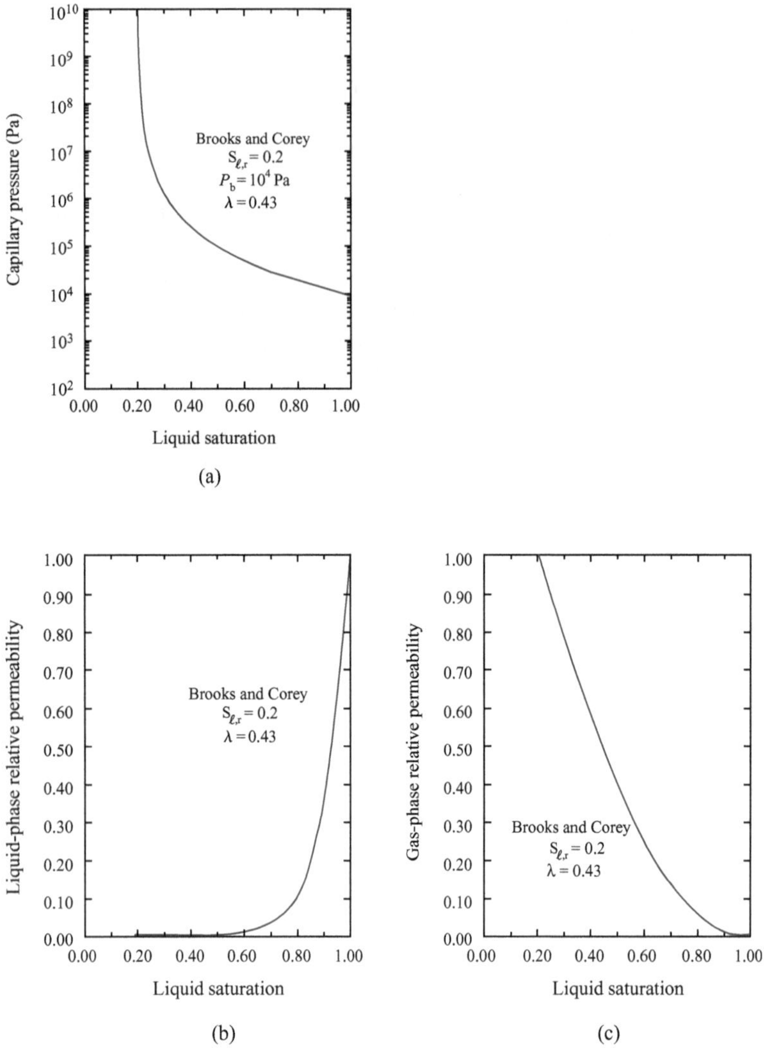

Figure 5.3. Representative Brooks and Corey two-phase characteristic curves (a) Capillary pressure, (b) Liquid relative permeability, (c) Gas relative permeability

The primary use of two-phase characteristic curves is for the flow of liquid in porous media; the flow of gas is not often investigated in detail. As mentioned above, the Parker et al. (1987) and Luckner et al. (1989) extensions for gas-phase relative permeability were made without comparison to experimental data. Brooks and Corey (1964, 1966) present some data-model comparisons for gas (non-wetting) that showed good agreement.

While the van Genuchten and Brooks and Corey two-phase characteristic curves may look similar, there are important differences when they are used in flow situations. McWhorter and Sunada (1990) and Webb (1992) evaluated the differences between the two sets of characteristic curves for analytical two-phase flow situations. The differences in the saturation profiles are significant, probably due to the different shape of the gas-phase relative permeability expressions. In general, the van Genuchten (and Parker) set of curves lead to a much sharper interface, while the Brooks and Corey predictions are much more blunt.

5.1.3 Brinkman Extension

The Brinkman extension for two-phase (unsaturated) flow has been analyzed by Whitaker (1994). The Brinkman correction terms are negligible for practical two-phase flow problems in porous media and are generally ignored.

5.1.4 Forchheimer Extension

The Forchheimer correction would be important for two-phase flow when the Reynolds number is sufficiently large compared to one. However, a formal analysis of the Forchheimer extension for two-phase (unsaturated) flow has not been performed.

5.1.5 Low Permeability Effects

Section 2.1.5 presented the low permeability effects for all-gas conditions. For unsaturated flow in porous media, Reinecke and Sleep (2002) investigated the variation of the Klinkenberg coefficient for air, b_{air}, including summarizing previous investigations. They developed a correlation for the Knudsen diffusion coefficient and the Klinkenberg parameter for unsaturated flow conditions as a function of effective gas permeability, which is simply the gas-saturated permeability times the gas relative permeability. Their Klinkenberg parameter correlation for air is

$$b_{\mathrm{air}} = 5.57\, k_{\mathrm{g,e}}^{-24}$$

where b_{air} is in Pa and $k_{g,e}$ is in m^2. Note that the units for the above equation for $k_{g,e}$ were specified as cm^2 in the original reference (equation 30). The correct units for $k_{g,e}$ are m^2 (Sleep, personal communication, 2005) as given above.

At an effective gas permeability of about 10^{-11} m^2, the above correlation is about the same as Heid et al. (1950). As the effective permeability decreases, the above correlation gives a smaller value of the Klinkenberg factor, or a smaller effect on the permeability, than either the Heid et al. (1950) or Jones and Owens (1980) correlations presented in Chapter 2. This difference may be due to the fact that liquid will preferentially block smaller pores, which have the most influence on gas slippage.

5.2 DIFFUSION

5.2.1 Ordinary Diffusion

As discussed in Section 2.2.1.1, Fick's law is often modified for application to porous media by the introduction of a porous media factor, β, which modifies the diffusion coefficient.

$$D^*_{AB} = \beta\, D_{AB}$$

The term β is defined as

$$\beta = \phi\, S_g\, \tau$$

where $D^*_{AB,pm}$ is the effective diffusion coefficient for the AB gas system in a porous media, D_{AB} is the effective diffusion coefficient of the AB gas system in a clear fluid, ϕ is the porosity, S_g is the gas saturation, and τ is the tortuosity. The Millington and Quirk (1961) correlation discussed in Chapter 2 is

$$\tau = \phi^{1/3} S_g^{7/3}$$

which can be rewritten as

$$\tau = \tau_o\, \tau_g = \phi^{1/3} S_g^{7/3}$$

where τ_o is the tortuosity due to the porosity and τ_g is the tortuosity due to the partial saturation of the porous medium. In Chapter 2, the gas saturation (S_g) was equal to 1.0. In this chapter, the gas saturation is generally not equal to 1.0 due to unsaturated conditions, and the τ_g factor ($S_g^{7/3}$) is retained.

5.2.2 Free-Molecule diffusion

Section 2.2.2 discusses the Knudsen diffusion coefficient, D_{iK}, for gas-only conditions. The gas-only equation based on the Klinkenberg coefficient is

$$D_{iK} = \frac{k_g\, b_i}{\mu_g}$$

For unsaturated conditions, the equation for the Knudsen diffusion coefficient needs to be modified due to the presence of liquid. Sleep (1998) used the Millington and Quirk (1961) tortuosity factor to evaluate the effect of unsaturated conditions, although the author notes that the approach might not be appropriate because molecular diffusion and Knudsen diffusion are two different processes. The porous media porosity, τ_o, is included twice with this approach, once in the tortuosity factor, τ, and once in the Klinkenberg factor as discussed in Chapter 2, and the resulting equation does not reduce to the proper form for all-gas conditions. Another approach for unsaturated flow was suggested by Webb (2001), who multiplied the Knudsen diffusion by the partial saturation tortuosity coefficient, τ_g, which is consistent with the arguments above and reduces to the proper expression for all-gas conditions.

In more recent work, Reinecke and Sleep (2002) have developed expressions for the Knudsen diffusion coefficient of air using the above Knudsen diffusion expression discussed in Section 5.1.5. Correction factors for the Klinkenberg coefficient for different gases and different conditions are discussed in Section 2.1.4. The gas permeability in the above equation is simply the effective value, or the gas-only value times the relative permeability. Any tortuosity factor is implicitly included in the Klinkenberg parameter.

Obviously this expression for the Knudsen diffusion coefficient is preferred to those proposed by Sleep (1998) and by Webb (2001), which were only estimates of the effect of unsaturated conditions.

5.3 COMBINED EFFECTS

Section 2.3 discusses combined effects for gas-only conditions. A number of parameters in the Dusty Gas Model need to be modified for application to unsaturated conditions as discussed above. The relative permeability should be used for the advection part of the transport, while tortuosity factors and a modified Klinkenberg parameter should be used for the ordinary (molecular) and free-molecule (Knudsen) diffusion components, respectively.

5.4 DISSOLVED GAS

In addition to the added complexity of unsaturated conditions on gas transport in the gas phase, the addition of a liquid phase adds another pathway for advection and diffusion of the gas. Gas will not only be transported in the gas phase; gas will also be transported as a dissolved species in the liquid phase. Due to the low concentration of dissolved gas in the liquid as discussed by Bird et al. (1960, page 538), Fick's law is adequate for dissolved gas diffusion. Note that advection of liquid will also transport dissolved gas. A simple addition of the two fluxes, similar to the ADM discussed in by Webb (Chapter 2 of this book), is probably adequate for dissolved gas flux in the liquid phase. In general, diffusion coefficients for liquids are about a factor of 10^4 lower than for gases at low pressure (Reid et al., 1987, Tables 11.2 and 11.5). However, the diffusion flux of gas in the liquid phase can be significant because of the concentration gradient in liquid phase, which can be larger than in the gas phase. The ratio of the gas concentration in the liquid and gas phases is given by the dimensionless Henry's constant as discussed by Ho (Chapter 3 of this book), or

$$K_H = \frac{c_g}{c_\ell}$$

where c_g is the concentration in the gas phase, and c_ℓ is the concentration in the liquid phase. A low value of the dimensionless Henry's constant indicates a significant concentration of dissolved gas in the liquid phase compared to the gas phase. Values of the dimensionless Henry's constant vary widely. For example, values for a number of trace organics range from 3.7×10^{-9} for bromacil to 1.4×10^{2} for n-octane (Jury et al., 1984).

A simplified approach to calculating the effective ordinary diffusion coefficient of gas under unsaturated conditions including dissolved gas diffusion was developed by Jury et al. (1983a). The model neglects advection and only considers ordinary (continuum) diffusion. The Millington and Quirk (1961) tortuosity expression is assumed to apply to both phases based on the saturation of that particular phase. The resulting expression for the unsaturated porous media diffusion coefficient can be written as

$$D^*_{\mathrm{AB}} = \phi^{4/3}\left(S_{\mathrm{g}}^{10/3} D_{\mathrm{AB,g}} + S_{\ell}^{10/3} D_{\mathrm{AB},\ell}/K_{\mathrm{H}}\right)$$

This expression uses the same approach presented by Jury et al. (1983a, 1991) but is slightly different. The above expression is based on the gradient of the concentration in the gas phase rather than the total concentration (gas, liquid, and solid phases). Note that the approach ignores the coordinate system issue discussed in Section 2.2.1.1. The above equation is a simple way to estimate the importance of dissolved gas diffusion in the liquid phase.

Figure 5.4 shows the results of an example calculation of the unsaturated porous medium diffusion coefficient for a porosity of 0.5, dimensionless Henry's constant of 10^{-4}, clear fluid gas diffusivity of 1.0×10^{-5} m^2/s at 1 bar, and a clear fluid dissolved gas diffusivity of 1.0×10^{-9} m^2/s. The figure shows the clear fluid gas diffusivity value as well as the reduction in the porous medium diffusivity due to the porous media. Diffusion in the gas phase dominates at low liquid saturation values, while dissolved gas diffusion in the liquid phase dominates at higher liquid saturations. Jury et al. (1991, 1984a) show that diffusion in the gas, liquid, or both phases may dominate depending on the dimensionless Henry's constant and the liquid saturation value.

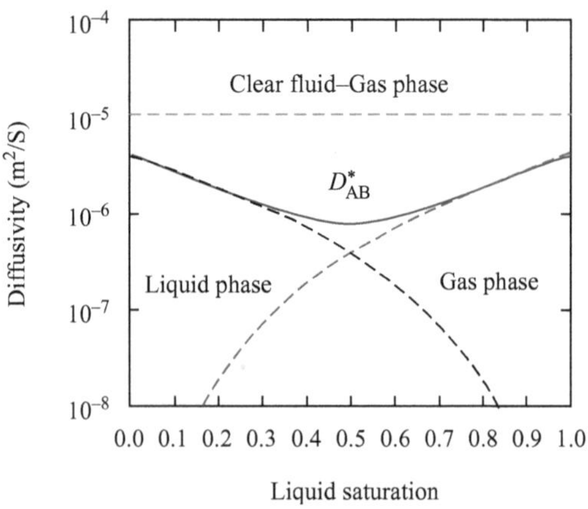

Figure 5.4. Unsaturated porous media diffusion coefficient

5.5 SIMPLIFIED MODELS FOR SOILS

There are simplified one-dimensional models for gas transport in soils that are often used. They use the effective diffusion coefficient discussed above as well as convection in the liquid phase due to evaporation or leaching and diffusion through a stagnant boundary layer at the soil surface. Because these models use simplifying assumptions such as constant water flux, constant atmospheric boundary layer thickness, negligible advection in the gas phase, etc., the models are generally used as screening models to investigate different scenarios.

The original screening models were developed by Jury et al. and are called the Behavioral Assessment Model (BAM) for the screening of various pesticides (Jury et al., 1983a, 1984a, b, c) and the Buried Chemical Model (BCM). The BCM is simply an extension of the BAM for buried organic chemicals (Jury et al., 1990). The BAM and BCM have been used to screen the behavior of various pesticides and volatile organic compounds by identifying the differences under typical environmental conditions. The BAM and BCM have been used in a study of landmine chemical signatures as discussed by Phelan (Chapter 21 of this book). Numerous extensions of Jury's approach have been made to include vapor-solid absorption such as Shoemaker et al. (1990).

The BAM and BCM are closed form solutions that are very amenable to fast calculations of the various effects. Note that the original equations in Jury et al. (1983a) have some errors in them. The equations are correct in Jury et al. (1983b, 1990).

Density effects may also influence unsaturated gas transport in soil. For example, evaporation of Volatile Organic Compounds (VOCs) in the subsurface, whose gas phase is heavier than air, may lead to gas stratification in the soil. This situation is discussed by Falta et al. (1989).

5.6 EXPERIMENTAL DATA

In contrast to the data presented in Section 2.4 for gas transport in porous media, there are no similar comprehensive data available for unsaturated flow conditions. As mentioned earlier, most investigations of flow and transport in the unsaturated (vadose) zone are concerned with liquid flow, not gas flow.

ACKNOWLEDGMENTS

Sandia is a multiprogram laboratory operated by Sandia Corporation, a Lockheed Martin Company, for the United States Department of Energy's National Nuclear Security Administration under Contract DE-AC04–94AL85000.

NOMENCLATURE

b	Klinkenberg coefficient
c	concentration
D	ordinary diffusion coefficient
D_{iK}	Knudsen diffusion coefficient

F	mass flux
$\overline{g}$	gravity
k	permeability
K_H	Henry's constant
n, m	van Genuchten fitting parameters
P	pressure
S	saturation

Greek

α	van Genuchten capillary pressure parameter
β	porous media factor
τ	tortuosity
ϕ	porosity
ρ	density
μ	viscosity

Subscript

A,B	components A, B
air	value for air
b	bubbling
c	capillary
e	effective
g	gas phase
ℓ	liquid phase
r	relative, residual
0	all-gas conditions

Superscript

D	ordinary diffusion
$*$	effective porous media value

REFERENCES

Bird, R.B., Stewart, W.E., and Lightfoot, E.N., 1960, Transport Phenomena, John Wiley & Sons, New York.

Brooks, R.H., and Corey, A.T., 1964, Hydraulic Properties of Porous Media, Hydrology Paper No. 3, Fort Collins, CO, Colorado State University.

Brooks, R.H., and Corey, A.T., 1966, "Properties of Porous Media Affecting Fluid Flow," Journal of the Irrigation and Drainage Division, Proceedings of the American Society of Civil Engineers (ASCE), Vol. 92, no. IR2, pp. 61–88.

Burdine, N.T., 1953, "Relative Permeability Calculations From Pore Size Distribution Data," Transactions of the American Institute of Mining and Metallurgical Engineers, Vol. 198, pp. 71–78.

Campbell, G.S., and Shiowaza, S., "Prediction of hydraulic properties of soils using particle-size distribution and bulk density data, International Workshop on Indirect Methods for Estimating the Hydraulic Properties of Unsaturated Soils, University of California, Riverside, CA, 1992.

Corey, A.T., 1986, Mechanics of Immiscible Fluids in Porous Media, Second Edition, Water Resources Publications.

deMarsily, G., 1986, Quantitative Hydrology, Academic Press, Inc.

Dullien, F.A.L., 1992, Porous Media: Fluid Transport and Pore Structure, Second Edition, Academic Press, Inc., San Diego.

Falta, R.W., Javandel, I., Pruess, K., and Witherspoon, P.A., 1989, "Density-Driven Flow of Gas in the Unsaturated Zone Due to the Evaporation of Volatile Organic Compounds," Water Resourc. Res., 25:2159–2169.

Heid, J.G., McMahon, J.J., Nielson, R.F., and Yuster, S.T., 1950, "Study of the Permeability of Rocks to Homogeneous Fluids," API Drilling and Production Practice, 230–244.

Jones, F.O., and Owens, W.W., 1980, "A Laboratory Study of Low-Permeability Gas Sands," J. Petroleum Techology (September 1980), pp. 1631–1640

Jury, W.A., Spencer, W.F., and Farmer, W.J., 1983a, "Behavior Assessment Model for Trace Organics in Soil: I. Model Development," J. Environ. Qual., Vol. 12, no. 4, pp. 558–564.

Jury, W.A., Spencer, W.F., and Farmer, W.J., 1983b, Use of Models for Assessing Relative Volatility, Mobility, and Persistence of Pesticides and Other Trace Organics in Soils, Hazard Assessment of Chemicals: Current Developments, Vol. 2, Academic Press.

Jury, W.A., Farmer, W.J., and Spencer, W.F., 1984a, "Behavior Assessment Model for Trace Organics in Soil: II. Chemical Classification and Parameter Sensitivity," J. Environ. Qual., Vol. 13, no. 4, pp. 567–572.

Jury, W.A., Spencer, W.F., and Farmer, W.J., 1984b, "Behavior Assessment Model for Trace Organics in Soil: III. Application of Screening Model," J. Environ. Qual., Vol. 13, no. 4, pp. 573–579.

Jury, W.A., Spencer, W.F., and Farmer, W.J., 1984c, "Behavior Assessment Model for Trace Organics in Soil: IV. Review of Experimental Evidence," J. Environ. Qual., Vol. 13, no. 4, pp. 580–586.

Jury, W.A., Russo, D., Streile, G., and El Abd, H., 1990, "Evaluation of Volatilization by Organic Chemicals Residing Below the Soil Surface," Water Resources Research, Vol. 26, no. 1, pp 13–20.

Jury, W.A. Gardner, W.R., and Gardner, W.H., 1991, Soil Physics, Fifth Edition, John Wiley & Sons.

Kaviany, M., 1995, Principles of Heat Transfer in Porous Media, Second Edition, Springer-Verlag New York, Inc., New York. Check 1995 Second Edition.

Luckner, L., van Genuchten, M.Th., and Nielsen, D.R., 1989, "A Consistent Set of Parametric Models for the Two-Phase Flow of Immiscible Fluids in the Subsurface," Water Resources Research, Vol. 25, no. 10, pp. 2187–2193.

McWhorter, D.B., and Sunada, D.K., 1990, "Exact Integral Solutions for Two-Phase Flow," Water Resources Research, Vol. 26, no. 3, pp. 399–413.

Millington, R.J., and Quirk, J.M., 1961, "Permeability of porous solids," Trans. Faraday Soc., 57:1200–1207.

Mualem, Y., 1976, "A New Model for Predicting the Hydraulic Conductivity of Unsaturated Porous Media," Water resources Research, Vol. 12, no. 3, pp. 513–522.

Nield, D.A., and Bejan, A., 1999, Convection in Porous Media, Second Edition, Springer-Verlag New York, Inc., New York.

Parker, J.C., Lenhard, R.J., and Kuppusamy, T., 1987, "A Parametric Model for Constitutive Properties Governing Multiphase Flow in Porous Media," Water Resources Research, Vol. 23, no. 4, pp. 618–624.

Reid, R.C., Prausnitz, J.M., and Poling, B.E., 1987, The Properties of Gases & Liquids, Fourth Edition, McGraw-Hill, Inc..

Reinecke, S.A., and Sleep, B.E., 2002, "Knudsen diffusion, gas permeability, and water content in an unconsolidated porous medium," Water Resources Research, Vol. 38, no. 12, p. 1280.

Schaap, M.G., and Leij, F.J., 2000, "Improved Prediction of Unsaturated Hydraulic Conductivity with the Mualem-van Genuchten Model," Soil Sci. Soc. Am. J., Vol. 64, pp. 843–851.

Shoemaker, C.A., Culver, T.B., Lion, L.W., and Peterson, M.G., 1990, "Analytical Models of the Impact of Two-Phase Sorption on Subsurface Transport of Volatile Chemicals," Water Resources Research, Vol. 26, No. 4, pp. 745–758.

Sleep, B.E., 1998, "Modeling transient organic vapor transport in porous media with the dusty gas model," Advances in Water Resources, 22:247–256.

———. 2005, Personal Communication to S.W. Webb.

Stephens, D.B., 1996, Vadose Zone Hydrology, Lewis Publishers.

Stephens, D.B., Lambert, K., and Watson, D., 1987, "Regression models for hydraulic conductivity and field test of the borehole permeameter," Water resources Research, 23:2207.

van Genuchten, R., 1978, Calculating The Unsaturated Hydraulic Conductivity With a New Closed-Form Analytical Model, Water Resources Bulletin 78-WR-08, Department of Civil Engineering, Princeton University.

van Genuchten, M.Th., 1980, "A Closed-Form Equation for Predicting the Hydraulic Conductivity of Unsaturated Soils," Soil Science Society of America Journal, Vol. 44, no. 5, pp. 892–898.

van Genuchten, M.Th., and Nielsen, D.R., 1985, "On Describing and Predicting the Hydraulic Properties of Unsaturated Soils," Annales Geophysicae, Vol. 3, no. 5, pp. 615–628.

van Genuchten, M.Th., Leij, F.J., and Yates, S.R., 1991, The RETC Code for Quantifying the Hydraulic Functions of Unsaturated Soils, Version 1.0. EPA Report 600/2–91/065, U.S. Salinity Laboratory, USDA, ARS, Riverside, California.

Webb, S.W., 1992, "Steady-State Saturation Profiles for Linear Immiscible Fluid Displacement in Porous Media," Heat and Mass Transfer in Porous Media, I. Catton, ed., ASME Winter Annual Meeting, Anaheim, CA, ASME, pp. 35–46.

Webb, S.W., 2000, "A simple extension of two-phase characteristic curves to include the dry region," Water Resources Research, Vol. 36, no. 6, pp. 1425–1430.

Webb, S.W., 2001, Modification of TOUGH2 to Include the Dusty Gas Model for Gas Diffusion, SAND2001-3214, Sandia National Laboratories, Albuquerque, NM.

Whitaker, S., 1994, "The closure problem for two-phase flow in homogeneous porous media", Chemical Engineering Science 49, 765–780.

CHAPTER 6

CONSERVATION EQUATIONS

STEPHEN WHITAKER
University of California at Davis, Davis, CA, USA

6.1 INTRODUCTION

In this section, the conservation equations for mass, momentum and energy of multicomponent systems are presented from the continuum point of view. A key concept associated with the continuum approach to multicomponent transport phenomena is the *species body* illustrated in Figure 6.1.

There we have illustrated the volume, $\mathscr{V}_A(t)$, occupied by species A in a binary system containing species A and B. For *single component* transport phenomena, a *motion* is described by

$$\mathbf{r} = \mathbf{r}(\mathbf{R}, t) \tag{6.1}$$

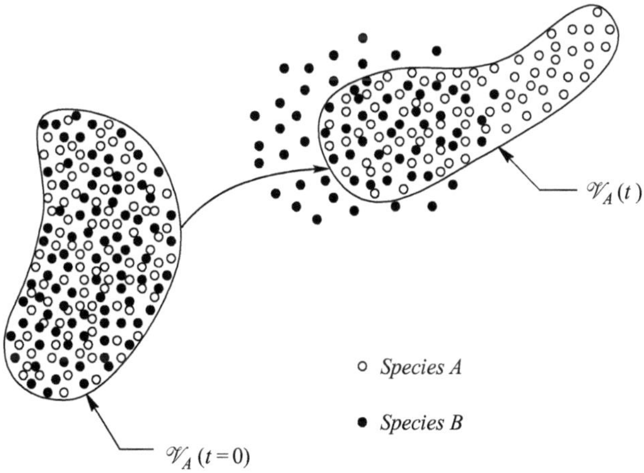

Figure 6.1. Motion of a species body

C. Ho and S. Webb (eds.), Gas Transport in Porous Media, 71–120.
© 2006 Springer.

in which $\mathbf{r}$ represents the time-dependent position of a *material element* whose reference position is $\mathbf{R}$. The velocity of any material element can be expressed as

$$\mathbf{v} = \left(\frac{\mathrm{d}\mathbf{r}}{\mathrm{d}t}\right)_{\mathbf{R}} = \frac{D\mathbf{r}}{Dt} \tag{6.2}$$

in which $D\mathbf{r}/Dt$ is referred to as the *material derivative*. The motion of a species body is described in an analogous manner, thus the motion of a material element of species A is represented by (Slattery, 1999)

$$\mathbf{r}_A = \mathbf{r}_A(\mathbf{R}_A, t) \tag{6.3}$$

and the velocity is given by

$$\mathbf{v}_A = \left(\frac{\mathrm{d}\mathbf{r}_A}{\mathrm{d}t}\right)_{\mathbf{R}_A} \tag{6.4}$$

In terms of the concept of a species body, we state the two axioms for the mass of multicomponent systems as

$$\text{AXIOM I:} \quad \frac{\mathrm{d}}{\mathrm{d}t} \int_{\mathscr{V}_A(t)} \rho_A \mathrm{d}V = \int_{\mathscr{V}_A(t)} r_A \mathrm{d}V, \quad A = 1, 2, \ldots, N \tag{6.5}$$

$$\text{AXIOM II:} \quad \sum_{A=1}^{A=N} r_A = 0 \tag{6.6}$$

Here ρ_A represents the mass density of species A while r_A represents the mass rate of production per unit volume of species A owing to chemical reaction. In order to extract a governing differential equation from Eq. (6.5), we first recall the general transport theorem (Whitaker, 1981) for some scalar ψ and an *arbitrary*, continuous velocity $\mathbf{w}$

$$\frac{\mathrm{d}}{\mathrm{d}t} \int_{\mathscr{V}_a(t)} \psi \mathrm{d}V = \int_{\mathscr{V}_a(t)} \frac{\partial \psi}{\partial t} \mathrm{d}V + \int_{\mathscr{A}_a(t)} \psi \mathbf{w} \cdot \mathbf{n} \mathrm{d}A \tag{6.7}$$

Here $\mathscr{V}_a(t)$ represents an arbitrary volume, the surface of which has a speed of displacement given by $\mathbf{w} \cdot \mathbf{n}$. Use of the divergence theorem allows us to express the general transport theorem in the form

$$\frac{\mathrm{d}}{\mathrm{d}t} \int_{\mathscr{V}_a(t)} \psi \mathrm{d}V = \int_{\mathscr{V}_a(t)} \left[\frac{\partial \psi}{\partial t} + \nabla \cdot (\psi \mathbf{w})\right] \mathrm{d}V \tag{6.8}$$

We can choose the *arbitrary velocity* $\mathbf{w}$ to be equal to the species velocity $\mathbf{v}_A$ so that this result takes the form of a Reynolds transport theorem for species A

$$\frac{\mathrm{d}}{\mathrm{d}t}\int_{\mathscr{V}_A(t)} \psi_A \mathrm{d}V = \int_{\mathscr{V}_A(t)} \left[\frac{\partial \psi_A}{\partial t} + \nabla\cdot(\psi_A \mathbf{v}_A)\right]\mathrm{d}V \tag{6.9}$$

Here ψ_A is any function associated with species A and if $\psi_A = \rho_A$ this result can be used with the axiom given by Eq. (6.5) to obtain

$$\int_{\mathscr{V}_A(t)} \left[\frac{\partial \rho_A}{\partial t} + \nabla\cdot(\rho_A \mathbf{v}_A) - r_A\right]\mathrm{d}V = 0, \quad A = 1,2,\ldots,N \tag{6.10}$$

Here we note that the volume $\mathscr{V}_A(t)$ is arbitrary in the sense that the *Euler cut principle* (Truesdell, 1968) suggests that we can identify *any region* in space as the species body. If we assume that the integrand in Eq (6.10) is continuous, the arbitrary nature of $\mathscr{V}_A(t)$ leads us to conclude that the integrand must be zero. Requiring that Eq. (6.10) be satisfied leads to the *species continuity equation*

$$\frac{\partial \rho_A}{\partial t} + \nabla\cdot(\rho_A \mathbf{v}_A) = r_A, \quad A = 1,2,\ldots,N \tag{6.11}$$

which forms the basis for much of the analysis presented in this monograph.

6.1.1 Continuity Equation

To derive the *continuity equation* from the species continuity equation, we sum Eq. (6.11) over all species and impose the axiom given by Eq. (6.6) to obtain

$$\frac{\partial}{\partial t}\sum_{A=1}^{A=N} \rho_A + \nabla\cdot\sum_{A=1}^{A=N} \rho_A \mathbf{v}_A = 0 \tag{6.12}$$

We define the *total mass density* as

$$\rho = \sum_{A=1}^{A=N} \rho_A \tag{6.13}$$

and note that the mass fraction is given by

$$\omega_A = \frac{\rho_A}{\rho} \tag{6.14}$$

In terms of the total mass density Eq. (6.12) takes the form

$$\frac{\partial \rho}{\partial t} + \nabla\cdot\sum_{A=1}^{A=N} \rho_A \mathbf{v}_A = 0 \tag{6.15}$$

and use of the definition of the mass average velocity

$$\mathbf{v} = \sum_{A=1}^{A=N} \omega_A \mathbf{v}_A, \quad \text{mass average velocity} \tag{6.16}$$

leads us to

$$\frac{\partial \rho}{\partial t} + \nabla \cdot (\rho \mathbf{v}) = 0 \tag{6.17}$$

This result has exactly the same form as the continuity equation for a single component when there is no chemical reaction.

6.1.2 Molar Forms

Often the molar form of Eq. (6.11) is preferred because both reaction rates and phase equilibria are expressed in molar quantities. We can divide Eq. (6.11) by the molecular mass of species A and make use of the definitions

$$c_A = \frac{\rho_A}{M_A}, \quad R_A = \frac{r_A}{M_A} \tag{6.18}$$

in order to express Eqs. (6.11) and (6.6) as

$$\frac{\partial c_A}{\partial t} + \nabla \cdot (c_A \mathbf{v}_A) = R_A, \qquad A = 1, 2, \ldots, N \tag{6.19}$$

$$\sum_{A=1}^{A=N} M_A R_A = 0 \tag{6.20}$$

This latter constraint on the molar rates of production owing to chemical reactions is not particularly useful, and one normally draws upon the concept that *atomic species are neither created nor destroyed by chemical reactions* to provide the framework for chemically reacting systems. This axiomatic statement can be expressed as

$$\text{AXIOM II:} \quad \sum_{A=1}^{A=N} N_{JA} R_A = 0, \quad J = 1, 2, \ldots, T \tag{6.21}$$

in which T represents the number of atomic species that take part in the reaction, and N_{JA} represents the number of J-type atoms contained in A-type molecules. The matrix of coefficients, N_{JA}, is often referred to as the *chemical composition matrix*. The reaction in which ethane and oxygen undergo complete combustion to form carbon dioxide and water can be expressed as

$$C_2H_6 + O_2 \rightarrow CO_2 + H_2O \tag{6.22}$$

and the chemical composition matrix for this reaction is represented by

$$\begin{array}{l} \text{Molecular species} \rightarrow \\ \\ \textit{carbon} \\ \textit{hydrogen} \\ \textit{oxygen} \end{array} \quad \begin{array}{c} C_2H_6 \quad O_2 \quad H_2O \quad CO_2 \\ \\ \begin{bmatrix} 2 & 0 & 0 & 1 \\ 6 & 0 & 2 & 0 \\ 0 & 2 & 1 & 2 \end{bmatrix} \end{array}$$

Equation (6.21) can be used to determine stoichiometric coefficients and the number of independent reactions that are needed to describe any given system (Cerro et al., 2004). In addition, it represents the *fundamental axiom* concerning conservation of mass for reacting systems since it can be used to derive Eq. (6.6).

6.1.3 Mechanics

To describe multicomponent systems we use four axiomatic statements concerning (1) the linear momentum of species A, (2) the angular momentum of species A, (3) the total diffusive force in a mixture, and (4) the rate of production of linear momentum owing to chemical reactions. These axioms take the form:

AXIOM I:

$$\frac{\mathrm{d}}{\mathrm{d}t} \int_{\mathscr{V}_A(t)} \rho_A \mathbf{v}_A \mathrm{d}V = \int_{\mathscr{V}_A(t)} \rho_A \mathbf{b}_A \mathrm{d}V + \int_{\mathscr{A}_A(t)} \mathbf{t}_{A(\mathbf{n})} \mathrm{d}A + \int_{\mathscr{V}_A(t)} \sum_{B=1}^{B=N} \mathbf{P}_{AB} \mathrm{d}V + \int_{\mathscr{V}_A(t)} r_A \mathbf{v}_A \mathrm{d}V \tag{6.23}$$

AXIOM II:

$$\frac{\mathrm{d}}{\mathrm{d}t} \int_{\mathscr{V}_A(t)} \mathbf{r} \times \rho_A \mathbf{v}_A \mathrm{d}V = \int_{\mathscr{V}_A(t)} \mathbf{r} \times \rho_A \mathbf{b}_A \mathrm{d}V + \int_{\mathscr{A}_A(t)} \mathbf{r} \times \mathbf{t}_{A(\mathbf{n})} \mathrm{d}A + \int_{\mathscr{V}_A(t)} \sum_{B=1}^{B=N} \mathbf{r} \times \mathbf{P}_{AB} \mathrm{d}V + \int_{\mathscr{V}_A(t)} \mathbf{r} \times r_A \mathbf{v}_A \mathrm{d}V \tag{6.24}$$

AXIOM III:

$$\sum_{A=1}^{A=N} \sum_{B=1}^{B=N} \mathbf{P}_{AB} = 0 \tag{6.25}$$

AXIOM IV:

$$\sum_{A=1}^{A=N} r_A \mathbf{v}_A = 0 \tag{6.26}$$

The first of these axioms represents the *linear momentum equation* for species A, and the point equation associated with Eq. (6.23) is the governing differential equation for $\mathbf{v}_A$. Axiom I simply indicates that the time rate of change of the linear momentum of a species A body is equal to the force acting on the body plus the rate of production of momentum owing to chemical reaction. In Eq. (6.23), the term $\mathbf{P}_{AB}$ represents the force per unit volume exerted *by species B on species A*, and the sum from $B = 1$ to $B = N$ represents the force exerted by all the other molecular species on species A.

The angular momentum equation given by Eq. (6.24) will lead to the symmetry of the species stress tensor, and should be thought of as simply a constraint on the contact force. The axiom given by Eq. (6.25) indicates that the total sum of the intermolecular forces is zero and it is satisfied by the condition $\mathbf{P}_{AB} = -\mathbf{P}_{BA}$ that appears in the Stefan-Maxwell equations. Axiom IV requires that linear momentum is neither created nor destroyed by chemical reactions.

In order to derive the governing differential equation associated with Eq. (6.23), one follows the classical developments concerning stress (Cauchy's lemma and Cauchy's fundamental theorem) to obtain

$$\mathbf{t}_{A(\mathbf{n})} = -\mathbf{t}_{A(-\mathbf{n})} \tag{6.27}$$

$$\mathbf{t}_{A(\mathbf{n})} = \mathbf{n} \cdot \mathbf{T}_A \tag{6.28}$$

One then makes use of the appropriate form of the general transport theorem so that the following differential equation can be extracted from Eq. (6.23):

$$\underbrace{\frac{\partial}{\partial t}(\rho_A \mathbf{v}_A)}_{\substack{\text{local}\\ \text{acceleration}}} + \underbrace{\nabla \cdot (\rho_A \mathbf{v}_A \mathbf{v}_A)}_{\substack{\text{convective}\\ \text{acceleration}}} = \underbrace{\rho_A \mathbf{b}_A}_{\substack{\text{body}\\ \text{force}}} + \underbrace{\nabla \cdot \mathbf{T}_A}_{\substack{\text{surface}\\ \text{force}}} \tag{6.29}$$

$$+ \underbrace{\sum_{B=1}^{B=N} \mathbf{P}_{AB}}_{\substack{\text{diffusive}\\ \text{force}}} + \underbrace{r_A \mathbf{v}_A}_{\substack{\text{source of momentum}\\ \text{owing to reaction}}} \qquad A = 1, 2, \ldots, N$$

Here we have identified the forces associated with $\mathbf{P}_{AB}$ as the *diffusive force* since this force is so closely related to the process of diffusion.

The analysis of the second axiom is algebraically complex but the final result simply indicates that the species stress tensor is symmetric

$$\mathbf{T}_A = \mathbf{T}_A^T, \quad A = 1, 2, \ldots, N \tag{6.30}$$

The third and fourth axioms remain unchanged and are ready for use.

6.1.4 Total Momentum Equation

In order to compare the results given by Eqs (6.29) and (6.30) with those for single component systems, we first sum Eq. (6.29) over all N species in order to develop the total momentum equation. This is given by

$$\frac{\partial}{\partial t}\sum_{A=1}^{A=N}(\rho_A \mathbf{v}_A) + \nabla\cdot\sum_{A=1}^{A=N}(\rho_A \mathbf{v}_A \mathbf{v}_A) = \sum_{A=1}^{A=N}\rho_A \mathbf{b}_A + \nabla\cdot\sum_{A=1}^{A=N}\mathbf{T}_A \tag{6.31}$$

in which Axioms III and IV have been used to eliminate the sum of the last two terms in Eq. (6.29). We can define the following quantities

$$\rho = \sum_{A=1}^{A=N}\rho_A, \quad \textit{total density} \tag{6.32}$$

$$\omega_A = \frac{\rho_A}{\rho}, \quad \textit{mass fraction} \tag{6.33}$$

$$\mathbf{v} = \sum_{A=1}^{A=N}\omega_A \mathbf{v}_A, \quad \textit{mass average velocity} \tag{6.34}$$

$$\mathbf{b} = \sum_{A=1}^{A=N}\omega_A \mathbf{b}_A, \quad \textit{mass average body force} \tag{6.35}$$

so that Eq. (6.31) takes the form

$$\frac{\partial}{\partial t}(\rho\mathbf{v}) + \nabla\cdot\sum_{A=1}^{A=N}\rho_A \mathbf{v}_A \mathbf{v}_A = \rho\mathbf{b} + \nabla\cdot\sum_{A=1}^{A=N}\mathbf{T}_A \tag{6.36}$$

In order to extract a simplified form of the convective acceleration, we need to introduce the important concept of a *diffusion velocity*.

6.1.5 Diffusion Velocity

There are two diffusion velocities that are commonly used; the *mass diffusion velocity* and the *molar diffusion velocity*. The first of these is defined according to the decomposition

$$\mathbf{v}_A = \mathbf{v} + \mathbf{u}_A \tag{6.37}$$

and one can use this to show that that the mass diffusion velocities are constrained by

$$\sum_{A=1}^{A=N}\rho_A \mathbf{u}_A = 0 \tag{6.38}$$

One can use the representation given by Eq. (6.37) in order to express the convective inertial term in Eq. (6.36) as

$$\sum_{A=1}^{A=N}\rho_A \mathbf{v}_A \mathbf{v}_A = \sum_{A=1}^{A=N}\rho_A \mathbf{v}_A(\mathbf{v} + \mathbf{u}_A) \tag{6.39}$$

and a little thought will indicate that this result takes the form

$$\sum_{A=1}^{A=N} \rho_A \mathbf{v}_A \mathbf{v}_A = \rho \mathbf{v}\mathbf{v} + \sum_{A=1}^{A=N} \rho_A \mathbf{v}_A \mathbf{u}_A \tag{6.40}$$

With the aid of Eq. (6.37) one can express the last term in this expression as

$$\sum_{A=1}^{A=N} \rho_A \mathbf{v}_A \mathbf{u}_A = \sum_{A=1}^{A=N} \rho_A \mathbf{v}\, \mathbf{u}_A + \sum_{A=1}^{A=N} \rho_A \mathbf{u}_A \mathbf{u}_A \tag{6.41}$$

and the constraint on the mass diffusion velocities given by Eq. (6.38) leads to the simplification

$$\sum_{A=1}^{A=N} \rho_A \mathbf{v}_A \mathbf{u}_A = \sum_{A=1}^{A=N} \rho_A \mathbf{u}_A \mathbf{u}_A \tag{6.42}$$

Use of this result in Eq. (6.40) provides

$$\sum_{A=1}^{A=N} \rho_A \mathbf{v}_A \mathbf{v}_A = \rho \mathbf{v}\mathbf{v} + \sum_{A=1}^{A=N} \rho_A \mathbf{u}_A \mathbf{u}_A \tag{6.43}$$

and this allows us to write Eq. (6.36) as

$$\frac{\partial}{\partial t}(\rho \mathbf{v}) + \nabla \cdot (\rho \mathbf{v}\mathbf{v}) + \nabla \cdot \sum_{A=1}^{A=N} \rho_A \mathbf{u}_A \mathbf{u}_A = \rho \mathbf{b} + \nabla \cdot \sum_{\mathbf{A=1}}^{\mathbf{A=N}} \mathbf{T}_A \tag{6.44}$$

At this point it is convenient to define a total stress tensor for multicomponent systems as

$$\mathbf{T} = \sum_{A=1}^{A=N} \mathbf{T}_A - \rho_A \mathbf{u}_A \mathbf{u}_A \tag{6.45}$$

in which the terms represented by $\rho_A \mathbf{u}_A \mathbf{u}_A$ are referred to as the *diffusive stresses* (Truesdell and Toupin, 1960). Use of this result in Eq. (6.44) leads to

$$\frac{\partial}{\partial t}(\rho \mathbf{v}) + \nabla \cdot (\rho \mathbf{v}\mathbf{v}) = \rho \mathbf{b} + \nabla \cdot \mathbf{T} \tag{6.46}$$

This result is identical in form to Cauchy's first equation for single component systems; however, in this case each term has a *greater physical significance* because of the definitions given by Eqs. (6.32), (6.34), (6.35), and (6.45). It is easy to see that one can use Eq. (6.30) along with Eq. (6.45) to produce the symmetry condition

$$\mathbf{T} = \mathbf{T}^T \tag{6.47}$$

which is identical to Cauchy's second equation. For a Stokesian fluid (Serrin, 1959) the stress tensor can be *decomposed* according to

$$\mathbf{T} = -p\mathbf{I} + \boldsymbol{\tau} \tag{6.48}$$

in which $\boldsymbol{\tau}$ is the *viscous stress tensor*. Use of Eq. (6.48) in Eq. (6.46) leads to the viscous stress equations of motion.

$$\frac{\partial}{\partial t}(\rho\mathbf{v}) + \nabla\cdot(\rho\mathbf{v}\mathbf{v}) = \rho\mathbf{b} - \nabla p + \nabla\cdot\boldsymbol{\tau} \tag{6.49}$$

To complete the analysis one needs a constitutive equation for the viscous stress tensor so that Eq. (6.49) can be used to determine the mass average velocity and this is an extremely important result for many mass transfer and fluid mechanics problems. For Newtonian fluids and incompressible flows, the viscous stress tensor takes the form

$$\boldsymbol{\tau} = \mu(\nabla\mathbf{v} + \nabla\mathbf{v}^T) \tag{6.50}$$

and one obtains the Navier-Stokes equations

$$\rho\frac{\partial\mathbf{v}}{\partial t} + \rho\mathbf{v}\nabla\cdot\mathbf{v} = \rho\mathbf{g} - \nabla p + \mu\nabla^2\mathbf{v} \tag{6.51}$$

Here body forces have been limited to the gravitational force in which the gravitational acceleration is represented by $\mathbf{g}$. In general, an equation of state is required to determine the pressure in Eqs. (6.49) and (6.51), and for an N-component system we suggest a generic form given by

$$p = p(T, c_A, c_B, \ldots, c_N) \tag{6.52}$$

When the flow can be treated as incompressible, this equation of state can be discarded in the classical manner.

6.1.6 Energy

The energy for a multicomponent system represents a complex matter for two reasons. First, the axiomatic statement for the thermal and kinetic energy of a species body is complex in its own right, and second, one must subtract the mechanical energy from the total energy in order to obtain the desired thermal energy equation. Because of the complexity of the subject, this presentation will only provide an outline of the steps involved. The axioms for the thermal and mechanical energy of a species body can be expressed as

AXIOM I:

$$\frac{\mathrm{d}}{\mathrm{d}t}\int\limits_{\mathscr{V}_A(t)}(\rho_A e_A+\tfrac{1}{2}\rho_A \mathrm{v}_A^2)\mathrm{d}V = -\int\limits_{\mathscr{A}_A(t)}\mathbf{q}_A\cdot\mathbf{n}\mathrm{d}A+\int\limits_{\mathscr{V}_A(t)}\sum_{B=1}^{B=N}Q_{AB}\mathrm{d}V$$
$$+\int\limits_{\mathscr{A}_A(t)}\mathbf{v}_A\cdot\mathbf{t}_{A(\mathbf{n})}\mathrm{d}A+\int\limits_{\mathscr{V}_A(t)}\rho_A\mathbf{b}_A\cdot\mathbf{v}_A\,\mathrm{d}V$$
$$+\int\limits_{\mathscr{V}_A(t)}\sum_{B=1}^{B=N}\mathbf{v}_A\cdot\mathbf{P}_{AB}\,\mathrm{d}V$$
$$+\int\limits_{\mathscr{V}_A(t)}r_A(\rho_A e_A+\tfrac{1}{2}\rho_A\mathrm{v}_A^2)\mathrm{d}V \tag{6.53}$$

AXIOM II:

$$\sum_{A=1}^{A=N} r_A(e_A+\tfrac{1}{2}\mathrm{v}_A^2)=0 \tag{6.54}$$

AXIOM III:

$$\sum_{A=1}^{A=N}\sum_{B=1}^{B=N}Q_{AB}=0 \tag{6.55}$$

AXIOM IV:

$$e_A \text{ is a function of the state of the system} \tag{6.56}$$

Here we have used e_A to represent the internal energy per unit mass of species A, we have used $\mathbf{q}_A$ to represent the heat flux (*conductive and radiative*) transferred from the surroundings to species A, and we have used Q_{AB} to represent the volumetric rate of transfer of thermal energy from species B to species A. Determination of e_A on the basis of Eq. (6.53) would be extremely difficult; however, the axiom given by Eq. (6.56) indicates that this is unnecessary and instead we need only determine the temperature, pressure, and composition of the system in order to determine e_A.

To produce a useful result from Eq. (6.53), we first derive the governing differential equations for $e_A+\frac{1}{2}\rho_A\mathrm{v}_A^2$. The species mechanical energy equation is then subtracted from the species total energy equation to obtain the species thermal energy equation. This is a rather lengthy process (Whitaker, 1989) that leads to the governing equation for the species energy.

$$\frac{\partial}{\partial t}(\rho_A\,e_A)+\nabla\cdot(\rho_A\,e_A\mathbf{v}_A)=-\nabla\cdot\mathbf{q}_A+\sum_{B=1}^{B=N}Q_{AB}+\nabla\mathbf{v}_A:\mathbf{T}_A+r_A\,e_A \tag{6.57}$$

There are a variety of paths that one can follow at this point; however, the development of an *enthalpy transport equation* seems to be the most productive route. The species enthalpy per unit mass is defined by

$$\rho_A h_A = \rho_A e_A + p_A \tag{6.58}$$

in which p_A is the partial pressure of species A. This is defined according to

$$p_A = \rho_A^2 \left(\frac{\partial e_A}{\partial \rho_A} \right)_{s, \rho_B, \rho_C, \ldots} \tag{6.59}$$

in which s is the total entropy per unit mass of the mixture. Following the approach used for a Stokesian fluid, the species stress tensor is decomposed according to

$$\mathbf{T}_A = -\mathbf{I} p_A + \boldsymbol{\tau}_A \tag{6.60}$$

and Eqs. (6.58) through (6.60) can be used in Eq. (6.57) to obtain

$$\begin{aligned} \frac{\partial}{\partial t}(\rho_A h_A) + \nabla \cdot (\rho_A h_A \mathbf{v}_A) = &- \nabla \cdot \mathbf{q}_A + \sum_{B=1}^{B=N} Q_{AB} \\ &+ \nabla \mathbf{v}_A : (-\mathbf{I} p_A) + \nabla \mathbf{v}_A : \boldsymbol{\tau}_A + r_A e_A \\ &+ \frac{\partial p_A}{\partial t} + \nabla \cdot (p_A \mathbf{v}_A) \end{aligned} \tag{6.61}$$

We now make use of the definitions,

$$\rho h = \sum_{A=1}^{A=N} \rho_A h_A, \qquad \mathbf{q} = \sum_{A=1}^{A=N} \mathbf{q}_A, \qquad p = \sum_{A=1}^{A=N} p_A \tag{6.62}$$

along with Axiom III, and sum Eq. (6.61) over all species to obtain

$$\begin{aligned} \frac{\partial}{\partial t}(\rho h) + \nabla \cdot \sum_{A=1}^{A=N} \rho_A h_A \mathbf{v}_A = &- \nabla \cdot \mathbf{q} + \sum_{A=1}^{A=N} \nabla \mathbf{v}_A : (-\mathbf{I} p_A) \\ &+ \sum_{A=1}^{A=N} \nabla \mathbf{v}_A : \boldsymbol{\tau}_A + \sum_{A=1}^{A=N} r_A e_A \\ &+ \frac{\partial p}{\partial t} + \nabla \cdot \sum_{A=1}^{A=N} p_A \mathbf{v}_A \end{aligned} \tag{6.63}$$

On the basis of Eqs. (6.6), (6.26), and (6.54), and the velocity decomposition given by Eq. (6.37) we can prove that

$$\sum_{A=1}^{A=N} r_A e_A + \sum_{A=1}^{A=N} r_A \tfrac{1}{2} u_A^2 = 0 \tag{6.64}$$

In addition, we can use the velocity decomposition given by Eq. (6.37) along with the stress decomposition given by Eq. (6.60) to obtain the following results:

$$\sum_{A=1}^{A=N} \rho_A h_A \mathbf{v}_A = \rho h \mathbf{v} + \sum_{A=1}^{A=N} \rho_A h_A \mathbf{u}_A \tag{6.65}$$

$$\sum_{A=1}^{A=N} \nabla \mathbf{v}_A : (-\mathbf{I}p_A) = -(\nabla \cdot \mathbf{v})p - \sum_{A=1}^{A=N} (\nabla \cdot \mathbf{u}_A)p_A \tag{6.66}$$

$$\sum_{A=1}^{A=N} \nabla \mathbf{v}_A : \boldsymbol{\tau}_A = \nabla \mathbf{v} : \boldsymbol{\tau} + \sum_{A=1}^{A=N} \nabla \mathbf{u}_A : \boldsymbol{\tau}_A \tag{6.67}$$

$$\sum_{A=1}^{A=N} p_A \mathbf{v}_A = p\mathbf{v} + \sum_{A=1}^{A=N} p_A \mathbf{u}_A \tag{6.68}$$

Use of Eqs. (6.64) through (6.68) in Eq. (6.63) leads to

$$\frac{\partial}{\partial t}(\rho h) + \nabla \cdot (\rho h \mathbf{v}) = -\nabla \cdot \mathbf{q} + \frac{Dp}{Dt} + \nabla \mathbf{v} : \boldsymbol{\tau} + \sum_{A=1}^{A=N} \rho_A h_A \mathbf{u}_A$$
$$+ \sum_{A=1}^{A=N} \mathbf{u}_A \cdot \nabla p_A + \sum_{A=1}^{A=N} \nabla \mathbf{u}_A : \boldsymbol{\tau}_A - \sum_{A=1}^{A=N} r_A \tfrac{1}{2} u_A^2 \tag{6.69}$$

The *first five terms* in this result have the same form as the enthalpy transport equation for single component systems (Whitaker, 1983), while the *last four terms* are all associated with diffusive fluxes. When convective transport is *significant*, the rate of work terms involving the diffusive fluxes can be neglected relative to Dp/Dt and $\nabla \mathbf{v} : \boldsymbol{\tau}$, and when convective transport is *insignificant* these rate of work terms can be discarded relative to the thermal terms in Eq. (6.69). When chemical reactions occur, we can show that the last term is negligible, thus we will discard the last three terms in Eq. (6.69) in order to express the enthalpy transport equation as

$$\frac{\partial}{\partial t}(\rho h) + \nabla \cdot (\rho h \mathbf{v}) = -\nabla \cdot \mathbf{q} + \frac{Dp}{Dt} + \nabla \mathbf{v} : \boldsymbol{\tau} + \sum_{A=1}^{A=N} \rho_A h_A \mathbf{u}_A \tag{6.70}$$

Given that h is a function of the state of the system, we can express the total enthalpy per unit mass in terms of a thermal equation of state given by

$$h = h(T, p, \omega_A, \omega_B, \ldots, \omega_{N-1}) \tag{6.71}$$

In order to transform Eq. (6.70) from an enthalpy transport equation to a transport equation for the temperature, one must develop the time and space derivatives of Eq. (6.71). This becomes algebraically tedious; however, the details are given elsewhere (Whitaker, 1989, Eqs. 50 through 67) and the result is given by

$$\rho c_p \frac{DT}{Dt} = -\nabla \cdot \mathbf{q} + T\beta \frac{Dp}{Dt} + \nabla \mathbf{v} : \boldsymbol{\tau} - \sum_{A=1}^{A=N} \rho_A \mathbf{u}_A \cdot \nabla h_A - \sum_{A=1}^{A=N} r_A h_A \tag{6.72}$$

At this point we can see that the last term in Eq. (6.69) can be discarded on the basis of

$$\sum_{A=1}^{A=N} r_A \tfrac{1}{2} u_A^2 \ll \sum_{A=1}^{A=N} r_A h_A \tag{6.73}$$

It is convenient to express the last term in Eq. (6.72) in terms of the partial molar enthalpy and the molar rate of reaction. These are given by

$$H_A = h_A M_A, \qquad R_A = \frac{r_A}{M_A} \tag{6.74}$$

and these definitions allow us to express the last term in Eq. (6.72) as

$$\sum_{A=1}^{A=N} r_A h_A = \sum_{A=1}^{A=N} R_A H_A \tag{6.75}$$

For the special case of a *single independent reaction*, the molar rates of reaction can be expressed in terms of the stoichiometric coefficients and the molar rate of reaction of the *pivot species*, N, to obtain

$$\sum_{A=1}^{A=N} R_A H_A = \left[\sum_{A=1}^{A=N} \nu_A H_A \right] R_N = \Delta H_{\text{reaction}} R_N \tag{6.76}$$

Use of this result in Eq. (6.72) provides

$$\rho c_p \frac{DT}{Dt} = -\nabla \cdot \mathbf{q} + T\beta \frac{Dp}{Dt} + \nabla \mathbf{v} : \boldsymbol{\tau} - \sum_{A=1}^{A=N} \rho_A \mathbf{u}_A \cdot \nabla h_A - \Delta H_{\text{reaction}} R_N \tag{6.77}$$

When there are multiple independent reactions (Cerro et al., 2004), one must identify those reactions and the heats of reaction associated with them. The diffusive flux of enthalpy, $\rho_A \mathbf{u}_A \cdot \nabla h_A$, can be ignored for most problems; however, combustion processes are a special case that must be considered with care. In addition, care must be taken to remember that $\mathbf{q}$ represents both the conductive and radiative transport, and for an isotropic medium we would express this quantity as

$$\mathbf{q} = -\lambda \nabla T + \mathbf{q}^R \tag{6.78}$$

In general, we wish to use the axioms for energy to determine a single temperature, T, and this means that the species energy equations represented by Eqs. (6.57) are not needed. Instead, it is the total energy equation given by Eq. (6.77) that provides us with information about the temperature. However, what is true for energy is *not true* for either mass or momentum, that is, we must make use of the species continuity equation given by Eq. (6.11) or by Eq. (6.19) and the species momentum equation given by Eq. (6.29).

6.1.7 Stefan-Maxwell Equations

The species momentum equation given by Eq. (6.29) and repeated here as Eq. (6.79), would appear to be overwhelmingly complex; however, some reasonable simplifications will lead us from that complex result to the relatively simple form given by the Stefan-Maxwell equations.

$$\underbrace{\frac{\partial}{\partial t}(\rho_A \mathbf{v}_A)}_{\substack{local\\acceleration}} + \underbrace{\nabla \cdot (\rho_A \mathbf{v}_A \mathbf{v}_A)}_{\substack{convective\\acceleration}} = \underbrace{\rho_A \mathbf{b}_A}_{\substack{body\\force}} + \underbrace{\nabla \cdot \mathbf{T}_A}_{\substack{surface\\force}} + \underbrace{\sum_{B=1}^{B=N} \mathbf{P}_{AB}}_{\substack{diffusive\\force}} + \underbrace{r_A \mathbf{v}_A}_{\substack{source\ of\ momentum\\owing\ to\ reaction}}, \quad A = 1, 2, \ldots, N \tag{6.79}$$

We begin our analysis of the species momentum equation by making use of the following representation for the species stress tensor

$$\mathbf{T}_A = -p_A \mathbf{I} + \boldsymbol{\tau}_A, \quad \textit{viscous fluid} \tag{6.80}$$

Here p_A is the partial pressure of species A and this representation is only valid for an isotropic, inelastic fluid. In addition to this limitation, we restrict our analysis to ideal solutions so that the partial pressure can be expressed as

$$p_A = x_A p, \quad \textit{ideal solution} \tag{6.81}$$

in which x_A is the mole fraction of species A and p is the total pressure. Use of Eqs. (6.80) and (6.81) in the species momentum equation given by Eq. (6.29) leads to

$$\frac{\partial}{\partial t}(\rho_A \mathbf{v}_A) + \nabla \cdot (\rho_A \mathbf{v}_A \mathbf{v}_A) = \rho_A \mathbf{b}_A - x_A \nabla p - p \nabla x_A + \nabla \cdot \boldsymbol{\tau}_A + \sum_{B=1}^{B=N} \mathbf{P}_{AB} + r_A \mathbf{v}_A, \quad A = 1, 2, \ldots, N \tag{6.82}$$

Our next step in the analysis of the species momentum equation is the use of Maxwell's representation for the force $\mathbf{P}_{AB}$ which we express as (Chapman and Cowling, 1970, pg. 109)

$$\mathbf{P}_{AB} = \frac{p x_A x_B (\mathbf{v}_B - \mathbf{v}_A)}{\mathscr{D}_{AB}}, \quad A, B = 1, 2, \ldots, N \tag{6.83}$$

Use of this result with Eq. (6.82) allows us to express the species momentum equation in the form

$$p^{-1}\left\{\frac{\partial}{\partial t}(\rho_A \mathbf{v}_A) + \nabla\cdot(\rho_A \mathbf{v}_A \mathbf{v}_A) - \rho_A \mathbf{b}_A + x_A \nabla p - \nabla\cdot\boldsymbol{\tau}_A - r_A \mathbf{v}_A\right\} \tag{6.84}$$

$$= -\nabla x_A + \sum_{B=1}^{B=N} \frac{x_A x_B(\mathbf{v}_B - \mathbf{v}_A)}{\mathscr{D}_{AB}}, \quad A = 1, 2, \ldots, N$$

The left hand side of this result can be simplified by the use of Eqs. (6.11) and the total momentum equation given by Eq. (6.49). The algebra is rather lengthy and the result is given by

$$p^{-1}\rho_A\left(\frac{\partial \mathbf{u}_A}{\partial t} + \mathbf{v}_A\cdot\nabla\mathbf{u}_A + \mathbf{u}_A\cdot\nabla\mathbf{v}\right) - p^{-1}(\omega_A - x_A)\nabla p \tag{6.85}$$

$$+p^{-1}(\omega_A \nabla\cdot\boldsymbol{\tau} - \nabla\cdot\boldsymbol{\tau}_A) - p^{-1}\rho_A(\mathbf{b} - \mathbf{b}_A) +$$

$$= -\nabla x_A + \sum_{B=1}^{B=N} \frac{x_A x_B(\mathbf{v}_B - \mathbf{v}_A)}{\mathscr{D}_{AB}}, \quad A = 1, 2, \ldots, N$$

When the two terms on the right hand side are dominant, all of the terms on the left hand side can be neglected. We express this idea as

$$p^{-1}\rho_A\left(\frac{\partial \mathbf{u}_A}{\partial t} + \mathbf{v}_A\cdot\nabla\mathbf{u}_A + \mathbf{u}_A\cdot\nabla\mathbf{v}\right) - p^{-1}(\omega_A - x_A)\nabla p \tag{6.86}$$

$$+p^{-1}(\omega_A \nabla\cdot\boldsymbol{\tau} - \nabla\cdot\boldsymbol{\tau}_A) - p^{-1}\rho_A(\mathbf{b} - \mathbf{b}_A)$$

$$\ll \nabla x_A, \quad A = 1, 2, \ldots, N$$

which leads us to the $N-1$ independent Stefan-Maxwell equations that take the form (Bird et al., 2002)

$$0 = -\nabla x_A + \sum_{B=1}^{B=N} \frac{x_A x_B(\mathbf{v}_B - \mathbf{v}_A)}{\mathscr{D}_{AB}}, \quad A = 1, 2, \ldots, N-1 \tag{6.87}$$

In this approach, one uses Eq. (6.49) to determine the mass average velocity, $\mathbf{v}$, and Eq. (6.87) to determine $N-1$ species velocities. There are numerous problems of mass transport in porous media for which the simplifications leading to Eq. (6.87) are *not valid*. Some of these cases are described in the classic work of Jackson (1977) on transport in porous catalyst and in the summary of the dusty gas model by Mason and Malinauskas (1983). More recent studies are described in the work of Kerkhof (1996, 1997).

6.1.8 Dilute Solution Convective-Diffusion Equation

If species A is dilute, that is, $x_A \ll 1$, one can make use of Eqs. (6.19) and (6.87) to obtain a classic form of the convective-diffusion equation. One begins with the use of Eq. (6.37) in order to express Eq. (6.87) in the form

$$0 = -\nabla x_A + \sum_{\substack{B=1 \\ B \neq A}}^{B=N} \frac{x_A x_B (\mathbf{u}_B - \mathbf{u}_A)}{\mathscr{D}_{AB}}, \quad A = 1, 2, \ldots, N-1 \tag{6.88}$$

which can be rearranged to obtain

$$0 = -\nabla x_A + \frac{x_A}{c} \sum_{\substack{B=1 \\ B \neq A}}^{B=N} \frac{c_B \mathbf{u}_B}{\mathscr{D}_{AB}} - \frac{1}{c} \left\{ \sum_{\substack{B=1 \\ B \neq A}}^{B=N} \frac{x_B}{\mathscr{D}_{AB}} \right\} c_A \mathbf{u}_A \tag{6.89}$$

The mass diffusion velocities are constrained by

$$0 = \sum_{B=1}^{B=N} \omega_B \mathbf{u}_B \tag{6.90}$$

and from this we can conclude that diffusion velocities tend to be of the same order of magnitude. This suggests that Eq. (6.89) can be simplified by imposing the restriction

$$\frac{x_A}{c} \sum_{\substack{B=1 \\ B \neq A}}^{B=N} \frac{c_B \mathbf{u}_B}{\mathscr{D}_{AB}} \ll \frac{1}{c} \left\{ \sum_{\substack{B=1 \\ B \neq A}}^{B=N} \frac{x_B}{\mathscr{D}_{AB}} \right\} c_A \mathbf{u}_A \tag{6.91}$$

whenever the mole fraction of species A is small compared to one, that is,

$$x_A \ll 1 \tag{6.92}$$

On the basis of Eq. (6.91) we simplify Eq. (6.89) to the form

$$0 = -\nabla x_A - \frac{1}{c} \left\{ \sum_{\substack{B=1 \\ B \neq A}}^{B=N} \frac{x_B}{\mathscr{D}_{AB}} \right\} c_A \mathbf{u}_A \tag{6.93}$$

and we define the mixture diffusivity for species A according to

$$\frac{1}{\mathscr{D}_{Am}} = \sum_{\substack{B=1 \\ B \neq A}}^{B=N} \frac{x_B}{\mathscr{D}_{AB}} \tag{6.94}$$

so that the *mixed-mode diffusive flux* can be expressed as

$$c_A \mathbf{u}_A = -\, c\,\mathscr{D}_{Am} \nabla x_A \tag{6.95}$$

We refer to $c_A \mathbf{u}_A$ as a mixed-mode diffusive flux because it is constructed in terms of a *molar* concentration, c_A, and a *mass* diffusion velocity, $\mathbf{u}_A$. Given the dilute solution constraint for the diffusing species indicated by Eq. (6.92), one can think of cases in which the temperature and pressure gradients are *small enough* so that the following restriction is satisfied.

$$x_A \nabla c \ll c \nabla x_A \tag{6.96}$$

Under these conditions the mixed-mode diffusive flux takes the form

$$c_A \mathbf{u}_A = -\mathscr{D}_{Am} \nabla c_A \tag{6.97}$$

and use of this result with Eqs. (6.19) and (6.37) yields

$$\frac{\partial c_A}{\partial t} + \nabla \cdot (c_A \mathbf{v}) = \nabla \cdot \left(\mathscr{D}_{Am} \nabla c_A \right) + R_A \tag{6.98}$$

This result is ubiquitous in the chemical engineering literature; however, the limitations imposed by Eqs. (6.86), (6.92) and (6.96) are not always made clear.

6.1.9 Non-Dilute Solutions

When the simplification indicated by Eq. (6.91) is *not valid*, we must work directly with Eq. (6.89) which can be arranged in the form

$$0 = -c\nabla x_A + x_A \sum_{\substack{B=1 \\ B \neq A}}^{B=N} \frac{\mathbf{J}_B}{\mathscr{D}_{AB}} - \left\{ \sum_{\substack{B=1 \\ B \neq A}}^{B=N} \frac{x_B}{\mathscr{D}_{AB}} \right\} \mathbf{J}_A, \tag{6.99}$$

$$A = 1, 2, \ldots, N-1$$

where the *mixed-mode* diffusive flux is given by

$$\mathbf{J}_B = c_B \mathbf{u}_B \tag{6.100}$$

In terms of this diffusive flux, Eq. (6.19) takes the form

$$\frac{\partial c_A}{\partial t} + \nabla \cdot (c_A \mathbf{v}) = -\nabla \cdot \mathbf{J}_A + R_A\,, \quad A = 1, 2, \ldots, N \tag{6.101}$$

in which the diffusive flux can be expressed as

$$\mathbf{J}_A = -c\mathscr{D}_{Am}\nabla x_A - x_A \sum_{\substack{B=1\\B\neq A}}^{B=N} \frac{\mathscr{D}_{Am}}{\mathscr{D}_{AB}}\mathbf{J}_B, \quad A = 1,\, 2,\, \ldots,\, N-1 \tag{6.102}$$

The mixed-mode diffusive fluxes are constrained by

$$\sum_{A=1}^{A=N} \mathbf{J}_A \left(\frac{M_A}{M_N}\right) = 0 \tag{6.103}$$

thus we have N equations that can be used to determine the N diffusive fluxes. At this point we define a matrix $[R]$ according to

$$[R] = \begin{bmatrix} 1 & - & \frac{x_A\mathscr{D}_{Am}}{\mathscr{D}_{AB}} & - & \frac{x_A\mathscr{D}_{Am}}{\mathscr{D}_{AC}} & - & \ldots\ldots & - & \frac{x_A\mathscr{D}_{Am}}{\mathscr{D}_{AN}} \\ -\ \frac{x_B\mathscr{D}_{Bm}}{\mathscr{D}_{BA}} & + & 1 & - & \frac{x_B\mathscr{D}_{Bm}}{\mathscr{D}_{BC}} & - & \ldots\ldots & - & \frac{x_B\mathscr{D}_{Bm}}{\mathscr{D}_{BN}} \\ -\ \frac{x_C\mathscr{D}_{Cm}}{\mathscr{D}_{CA}} & - & \frac{x_C\mathscr{D}_{Cm}}{\mathscr{D}_{CB}} & + & 1 & - & \ldots\ldots & - & \frac{x_C\mathscr{D}_{Cm}}{\mathscr{D}_{CN}} \\ . & . & . & . & . & - & \ldots\ldots & - & . \\ . & . & . & . & . & - & \ldots\ldots & - & . \\ \frac{M_A}{M_N} & + & \frac{M_B}{M_N} & + & \frac{M_C}{M_N} & + & \ldots\ldots & + & 1 \end{bmatrix} \tag{6.104}$$

and use Eqs. (6.102) and (6.103) to express the N diffusive fluxes according to

$$[R]\begin{bmatrix} \mathbf{J}_A \\ \mathbf{J}_B \\ \mathbf{J}_C \\ \cdots \\ \cdots \\ \mathbf{J}_N \end{bmatrix} = -c \begin{bmatrix} \mathscr{D}_{Am}\nabla x_A \\ \mathscr{D}_{Bm}\nabla x_B \\ \mathscr{D}_{Cm}\nabla x_C \\ \cdots \\ \mathscr{D}_{N-1m}\nabla x_{N-1} \\ 0 \end{bmatrix} \tag{6.105}$$

We assume that the inverse of $[R]$ exists in order to express the column matrix of diffusive flux vectors in the form

$$\begin{bmatrix} \mathbf{J}_A \\ \mathbf{J}_B \\ \mathbf{J}_C \\ \cdots \\ \cdots \\ \mathbf{J}_N \end{bmatrix} = -c\,[R]^{-1} \begin{bmatrix} \mathscr{D}_{Am}\nabla x_A \\ \mathscr{D}_{Bm}\nabla x_B \\ \mathscr{D}_{Cm}\nabla x_C \\ \cdots \\ \mathscr{D}_{N-1m}\nabla x_{N-1} \\ 0 \end{bmatrix} \tag{6.106}$$

in which the column matrix on the right hand side of this result can be expressed as

$$\begin{bmatrix} \mathscr{D}_{Am}\nabla x_A \\ \mathscr{D}_{Bm}\nabla x_B \\ \mathscr{D}_{Cm}\nabla x_C \\ \cdots \\ \mathscr{D}_{N-1m}\nabla x_{N-1} \\ 0 \end{bmatrix} = \begin{bmatrix} \mathscr{D}_{Am} & 0 & 0 & \cdots\cdots & 0 & 0 \\ 0 & \mathscr{D}_{Bm} & 0 & \cdots\cdots & 0 & 0 \\ 0 & 0 & \mathscr{D}_{Cm} & \cdots\cdots & 0 & 0 \\ . & . & . & \cdots\cdots & . & . \\ 0 & 0 & 0 & \cdots\cdots & \mathscr{D}_{N-1m} & 0 \\ 0 & 0 & 0 & \cdots\cdots & 0 & \mathscr{D}_{Nm} \end{bmatrix} \begin{bmatrix} \nabla x_A \\ \nabla x_B \\ \nabla x_C \\ \cdot\cdot \\ \nabla x_{N-1} \\ 0 \end{bmatrix} \tag{6.107}$$

The *diffusivity matrix* is now defined by

$$[D] = [R]^{-1} \begin{bmatrix} \mathscr{D}_{Am} & 0 & 0 & \cdots\cdots & 0 & 0 \\ 0 & \mathscr{D}_{Bm} & 0 & \cdots\cdots & 0 & 0 \\ 0 & 0 & \mathscr{D}_{Cm} & \cdots\cdots & 0 & 0 \\ . & . & . & \cdots\cdots & . & . \\ 0 & 0 & 0 & \cdots\cdots & \mathscr{D}_{N-1m} & 0 \\ 0 & 0 & 0 & \cdots\cdots & 0 & \mathscr{D}_{Nm} \end{bmatrix} \tag{6.108}$$

so that Eq. (6.106) takes the form

$$\begin{bmatrix} \mathbf{J}_A \\ \mathbf{J}_B \\ \mathbf{J}_C \\ \cdots \\ \cdots \\ \mathbf{J}_N \end{bmatrix} = -c\,[D] \begin{bmatrix} \nabla x_A \\ \nabla x_B \\ \nabla x_C \\ \cdots \\ \nabla x_{N-1} \\ 0 \end{bmatrix} \tag{6.109}$$

This result can be expressed in a form analogous to that given by Eq. (6.102) leading to

$$\mathbf{J}_A = -c \sum_{E=1}^{E=N-1} D_{AE} \nabla x_E, \qquad A = 1,\ 2,\ \ldots, N \tag{6.110}$$

In the general case, the elements of the diffusivity matrix, D_{AE}, will depend on the mole fractions in a non-trivial manner. When this result is used in Eq. (6.101), we obtain the non-linear, coupled governing differential equation for given by

$$\frac{\partial c_A}{\partial t} + \nabla \cdot (c_A \mathbf{v}) = \nabla \cdot \left(c \sum_{E=1}^{E=N-1} D_{AE} \nabla x_E \right) + R_A, \quad A = 1, 2, \ldots, N \tag{6.111}$$

Use of the transport equations presented in this section to describe transport phenomena in porous media requires that we *upscale* the equations from the so-called point scale to an appropriate larger scale (Cushman, 1990, 1997). This analysis is described in the following section.

6.2 VOLUME AVERAGED CONSERVATION EQUATIONS

To capture the essential features of mass transfer in porous media, we develop the local volume averaged forms of the governing equations. Volume averaging can occur at many levels and some of those levels associated with a study of biofilm growth in porous media are illustrated in Figure 6.2. In that figure, Level I represents an heterogeneous aquifer in which mass transport and biological reactions occur. The averaging volume used to develop transport equations in the aquifer is illustrated as Level II. There it is apparent that a two-region model is being used to model the aquifer. In order to describe mass transport and reaction in one of those two regions, we make use of the averaging volume illustrated as Level III in Figure 6.2. There we have shown a fluid phase, a solid phase, and the biofilm phase. Continuing on to Level IV we indicate that the solid phase at Level III is actually a micro-porous medium in which the diffusive transport may play an important role in the biofilm growth. Although not shown, one must make use of another averaging volume to analysis the process of diffusion and reaction in the biofilms (Wood and Whitaker, 1998). In this work, we consider volume averaging at Level III, which is often referred to as the

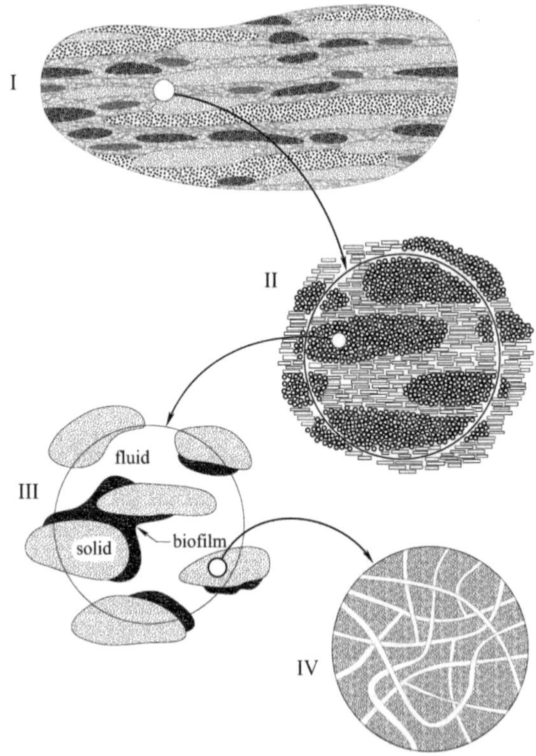

Figure 6.2. Hierarchical porous medium containing biofilms

Darcy scale. The identification with Darcy's law comes from the fact that the no-slip condition is typically applied at the fluid-solid interface illustrated in Level III. The system at Level III in Figure 6.2 is illustrated in Figure 6.3 without the biolfilm. There the fluid is identified as the γ-phase and the solid as the impermeable κ-phase. The development of local volume averaged equations requires that we define two types of averages in terms of the averaging volume, $\mathscr{V}$, illustrated in Figure 6.3. The first of these is the *superficial average* of some function ψ_γ defined according to

$$\langle \psi_\gamma \rangle = \frac{1}{\mathscr{V}} \int_{V_\gamma} \psi_\gamma \mathrm{d}V \tag{6.112}$$

while the second is the *intrinsic average* is defined by

$$\langle \psi_\gamma \rangle^\gamma = \frac{1}{V_\gamma} \int_{V_\gamma} \psi_\gamma \mathrm{d}V \tag{6.113}$$

These two averages are related according to

$$\langle \psi_\gamma \rangle = \varepsilon_\gamma \langle \psi_\gamma \rangle^\gamma \tag{6.114}$$

in which ε_γ is the volume fraction of the γ-phase defined explicitly as

$$\varepsilon_\gamma = \frac{V_\gamma}{\mathscr{V}} \tag{6.115}$$

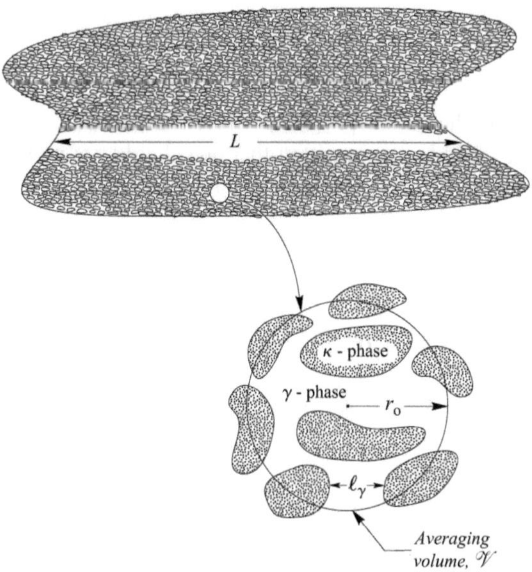

Figure 6.3. Two-phase system

Since the intrinsic and superficial averages differ by a factor of ε_γ, it is essential to make use of a notation that clearly distinguishes between the two averages.

When we form the volume average of any transport equation, we are immediately confronted with the average of a gradient (or divergence) whereas it is the gradient (or divergence) of the average that we are seeking. In order to interchange integration and differentiation, we will make use of the spatial averaging theorem (Whitaker, 1999). For the two-phase system illustrated in Figure 6.3 this theorem can be expressed as

$$\langle \nabla \psi_\gamma \rangle = \nabla \langle \psi_\gamma \rangle + \frac{1}{\mathscr{V}} \int_{A_{\gamma\kappa}} \mathbf{n}_{\gamma\kappa} \psi_\gamma \, \mathrm{d}A \tag{6.116}$$

in which ψ_γ is any function associated with the γ-phase. Here $A_{\gamma\kappa}$ represents the interfacial area contained within the averaging volume, and we have used $\mathbf{n}_{\gamma\kappa}$ to represent the unit normal vector pointing *from* the γ-phase *toward* the κ-phase. In order for the method of volume averaging to lead to a successful result, the length scales illustrated in Figure 6.3 must be disparate as indicated by

$$\ell_\gamma \ll r_0 \ll L \tag{6.117}$$

When these constraints are satisfied, Eq. (6.116) can be used to derive an alternate form given by

$$\langle \nabla \psi_\gamma \rangle = \varepsilon_\gamma \nabla \langle \psi_\gamma \rangle^\gamma + \frac{1}{\mathscr{V}} \int_{A_{\gamma\kappa}} \mathbf{n}_{\gamma\kappa} \tilde{\psi}_\gamma \, \mathrm{d}A \tag{6.118}$$

in which $\tilde{\psi}_\gamma$ is the *spatial deviation* defined according to

$$\tilde{\psi}_\gamma = \psi_\gamma - \langle \psi_\gamma \rangle^\gamma \tag{6.119}$$

To avoid errors in the development of closed-form volume averaged transport equations, Gray (1975) has pointed out that one should define the spatial deviation in terms of the *intrinsic average* as indicated by Eq. (6.119).

6.2.1 Mass Transfer

In order to explore the transport of species A in the porous medium illustrated in Figure 6.3, we begin with Eq. (6.19) in the form

$$\frac{\partial c_{A\gamma}}{\partial t} + \nabla \cdot (c_{A\gamma} \mathbf{v}_{A\gamma}) = R_{A\gamma}, \qquad A = 1, 2, \ldots, N \tag{6.120}$$

If the γ–κ system is rigid and surface transport can be neglected, one can use Eq. (6.120) to derive a jump condition (Slattery, 1990; Edwards et al., 1991; Whitaker, 1992) of the form

$$\frac{\partial c_{As}}{\partial t} = \left(c_{A\gamma} \mathbf{v}_{A\gamma} \right) \cdot \mathbf{n}_{\gamma\kappa} + R_{As}, \text{ at the } \gamma - \kappa \text{ interface}, \quad A = 1, 2, \ldots, N \tag{6.121}$$

Here we have used c_{As} to represent the *surface concentration* of species A (moles per unit area) and R_{As} to represent the molar rate of production of species A owing to heterogeneous reaction (moles per unit area per unit time). In order to obtain a solution for $c_{A\gamma}$, and therefore $\langle c_{A\gamma}\rangle^{\gamma}$, we require a *connection* between the surface concentration, c_{As}, and the bulk concentration, $c_{A\gamma}$. The two classic connections are the linear interfacial flux relation given by

$$\left(c_{A\gamma}\mathbf{v}_{A\gamma}\right)\cdot\mathbf{n}_{\gamma\kappa} = k_{A1}\,c_{A\gamma} - k_{-A1}\,c_{As}, \text{at the } \gamma\text{–}\kappa \text{ interface}, \quad A = 1, 2, \ldots, N \tag{6.122}$$

and the approximation of local adsorption equilibrium. For a linear adsorption isotherm, this latter connection is given by

$$c_{As} = K_{\text{eq}}\,c_{A\gamma}, \quad \text{at the } \gamma\text{–}\kappa \text{ interface}, \quad A = 1,\ 2,\ \ldots,\ N \tag{6.123}$$

While this result is strictly true at the condition of equilibrium, the general form can often be used as a reasonable approximation for dynamic systems (Whitaker, 1986b, 1999).

We begin the averaging procedure with Eq. (6.120) and express the superficial average of that result as

$$\left\langle\frac{\partial c_{A\gamma}}{\partial t}\right\rangle + \left\langle\nabla\cdot\left(c_{A\gamma}\mathbf{v}_{A\gamma}\right)\right\rangle = \langle R_{A\gamma}\rangle, \quad A = 1, 2, \ldots, N \tag{6.124}$$

For a rigid porous medium, one can use the transport theorem and the averaging theorem to obtain

$$\frac{\partial\langle c_{A\gamma}\rangle}{\partial t} + \nabla\cdot\langle c_{A\gamma}\mathbf{v}_{A\gamma}\rangle + \frac{1}{\mathscr{V}}\int\limits_{A_{\gamma\kappa}}\mathbf{n}_{\gamma\kappa}\cdot\left(c_{A\gamma}\mathbf{v}_{A\gamma}\right)\mathrm{d}A = \langle R_{A\gamma}\rangle \tag{6.125}$$

where it is understood that this applies to all N species. Since we seek a transport equation for the intrinsic average concentration, we make use of Eq. (6.114) to express Eq. (6.125) in the form

$$\varepsilon_\gamma\frac{\partial\langle c_{A\gamma}\rangle^{\gamma}}{\partial t} + \nabla\cdot\langle c_{A\gamma}\mathbf{v}_{A\gamma}\rangle + \frac{1}{\mathscr{V}}\int\limits_{A_{\gamma\kappa}}\mathbf{n}_{\gamma\kappa}\cdot\left(c_{A\gamma}\mathbf{v}_{A\gamma}\right)\mathrm{d}A = \varepsilon_\gamma\langle R_{A\gamma}\rangle^{\gamma} \tag{6.126}$$

At this point, it is convenient to make use of the jump condition given by Eq. (6.121) in order to obtain

$$\varepsilon_\gamma\frac{\partial\langle c_{A\gamma}\rangle^{\gamma}}{\partial t} + \nabla\cdot\langle c_{A\gamma}\mathbf{v}_{A\gamma}\rangle = \langle R_{A\gamma}\rangle^{\gamma} - \frac{1}{\mathscr{V}}\int\limits_{A_{\gamma\kappa}}\frac{\partial c_{As}}{\partial t}\mathrm{d}A + \frac{1}{\mathscr{V}}\int\limits_{A_{\gamma\kappa}}R_{As}\mathrm{d}A \tag{6.127}$$

We now define the *intrinsic interfacial area average* according to

$$\langle \psi_\gamma \rangle_{\gamma\kappa} = \frac{1}{A_{\gamma\kappa}} \int_{A_{\gamma\kappa}} \psi_\gamma \, \mathrm{d}A \tag{6.128}$$

and use this definition with c_{As} and R_{As} so that Eq. (6.127) takes the convenient form given by

$$\underbrace{\varepsilon_\gamma \frac{\partial \langle c_{A\gamma} \rangle^\gamma}{\partial t}}_{accumulation} + \underbrace{\nabla \cdot \langle c_{A\gamma} \mathbf{v}_{A\gamma} \rangle}_{transport} = \underbrace{\langle R_{A\gamma} \rangle^\gamma}_{\substack{homogeneous \\ reaction}} \underbrace{- a_\mathrm{v} \frac{\partial \langle c_{As} \rangle_{\gamma\kappa}}{\partial t}}_{adsorption} + \underbrace{a_\mathrm{v} \langle R_{As} \rangle_{\gamma\kappa}}_{\substack{heterogeneous \\ reaction}} \tag{6.129}$$

One must keep in mind that this is a *general result* based on Eqs. (6.120) and (6.121); however, *only* the first term in Eq. (6.129) is in a form that is ready for applications. On the right hand side of Eq. (6.129) we are confronted with the necessity of chemical kinetic constitutive equations for the reaction rate terms and a suitable connection between the surface concentration, c_{As}, and the bulk concentration, $c_{A\gamma}$. In this development, we will avoid these problems and concentration on a *single aspect* of Eq. (6.129), the determination of the species velocity, $\mathbf{v}_{A\gamma}$.

The manner in which $\mathbf{v}_{A\gamma}$ is determined depends on the particular characteristics of the species momentum equation given by Eq. (6.84). When Eqs. (6.49) and (6.87) are a valid representation of the N species momentum equations, we make use of the total momentum equation to determine the mass average velocity, $\mathbf{v}_\gamma$, and the Stefan-Maxwell equations to determine the $N - 1$ independent species velocities. Under these circumstances, Eq. (6.129) takes the form

$$\varepsilon_\gamma \frac{\partial \langle c_{A\gamma} \rangle^\gamma}{\partial t} + \nabla \cdot \langle c_{A\gamma} \mathbf{v}_\gamma \rangle = \langle \nabla \cdot \mathbf{J}_{A\gamma} \rangle + \langle R_{A\gamma} \rangle^\gamma - a_\mathrm{v} \frac{\partial \langle c_{As} \rangle_{\gamma\kappa}}{\partial t} + a_\mathrm{v} \langle R_{As} \rangle_{\gamma\kappa} \tag{6.130}$$

and we are confronted with the average of a product, $c_{A\gamma} \mathbf{v}_\gamma$, whereas it is the product of the averages that we desire. Here we follow the approach used in turbulence modeling and decompose the point values of the concentration and velocity into the intrinsic average values and the spatial deviation values. These decompositions are given by

$$c_{A\gamma} = \langle c_{A\gamma} \rangle^\gamma + \tilde{c}_{A\gamma}, \quad \mathbf{v}_\gamma = \langle \mathbf{v}_\gamma \rangle^\gamma + \tilde{\mathbf{v}}_\gamma \tag{6.131}$$

so that the volume averaged convective transport is described as

$$\langle \mathbf{v}_\gamma c_{A\gamma} \rangle = \big\langle \langle \mathbf{v}_\gamma \rangle^\gamma \langle c_{A\gamma} \rangle^\gamma + \tilde{\mathbf{v}}_\gamma \langle c_{A\gamma} \rangle^\gamma + \langle \mathbf{v}_\gamma \rangle^\gamma \tilde{c}_{A\gamma} + \tilde{\mathbf{v}}_\gamma \tilde{c}_{A\gamma} \big\rangle \tag{6.132}$$

The analysis of the convective transport is not simple; however, reasonable arguments are available (Whitaker, 1999, Sec. 3.2.3) indicating that when the length-scale constraint indicated by Eq. (6.117) is valid Eq. (6.132) simplifies to

$$\langle \mathbf{v}_\gamma c_{A\gamma} \rangle = \varepsilon_\gamma \langle \mathbf{v}_\gamma \rangle^\gamma \langle c_{A\gamma} \rangle^\gamma + \langle \tilde{\mathbf{v}}_\gamma \tilde{c}_{A\gamma} \rangle \tag{6.133}$$

Use of this result in Eq. (6.130) leads to

$$\varepsilon_\gamma \frac{\partial \langle c_{A\gamma} \rangle^\gamma}{\partial t} + \nabla \cdot \left(\langle \mathbf{v}_\gamma \rangle \langle c_{A\gamma} \rangle^\gamma \right) = \langle \nabla \cdot \mathbf{J}_{A\gamma} \rangle + \underbrace{\nabla \cdot \langle \tilde{\mathbf{v}}_\gamma \tilde{c}_{A\gamma} \rangle}_{\substack{\textit{dispersive} \\ \textit{transport}}} + \langle R_{A\gamma} \rangle^\gamma \tag{6.134}$$

$$- a_{\mathrm{v}} \frac{\partial \langle c_{As} \rangle_{\gamma\kappa}}{\partial t} + a_{\mathrm{v}} \langle R_{As} \rangle_{\gamma\kappa}$$

where the volume average velocity has been expressed in terms of the superficial average rather than the intrinsic average since it is the superficial average that appears in the traditional form of Darcy's law. In the following section we indicate how $\langle \mathbf{v}_\gamma \rangle$ can be determined and we then return to the problem of expressing the diffusive and dispersive flux in some useful form.

6.3 MOMENTUM TRANSFER

An analysis of momentum transfer is based on the continuity equation and the linear momentum equation. The former was given earlier as Eq. (6.17) and represented here in the form

$$\frac{\partial \rho_\gamma}{\partial t} + \nabla \cdot (\rho_\gamma \mathbf{v}_\gamma) = 0 \tag{6.135}$$

When the density can be treated as a constant, this reduces to

$$\nabla \cdot \mathbf{v}_\gamma = 0 \tag{6.136}$$

and the local volume averaged form is given by

$$\nabla \cdot \langle \mathbf{v}_\gamma \rangle = 0 \tag{6.137}$$

This form of the continuity equation is an attractive approximation for the case of dilute solution mass transfer described by Eqs. (6.88) through (6.98). For non-dilute solutions, the approximation given by Eq. (6.136) may not be acceptable.

In the total momentum equation given by Eq. (6.49), it is often reasonable to ignore variations in the density and viscosity *in the viscous term* so that the governing equation for $\mathbf{v}_\gamma$ is given by

$$\frac{\partial}{\partial t} \left(\rho_\gamma \mathbf{v}_\gamma \right) + \nabla \cdot \left(\rho_\gamma \mathbf{v}_\gamma \mathbf{v}_\gamma \right) = \rho_\gamma \mathbf{g} - \nabla p_\gamma + \mu \nabla^2 \mathbf{v}_\gamma \tag{6.138}$$

In addition, there is a class of problems for which both the normal and tangential components of the mass average velocity are zero at the fluid solid interface leading to the boundary condition given by

$$\text{BC.} \qquad \mathbf{v}_\gamma = 0\,, \quad \text{at the } \gamma\text{–}\kappa \text{ interface} \tag{6.139}$$

This condition requires that the mean free path be small compared to the pore diameter and that the effect of adsorption and reaction at the γ–κ interface be neglected. When these two conditions are not satisfied, one is confronted with Knudsen flow, slip owing to multicomponent effects (Graham, 1833; Kramers and Kistemaker, 1943), and coupling of the adsorption/reaction process with the momentum transfer process. Within the framework of the method of volume averaging, some of these conditions have been examined by Whitaker (1987) and more recently by Altevogt et al. (2003a, 2003b). Detailed closure problems have not been developed for either case, thus a comparison between theory and experiment *in the absence of adjustable parameters* has not been made. In the absence of this type of comparison between theory and experiment, the dusty gas remains as a popular model of this complex process (Mason and Malinauskas, 1983).

The superficial average of Eq. (6.138) can be expressed as

$$\left\langle \frac{\partial}{\partial t}(\rho_\gamma \mathbf{v}_\gamma) \right\rangle + \langle \nabla \cdot (\rho_\gamma \mathbf{v}_\gamma \mathbf{v}_\gamma) \rangle = -\,\langle \nabla p_\gamma \rangle + \langle \rho_\gamma \mathbf{g} \rangle + \langle \mu_\gamma \nabla \cdot \nabla \mathbf{v}_\gamma \rangle \tag{6.140}$$

and if we neglect variations of the density in both the body force and the inertial terms, we can express this result as

$$\rho_\gamma \left\langle \frac{\partial \mathbf{v}_\gamma}{\partial t} \right\rangle + \rho_\gamma \langle \nabla \cdot (\mathbf{v}_\gamma \mathbf{v}_\gamma) \rangle = -\,\langle \nabla p_\gamma \rangle + \varepsilon_\gamma \rho_\gamma \mathbf{g} + \mu_\gamma \langle \nabla \cdot \nabla \mathbf{v}_\gamma \rangle \tag{6.141}$$

Here we see the need to interchange integration and differentiation in every term except the one representing the gravitational force. Use of the general transport theorem, the averaging theorem, and the second of Eqs. (6.131) eventually leads to

$$\begin{aligned} \rho_\gamma \frac{\partial \langle \mathbf{v}_\gamma \rangle^\gamma}{\partial t} + \rho_\gamma \langle \mathbf{v}_\gamma \rangle^\gamma \cdot \nabla \langle \mathbf{v}_\gamma \rangle^\gamma + \rho_\gamma \varepsilon_\gamma^{-1} \nabla \cdot \langle \tilde{\mathbf{v}}_\gamma \tilde{\mathbf{v}}_\gamma \rangle \qquad\qquad (6.142) \\ = -\,\nabla \langle p_\gamma \rangle^\gamma + \rho_\gamma \mathbf{g} + \mu_\gamma \nabla^2 \langle \mathbf{v}_\gamma \rangle^\gamma + \frac{1}{V_\gamma} \int_{A_{\gamma\kappa}} \mathbf{n}_{\gamma\kappa} \cdot \left(-\mathbf{I} \tilde{p}_\gamma + \mu_\gamma \nabla \tilde{\mathbf{v}}_\gamma \right) \mathrm{d}A \end{aligned}$$

To obtain this result, one must impose the length scale constraints given by Eq. (6.117) and a step-by-step development is given by Whitaker (1997). Although we seek a governing equation for $\langle \mathbf{v}_\gamma \rangle$, the form given by Eq. (6.142) is a necessary step since we can use it in conjunction with the decomposition

$$\mathbf{v}_\gamma = \langle \mathbf{v}_\gamma \rangle^\gamma + \tilde{\mathbf{v}}_\gamma \tag{6.143}$$

and Eq. (6.138) to determine the spatial deviation velocity, $\tilde{\mathbf{v}}_\gamma$. The boundary value problem for $\tilde{\mathbf{v}}_\gamma$ and $\tilde{p}_\gamma$ is known as the *closure problem* and the development of this problem is described in detail by Whitaker (1996). The analysis leads to the *mappings* given by

$$\tilde{\mathbf{v}}_\gamma = \mathbf{M} \cdot \langle \mathbf{v}_\gamma \rangle^\gamma \tag{6.144a}$$

$$\tilde{p}_\gamma = \mu_\gamma \mathbf{m} \cdot \langle \mathbf{v}_\gamma \rangle^\gamma \tag{6.144b}$$

The mapping tensor **M** and the mapping vector **m** are determined by a closure problem that must be solved in some representative region which is necessarily a spatially periodic model of a porous medium (Quintard and Whitaker, 1994a–e). Such a model is illustrated in Figure 6.4. Use of Eqs. (6.144) in Eq. (6.142) and imposition of the length scale constraints given by Eq. (6.117) leads to

$$\begin{aligned} &\rho_\gamma \frac{\partial \langle \mathbf{v}_\gamma \rangle^\gamma}{\partial t} + \rho_\gamma \langle \mathbf{v}_\gamma \rangle^\gamma \nabla \cdot \langle \mathbf{v}_\gamma \rangle^\gamma + \rho_\gamma \varepsilon_\gamma^{-1} \nabla \cdot \left(\langle \mathbf{v}_\gamma \rangle^\gamma \cdot \langle \mathbf{M}^{\mathrm{T}} \mathbf{M} \rangle \cdot \langle \mathbf{v}_\gamma \rangle^\gamma \right) \\ &= -\nabla \langle p_\gamma \rangle^\gamma + \rho_\gamma \mathbf{g} + \mu_\gamma \nabla^2 \langle \mathbf{v}_\gamma \rangle^\gamma \\ &\quad + \mu_\gamma \left\{ \frac{1}{V_\gamma} \int_{A_{\gamma\kappa}} \mathbf{n}_{\gamma\kappa} \cdot (-\mathbf{I}\mathbf{m} + \nabla \mathbf{M}) \, dA \right\} \cdot \langle \mathbf{v}_\gamma \rangle^\gamma \end{aligned} \tag{6.145}$$

It is shown elsewhere (Whitaker, 1996) that the solution for the closure variables is analogous to the solution of the Navier-Stokes equations, thus the mapping variables **M** and **m** can be determined by classical means. In general, the terms on the left hand side of Eq. (6.145) are negligible and the volume averaged form of the Navier-Stokes

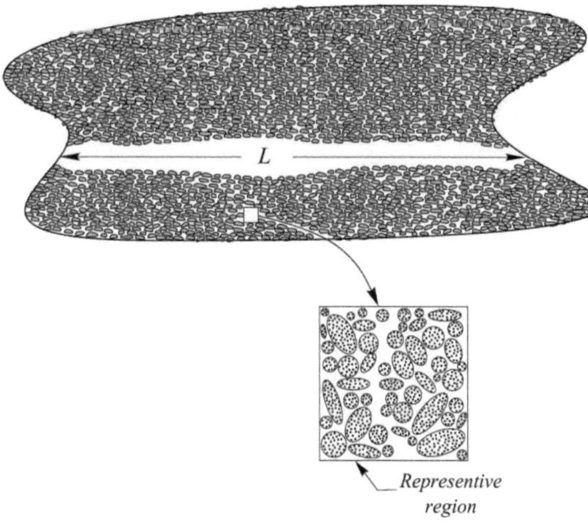

Figure 6.4. Spatially periodic model of a porous medium

equations simplifies to

$$0 = -\nabla\langle p_\gamma\rangle^\gamma + \rho_\gamma \mathbf{g} + \mu_\gamma \nabla^2\langle \mathbf{v}_\gamma\rangle^\gamma + \mu_\gamma \left\{ \frac{1}{V_\gamma} \int_{A_{\gamma\kappa}} \mathbf{n}_{\gamma\kappa} \cdot (-\mathbf{Im} + \nabla\mathbf{M})\, dA \right\} \cdot \langle \mathbf{v}_\gamma\rangle^\gamma \tag{6.146}$$

Rather than attack the general closure problem for **m** and **M**, it is convenient to *decompose* the problem into two parts. The first part will produce the Darcy's law (Darcy, 1856) permeability tensor that depends only on the geometry of the porous medium under consideration, and the second part will lead to an inertial correction, i.e., the Forchheimer equation (Forchheimer, 1901). To accomplish this decomposition, we represent **m** and **M** as

$$\mathbf{m} = \mathbf{b} + \mathbf{c}, \qquad \mathbf{M} = \mathbf{B} + \mathbf{C} \tag{6.147}$$

The details of the closure problems for **b** and **B** and for **c** and **C** are given elsewhere (Whitaker, 1996), and here we only note that the use of Eq. (6.147) leads to

$$\begin{aligned} 0 = &-\nabla\langle p_\gamma\rangle^\gamma + \rho_\gamma \mathbf{g} + \mu_\gamma \nabla^2\langle \mathbf{v}_\gamma\rangle^\gamma \\ &+ \mu_\gamma \left\{ \frac{1}{V_\gamma} \int_{A_{\gamma\kappa}} \mathbf{n}_{\gamma\kappa} \cdot (-\mathbf{Ib} + \nabla\mathbf{B})\, dA \right\} \cdot \langle \mathbf{v}_\gamma\rangle^\gamma \\ &+ \mu_\gamma \left\{ \frac{1}{V_\gamma} \int_{A_{\gamma\kappa}} \mathbf{n}_{\gamma\kappa} \cdot (-\mathbf{Ic} + \nabla\mathbf{C})\, dA \right\} \cdot \langle \mathbf{v}_\gamma\rangle^\gamma \end{aligned} \tag{6.148}$$

in which the *Darcy's law permeability tensor* is defined by

$$\frac{1}{V_\gamma} \int_{A_{\gamma\kappa}} \mathbf{n}_{\gamma\kappa} \cdot [-\mathbf{Ib} + \nabla\mathbf{B}]\, dA = -\,\varepsilon_\gamma\, \mathbf{K}^{-1} \tag{6.149}$$

and the *Forchheimer correction tensor* is defined by

$$\frac{1}{V_\gamma} \int_{A_{\gamma\kappa}} \mathbf{n}_{\gamma\kappa} \cdot [-\mathbf{Ic} + \nabla\mathbf{C}]\, dA = -\,\varepsilon_\gamma\, \mathbf{K}^{-1} \cdot \mathbf{F} \tag{6.150}$$

Here we note that the definitions of **K** and **F** have been deliberately chosen to produce a momentum equation containing the *superficial average* velocity rather than the *intrinsic average* velocity that appears in Eq. (6.148). Use of these two definitions in Eq. (6.148) leads to a result

$$0 = -\nabla\langle p_\gamma\rangle^\gamma + \rho_\gamma \mathbf{g} + \mu_\gamma \nabla^2\langle \mathbf{v}_\gamma\rangle^\gamma - \mu_\gamma\, \mathbf{K}^{-1} \cdot \langle \mathbf{v}_\gamma\rangle - \mu_\gamma\, \mathbf{K}^{-1} \cdot \mathbf{F} \cdot \langle \mathbf{v}_\gamma\rangle \tag{6.151}$$

that can be arranged in the form

$$\langle \mathbf{v}_\gamma \rangle = -\frac{\mathbf{K}}{\mu_\gamma} \cdot \left(\nabla \langle p_\gamma \rangle^\gamma - \rho_\gamma \mathbf{g} - \underbrace{\mu_\gamma \nabla^2 \langle \mathbf{v}_\gamma \rangle^\gamma}_{\substack{\textit{Brinkman} \\ \textit{correction}}} \right) - \underbrace{\mathbf{F} \cdot \langle \mathbf{v}_\gamma \rangle}_{\substack{\textit{Fochheimer} \\ \textit{correction}}} \tag{6.152}$$

It is important to recognize that the Brinkman correction (Brinkman, 1947) appears naturally in terms of the *intrinsic average* velocity and not the *superficial average* velocity that appears in Darcy's law and the Forchheimer correction. For application purposes, one needs to work with the superficial velocity so that Eq. (6.152) takes the form

$$\langle \mathbf{v}_\gamma \rangle = -\frac{\mathbf{K}}{\mu_\gamma} \left[\nabla \langle p_\gamma \rangle^\gamma - \rho_\gamma \mathbf{g} - \underbrace{\left(\mu_\gamma / \varepsilon_\gamma \right) \nabla^2 \langle \mathbf{v}_\gamma \rangle}_{\substack{\textit{Brinkman} \\ \textit{correction}}} \right] - \underbrace{\mathbf{F} \cdot \langle \mathbf{v}_\gamma \rangle}_{\substack{\textit{Forchheimer} \\ \textit{correction}}} \tag{6.153}$$

in which $\mu_\gamma / \varepsilon_\gamma$ is *sometimes* referred to as the *Brinkman viscosity*. When values of the Brinkman viscosity different than $\mu_\gamma / \varepsilon_\gamma$ are encountered, it means that this term is being used as an *empirical* correlating factor.

6.4 CLOSURE FOR MASS TRANSFER

Given a means of determining the superficial mass average velocity, we are ready to return to Eq. (6.134) and list some of the results that are available. For dilute solutions, the development given by Eqs. (6.88) through (6.98) provides

$$\varepsilon_\gamma \frac{\partial \langle c_{A\gamma} \rangle^\gamma}{\partial t} + \nabla \cdot \left(\langle \mathbf{v}_\gamma \rangle \langle c_{A\gamma} \rangle^\gamma \right) = \left\langle \nabla \cdot \left(\mathscr{D}_{Am} \nabla c_{A\gamma} \right) \right\rangle + \underbrace{\nabla \cdot \langle \tilde{\mathbf{v}}_\gamma \tilde{c}_{A\gamma} \rangle}_{\substack{\textit{dispersive} \\ \textit{transport}}} - a_v \frac{\partial \langle c_{As} \rangle_{\gamma\kappa}}{\partial t} + a_v \langle R_{As} \rangle_{\gamma\kappa} \tag{6.154}$$

Here we have discarded the homogeneous reaction rate term since homogeneous reactions are generally unimportant in porous media processes. Use of the averaging theorem and the decomposition given by the first of Eqs. (6.131) allows us to express

Eq. (6.154) in the form (Whitaker, 1999)

$$\varepsilon_\gamma \frac{\partial \langle c_{A\gamma}\rangle^\gamma}{\partial t} + \nabla \cdot \left(\langle \mathbf{v}_\gamma\rangle \langle c_{A\gamma}\rangle^\gamma\right) \tag{6.155}$$

$$= \underbrace{\nabla \cdot \left[\varepsilon_\gamma \mathscr{D}_{Am}\left(\nabla \langle c_{A\gamma}\rangle^\gamma + \frac{1}{V_\gamma}\int\limits_{A_{\gamma\kappa}} \mathbf{n}_{\gamma\kappa}\tilde{c}_{A\gamma}\, dA\right)\right]}_{\textit{diffusion}}$$

$$+ \underbrace{\nabla \cdot \langle \tilde{\mathbf{v}}_\gamma \tilde{c}_{A\gamma}\rangle}_{\substack{\textit{dispersive}\\ \textit{transport}}} - \underbrace{a_v \frac{\partial \langle c_{As}\rangle_{\gamma\kappa}}{\partial t}}_{\textit{adsorption}} + \underbrace{a_v \langle R_{As}\rangle_{\gamma\kappa}}_{\substack{\textit{heterogeneous}\\ \textit{reaction}}}$$

The first two terms on the left hand side of this result require only the knowledge of the superficial mass average velocity, and when Eqs. (6.138) and (6.139) are applicable this represents a straightforward matter. The last four terms all depend, directly or indirectly, on the spatial deviation concentration, $\tilde{c}_{A\gamma}$, and are thus sensitive to the approximations made in the closure problem. In the absence of a closure problem, these four terms become sensitive to the intuition associated with a particular model.

6.4.1 Passive Transport

When there is negligible adsorption and/or heterogeneous reaction, the spatial deviation concentration takes the form (Whitaker, 1999)

$$\tilde{c}_{A\gamma} = \mathbf{b}_\gamma \cdot \nabla \langle c_{A\gamma}\rangle^\gamma \tag{6.156}$$

and this leads to a transport equation of the form

$$\varepsilon_\gamma \frac{\partial \langle c_{A\gamma}\rangle^\gamma}{\partial t} + \nabla \cdot \left(\langle \mathbf{v}_\gamma\rangle \langle c_{A\gamma}\rangle^\gamma\right) = \nabla \cdot \left[\left(\varepsilon_\gamma \mathbf{D}_{\text{eff}} + \varepsilon_\gamma \mathbf{D}_\gamma\right) \cdot \nabla \langle c_{A\gamma}\rangle^\gamma\right] \tag{6.157}$$

Here $\mathbf{D}_{\text{eff}}$ is the effective diffusivity tensor defined by

$$\mathbf{D}_{\text{eff}} = \mathscr{D}_{Am}\left(\mathbf{I} + \frac{1}{V_\gamma}\int\limits_{A_{\gamma\kappa}} \mathbf{n}_{\gamma\kappa}\mathbf{b}_\gamma\, dA\right) \tag{6.158}$$

For purely diffusive transport in porous media that are isotropic at the Darcy scale, good agreement between theory and experiment can be obtained using very simple geometrical models (Quintard, 1993). For porous media that are anisotropic at the Darcy scale, simple models are not satisfactory and more work is required in order to develop a predictive theory of diffusion (Ochoa-Tapia et al., 1994). In Eq. (6.157) we have used $\mathbf{D}_\gamma$ to represent the hydrodynamic dispersion tensor defined by

$$\mathbf{D}_\gamma = -\frac{1}{V_\gamma} \int_{V_\gamma} \tilde{\mathbf{v}}_\gamma \mathbf{b}_\gamma \, \mathrm{d}V \tag{6.159}$$

In general, these two tensor coefficients are combined so that Eq. (6.157) takes the form

$$\varepsilon_\gamma \frac{\partial \langle c_{A\gamma}\rangle^\gamma}{\partial t} + \nabla \cdot \left(\langle \mathbf{v}_\gamma \rangle \langle c_{A\gamma}\rangle^\gamma \right) = \nabla \cdot \left(\varepsilon_\gamma \mathbf{D}^*_\gamma \cdot \nabla \langle c_{A\gamma}\rangle^\gamma \right) \tag{6.160}$$

in which $\mathbf{D}^*_\gamma$ is referred to as the total dispersion tensor. Simple geometric models fail to provide good agreement with experimental results for both lateral and longitudinal dispersion (Eidsath et al., 1983); however, complex, two-dimensional unit cells for a spatially periodic model of a porous medium can be used to provide attractive agreement (Whitaker, 1999, Sec. 3.4). At this point in time, it would appear that three-dimensional unit cells containing an appropriate degree of randomness will be required to accurately predict the dispersion tensor for porous media that are isotropic at the Darcy scale.

6.4.2 Active Transport: Adsorption

When adsorption occurs in the absence of heterogeneous reaction, Eq. (6.155) reduces to

$$\begin{aligned} \varepsilon_\gamma \frac{\partial \langle c_{A\gamma}\rangle^\gamma}{\partial t} &+ \nabla \cdot \left(\langle \mathbf{v}_\gamma \rangle \langle c_{A\gamma}\rangle^\gamma \right) \\ &= \underbrace{\nabla \cdot \left[\varepsilon_\gamma \mathscr{D}_{Am} \left(\nabla \langle c_{A\gamma}\rangle^\gamma + \frac{1}{V_\gamma} \int_{A_{\gamma\kappa}} \mathbf{n}_{\gamma\kappa} \tilde{c}_{A\gamma} \, , \mathrm{d}A \right) \right]}_{\text{diffusion}} \\ &\quad + \underbrace{\nabla \cdot \langle \tilde{\mathbf{v}}_\gamma \tilde{c}_{A\gamma} \rangle}_{\substack{\text{dispersive} \\ \text{transport}}} - \underbrace{a_{\mathrm{v}} \frac{\partial \langle c_{As}\rangle_{\gamma\kappa}}{\partial t}}_{\text{adsorption}} \end{aligned} \tag{6.161}$$

For many dynamic processes, the condition of local adsorption equilibrium (Whitaker, 1999, Problem 1-3) can be used as a reasonable approximation (Wood et al., 2004). For a linear adsorption isotherm, this allows us to use

$$c_{As} = K_{\mathrm{eq}} c_{A\gamma} \tag{6.162}$$

in Eq. (6.161) to obtain

$$\varepsilon_\gamma \frac{\partial \langle c_{A\gamma}\rangle^\gamma}{\partial t} + \nabla \cdot \left(\langle \mathbf{v}_\gamma \rangle \langle c_{A\gamma}\rangle^\gamma\right) \tag{6.163}$$

$$= \underbrace{\nabla \cdot \left[\varepsilon_\gamma \mathscr{D}_{Am} \left(\nabla \langle c_{A\gamma}\rangle^\gamma + \frac{1}{V_\gamma} \int_{A_{\gamma\kappa}} \mathbf{n}_{\gamma\kappa} \tilde{c}_{A\gamma} \, \mathrm{d}A\right)\right]}_{\textit{diffusion}}$$

$$+ \underbrace{\nabla \cdot \langle \tilde{\mathbf{v}}_\gamma \tilde{c}_{A\gamma}\rangle}_{\substack{\textit{dispersive} \\ \textit{transport}}} - \underbrace{a_\mathrm{v} K_{eq} \frac{\partial \langle c_{A\gamma}\rangle_{\gamma\kappa}}{\partial t}}_{\textit{adsorption}}$$

Use of the first of Eqs. (6.131) along with the approximation (Whitaker, 1999, Sec. 1.3.3) given by

$$\left\langle \langle c_{A\gamma}\rangle^\gamma \right\rangle_{\gamma\kappa} = \langle c_{A\gamma}\rangle^\gamma \tag{6.164}$$

allows us to express Eq. (6.163) as

$$\varepsilon_\gamma \left(1 + \frac{a_\mathrm{v} K_\mathrm{eq}}{\varepsilon_\gamma}\right) \frac{\partial \langle c_{A\gamma}\rangle^\gamma}{\partial t} + \nabla \cdot \left(\langle \mathbf{v}_\gamma \rangle \langle c_{A\gamma}\rangle^\gamma\right) \tag{6.165}$$

$$= \underbrace{\nabla \cdot \left[\varepsilon_\gamma \mathscr{D}_{Am} \left(\nabla \langle c_{A\gamma}\rangle^\gamma + \frac{1}{V_\gamma} \int_{A_{\gamma\kappa}} \mathbf{n}_{\gamma\kappa} \tilde{c}_{A\gamma} \mathrm{d}A\right)\right]}_{\textit{diffusion}}$$

$$+ \underbrace{\nabla \cdot \langle \tilde{\mathbf{v}}_\gamma \tilde{c}_{A\gamma}\rangle}_{\substack{\textit{dispersive} \\ \textit{transport}}} - a_\mathrm{v} K_\mathrm{eq} \frac{\partial \langle \tilde{c}_{A\gamma}\rangle_{\gamma\kappa}}{\partial t}$$

When the length scale constraints indicated by Eq. (6.117) are valid one can show that the spatial deviation concentration is constrained by

$$\tilde{c}_{A\gamma} \ll \langle c_{A\gamma}\rangle^\gamma \tag{6.166}$$

This, in turn, leads to the inequalities given by

$$\langle \tilde{c}_{A\gamma}\rangle_{\gamma\kappa} \lll \langle c_{A\gamma}\rangle^\gamma, \qquad \frac{\partial \langle \tilde{c}_{A\gamma}\rangle_{\gamma\kappa}}{\partial t} \lll \frac{\partial \langle c_{A\gamma}\rangle^\gamma}{\partial t} \tag{6.167}$$

The second of these allows us to discard the last term in Eq. (6.165) so that the local volume averaged mass transport equation for linear adsorption takes the form

$$\varepsilon_\gamma\left(1+\frac{a_v K_{eq}}{\varepsilon_\gamma}\right)\frac{\partial\langle c_{A\gamma}\rangle^\gamma}{\partial t}+\nabla\cdot\left(\langle\mathbf{v}_\gamma\rangle\langle c_{A\gamma}\rangle^\gamma\right) \tag{6.168}$$

$$=\underbrace{\nabla\cdot\left[\varepsilon_\gamma\mathscr{D}_{Am}\left(\nabla\langle c_{A\gamma}\rangle^\gamma+\frac{1}{V_\gamma}\int\limits_{A_{\gamma\kappa}}\mathbf{n}_{\gamma\kappa}\tilde{c}_{A\gamma}\,dA\right)\right]}_{\textit{diffusion}}+\underbrace{\nabla\cdot\langle\tilde{\mathbf{v}}_\gamma\tilde{c}_{A\gamma}\rangle}_{\substack{\textit{dispersive}\\ \textit{transport}}}$$

The closure problem associated with this transport process has been explored in detail elsewhere (Whitaker, 1997) where it is shown that the spatial deviation concentration can be expressed as

$$\tilde{c}_{A\gamma}=\mathbf{b}_\gamma\cdot\nabla\langle c_{A\gamma}\rangle^\gamma+s_\gamma K_{eq}\frac{\partial\langle c_{A\gamma}\rangle^\gamma}{\partial t} \tag{6.169}$$

Here the vector $\mathbf{b}_\gamma$ and the scalar s_γ are the *closure variables* or the *mapping variables*. The latter description is often used because $\mathbf{b}_\gamma$ maps $\nabla\langle c_{A\gamma}\rangle^\gamma$ onto $\tilde{c}_{A\gamma}$, and s_γ maps $K_{eq}\partial\langle c_{A\gamma}\rangle^\gamma\partial t$ onto $\tilde{c}_{A\gamma}$. Use of Eq. (6.169) in Eq. (6.168) leads to

$$\varepsilon_\gamma\left(1+\frac{a_v K_{eq}}{\varepsilon_\gamma}\right)\frac{\partial\langle c_{A\gamma}\rangle^\gamma}{\partial t}+\nabla\cdot\left(\langle\mathbf{v}_\gamma\rangle\langle c_{A\gamma}\rangle^\gamma\right) \tag{6.170}$$

$$+\varepsilon_\gamma\mathbf{d}_\gamma\cdot\nabla\left(\frac{K_{eq}\partial\langle c_{A\gamma}\rangle^\gamma}{\partial t}\right)=\nabla\cdot\left(\varepsilon_\gamma\mathbf{D}_\gamma^*\cdot\nabla\langle c_{A\gamma}\rangle^\gamma\right)$$

in which $\mathbf{d}_\gamma$ is a dimensionless vector and $\mathbf{D}_\gamma^*$ is the total dispersion tensor. These quantities are defined by

$$\varepsilon_\gamma\mathbf{d}_\gamma=\frac{1}{\mathscr{V}}\int\limits_{V_\gamma}s_\gamma\tilde{\mathbf{v}}_\gamma dV-\frac{1}{\mathscr{V}}\int\limits_{A_{\gamma\kappa}}\mathbf{n}_{\gamma\kappa}s_\gamma\mathscr{D}_{Am}dA \tag{6.171}$$

$$\mathbf{D}_\gamma^*=\mathscr{D}_{Am}\left(\mathbf{I}+\frac{1}{V_\gamma}\int\limits_{A_{\gamma\kappa}}\mathbf{n}_{\gamma\kappa}\mathbf{b}_\gamma\,dA\right)-\frac{1}{V_\gamma}\int\limits_{V_\gamma}\tilde{\mathbf{v}}_\gamma\mathbf{b}_\gamma\,dV \tag{6.172}$$

and can be calculated by means of the closure problems for $\mathbf{b}_\gamma$ and s_γ. If the flow can be approximated as incompressible, the continuity equation is given by

$$\nabla\cdot\mathbf{v}_\gamma=0 \tag{6.173}$$

and the volume averaged form is

$$\nabla\cdot\left(\varepsilon_\gamma\langle\mathbf{v}_\gamma\rangle^\gamma\right)=0 \tag{6.174}$$

For the special case of a homogeneous porous medium, Eq. (6.170) can be expressed as

$$\varepsilon_\gamma \left(1 + \frac{a_\mathrm{v} K_\mathrm{eq}}{\varepsilon_\gamma}\right) \frac{\partial \langle c_{A\gamma}\rangle^\gamma}{\partial t} + \varepsilon_\gamma \langle \mathbf{v}_\gamma \rangle^\gamma \cdot \nabla \langle c_{A\gamma}\rangle^\gamma \\ + \varepsilon_\gamma \mathbf{d}_\gamma \cdot \nabla \left(\frac{K_\mathrm{eq} \partial \langle c_{A\gamma}\rangle^\gamma}{\partial t}\right) = \varepsilon_\gamma \mathbf{D}_\gamma^* : \nabla\nabla \langle c_{A\gamma}\rangle^\gamma \tag{6.175}$$

provided one is willing to ignore the variations in the total dispersion tensor that result from variations in the velocity field. For a homogeneous porous medium and a constant velocity, Eq. (6.175) is exact. If the adsorption isotherm is *nonlinear*, K_eq is a function of $\langle c_{A\gamma}\rangle^\gamma$; however, for linear adsorption K_eq is a constant and Eq. (6.175) simplifies to

$$\varepsilon_\gamma \left(1 + \frac{a_\mathrm{v} K_\mathrm{eq}}{\varepsilon_\gamma}\right) \frac{\partial \langle c_{A\gamma}\rangle^\gamma}{\partial t} + \varepsilon_\gamma \langle \mathbf{v}_\gamma \rangle^\gamma \cdot \nabla \langle c_{A\gamma}\rangle^\gamma \\ + \varepsilon_\gamma \mathbf{d}_\gamma K_\mathrm{eq} \cdot \nabla \left(\frac{\partial \langle c_{A\gamma}\rangle^\gamma}{\partial t}\right) = \varepsilon_\gamma \mathbf{D}_\gamma^* : \nabla\nabla \langle c_{A\gamma}\rangle^\gamma \tag{6.176}$$

For *chromatographic processes*, this result can be simplified following the original analysis of Golay (1958). For pulsed systems, we identify the pulse velocity by $\mathbf{u}_p$ and express the time derivative of the concentration as

$$\frac{\partial \langle c_{A\gamma}\rangle^\gamma}{\partial t} = \left.\frac{\mathrm{d} \langle c_{A\gamma}\rangle^\gamma}{\mathrm{d}t}\right|_{\mathbf{u}_p} - \mathbf{u}_p \cdot \nabla \langle c_{A\gamma}\rangle^\gamma \tag{6.177}$$

Here the subscript $\mathbf{u}_p$ is used to indicate the time derivative as determined by an observer moving at the velocity $\mathbf{u}_p$. When the following restriction is valid

$$\frac{\partial \langle c_{A\gamma}\rangle^\gamma}{\partial t} \gg \left.\frac{\mathrm{d} \langle c_{A\gamma}\rangle^\gamma}{\mathrm{d}t}\right|_{\mathbf{u}_p} \tag{6.178}$$

the mixed derivative term in Eq. (6.176) can be expressed as

$$\varepsilon_\gamma \mathbf{d}_\gamma \, K_\mathrm{eq} \cdot \nabla \left(\frac{\partial \langle c_{A\gamma}\rangle^\gamma}{\partial t}\right) = -\varepsilon_\gamma K_\mathrm{eq} \mathbf{d}_\gamma \mathbf{u}_p : \nabla\nabla \langle c_{A\gamma}\rangle^\gamma \tag{6.179}$$

Use of this result in Eq. (6.175) leads to a *chromatographic equation* in the form

$$\frac{\partial \langle c_{A\gamma}\rangle^\gamma}{\partial t} + \frac{\langle \mathbf{v}_\gamma \rangle^\gamma}{\left(1 + \varepsilon_\gamma^{-1} a_\mathrm{v} K_\mathrm{eq}\right)} \cdot \nabla \langle c_{A\gamma}\rangle^\gamma \\ = \left(\frac{\mathbf{D}_\gamma^* + K_\mathrm{eq} \mathbf{d}_\gamma \mathbf{u}_p}{1 + \varepsilon_\gamma^{-1} a_\mathrm{v} K_\mathrm{eq}}\right) : \nabla\nabla \langle c_{A\gamma}\rangle^\gamma \tag{6.180}$$

This clearly indicates that the pulse velocity is given by

$$\mathbf{u}_p = \frac{\langle \mathbf{v}_\gamma \rangle^\gamma}{\left(1 + \varepsilon_\gamma^{-1} a_v K_{\text{eq}}\right)} \tag{6.181}$$

and that the dispersion tensor in this intrinsic averaged mass transport equation is a complex function of the equilibrium coefficient K_{eq}. It is convenient to express Eq. (6.180) as

$$\frac{\partial \langle c_{A\gamma} \rangle^\gamma}{\partial t} + \frac{\langle \mathbf{v}_\gamma \rangle^\gamma}{\left(1 + \varepsilon_\gamma^{-1} a_v K_{\text{eq}}\right)} \cdot \nabla \langle c_{A\gamma} \rangle^\gamma = \mathbf{D}_\gamma^{**} : \nabla\nabla \langle c_{A\gamma} \rangle^\gamma \tag{6.182}$$

in which the dispersion tensor takes the form

$$\mathbf{D}_\gamma^{**} = \frac{\mathbf{D}_\gamma^{*} \left(1 + \varepsilon_\gamma^{-1} a_v K_{\text{eq}}\right) + K_{\text{eq}}\, \mathbf{d}_\gamma \, \langle \mathbf{v}_\gamma \rangle^\gamma}{\left(1 + \varepsilon_\gamma^{-1} a_v K_{\text{eq}}\right)^2} \tag{6.183}$$

In order to determine $\mathbf{D}_\gamma^{**}$ one must solve the two closure problems given by Whitaker (1997), and methods of solution are described by various authors (Eidsath et al., 1983; Sahraoui and Kaviany, 1994; Quintard and Whitaker, 1994f, 1995).

In the previous sections we have given several examples of mass and momentum transport for dilute solutions, and there are many other processes involving heterogeneous reaction, non-linear adsorption, slightly compressible flow, etc., that can be analyzed in the same manner. The more general mass transfer problem based on Eq. (6.111) is much more complex, and in the following section we treat the simplest case of non-dilute solution transport in porous media.

6.5 CLOSURE FOR NON-DILUTE DIFFUSION

In this section we consider the simplest possible case of non-dilute mass transfer in porous media. This occurs when there is no homogeneous reaction, no heterogeneous reaction, no adsorption, negligible convective transport, and the conditions are such that the total molar concentration can be treated as a specified constant. Even with these severe simplifications, the analysis is complex; however, the result is relatively simple, i.e., there is a single tortuosity tensor that can be used to determine all the effective diffusivity tensors. For negligible convective transport and no homogeneous reaction, Eq. (6.111) simplifies to

$$\frac{\partial c_{A\gamma}}{\partial t} = \nabla \cdot \left(c_\gamma \sum_{E=1}^{E=N-1} D_{AE} \nabla x_{E\gamma} \right), \qquad A = 1, 2, \ldots, N \tag{6.184}$$

and we are confronted with a purely diffusive problem. Including heterogeneous reaction is straightforward and the details are given by Arce et al. (2005). When the total molar concentration is constrained by

$$x_{A\gamma} \nabla c_\gamma \ll c_\gamma \nabla x_{A\gamma}, \qquad A = 1, 2, \ldots, N \tag{6.185}$$

we can impose the simplification

$$c_\gamma = \langle c_\gamma \rangle^\gamma = \text{constant} \tag{6.186}$$

Imposition of this condition means that there are only $N - 1$ independent transport equations of the form given by

$$\frac{\partial c_{A\gamma}}{\partial t} = \nabla \cdot \sum_{E=1}^{E=N-1} D_{AE} \nabla c_{E\gamma}, \qquad A = 1, 2, \ldots, N-1 \tag{6.187}$$

and for the case of passive transport the boundary conditions at the fluid-solid boundary are given by

$$-\mathbf{n}_{\gamma\kappa} \cdot \sum_{E=1}^{E=N-1} D_{AE} \nabla c_{E\gamma} = 0, \quad \text{at the } \gamma\text{–}\kappa \text{ interface} \tag{6.188}$$

$$A = 1, 2, \ldots, N-1,$$

One can make use of the averaging theorem and the no-flux boundary condition to show that the volume averaged form of Eq. (6.187) is given by

$$\varepsilon_\gamma \frac{\partial \langle c_{A\gamma} \rangle^\gamma}{\partial t} = \nabla \cdot \sum_{E=1}^{E=N-1} \langle D_{AE} \nabla c_{E\gamma} \rangle, \qquad A = 1, 2, \ldots, N-1 \tag{6.189}$$

At this point we decompose the elements of the diffusion matrix according to

$$D_{AE} = \langle D_{AE} \rangle^\gamma + \tilde{D}_{AE} \tag{6.190}$$

and neglect $\tilde{D}_{AE}$ relative to $\langle D_{AE} \rangle^\gamma$ so that Eq. (6.189) simplifies to

$$\varepsilon_\gamma \frac{\partial \langle c_{A\gamma} \rangle^\gamma}{\partial t} = \nabla \cdot \sum_{E=1}^{E=N-1} \langle D_{AE} \rangle^\gamma \langle \nabla c_{E\gamma} \rangle, \quad A = 1, 2, \ldots, N-1 \tag{6.191}$$

We can represent this simplification as,

$$\tilde{D}_{AE} \ll \langle D_{AE} \rangle^\gamma \tag{6.192}$$

and when it is *not satisfactory* it may be possible to develop a *correction* based on the retention of the spatial deviation, $\tilde{D}_{AE}$. The inequality represented by Eq. (6.192) is equivalent to ignoring variations of D_{AE} within the averaging volume.

The volume averaging theorem can be used with the average of the gradient in Eq. (6.191) in order to obtain

$$\langle \nabla c_{E\gamma} \rangle = \nabla \langle c_{E\gamma} \rangle + \frac{1}{\mathscr{V}} \int_{A_{\gamma\kappa}} \mathbf{n}_{\gamma\kappa} c_{E\gamma} \, \mathrm{d}A \tag{6.193}$$

and one can follow an established analysis (Whitaker, 1999, Chapter 1) in order to express this result as

$$\langle \nabla c_{E\gamma} \rangle = \varepsilon_\gamma \nabla \langle c_{E\gamma} \rangle^\gamma + \frac{1}{\mathscr{V}} \int_{A_{\gamma\kappa}} \mathbf{n}_{\gamma\kappa} \tilde{c}_{E\gamma} \mathrm{d}A \tag{6.194}$$

Use of this result in Eq. (6.191) provides

$$\varepsilon_\gamma \frac{\partial \langle c_{A\gamma} \rangle^\gamma}{\partial t} = \nabla \cdot \left[\sum_{E=1}^{E=N-1} \langle D_{AE} \rangle^\gamma \left(\varepsilon_\gamma \nabla \langle c_{E\gamma} \rangle^\gamma + \underbrace{\frac{1}{\mathscr{V}} \int_{A_{\gamma\kappa}} \mathbf{n}_{\gamma\kappa} \tilde{c}_{E\gamma} \mathrm{d}A}_{\textit{filter}} \right) \right] \tag{6.195}$$

in which the area integral of $\mathbf{n}_{\gamma\kappa}\tilde{c}_{E\gamma}$ has been identified as a *filter*. Not all the information available at the length scale associated with $\tilde{c}_{E\gamma}$ will pass through this filter to influence the transport equation for $\langle c_{A\gamma} \rangle^\gamma$, and the existence of filters of this type is a recurring theme in the method of volume averaging (Whitaker, 1999).

In order to obtain a closed form of Eq. (6.195), we need a representation for the spatial deviation concentration, $\tilde{c}_{A\gamma}$, and this requires the development of the *closure problem*. To develop this closure problem, we return to Eq. (6.187) and make use of Eqs. (6.190) and (6.192) to obtain

$$\frac{\partial c_{A\gamma}}{\partial t} = \nabla \cdot \sum_{E=1}^{E=N-1} \langle D_{AE} \rangle^\gamma \nabla c_{E\gamma}, \qquad A = 1, 2, \ldots, N-1 \tag{6.196}$$

If we ignore variations in ε_γ and subtract Eq. (6.195) from Eq. (6.196), we can arrange the result as

$$\frac{\partial \tilde{c}_{A\gamma}}{\partial t} = \nabla \cdot \left[\sum_{E=1}^{E=N-1} \langle D_{AE} \rangle^\gamma \nabla \tilde{c}_{E\gamma} \right] - \nabla \cdot \left[\sum_{E=1}^{E=N-1} \frac{\langle D_{AE} \rangle^\gamma}{\varepsilon_\gamma} \frac{1}{\mathscr{V}} \int_{A_{\gamma\kappa}} \mathbf{n}_{\gamma\kappa} \tilde{c}_{E\gamma} \mathrm{d}A \right] \tag{6.197}$$

in which it is understood that this result applies to all $N-1$ species. In order to develop a boundary condition for the spatial deviation concentration, we again make use of the concentration decomposition given by the first of Eqs. (6.131) along with Eqs. (6.190) and (6.192) to obtain

$$\begin{aligned} &- \sum_{E=1}^{E=N-1} \mathbf{n}_{\gamma\kappa} \cdot \langle D_{AE} \rangle^\gamma \, \nabla \tilde{c}_{E\gamma} \\ &\quad = \sum_{E=1}^{E=N-1} \mathbf{n}_{\gamma\kappa} \cdot \langle D_{AE} \rangle^\gamma \, \nabla \langle c_{E\gamma} \rangle^\gamma, \quad \text{at the } \gamma\text{–}\kappa \text{ interface} \end{aligned} \tag{6.198}$$

With this result we can construct a boundary value problem for $\tilde{c}_{A\gamma}$ in which the governing differential equation is given by

$$\underbrace{\frac{\partial \tilde{c}_{A\gamma}}{\partial t}}_{accumulation} = \underbrace{\nabla \cdot \left[\sum_{E=1}^{E=N-1} \langle D_{AE} \rangle^{\gamma} \nabla \tilde{c}_{E\gamma} \right]}_{diffusion} - \underbrace{\nabla \cdot \left[\sum_{E=1}^{E=N-1} \frac{\langle D_{AE} \rangle^{\gamma}}{\varepsilon_{\gamma}} \frac{1}{\mathscr{V}} \int_{A_{\gamma\kappa}} \mathbf{n}_{\gamma\kappa} \tilde{c}_{E\gamma} \mathrm{d}A \right]}_{non\text{-}local\ diffusion} \tag{6.199}$$

while the interfacial boundary condition takes the form

$$\text{BC.1} \qquad \underbrace{-\sum_{E=1}^{E=N-1} \mathbf{n}_{\gamma\kappa} \cdot \langle D_{AE} \rangle^{\gamma} \nabla \tilde{c}_{E\gamma}}_{diffusive\ flux} = \underbrace{\sum_{E=1}^{E=N-1} \mathbf{n}_{\gamma\kappa} \cdot \langle D_{AE} \rangle^{\gamma} \nabla \langle c_{E\gamma} \rangle^{\gamma}}_{diffusive\ source}, \quad \text{at the } \gamma\text{–}\kappa \text{ interface} \tag{6.200}$$

In addition to the interfacial boundary condition, we require a boundary condition at the entrances and exits of the macroscopic system illustrated in Figure 6.3 and we express this boundary condition as

$$\text{BC.2} \qquad \tilde{c}_{A\gamma} = \mathscr{F}(\mathbf{r}, t), \qquad \text{at } \mathscr{A}_{\gamma e} \tag{6.201}$$

To complete our statement of the boundary value problem, we propose an initial condition of the form

$$\text{IC.} \qquad \tilde{c}_{A\gamma} = \mathscr{F}(\mathbf{r}), \quad \text{at } t = 0 \tag{6.202}$$

In general, both the boundary condition at the entrances and exits of the macroscopic system and the initial condition are unknown in terms of the spatial deviation concentration, $\tilde{c}_{A\gamma}$. However, neither of these is important when the separation of length scales indicated by Eq. (6.117) is valid. Under these circumstances, the boundary condition imposed at $\mathscr{A}_{\gamma e}$ influences the $\tilde{c}_{A\gamma}$ field only over a negligibly small region, and the initial condition given by Eq. (6.202) can be discarded because the closure problem is quasi-steady. Under these circumstances, the closure problem can be solved in some representative, local region (Quintard and Whitaker, 1994a–e).

In the governing differential equation for $\tilde{c}_{A\gamma}$, we have identified the accumulation term, the diffusion term, and the so-called non-local diffusion term. In the boundary condition imposed at the γ–κ interface, we have identified the diffusive flux, and a non-homogeneous term referred to as the *diffusion source*. If the diffusion source in Eq. (6.200) were zero, the $\tilde{c}_{A\gamma}$ field would be generated only by the non-homogeneous terms that might appear in the boundary condition imposed at $\mathscr{A}_{\gamma e}$ or in the initial condition given by Eq. (6.202). One can easily develop arguments indicating that the closure problem for $\tilde{c}_{A\gamma}$ is quasi-steady, thus the initial condition is of no importance (Whitaker, 1999, Chapter 1). In addition, one can develop arguments indicating that the boundary condition imposed at $\mathscr{A}_{\gamma e}$ will influence the $\tilde{c}_{A\gamma}$ field over a negligibly small portion of the field of interest. Because of this, any useful solution to the closure problem must be developed for some *representative region* which is most often conveniently described in terms of a unit cell in a spatially periodic system. These ideas lead to a closure problem of the form

$$0 = \nabla \cdot \left[\underbrace{\sum_{E=1}^{E=N-1} \langle D_{AE}\rangle^\gamma \nabla \tilde{c}_{E\gamma}}_{\textit{diffusion}} \right] - \nabla \cdot \left[\underbrace{\sum_{E=1}^{E=N-1} \frac{\langle D_{AE}\rangle^\gamma}{\varepsilon_\gamma \mathscr{V}} \int_{A_{\gamma\kappa}} \mathbf{n}_{\gamma\kappa} \tilde{c}_{E\gamma} \, dA}_{\textit{non-local diffusion}} \right] \tag{6.203}$$

$$\text{BC.1} \qquad - \underbrace{\sum_{E=1}^{E=N-1} \mathbf{n}_{\gamma\kappa} \cdot \langle D_{AE}\rangle^\gamma \, \nabla \tilde{c}_{E\gamma}}_{\textit{diffusive flux}} = \underbrace{\sum_{E=1}^{E=N-1} \mathbf{n}_{\gamma\kappa} \cdot \langle D_{AE}\rangle^\gamma \, \nabla \langle c_{E\gamma}\rangle^\gamma}_{\textit{diffusive source}}, \quad \text{at the } \gamma\text{–}\kappa \text{ interface} \tag{6.204}$$

$$\text{BC.2} \qquad \tilde{c}_{A\gamma}(\mathbf{r} + \ell_i) = \tilde{c}_{A\gamma}(\mathbf{r}), \quad i = 1, 2, 3 \tag{6.205}$$

Here we have used ℓ_i to represent the three base vectors needed to characterize a spatially periodic system. The use of a spatially periodic system does not limit this analysis to simple systems since a periodic system can be arbitrary complex (Quintard and Whitaker, 1994a–e). However, the periodicity condition imposed by Eq. (6.205) can only be strictly justified when $\langle D_{AE}\rangle^\gamma$, $\langle c_{A\gamma}\rangle^\gamma$, and $\nabla \langle c_{A\gamma}\rangle^\gamma$ are constants and this does not occur for the types of systems under consideration. This matter has been examined elsewhere (Whitaker, 1986a) and the analysis suggests that the traditional separation of length scales allows one to treat $\langle D_{AE}\rangle^\gamma$, $\langle c_{A\gamma}\rangle^\gamma$, and $\nabla \langle c_{A\gamma}\rangle^\gamma$ as constants within the framework of the closure problem.

One can show (Whitaker, 1999) that the non-local diffusion term is negligible when the length scales are separated as indicated in Eq. (6.117), and under these circumstances the boundary value problem for the spatial deviation concentration takes the form

$$0 = \nabla \cdot \left[\sum_{E=1}^{E=N-1} \langle D_{AE} \rangle^{\gamma} \, \nabla \tilde{c}_{E\gamma} \right] \tag{6.206}$$

$$\text{BC.1} \qquad - \sum_{E=1}^{E=N-1} \mathbf{n}_{\gamma\kappa} \cdot \langle D_{AE} \rangle^{\gamma} \, \nabla \tilde{c}_{E\gamma} = \sum_{E=1}^{E=N-1} \mathbf{n}_{\gamma\kappa} \cdot \langle D_{AE} \rangle^{\gamma} \nabla \langle c_{E\gamma} \rangle^{\gamma} , \quad \text{at } A_{\gamma\kappa} \tag{6.207}$$

$$\text{BC.2} \qquad \tilde{c}_{A\gamma}(\mathbf{r} + \ell_i) = \tilde{c}_{A\gamma}(\mathbf{r}), \quad i = 1, 2, 3 \tag{6.208}$$

Here one must remember that the subscript A represents species $A, B, C, \ldots, N-1$.

In this boundary value problem, there is only a single non-homogeneous term represented by $\nabla \langle c_{E_\gamma} \rangle^{\gamma}$ in the boundary condition imposed at the γ–κ interface. If this source term were zero, the solution to this boundary value problem would be given by $\tilde{c}_{A\gamma} = constant$. Any constant associated with $\tilde{c}_{A\gamma}$ will not pass through the filter in Eq. (6.195), and this suggests that a solution can be expressed in terms of the gradients of the volume averaged concentration. Since the system is linear in the $N-1$ independent gradients of the average concentration, we are led to a solution of the form

$$\tilde{c}_{E\gamma} = \mathbf{b}_{EA} \cdot \nabla \langle c_{A\gamma} \rangle^{\gamma} + \mathbf{b}_{EB} \cdot \nabla \langle c_{B\gamma} \rangle^{\gamma} + \mathbf{b}_{EC} \cdot \nabla \langle c_{C\gamma} \rangle^{\gamma} + \ldots.. + \mathbf{b}_{E,N-1} \cdot \nabla \langle c_{N-1\gamma} \rangle^{\gamma} \tag{6.209}$$

Here the vectors, $\mathbf{b}_{EA}, \mathbf{b}_{EB}$, etc., are referred to as the *closure variables* or the *mapping variables* since they map the gradients of the volume averaged concentrations onto the spatial deviation concentrations. In this representation for $\tilde{c}_{A\gamma}$, we can ignore the spatial variations of $\nabla \langle c_{A\gamma} \rangle^{\gamma}, \nabla \langle c_{B\gamma} \rangle^{\gamma}$ etc., within the framework of a local closure problem, and we can use Eq. (6.209) in Eq. (6.206) to obtain

$$0 = \nabla \cdot \left[\sum_{E=1}^{E=N-1} \langle D_{AE} \rangle^{\gamma} \sum_{D=1}^{D=N-1} \nabla \mathbf{b}_{ED} \cdot \nabla \langle c_{D\gamma} \rangle^{\gamma} \right] \tag{6.210}$$

$$\text{BC.1} \quad - \sum_{E=1}^{E=N-1} \mathbf{n}_{\gamma\kappa} \cdot \langle D_{AE} \rangle^{\gamma} \sum_{D=1}^{D=N-1} \nabla \mathbf{b}_{ED} \cdot \nabla \langle c_{D\gamma} \rangle^{\gamma} = \sum_{E=1}^{E=N-1} \mathbf{n}_{\gamma\kappa} \cdot \langle D_{AE} \rangle^{\gamma} \, \nabla \langle c_{E\gamma} \rangle^{\gamma} , \quad \text{at } A_{\gamma\kappa} \tag{6.211}$$

BC.2 $$\mathbf{b}_{AE}(\mathbf{r}+\ell_i)=\mathbf{b}_{AE}(\mathbf{r}) , \quad i=1,2,3 , \quad A=1,\ 2,\ \dots,N-1 \tag{6.212}$$

The derivation of Eqs. (6.210) and (6.211) requires the use of simplifications of the form

$$\nabla\left(\mathbf{b}_{EA}\cdot\nabla\langle c_{A\gamma}\rangle^{\gamma}\right)=\nabla\mathbf{b}_{EA}\cdot\nabla\langle c_{A\gamma}\rangle^{\gamma} \tag{6.213}$$

which result from the inequality

$$\mathbf{b}_{EA}\cdot\nabla\nabla\langle c_{A\gamma}\rangle^{\gamma} \quad \ll \quad \nabla\mathbf{b}_{EA}\cdot\nabla\langle c_{A\gamma}\rangle^{\gamma} \tag{6.214}$$

The basis for this inequality is the separation of length scales indicated by Eq. (6.117), and a detailed discussion is available elsewhere (Whitaker, 1999). One should keep in mind that the boundary value problem given by Eqs. (6.210) through (6.212) applies to all $N-1$ species and that the $N-1$ concentration gradients are independent. This latter condition allows us to obtain

$$0=\nabla\cdot\left[\sum_{E=1}^{E=N-1}\langle D_{AE}\rangle^{\gamma}\ \nabla\mathbf{b}_{ED}\right], \quad D=1,\ 2,\ \dots,N-1 \tag{6.215}$$

BC.1 $$-\sum_{E=1}^{E=N-1}\mathbf{n}_{\gamma\kappa}\cdot\langle D_{AE}\rangle^{\gamma}\ \nabla\mathbf{b}_{ED}=\mathbf{n}_{\gamma\kappa}\langle D_{AD}\rangle^{\gamma} , \quad \text{at } A_{\gamma\kappa} \tag{6.216}$$

$$D=1,\ 2,\ \dots,N-1 ,$$

Periodicity: $$\mathbf{b}_{AD}(\mathbf{r}+\ell_i)=\mathbf{b}_{AD}(\mathbf{r}), \quad i=1,2,3, \quad D=1,\ 2,\ \dots,N-1 \tag{6.217}$$

At this point it is convenient to expand the closure problem for species A in order to obtain

First Problem for Species A

$$0=\nabla\cdot\Big\{\langle D_{AA}\rangle^{\gamma}\Big[\nabla\mathbf{b}_{AA}+\left(\langle D_{AA}\rangle^{\gamma}\right)^{-1}\langle D_{AB}\rangle^{\gamma}\ \nabla\mathbf{b}_{BA} \tag{6.218a}$$
$$+\left(\langle D_{AA}\rangle^{\gamma}\right)^{-1}\langle D_{AC}\rangle^{\gamma}\nabla\mathbf{b}_{CA}+\ldots+\left(\langle D_{AA}\rangle^{\gamma}\right)^{-1}\langle D_{A,N-1}\rangle^{\gamma}\nabla\mathbf{b}_{N-1,A}\Big]\Big\}$$

BC. $$-\mathbf{n}_{\gamma\kappa}\cdot\nabla\mathbf{b}_{AA}-\mathbf{n}_{\gamma\kappa}\cdot\left(\langle D_{AA}\rangle^{\gamma}\right)^{-1}\langle D_{AB}\rangle^{\gamma}\ \nabla\mathbf{b}_{BA} \tag{6.218b}$$
$$-\mathbf{n}_{\gamma\kappa}\cdot\left(\langle D_{AA}\rangle^{\gamma}\right)^{-1}\langle D_{AC}\rangle^{\gamma}\ \nabla\mathbf{b}_{CA}-\dots.$$
$$-\mathbf{n}_{\gamma\kappa}\cdot\left(\langle D_{AA}\rangle^{\gamma}\right)^{-1}\langle D_{A,N-1}\rangle^{\gamma}\ \nabla\mathbf{b}_{N-1,A}=\mathbf{n}_{\gamma\kappa} , \quad \text{at } A_{\gamma\kappa}$$

Periodicity: $$\mathbf{b}_{DA}(\mathbf{r}+\ell_i)=\mathbf{b}_{DA}(\mathbf{r}) , \quad i-1,2,3 , \quad D=1,2,\dots,N-1 \tag{6.218c}$$

Second Problem for Species A

$$0 = \nabla \cdot \Big\{ \langle D_{AB} \rangle^{\gamma} \Big[\big(\langle D_{AB} \rangle^{\gamma} \big)^{-1} \langle D_{AA} \rangle^{\gamma} \nabla \mathbf{b}_{AB} + \nabla \mathbf{b}_{BB} \tag{6.219a}$$

$$+ \big(\langle D_{AB} \rangle^{\gamma} \big)^{-1} \langle D_{AC} \rangle^{\gamma} \nabla \mathbf{b}_{CB} + \ldots . + \big(\langle D_{AB} \rangle^{\gamma} \big)^{-1} \langle D_{A,N-1} \rangle^{\gamma} \nabla \mathbf{b}_{N-1,B} \Big] \Big\}$$

$$\text{BC.} \qquad - \mathbf{n}_{\gamma\kappa} \cdot \nabla \mathbf{b}_{AA} - \mathbf{n}_{\gamma\kappa} \cdot \big(\langle D_{AA} \rangle^{\gamma} \big)^{-1} \langle D_{AB} \rangle^{\gamma} \nabla \mathbf{b}_{BA} \tag{6.219b}$$

$$- \mathbf{n}_{\gamma\kappa} \cdot \big(\langle D_{AA} \rangle^{\gamma} \big)^{-1} \langle D_{AC} \rangle^{\gamma} \nabla \mathbf{b}_{CA} - \ldots .$$

$$- \mathbf{n}_{\gamma\kappa} \cdot \big(\langle D_{AA} \rangle^{\gamma} \big)^{-1} \langle D_{A,N-1} \rangle^{\gamma} \nabla \mathbf{b}_{N-1,A} = \mathbf{n}_{\gamma\kappa} \,, \quad \text{at } A_{\gamma\kappa}$$

$$\text{Periodicity:} \quad \mathbf{b}_{DB}(\mathbf{r} + \ell_i) = \mathbf{b}_{DB}(\mathbf{r}) \,, \quad i = 1,2,3 \,, \quad D = 1,2,\ldots,N-1 \tag{6.219c}$$

Third Problem for Species A

etc. (6.220)

$N - 1$ *Problem for Species A*

etc. (6.221)

Here it is convenient to define a new set of *closure variables* or *mapping variables* according to

$$\mathbf{d}_{AA} = \mathbf{b}_{AA} + \big(\langle D_{AA} \rangle^{\gamma} \big)^{-1} \langle D_{AB} \rangle^{\gamma} \mathbf{b}_{BA} + \big(\langle D_{AA} \rangle^{\gamma} \big)^{-1} \langle D_{AC} \rangle^{\gamma} \mathbf{b}_{CA} \tag{6.222a}$$

$$+ \ldots . + \big(\langle D_{AA} \rangle^{\gamma} \big)^{-1} \langle D_{A,N-1} \rangle^{\gamma} \mathbf{b}_{N-1,A}$$

$$\mathbf{d}_{AB} = \big(\langle D_{AB} \rangle^{\gamma} \big)^{-1} \langle D_{AA} \rangle^{\gamma} \mathbf{b}_{AB} + \mathbf{b}_{BB} + \big(\langle D_{AB} \rangle^{\gamma} \big)^{-1} \langle D_{AC} \rangle^{\gamma} \mathbf{b}_{CB} \tag{6.222b}$$

$$+ \ldots . + \big(\langle D_{AB} \rangle^{\gamma} \big)^{-1} \langle D_{A,N-1} \rangle^{\gamma} \mathbf{b}_{N-1,B}$$

$$\mathbf{d}_{AC} = \big(\langle D_{AC} \rangle^{\gamma} \big)^{-1} \langle D_{AA} \rangle^{\gamma} \mathbf{b}_{AC} + \big(\langle D_{AC} \rangle^{\gamma} \big)^{-1} \langle D_{AB} \rangle^{\gamma} \mathbf{b}_{BC} \tag{6.222c}$$

$$+ \mathbf{b}_{CC} + \ldots . + \big(\langle D_{AC} \rangle^{\gamma} \big)^{-1} \langle D_{A,N-1} \rangle^{\gamma} \mathbf{b}_{N-1,C}$$

etc. (6.222n − 1)

With these definitions, the closure problems take the following simplified forms:

First Problem for Species A

$$0 = \nabla^2 \mathbf{d}_{AA} \tag{6.223a}$$

$$\text{BC.} \qquad -\mathbf{n}_{\gamma\kappa}\cdot\nabla\mathbf{d}_{AA} = \mathbf{n}_{\gamma\kappa}\,, \quad \text{at } A_{\gamma\kappa} \tag{6.223b}$$

$$\text{Periodicity:} \qquad \mathbf{d}_{AA}(\mathbf{r}+\ell_i) = \mathbf{d}_{AA}(\mathbf{r})\,, \quad i = 1,\ 2,\ 3 \tag{6.223c}$$

Second Problem for Species A

$$0 = \nabla^2\mathbf{d}_{AB} \tag{6.224a}$$

$$\text{BC.} \qquad -\ \mathbf{n}_{\gamma\kappa}\cdot\nabla\mathbf{d}_{AB} = \mathbf{n}_{\gamma\kappa}\,, \quad \text{at } A_{\gamma\kappa} \tag{6.224b}$$

$$\text{Periodicity:} \qquad \mathbf{d}_{AB}(\mathbf{r}+\ell_i) = \mathbf{d}_{AB}(\mathbf{r})\,, \qquad i = 1,2,3 \tag{6.224c}$$

Third Problem for Species A

$$\text{etc.} \tag{6.225}$$

$N-1$ *Problem for Species A*

$$\text{etc.} \tag{6.226}$$

To obtain these simplified forms, one must make repeated use of inequalities of the form given by Eq. (6.214). Each one of these closure problems is identical to that obtained by Ryan et al. (1981) and solutions have been developed by several workers (Ryan et al., 1981; Ochoa-Tapia et al., 1994; Chang, 1982, 1983; Quintard, 1993; Quintard and Whitaker, 1993a,b). In each case, the closure problem determines the *closure variable* to within an arbitrary constant, and this constant can be specified by imposing the condition

$$\langle\tilde{c}_{D\gamma}\rangle^\gamma = 0\,, \quad \text{or} \quad \langle\mathbf{d}_{GD}\rangle^\gamma = 0\,, \qquad \begin{cases} G = 1,2,\ldots,N-1 \\ D = 1,2,\ldots,N-1 \end{cases} \tag{6.227}$$

However, any constant associated with a closure variable will not pass through the filter in Eq. (6.195), thus this constraint on the average is not necessary.

6.5.1 Closed Form for Non-Dilute Diffusion

The closed form of Eq. (6.195) can be obtained by use of the representation for $\tilde{c}_{E\gamma}$ given by Eq. (6.209), along with the definitions represented by Eqs. (6.222). After some algebraic manipulation, one obtains

$$\varepsilon_\gamma \frac{\partial \langle c_{A\gamma}\rangle^\gamma}{\partial t} = \nabla \cdot \left[\varepsilon_\gamma \langle D_{AA}\rangle^\gamma \left(\mathbf{I} + \frac{1}{V_\gamma} \int_{A_{\gamma\kappa}} \mathbf{n}_{\gamma\kappa} \mathbf{d}_{AA} \, \mathrm{d}A \right) \cdot \nabla \langle c_{A\gamma}\rangle^\gamma \right. \tag{6.228}$$

$$+ \varepsilon_\gamma \langle D_{AB}\rangle^\gamma \left(\mathbf{I} + \frac{1}{V_\gamma} \int_{A_{\gamma\kappa}} \mathbf{n}_{\gamma\kappa} \mathbf{d}_{AB} \, \mathrm{d}A \right) \cdot \nabla \langle c_{B\gamma}\rangle^\gamma$$

$$+ \varepsilon_\gamma \langle D_{AC}\rangle^\gamma \left(\mathbf{I} + \frac{1}{V_\gamma} \int_{A_{\gamma\kappa}} \mathbf{n}_{\gamma\kappa} \mathbf{d}_{AC} \, \mathrm{d}A \right) \cdot \nabla \langle c_{C\gamma}\rangle^\gamma$$

$$\left. + \varepsilon_\gamma \langle D_{A,N-1}\rangle^\gamma \left(\mathbf{I} + \frac{1}{V_\gamma} \int_{A_{\gamma\kappa}} \mathbf{n}_{\gamma\kappa} \mathbf{d}_{A,N-1} \, \mathrm{d}A \right) \cdot \nabla \langle c_{N-1\gamma}\rangle^\gamma \right]$$

On the basis of the closure problems given by Eqs. (6.223a) through (6.226), we conclude that there is a single tensor that describes the tortuosity for species A. This means that Eq. (6.228) can be expressed as

$$\varepsilon_\gamma \frac{\partial \langle c_{A\gamma}\rangle^\gamma}{\partial t} = \nabla \cdot \left[\varepsilon_\gamma \mathbf{D}_{AA}^{\mathrm{eff}} \cdot \nabla \langle c_{A\gamma}\rangle^\gamma + \varepsilon_\gamma \mathbf{D}_{AB}^{\mathrm{eff}} \cdot \nabla \langle c_{B\gamma}\rangle^\gamma \right. \tag{6.229}$$

$$\left. + \ \varepsilon_\gamma \mathbf{D}_{AC}^{\mathrm{eff}} \cdot \nabla \langle c_{C\gamma}\rangle^\gamma + \ \ldots\ldots + \ \varepsilon_\gamma \mathbf{D}_{A,N-1}^{\mathrm{eff}} \cdot \nabla \langle c_{N-1\gamma}\rangle^\gamma \right]$$

in which the effective diffusivity tensors are related according to

$$\frac{\mathbf{D}_{AA}^{\mathrm{eff}}}{\langle D_{AA}\rangle^\gamma} = \frac{\mathbf{D}_{AB}^{\mathrm{eff}}}{\langle D_{AB}\rangle^\gamma} = \frac{\mathbf{D}_{AC}^{\mathrm{eff}}}{\langle D_{AC}\rangle^\gamma} = \ldots.. = \frac{\mathbf{D}_{A,N-1}^{\mathrm{eff}}}{\langle D_{A,N-1}\rangle^\gamma} \tag{6.230}$$

The remaining diffusion equations for species $B, C, \ldots, N-1$ have precisely the same form as Eq. (6.229), and the various effective diffusivity tensors are related to each other in the manner indicated by Eqs. (6.230). The generic closure problem can be expressed as

Generic Closure Problem

$$0 = \nabla^2 \mathbf{d} \tag{6.231a}$$

$$\text{BC.} \qquad -\mathbf{n}_{\gamma\kappa} \cdot \nabla \mathbf{d} = \mathbf{n}_{\gamma\kappa} \,, \qquad \text{at } A_{\gamma\kappa} \tag{6.231b}$$

$$\text{Periodicity:} \qquad \mathbf{d}(\mathbf{r} + \ell_i) = \mathbf{d}(\mathbf{r}) \,, \qquad i = 1,\ 2,\ 3 \tag{6.231c}$$

and solution of this boundary value problem is relatively straightforward. The existence of a single, generic closure problem that allows for the determination of all the effective diffusivity tensors represents the principle result of this section. On the basis of this single closure problem, the tortuosity tensor is defined according to

$$\boldsymbol{\tau} = \mathbf{I} + \frac{1}{V_\gamma} \int_{A_{\gamma\kappa}} \mathbf{n}_{\gamma\kappa} \mathbf{d} \, \mathrm{d}A \tag{6.232}$$

and we can express Eqs. (6.230) in the form

$$\mathbf{D}_{AA}^{\text{eff}} = \boldsymbol{\tau} \langle D_{AA} \rangle^\gamma \,, \quad \mathbf{D}_{AB}^{\text{eff}} = \boldsymbol{\tau} \langle D_{AB} \rangle^\gamma \,, \ldots., \quad \mathbf{D}_{A,N-1}^{\text{eff}} = \boldsymbol{\tau} \langle D_{A,N-1} \rangle^\gamma \tag{6.233}$$

Substitution of these results into Eq. (6.229) allows us to represent the local volume averaged diffusion equations as

$$\varepsilon_\gamma \frac{\partial \langle c_{A\gamma} \rangle^\gamma}{\partial t} = \nabla \cdot \left[\sum_{E=1}^{E=N-1} \varepsilon_\gamma \boldsymbol{\tau} \langle D_{AE} \rangle^\gamma \cdot \nabla \langle c_{E\gamma} \rangle^\gamma \right], \quad A = 1, 2, \ldots, N-1 \tag{6.234}$$

It is important to remember that this analysis has been simplified on the basis of Eq. (6.186) and the problem becomes more complex when the total molar concentration cannot be treated as a constant. For a porous medium that is isotropic in the volume averaged sense, the tortuosity tensor takes the classical form

$$\boldsymbol{\tau} = \mathbf{I} \, \tau^{-1} \tag{6.235}$$

in which $\mathbf{I}$ is the unit tensor and τ is *the* tortuosity. For isotropic porous media, we can express Eq. (6.234) as

$$\varepsilon_\gamma \frac{\partial \langle c_{A\gamma} \rangle^\gamma}{\partial t} = \nabla \cdot \left[\sum_{E=1}^{E=N-1} \left(\frac{\varepsilon_\gamma}{\tau} \right) \langle D_{AE} \rangle^\gamma \, \nabla \langle c_{E\gamma} \rangle^\gamma \right], \qquad A = 1, 2, \ldots, N-1 \tag{6.236}$$

Often ε_γ and τ can be treated as constants; however, the diffusion coefficients in this transport equation will be functions of the local volume averaged mole fractions, and this means that we are faced with a coupled, non-linear diffusion problem. Future studies of non-dilute solutions need to include convective effects along with adsorption and heterogeneous reaction. In addition, the restriction imposed by Eq. (6.186) needs to be removed.

NOMENCLATURE

$\mathscr{A}_a(t)$	surface area of an arbitrary, moving volume, m^2
$\mathscr{A}_{\gamma e}$	area of entrances and exits of the γ-phase contained in the macroscopic region, m^2
$A_{\gamma\kappa}$	area of the γ–κ interface contained within the averaging volume, m^2
a_v	$A_{\gamma\kappa}/\mathscr{V}$ area per unit volume, m^{-1}
$\mathbf{b}_A$	species A body force per unit mass, N/kg
$\mathbf{b}$	total body force per unit mass, N/kg
$\mathbf{b}_\gamma$	vector that maps $\nabla\langle c_{A\gamma}\rangle^\gamma$ onto $\tilde{c}_{A\gamma}$ for a dilute solution, m
$\mathbf{b}_{EA}$	closure variable that maps $\nabla\langle c_{A\gamma}\rangle^\gamma$ onto $\tilde{c}_{A\gamma}$ for a non-dilute solution, m
c_A	bulk concentration of species A, moles/m^3
$c_{A\gamma}$	bulk concentration of species A in the γ-phase, moles/m^3
$\langle c_{A\gamma}\rangle$	superficial average bulk concentration of species A in the γ-phase, moles/m^3
$\langle c_{A\gamma}\rangle^\gamma$	intrinsic average bulk concentration of species A in the γ-phase, moles/m^3
$\langle c_{A\gamma}\rangle_{\gamma\kappa}$	intrinsic area average bulk concentration of species A at the γ–κ interface, moles/m^3
$\tilde{c}_{A\gamma}$	$c_{A\gamma} - \langle c_{A\gamma}\rangle^\gamma$, spatial deviation concentration of species A, moles/m^3
c_γ	$\sum_{A=1}^{A=N} c_{A\gamma}$, total molar concentration, moles/m^3
c_{As}	surface concentration of species A associated with the γ–κ interface, moles/m^2
$\langle c_{As}\rangle_{\gamma\kappa}$	area averaged surface concentration of species A associated with the $\gamma-\kappa$ interface, moles/m^2
$\mathscr{D}_{AB}$	binary diffusion coefficient for species A and B, m^2/s
$\mathscr{D}_{Am}$	$\mathscr{D}_{Am}^{-1} = \sum_{\substack{B=1\\B\neq A}}^{B=N} x_B/\mathscr{D}_{AB}$, mixture diffusivity, m^2/s
$\mathbf{D}_{\text{eff}}$	effective diffusivity tensor for the γ-phase, m^2/s
$\mathbf{D}_\gamma$	hydrodynamic dispersion tensor for the γ-phase, m^2/s
$\mathbf{D}_\gamma^*$	$\mathbf{D}_{\text{eff}} + \mathbf{D}_\gamma$, total dispersion tensor for the γ-phase, m^2/s
$[D]$	diffusivity matrix, m^2/s
D_{AE}	element of the diffusivity matrix, m^2/s
$\langle D_{AE}\rangle^\gamma$	intrinsic average element of the diffusivity matrix, m^2/s
$\tilde{D}_{AE}$	$D_{AE} - \langle D_{AE}\rangle^\gamma$, spatial deviation of an element of the diffusivity matrix, m^2/s
e_A	species A internal energy, kcal/kg m^3
$\mathbf{F}$	Forchheimer correction tensor
$\mathbf{g}$	gravitational acceleration vector, m^2/s
h_A	species A enthalpy or *partial mass enthalpy*, kcal/kmol m^3
H_A	species A partial molar enthalpy, kcal/kmol m^3

$\Delta H_{\text{reaction}}$	molar heat of reaction for a single independent reaction, kcal/kmol m^3
h	enthalpy per unit mass, kcal/kg m^3
$\mathbf{I}$	unit tensor
$\mathbf{J_B}$	$c_B\mathbf{u}_B$, mixed-mode diffusive flux for species B, mole/m^{2}s
$\mathbf{J}_{\mathbf{A}\gamma}$	$c_{A\gamma}\mathbf{u}_{A\gamma}$ mixed-mode diffusive flux for species A in the γ-phase, mole/m^{2}s
$\mathbf{K}$	Darcy's law permeability tensor, m^2
K_{eq}	linear adsorption equilibrium coefficient for species A, m
k_{Al}	adsorption rate coefficient for species A, m/s
k_{-Al}	desorption rate coefficient for species A, s^{-1}
ℓ_i	$i = 1, 2, 3$, lattice vectors for a spatially periodic model of a porous medium, m
ℓ_γ	small length scale associated with the γ-phase, m
L	large length scale associated with the porous medium, m
$\mathbf{M}$	second order tensor that maps $\langle \mathbf{v}_\gamma \rangle^\gamma$ onto $\tilde{\mathbf{v}}_\gamma$
$\mathbf{m}$	vector that maps $\mu_\gamma \langle \mathbf{v}_\gamma \rangle^\gamma$ on to $\tilde{p}_\gamma$, m^{-1}
M_A	molecular mass of species A, kg/kg mole
N_{JA}	elements of the composition matrix
$\mathbf{n}$	unit normal vector
$\mathbf{n}_{\gamma\kappa}$	unit normal vector directed from the γ-phase to the κ-phase
p	total pressure, N/m^2
p_γ	total pressure in the γ-phase, N/m^2
$\tilde{p}_\gamma$	$p_\gamma - \langle p_\gamma \rangle^\gamma$, spatial deviation pressure, N/m^2
p_A	partial pressure for species A, N/m^2
$\mathbf{P}_{AB}$	diffusive force per unit volume exerted by species B on species A, N/m^3
q_A	energy flux (conduction and radiation) vector for species A, kcal/m^{2}s
$\mathbf{q}$	energy flux (conduction and radiation) vector, kcal/m^{2}s
$\mathbf{q}^R$	radiation energy flux vector, kcal/m^{2}s
Q_{AB}	energy exchange between species A and species B, kcal/m^{3}s
r_o	radius of averaging volume, m
$\mathbf{r}$	position vector, m
$\mathbf{R}$	reference position vector, m
$\mathbf{r}_A$	position vector for species A, m
$\mathbf{R}_A$	reference position vector for species A, m
r_A	mass rate of homogeneous chemical reaction, kg/m^{3}s
R_A	molar rate of homogeneous chemical reaction, moles/m^{3}s
$R_{A\gamma}$	molar rate of homogeneous chemical reaction in the γ-phase, moles/m^{3}s
R_{As}	molar rate of heterogeneous chemical reaction associated with the γ–κ interface, moles/m^{2}s
$\langle R_{As} \rangle_{\gamma\kappa}$	area average molar rate of heterogeneous chemical reaction for species A, moles/m^{2}s
s	total entropy per unit mass, kcal/kg K m^3

s_γ	scalar that maps $K_{eq}\partial\langle c_{A\gamma}\rangle^\gamma/\partial t$ onto $\tilde{c}_{A\gamma}$, s/m
t	time, s
T	temperature, K
$\mathbf{t}_{A(\mathbf{n})}$	species A stress vector, N/m^2
$\mathbf{T}_A$	species A stress tensor, N/m^2
$\mathbf{T}$	total stress tensor, N/m^2
$\mathbf{u}_p$	pulse velocity, m/s
$\mathbf{u}_A$	$\mathbf{v}_A - \mathbf{v}_\gamma$, mass diffusion velocity for species A, m/s
$\mathbf{u}_{A\gamma}$	$\mathbf{v}_{A\gamma} - \mathbf{v}_\gamma$, species A mass diffusion velocity in the γ-phase, m/s
$\mathbf{v}_A$	velocity of species A, m/s
$\mathbf{v}_{A\gamma}$	velocity of species A in the γ-phase, m/s
$\mathbf{v}$	$\sum_{A=1}^{A=N} \omega_A \mathbf{v}_A$, mass average velocity, m/s
$\mathbf{v}_\gamma$	$\sum_{A=1}^{A=N} \omega_{A\gamma} \mathbf{v}_{A\gamma}$, mass average velocity in the γ-phase, m/s
$\langle \mathbf{v}_\gamma \rangle^\gamma$	intrinsic mass average velocity in the γ-phase, m/s
$\langle \mathbf{v}_\gamma \rangle$	superficial mass average velocity in the γ-phase, m/s
$\tilde{\mathbf{v}}_\gamma$	$\mathbf{v}_\gamma - \langle \mathbf{v}_\gamma \rangle^\gamma$, spatial deviation velocity in the γ-phase, m/s
$\mathscr{V}$	averaging volume, m^3
$\mathscr{V}_A(t)$	volume of a species A body, m^3
$\mathscr{V}_a(t)$	volume of an arbitrary, moving volume, m^3
V_γ	volume of the γ-phase contained within the averaging volume, m^3
V_κ	volume of the κ-phase contained within the averaging volume, m^3
$\mathbf{w}$	arbitrary velocity, m/s
x_A	c_A/c, mole fraction of species A
$x_{A\gamma}$	$c_{A\gamma}/c_\gamma$, mole fraction of species A in the γ-phase
$\mathbf{x}$	position vector locating the center of the averaging volume, m
$\mathbf{y}$	position vector locating points on the γ–κ interface relative to the center of the averaging volume, m

Greek Letters

β	coefficient of expansion, K^{-1}
ε_γ	volume fraction of the γ-phase (porosity)
ρ_A	mass density of species A, kg/m^3
$\rho_{A\gamma}$	mass density of species A in the γ-phase, kg/m^3
ρ	total mass density, kg/m^3
ρ_γ	total mass density for the γ-phase, kg/m^3
ω_A	ρ_A/ρ_γ, mass fraction of species A
$\omega_{A\gamma}$	$\rho_{A\gamma}/\rho_\gamma$, mass fraction of species A in the γ-phase
ν_A	stoichiometric coefficient for species A for a single independent reaction
λ	thermal conductivity, kcal/s m K
$\boldsymbol{\tau}$	viscous stress tensor, N/m^2, tortuosity tensor

$\boldsymbol{\tau}_A$ viscous stress tensor for species A, N/m^2
τ tortuosity
μ viscosity, Ns/m^2
μ_γ viscosity of the γ-phase, Ns/m^2

REFERENCES

Altevogt, A.S., Rolston, D.E. and Whitaker, S. 2003a, New equations for binary gas transport in porous media I: Equation development, Water Resources Research **26**, 695–715.

Altevogt, A.S., Rolston, D.E. and Whitaker, S. 2003b, New equations for binary gas transport in porous media II: Experimental validation, Water Resources Research **26**, 717–723.

Arce, P., Quintard, M. and Whitaker, S. 2005, The art and science of upscaling, in *Chemical Engineering: Trends and Developments*, edited by M.A. Galan and E.M de Valle, John Wiley & Sons, Ltd, London.

Bird, R.B., Stewart, W.E. and Lightfoot, E.N. 2002, *Transport Phenomena*, Second Edition, John Wiley and Sons, Inc., New York.

Brinkman, H.C. 1947, A calculation of the viscous force exerted by a flowing fluid on a dense swarm of particles, App. Sci. Res. **A1**, 27–34.

Chang, H.C. 1982, Multiscale analysis of effective transport in periodic heterogeneous media, Chem. Engng. Commun. **15**, 83–91.

Chang, H.C. 1983, Effective diffusion and conduction in two-phase media: A unified approach, AIChE Journal **29**, 846–853.

Cerro, R.L., Higgins, B.G. and Whitaker, S. 2004, *Material Balances for Chemical Engineers* available at http://www.higgins.ucdavis.edu/matbalance.html.

Chapman, S. and Cowling, T.G. 1970, *The Mathematical Theory of Nonuniform Gases*, third edition, Cambridge University Press.

Cushman, J.H. 1990, *Dynamics of Fluids in Hierarchical Porous Media*, Academic Press, London.

Cushman, J.H. 1997, *The Physics of Fluids in Hierarchical Porous Media: Angstroms to Miles*, Kluwer Academic Publishers, Dordrecht.

Darcy, H. 1856, Les Fontaines Publiques de la Ville de Dijon, Dalmont, Paris.

Edwards, D.A., Brenner, H. and Wasan, D.T. 1991, *Interfacial Transport Processes and Rheology*, Butterworth-Heinemann, Boston.

Eidsath, A.B., Carbonell, R.G., Whitaker, S. and Herrmann, L.R. 1983, Dispersion in pulsed systems III: Comparison between theory and experiments for packed beds, Chem. Engng. Sci. **38,** 1803–1816.

Forchheimer, P. 1901, Wasserbewegung durch Boden, Z. Ver. Deutsch. Ing. **45**, 1782–1788.

Golay, M.J.E. 1958, in *Gas Chromatography*, edited by D.H. Desty, Butterworths, London.

Graham, T. 1833, On the law of the diffusion of gases, Phil. Mag. **2**, 175–190, 269–276, 351–358.

Gray, W.G. 1975, A derivation of the equations for multiphase transport, Chem. Engng. Sci. **30**, 229–233.

Jackson, R. 1977, *Transport in Porous Catalysts*, Elsevier Publishing Co., New York.

Kerkhof, P.J.A.M. 1996, A modified Maxwell-Stefan Model for transport through inert membranes: The binary friction model, The Chemical Engineering Journal **64**, 319–343.

Kerkhof, P.J.A.M. 1997, New light on some old problems: Revisiting the Stefan tube, Graham's law, and the Bosanquet equation, Industrial & Engineering Chemistry Research **36**, 915–922.

Kramers, H.A. and Kistemaker, J. 1943, On the slip of a diffusing gas mixture along a wall, Physica **10**, 699–713.

Mason, E.A. and Malinauskas, A.P. 1983, *Gas Transport in Porous Media: The Dusty-Gas Model*, Elsevier, New York.

Ochoa-Tapia, J.A., Stroeve, P. and Whitaker, S. 1994, Diffusive transport in two-phase media: Spatially periodic models and Maxwell's theory for isotropic and anisotropic systems, Chemical Engineering Science **49**, 709–726.

Quintard, M. 1993, Diffusion in isotropic and anisotropic porous systems: Three-dimensional calculations, Transport in Porous Media **11**, 187–199.

Quintard, M. and Whitaker, S. 1993a, Transport in ordered and disordered porous media: Volume averaged equations, closure problems, and comparison with experiment, Chem. Engng. Sci. **48**, 2537–2564.

Quintard, M. and Whitaker, S. 1993b, One and two-equation models for transient diffusion processes in two-phase systems, pages 369-465 in *Advances in Heat Transfer*, Vol. 23, Academic Press, New York.

Quintard, M. and Whitaker, S. 1994a, Transport in ordered and disordered porous media I: The cellular average and the use of weighting functions, Transport in Porous Media **14**, 163–177.

Quintard, M. and Whitaker, S. 1994b, Transport in ordered and disordered porous media II: Generalized volume averaging, Transport in Porous Media **14**, 179–206.

Quintard, M. and Whitaker, S. 1994c, Transport in ordered and disordered porous media III: Closure and comparison between theory and experiment, Transport in Porous Media **15**, 31–49.

Quintard, M. and Whitaker, S. 1994d, Transport in ordered and disordered porous media IV: Computer generated porous media, Transport in Porous Media **15**, 51–70.

Quintard, M. and Whitaker, S. 1994e, Transport in ordered and disordered porous media V: Geometrical results for two-dimensional systems, Transport in Porous Media **15**, 183–196.

Quintard, M. and Whitaker, S. 1994f, Convection, dispersion, and interfacial transport of contaminants: Homogeneous porous media, Adv. Water Resour. **17**, 221–239.

Quintard, M. and Whitaker, S. 1995, Aerosol filtration: An analysis using the method of volume averaging, J. Aerosol Sci. **26**, 1227–1255.

Ryan, D., Carbonell, R.G. and Whitaker, S. 1981, A theory of diffusion and reaction in porous media, AIChE Symposium Series, #202, Vol. 71, 46–62.

Sahraoui, M. and Kaviany, M. 1994, Slip and no-slip temperature boundary conditions at the interface of porous, plain media: Convection, Int. J. Heat Mass Trans. **37**, 1029–1044.

Serrin, J. 1959, Mathematical Principles of Classical Fluid Mechanics, in *Handbuch der Physik*, Vol. VIII, Part 1, edited by S. Flugge and C. Truesdell, Springer Verlag, New York.

Slattery, J.C. 1990, *Interfacial Transport Phenomena*, Springer-Verlag, New York.

Slattery, J.C. 1999, *Advanced Transport Phenomena*, Cambridge University Press, Cambridge.

Truesdell, C. 1968, *Essays in the History of Mechanics*, Springer-Verlag, New York.

Truesdell, C. and Toupin, R. 1960, The Classical Field Theories, in *Handbuch der Physik*, Vol. III, Part 1, edited by S. Flugge, Springer Verlag, New York.

Whitaker, S. 1981, *Introduction to Fluid Mechanics*, R.E. Krieger Pub. Co., Malabar, Florida.

Whitaker, S. 1983, *Fundamental Principles of Heat Transfer*, R.E. Krieger Pub. Co., Malabar, Florida.

Whitaker, S. 1986a, Transport processes with heterogeneous reaction, pages 1 to 94 in *Concepts and Design of Chemical Reactors,* edited by S. Whitaker and A.E. Cassano, Gordon and Breach Publishers, New York.

Whitaker, S. 1986b, Transient diffusion, adsorption and reaction in porous catalysts: The reaction controlled, quasi-steady catalytic surface, Chem. Engng. Sci. **41**, 3015–3022.

Whitaker, S. 1987, Mass transport and reaction in catalyst pellets, Transport in Porous Media **22**, 269–299.

Whitaker, S. 1989, Heat transfer in catalytic packed bed reactors, in *Handbook of Heat and Mass Transfer,* Vol. 3, Chapter 10, *Catalysis, Kinetics & Reactor Engineering,* edited by N.P. Cheremisinoff, Gulf Publishers, Matawan, New Jersey.

Whitaker, S. 1992, The species mass jump condition at a singular surface, Chem. Engng. Sci. **47,** 1677–1685.

Whitaker, S. 1996, The Forchheimer equation: A theoretical development, Transport in Porous Media **25**, 27–61.

Whitaker, S. 1997, Volume averaging of transport equations, Chapter 1 in *Flow in Porous Media*, edited by J.P. Du Plessis, Computational Mechanics Publications, Southampton, U.K.

Whitaker, S. 1999, *The Method of Volume Averaging*, Kluwer Academic Publishers, Dordrecht.

Wood, B.D. and Whitaker, S. 1998, Diffusion and reaction in biofilms, Chem. Engng. Sci. **53**, 397–425.

Wood, B.D., Quintard, M. and Whitaker, S. 2004, Estimation of adsorption rate coefficients based on the Smoluchowski equation, Chem. Engng. Sci. **59**, 1905–1921.

CHAPTER 7

GAS-PHASE DISPERSION IN POROUS MEDIA

MOLLY S. COSTANZA-ROBINSON[1,2] AND MARK L. BRUSSEAU[3,4]
[1] *Program of Environmental Studies, Middlebury College, Middlebury, Vermont 05753*
mcostanz@middlebury.edu
[2] *Department of Chemistry and Biochemistry, Middlebury College, VT, USA*
[3] *Soil, Water and Environmental Science Department, The University of Arizona, Tucson, AZ, USA*
[4] *Hydrology and Water Resources Department, The University of Arizona, Tucson, AZ, USA*

7.1 INTRODUCTION

Solute dispersion refers to the spatial spreading of a solute plume over time. The spreading is essentially a mixing and consequent dilution of the solute plume with the resident fluid, as depicted in Figure 7.1. Consideration of dispersion is critical to understanding gas-phase transport in porous media.

Many transport principles that were originally developed to describe behavior in saturated porous media, and later for unsaturated water flow, can also be readily

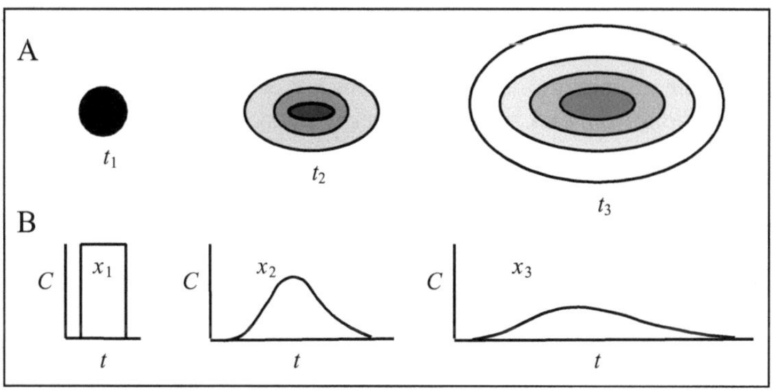

Figure 7.1. Spreading of a solute plume from an instantaneous point source. (A) Two-dimensional spatial "snapshots" (concentration versus *x-y* coordinates) as a function of time (*t*). (B) Temporal breakthough curves (concentration versus time) as a function of distance along axis of flow (*x*)

C. Ho and S. Webb (eds.), Gas Transport in Porous Media, 121–132.
© 2006 *Springer.*

applied to the transport of gases in unsaturated systems, including the concepts of dispersion. However, any rigorous or quantitative analysis of gas-phase systems requires consideration of the unique complexities of unsaturated systems and the properties of gases themselves. For example, unsaturated porous media have air-filled porosities that may vary both spatially- and temporally, and depend on such factors as soil-water content and particle-size distribution. Gas-phase diffusion coefficients are generally four to six orders of magnitude larger than aqueous-phase values, and in contrast to water, gases are significantly affected by pressure-temperature relationships. Gases experience slip-flow along pore walls, often termed the Klinkenberg effect, while water does not. In the following discussion, the authors use the terms "gas" and "vapor" interchangeably, while the term "solute" refers very broadly to the gas/vapor of interest.

7.2 THEORY

7.2.1 Diffusion

Dispersion includes diffusive and mechanical mixing components. Gas-phase diffusion is often assumed to be dominated by molecular diffusion, the random spreading of a solute along concentration gradients over time, described here by a one-dimensional Fick's second law:

$$\frac{\partial C}{\partial t} = D_a \frac{\partial^2 C}{\partial x^2} \tag{7.1}$$

where C is the gas concentration ($M{\cdot}L^{-3}$), t is time (T), D_a is the binary molecular diffusion coefficient in air ($L^2{\cdot}T^{-1}$), and x is the distance along the axis of flow (L). For molecular diffusion, molecule-molecule collisions are the only type of collisions that occur, implying a system without walls. In some cases, more complex gas-phase diffusion processes may also occur, including viscous, Knudsen, and nonequimolar diffusion (e.g., Scanlon et al., 2000). The former two processes occur due to the presence of pore walls and consequent molecule-wall collisions (Cunningham and Williams, 1980), while the latter requires both the presence of walls and a multicomponent gas (i.e., a mixture). Such conditions are present in porous media and may lead to deviations from Fick's law (e.g., Sleep, 1998). Baehr and Bruell (1990) report that high vapor pressures, such as those achieved near organic liquid sources, also cause deviation from Fick's law. In accordance with the bulk of the literature, this chapter will focus on molecular diffusion (e.g., assuming a system without walls), while other diffusion processes are discussed in detail elsewhere.

Diffusion is a solute-dependent component of dispersion, due to the relationships among average kinetic energy, velocity, and molecular mass. At a given temperature the average kinetic energy of all gases will be equal and described as:

$$E_k = \frac{3}{2}kT = \frac{1}{2}mv_{\text{rms}}^2 \tag{7.2}$$

where k is Boltzmann's constant ($J \cdot K^{-1}$), T is temperature (K), m is the solute mass (M), and v_{rms} is the root-mean-square velocity of the gas particles ($L \cdot T^{-1}$). Thus, given thermal equilibrium and consequent equal kinetic energy, lower molecular weight gases will exhibit higher average velocities relative to higher molecular weight gases. This higher velocity translates into larger diffusion coefficients and larger contributions to overall dispersion.

Diffusion processes often dominate transport in low permeability zones, such as within aggregates or fine-textured lenses. In the case of volatile organic compounds (VOCs), the occurrence of diffusion can both aid and present additional challenges to remediation efforts. Specifically, the larger spatial distribution of the VOC caused by diffusion may increase the probability of detecting the VOC in the subsurface, such that a remediation plan can be implemented. Conversely, it is this same diffusion process that is often largely responsible for the transport of VOCs in the vadose-zone to the water table resulting in groundwater contamination (e.g., Lupo, 1989). Moreover, contaminant diffusion can influence the fate of contamination in a system by altering its bioavailability. Diffusion largely governs gas-exchange between the atmosphere and soil, including the exchange of carbon dioxide and oxygen, and atmospheric pollutants, such as fluorocarbons (Weeks et al., 1982). Jury et al. (1991) estimate that only 0.5%, 1%, 0.1%, and 7–9% of the overall subsurface gas exchange is induced by temperature and barometric pressure changes, wind, and precipitation, respectively. Thus, they conclude that diffusion is the primary gas transport mechanism in soil systems. Little et al. (1992) review the critical role diffusion plays in transporting subsurface VOCs into homes and buildings.

7.2.2 Mechanical Mixing

Mechanical mixing is a solute-independent component of dispersion, governed by the physical properties of the porous medium and carrier gas velocity. Mechanical mixing is a lumped term, incorporating a number of sources of velocity variations that result in solute mixing and dilution. Such velocity variations may be caused by (A) non-uniform velocity profiles along the cross section of individual pores (e.g., velocities are higher in the center of the pore relative to near-wall velocities); (B) distributions in pore sizes (e.g., large pores promote higher velocities than smaller pores); and (C) the tortuosity of flow paths, as shown in Figure 7.2. The effective pore-size distribution and tortuosity are influenced by the presence of soil-water. At larger scales, dispersion may be caused by the presence of lenses of material with different permeabilities. Larger-scale differences in permeability may further influence capillary forces and the resultant large-scale water saturation, although much less is known about large-scale mechanical mixing processes (Freeze and Cherry, 1979; Selker et al., 1999). Because the magnitude of mechanical mixing depends on the degree of heterogeneity, it is expected that the magnitude of gas-phase dispersivity will be proportional to the system-scale as has been demonstrated for aqueous-phase dispersion (Pickens and Grisak, 1981; Gelhar et al., 1992).

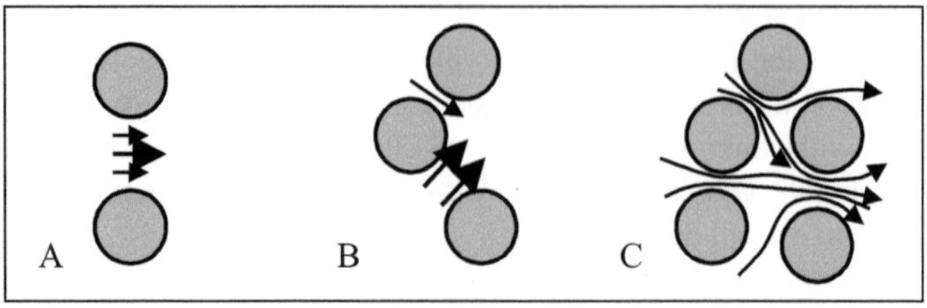

Figure 7.2. Schematic of sources of pore-scale velocity variations resulting in mechanical mixing: (A) Velocity variations within a single pore due to wall-effects; (B) Pore-size distributions; (C) tortuosity effects

7.2.3 Dispersion Coefficient and Peclet Number

Gas-transport in porous media is often described using an advection-dispersion equation (A-D Equation), as described in more detail elsewhere. In a one-dimensional A-D Equation, the effects of dispersion are represented by the longitudinal dispersion coefficient, D ($L^2 \cdot T^{-1}$), defined as:

$$D = D_a \tau + \alpha v \tag{7.3}$$

where D_a is the binary molecular diffusion coefficient in air ($L^2 \cdot T^{-1}$); τ is the tortuosity factor defined between 0 and 1 and inversely proportional to the tortuosity of the gas phase in the porous medium (dimensionless); α is the gas-phase longitudinal dispersivity, a measure of the physical heterogeneity of the media (L); and v is the average linear velocity of the gas ($L \cdot T^{-1}$). The first term on the right-hand side of Equation 7.3 represents the solute-dependent diffusive contributions to dispersion, while the second term represents the mechanical mixing dispersion component. Transverse dispersion also occurs and is described by a transverse dispersion coefficient; however, under most conditions transverse dispersion is observed to be much less significant than longitudinal dispersion and is not considered further here.

Diffusion coefficients (in air) are typically obtained from the literature. The tortuosity factor is estimated using empirical literature correlations incorporating total and air-filled porosity (Penman, 1940; Currie, 1961; Millington and Quirk, 1961; Millington and Shearer, 1971; Sallam et al., 1984; Karimi et al., 1987; Shimamura, 1992; Moldrup et al., 1996; Schaefer et al., 1997; Poulsen et al., 1998). The product of the diffusion coefficient and the tortuosity factor is often termed the effective diffusion coefficient, D_a^*. The reduction in D_a^* relative to the diffusion coefficient in air is due to the presence of the solid-phase media, resulting in smaller cross-sectional area available for diffusion, the tortuosity of the gas pathways, the presence of disconnected or "dead-end" pores, and at least in dry porous media, the geometry of the pores, as influenced by particle shape (Currie, 1960). Thus, D_a^* decreases in response to the presence of soil-water or greater bulk densities. Theoretically, the distribution of water due to pore-size distributions also influences measured D_a^* values (Bruce and Webber, 1953). Effective diffusion coefficients can be estimated using tortuosities (see above) or can measured directly in the laboratory (Taylor, 1949; Bruce and Webber, 1953; Currie, 1961; Shearer et al., 1973; Sallam et al., 1984; Johnson and Perrott, 1991; Barone et al., 1992; Jin and Jury, 1996; Batterman et al., 1996; Schaefer et al., 1997; Moldrup et al., 1998, 2000) and field (Raney, 1949; Lai et al., 1976; Weeks et al., 1982; Kreamer et al., 1988).

The longitudinal dispersivity term scales with the degree of heterogeneity of the physical system and is often measured using column-scale nonreactive tracer tests. Gas-phase longitudinal dispersivities have been measured in laboratory porous media systems and are found to range approximately between 0.2 and 2.9 cm (Popovičová and Brusseau, 1997; Ruiz et al., 1999; Garcia-Herruzo et al., 2000; Costanza-Robinson and Brusseau, 2002). Dispersivities measured in the field tend to be larger due to increased system heterogeneity. Ideally, dispersivities should be solute-independent and insensitive to changes in carrier gas velocity or nonequilibrium effects. However, if the data analysis fails to consider all relevant transport processes, dispersivity can become a "lumped" solute-dependent parameter that no longer solely represents physical heterogeneity of the porous medium (Costanza-Robinson and Brusseau, 2002).

The Peclet number, P_e, is a dimensionless measure of the degree of dispersion experienced by a solute, defined alternatively as the ratio between the advective and dispersive processes or the ratio of advective to diffusive processes (Rose, 1973). The former definition is more encompassing and will be used here. The Peclet number is usually obtained by fitting a solute breakthrough curve with an advective-dispersive solute transport model. The magnitude of the Peclet number is inversely proportional to the degree of dispersion. Thus, low Peclet numbers correspond to a large degree of solute spreading. The Peclet number, also termed the Brenner number (Rose and Passioura, 1971), can be related to the dispersion coefficient as follows:

$$P_e = \frac{vL}{D} \tag{7.4}$$

where L is a characteristic length of the system (L). The characteristic length can be defined at small- or macro-scales (e.g., grain diameter or column length) (Rose, 1973). The specific formulation of the Peclet number varies by application and field of study. The macroscale length is used commonly in the fields of soil physics and hydrology, while grain-scale lengths are often used in engineering disciplines. Thus, caution should be exercised when interpreting absolute values of Peclet numbers or when comparing values from different studies.

7.2.4 Functional Dependence of Dispersion and the Dispersion Coefficient

The dispersion coefficient is a function of the solute velocity, v. While the diffusion term is independent of v, the mechanical mixing component of dispersion is proportional to velocity (see Equation 7.3). Thus, D decreases with decreases in velocity. However, slower velocities lead to larger residence times, thereby allowing more time for diffusion to occur. This results in larger magnitudes of dispersion. When velocities are sufficiently large, the contribution of diffusion to dispersion will be negligible. Under these conditions, while the magnitude of the *dispersion coefficient* will change with changes in velocity, the magnitude of *dispersion* will be velocity-independent. This is shown by consideration of the Peclet number (Section 2.3). If diffusion is considered as negligible, the dispersion coefficient in Equation 7.4 can be replaced with the mechanical mixing term from Equation 7.3. As shown below, this results in a velocity-independent Peclet number:

$$P_e = \frac{vL}{D} = \frac{vL}{\alpha v} = \frac{L}{\alpha} \tag{7.5}$$

Thus, at high velocities P_e, a measure of the magnitude of dispersion, depends solely on properties of the porous medium, as represented by the dispersivity (α) and the characteristic length of the system (L). Gas retention in the system (e.g., adsorption, dissolution) results in longer residence time and to an increase in apparent dispersion. However, the magnitude of the dispersion coefficient, a measure of the dispersion *per unit time*, does not actually increase (Jury et al., 1991).

Inspection of Equation 7.3 shows that the diffusion contribution to D is solute-dependent, while that of mechanical mixing is not. Thus, the magnitude of dispersion will vary as a function of the solute when diffusion provides a significant contribution, with lower molecular-weight gases exhibiting greater dispersion. Conversely, dispersion will be solute-independent for larger velocities, when mechanical mixing dominates dispersion.

7.3 LABORATORY INVESTIGATIONS

7.3.1 Variables Affecting Diffusion

The magnitude of diffusion and its contribution to overall dispersion depends on properties of the solute and porous medium, as well as transport parameters, such as carrier velocity, as noted above. Ehlers et al. (1969) found diffusion contributions to gas-phase dispersion be directly and inversely related to temperature and bulk density of the medium, respectively. Others have reported similar bulk-density or total-porosity effects on diffusion rates (Taylor, 1949; Sallam et al., 1984; Karimi et al., 1987; Shimamura, 1992; Abu-El-Sha'r and Abriola, 1997). Lower bulk densities correspond to larger pores and less tortuosity; thus, these results are consistent with the discussion above.

At soil-water contents higher than 4 to 5% (wt), Ehlers et al. (1969) found soil-water content did not influence effective diffusion coefficients, although the technique used could not differentiate between gas-phase and aqueous-phase diffusion. Thus, at higher water contents, the expected decrease in gas-phase diffusion was likely offset by increased aqueous-phase diffusion. Karimi et al. (1987) examined the role of soil-water content on diffusion of benzene in a simulated landfill scenario and was able to isolate the process of vapor diffusion. Increasing soil-water content in the range 8 to 11% (wt) was observed to decrease the effective benzene diffusion coefficient. Similar inverse relationships between soil-water content and diffusion are reported for graded (Taylor, 1949) and aggregated porous media (Millington and Shearer, 1971; Arands et al., 1997). Shimamura (1992) presented similar results for a number of sandy soils with controlled grain-size distributions.

Taylor (1949) presented diffusion rates as a function of matric potential, a variable that is inversely related to soil-water content. The plot of effective diffusion distance (inversely proportional to tortuosity) versus matric potential displayed significantly less variability for the four natural and graded porous media studied, as compared to the plot of effective diffusion distance versus soil-water content. Viewing diffusion as a function of matric potential, incorporates both the effects of soil-water content and pore- and grain-size distributions, thereby allowing more general conclusions to be drawn for a variety of porous media.

7.3.2 Variables Affecting Mechanical Mixing and Total Dispersion

The total magnitude of dispersion depends on several factors, including physical properties of the porous medium, physicochemical properties of the gaseous solute, and flow conditions. Edwards and Richardson (1968) measured dispersion coefficients by varying argon velocity in a dry packed system, demonstrating that the dispersion coefficient is relatively constant at low Reynolds numbers (e.g., 0.01–0.5) and increases approximately linearly for higher Reynolds numbers. The Reynolds number is a measure of the turbulence of flow and for the same fluid and porous medium is directly proportional to average linear velocity. Thus, for the low velocity experiments, the magnitude of dispersion remained relatively constant, but increased linearly at higher velocities. The authors interpret this as indicating that molecular diffusion, a velocity-independent term (see Equation 7.3), dominates dispersion at low velocities (e.g., low Reynolds numbers). At higher velocities, mechanical mixing, which is directly proportional to velocity, dominates dispersion. Using expressions similar to Equation 7.3, Edwards and Richardson (1968) define three regions of dispersion: (1) low Reynolds numbers where the mechanical mixing term (in Equation 7.3) is negligible; (2) intermediate Reynolds numbers where both mechanical mixing and diffusion terms are significant; and (3) high Reynolds numbers where the diffusion term becomes negligible. This three-region approach had been previously applied to the case of saturated flow.

Popovičová and Brusseau (1997) also examined the role of carrier gas velocity on the magnitude of dispersion and the relative contributions of diffusion and mechanical mixing to methane transport in a dry, homogeneous, glass-bead column. At low velocities, virtually all methane dispersion was due to diffusion, while at larger pore velocities, mechanical mixing contributed more than 80% of the observed dispersion. Similar velocity-dependence of dispersion-contributions was observed for a heterogeneous glass-bead column, and total dispersion increased relative to the homogeneous system. The heterogeneous column had a macropore located in the center of the otherwise homogeneous column. At the highest velocities in the heterogeneous column, rate-limited diffusion (e.g., nonequilibrium effects) between the macro- and micropore domains became much more significant than dispersion processes.

Garges and Baehr (1998) simulated gas-phase miscible displacement breakthrough curves using a one-dimensional advection-dispersion transport model and varying the degree of dispersion, as represented by varying Peclet numbers. The paper provides excellent plots displaying the effect of different magnitudes of dispersion on the resulting breakthrough curves. As expected, larger dispersion coefficients results in breakthrough curves with shallower slopes for both the arrival and elution waves.

Batterman et al. (1995) examined the influence of porous media properties and the relative humidity of the gas-phase on diffusion rates and total dispersion coefficients. Effective diffusion coefficients were measured for a number of dry natural and synthetic media and were found to be consistent with predictions from empirical correlations (see Section 2.3). Total dispersion coefficients were measured for the same media with carrier relative gas humidities ranging from 0 to 90% (generally corresponding to very dry soils with gravimetric soil-water contents <1%). The authors concluded that mechanical mixing and diffusion contributed about equally to dispersion under these conditions.

Batterman et al. (1995) report that methane experienced greater overall dispersion than did trichloroethene (TCE) in column studies performed over a range of soil humidities. This trend is expected due to the much higher diffusivity of methane relative to TCE. However, additional analysis shows that the difference in reported dispersion coefficients for TCE and methane is too large to be explained by differences in diffusion coefficients alone. After correcting the total dispersion coefficients reported by Batterman et al. (1995) for diffusion, the absolute dispersion due to mechanical mixing is almost three times greater for methane than for TCE. Theoretically, mechanical mixing should be a solute-independent term. This indicates that additional transport processes were likely occurring in the experiments that were not considered in the original data analysis.

Costanza-Robinson and Brusseau (2002) observed that the lowest molecular weight compound studied, methane, had the largest diffusion contribution in a wetted homogeneous natural sand system, comprising approximately 60% of the total observed dispersion. For higher molecular weight compounds (e.g., difluoromethane and TCE) mechanical mixing dominated dispersion, contributing between 50 and 100% of the dispersion, depending on soil-water content. Relative contributions from mechanical mixing increased at higher soil-water contents due to the consequent increase in tortuosity and decrease in diffusion.

The influence of specific system properties on experimentally-determined dispersivity values, including soil-water content and particle- and pore-size distributions have not been fully studied. Furthermore, laboratory-measured dispersivity values are often derived from data from nonreactive tracer tests and are assumed to be representative of dispersivities for reactive compounds under various system conditions. Because dispersivity is a measure of the heterogeneity of the physical system, this assumption is theoretically justifiable. However, for this assumption to hold in a real system, all relevant transport processes must be appropriately accounted for in the analysis. For example, if additional transport processes, such as rate-limited mass-transfer are occurring, but are not considered in the data analysis, the dispersivity values will become lumped, solute-dependent values.

Costanza-Robinson and Brusseau (2002) performed experiments to examine these issues. Diffusional dispersion contributions were explicitly accounted for using properties of the porous medium and literature tortuosity correlations and diffusion constants. The total dispersion coefficient was obtained by fitting the experimental breakthrough curves with an equilibrium, one-dimensional, transport model. The difference between the total dispersion coefficient and the diffusional contributions was taken as the mechanical mixing contribution. The dispersivity values calculated in this manner were observed to be constant for the porous medium studied at soil-water contents ranging from 2 to 14%. This indicates that it may be appropriate to use a single dispersivity value to represent a given porous media over a range of natural conditions. Moreover, the calculated dispersivities were the same for all compounds studied, which included a nonreactive (methane), a water-soluble (difluoromethane), and a water-soluble and sorbing solute (TCE). This indicates that in the 2 to 14% soil-water content range, the data analysis appropriately accounted for transport processes, such that the dispersivity value was a true measure of the porous medium heterogeneity, rather than a solute-dependent lumped term. At soil-water contents greater than 14%, dispersivities became solute-dependent, indicating that additional transport processes were being lumped into the dispersivity term. The authors attributed this to rate-limited diffusion of the soluble solutes through water films, which was not considered in the data analysis.

7.4 FIELD AND MODELING INVESTIGATIONS

The relatively few experimental field investigations of gas-phase dispersion have focused on diffusion. Raney (1949) and Lai et al. (1976) have presented methods for measuring effective diffusion coefficients in-situ that are applicable to surface soils. Weeks et al. (1982) examined the use of atmospheric pollutants, fluorocarbons F-11 and F-12, in measuring effective vadose-zone diffusion rates, concluding that gas-phase diffusion is likely the most important transport mechanism in regions where groundwater recharge is small. As expected, soil tortuosity and the solubility and sorption of the gases resulted in measured effective diffusion coefficients that were much less than those estimated for free-air diffusion. Numerical-modeling results indicated that the near-surface region had lower tortuosities, while deeper layers contributed most significantly to reduced diffusion rates. Finally, Weeks et al. (1982) concluded that the relative agreement between their optimized tortuosity factors and tortuosity factors estimated via a number of theoretical and empirical approaches lends support to the use of diffusion theory in predicting soil gas concentrations, even in large-scale, heterogeneous natural systems.

Mathematical modeling of gas-phase transport has also focused largely on diffusional processes, citing barometric pressure gradients and consequent advection as minimal. Mendoza and Frind (1990) demonstrate that in dilute (e.g., low vapor pressure) vadose zone systems, diffusion from the soil to the atmosphere results in removal of 95% and 69% of the contaminant mass from systems having gas-phase permeabilities of 1.0×10^{-11} and 1.0×10^{-10} m^2, respectively. At the higher permeabilities other transport mechanisms, such as density-driven advection, contribute more significantly to gas transport. Lupo (1989) also demonstrates the importance of diffusion in the transport of aromatic contaminants in a simulated landfill scenario. Diffusion of chlorobenzene and benzene is calculated to be most critical in coarser soils and under conditions of lower groundwater recharge. Similarly, Baehr (1987) reports the important role diffusion plays in the transport of hydrocarbons in the vadose zone.

7.5 SUMMARY

Gas-phase dispersion is caused by mechanical mixing and diffusion processes. The magnitude of diffusion is inversely proportional to compound molecular weight, porous media bulk density, and soil-water content and directly proportional to temperature. Under natural-gradient conditions, diffusion will likely be the dominant transport mechanism. Mechanical mixing is likely to be dominant only under conditions of induced gas advection (e.g., miscible displacement experiments and soil vapor extraction systems) or relatively extreme changes in barometric pressure. The magnitude of dispersion depends on the degree of heterogeneity of the physical system. Experimental investigation of gas-phase dispersion has focused almost exclusively on laboratory-scale systems. Understanding of gas-phase dispersion is important for contaminant transport applications, as well as atmosphere-soil gas exchange processes.

REFERENCES

Abu-El-Sha'r, W., Abriola, L.M. Experimental Assessment of Gas Transport Mechanisms in Natural Porous Media: Parameter Evaluation. Water Resources Research, 33(4):505–516, 1997.

Arands, R., Lam, T., Massry, I., Berler, D.H., Muzzio, F.J., Kosson, D.S. Modeling and Experimental Validation of Volatile Organic Contaminant Diffusion through an Unsaturated Soil. Water Resources Research, 33(4):599–609, 1997.

Baehr, A.L. Selective Transport of Hydrocarbons in the Unsaturated Zone due to Aqueous and Vapor Phase Partitioning. Water Resources Research, 23(10):1926–1938, 1987.

Baehr, A.L. Bruell, C.J. Application of the Stefan-Maxwell Equations to Determine Limitations of Fick's Law when Modeling Organic Vapor Transport in Sand Columns. Water Resources Research, 26(6):1155–1163, 1990.

Barone, F.S., Rowe, R.K., Quickley, R.M. A Laboratory Estimation of Diffusion and Adsorption Coefficients for Several Volatile Organics in a Natural Clayey Soil. Journal of Contaminant Hydrology, 10:225–250, 1992.

Batterman, S., Kulshrestha, A., Cheng, H.Y. Hydrocarbon Vapor Transport in Low Moisture Soils. Environmental Science and Technology, 29(1):171–180, 1995.

Batterman, S., Padmanabham, I., Milne, P. Effective Gas-Phase Diffusion Coefficients in Soils at Varying Water Content Measured using a One-Flow Sorbent-Based Technique. Environmental Science and Technology, 30(3):770–778, 1996.

Bruce, R.R., Webber, L.R. The Use of a Diffusion Chamber as a Measure of the Rate of Oxygen Supplied by a Soil. Canadian Journal of Agricultural Science, 33:430–736, 1953.

Costanza-Robinson, M.S., Brusseau, M.L. Gas Phase Advection and Dispersion in Unsaturated Porous Media. Water Resources Research, 38(4):7/1–7/10, 2002.

Cunningham, R.E., Williams, R.J.J., *Diffusion in Gases and Porous Media.* New York: Plenum Press, 1980.

Currie, J.A. Gaseous Diffusion in Porous Media. Part 2. Dry Granular Materials. British Journal of Applied Physics, 11:318–324, 1960.

Currie, J.A. Gaseous Diffusion in Porous Media. Part 3. Wet Granular Materials. British Journal of Applied Physics, 12:275–281, 1961.

Edwards, M.F., Richardson, J.F. Gas Dispersion in Packed Beds. Chemical Engineering Science, 23: 109–123, 1968.

Ehlers, W., Farmer, W.J., Spencer, W.F., Letey, J. Lindane Diffusion in Soils: II. Water Content, Bulk Density, and Temperature Effects. Soil Science Society of America Proceedings, 33:505–508, 1969.

Freeze, R.A., Cherry, J.A., *Groundwater.* Englewood Cliffs: Prentice-Hall, Inc., 1979.

García-Herruzo, F., Rodríguez-Moroto, J.M., García-Delgado, R.A., Gómez-Lahoz, C., Vereda-Alonso, C. Column Study of the Influence of Air Humidity on the Retention of Hydrocarbons on Soil. Chemosphere, 41:1167–1172, 2000.

Garges, J.A. and Baehr, A.L. Type Curves to determine the Relative Importance of Advection and Dispersion for Solute and Vapor Transport. Ground Water, 36(6):959–965, 1998.

Gelhar, L.W., Welty, C., Rehfeldt, K.R. A Critical Review of Data on Field-Scale Dispersion in Aquifers. Water Resources Research, 28(7):1955–1974, 1992.

Jin, Y., Jury, W.A. Characterizing the Dependence of Gas Diffusion Coefficient on Soil Properties. Soil Science Society of America Journal, 60:66–71, 1996.

Johnson, R.L., Perrott, M. Gasoline Vapor Transport through a High-Water-Content Soil. Journal of Contaminant Hydrology, 8:317–334, 1991.

Jury, W.A., Gardner, W.R., Gardner, W.H., *Soil Physics*, 5th Edition. New York: John Wiley & Sons, Inc., 1991.

Karimi, A.A., Farmer, W.J., Cliath, M.M. Vapor-Phase Diffusion of Benzene in Soil. Journal of Environmental Quality, 16(1):38–43, 1987.

Kreamer, D.K., Weeks, E.P., Thompson, G.M. A Field Technique to Measure the Tortuosity and Sorption-Affected Porosity for Gaseous Diffusion of Materials in the Unsaturated Zone with Experimental Results from Near Barnwell, South Caroline. Water Resources Research, 24(3):331–341, 1988.

Lai, S.H., Tiedje, J.M., Erickson, A.E. In situ Measurement of Gas Diffusion Coefficient in Soils. Soil Science Society of America Journal, 40(1):3–6, 1976.

Little, J.C., Daisey, J.M., Nazaroff, W.W. Transport of Subsurface Contaminants into Buildings: An Exposure Pathway for Volatile Organics. Environmental Science and Technology, 26(11):2058–2065, 1992.

Lupo, M.J. Mathematical Evaluation of Volatile Organic Compound Transport via Pore-Space Dispersion versus Advection. National Superfund Conference, Washington, D.C., 1989.

Mendoza, C.A., Frind, E.O. Advective-Dispersive Transport of Dense Organic Vapors in the Unsaturated Zone – 2. Sensitivity Analysis. Water Resources Research, 26(3):388–398, 1990.

Millington, R.J., Quirk, J.P. Permeability of Porous Solids. Transactions of the Faraday Society, 53: 1200–1207, 1961.

Millington, R.G., Shearer, R.C. Diffusion in Aggregated Porous Media. Soil Science, 111(6):372–378, 1971.

Moldrup, P., Kruse, C.W., Tolston, D.E., Yamaguchi, T. Modeling Diffusion and Reaction in Soils: III. Predicting Gas Diffusivity from the Campbell Soil-Water Retention Model. Soil Science, 161(6): 366–375, 1996.

Moldrup, P., Poulsen, T.G., Schjønning, Olesen, T., Yamaguchi, T. Gas Permeability in Undisturbed Soils: Measurements and Predictive Models. Soil Science, 163(3):180–189, 1998.

Moldrup, P., Olesen, T., Schjønning, Yamaguchi, T., Rolston, D.E. Predicting the Gas Diffusion Coefficient in Undisturbed Soil from Soil Water Characteristics. Soil Science Society of America Journal, 64:94–100, 2000.

Penman, H.L. Gas and Vapour Movements in the Soil – I. The Diffusion of Vapours through Porous Solids. Journal of Agricultural Science, 30:437–462, 1940.

Pickens, J.F., Grisak, G.E. Scale-Dependent Dispersion in a Stratified Granular Aquifer. Water Resources Research, 17(4):1191–1211, 1981.

Popovičová, J., Brusseau, M.L. Dispersion and Transport of Gas-Phase Contaminants in Dry Porous Media: Effect of Heterogeneity and Gas Velocity. Journal of contaminant Hydrology, 28:157–169, 1997.

Poulsen, T.G., Moldrup, P., Schjønning, P., Massmann, J.W., Hansen, J.Å. Gas Permeability and Diffusivity in Undisturbed Soil: SVE Implications. Journal of Environmental Engineering, 124(10):979–986, 1998.

Raney, W.A. Field Measurement of Oxygen Diffusion through Soil. Soil Science Society of America Proceedings, 14:61–64, 1949.

Rose, D.A., Passioura, J.B. The Analysis of Experiments on Hydrodynamic Dispersion. Soil Science, 111:252–257, 1971.

Rose, D.A. Some Aspects of the Hydrodynamic Dispersion of Solutes in Porous Materials. Journal of Soil Science, 24(3):284–295, 1973.

Ruiz, J., Bilbao, Rafael, Murillo, M.B. Convective Transport and Removal of Vapors of Two Volatile Compounds in Sand Columns under Different Air Humidity Conditions. Environmental Science and Technology, 33(21):3774–3780, 1999.

Sallam, A., Jury, W.A., Letey, J. Measurement of Gas Phase Diffusion Coefficient under Relatively Low Air-filled Porosity. Soil Science Society of America Journal, 48:3–6, 1984.

Scanlon, B.R., Nicot, J.P., Massmann, J.M. "Soil Gas Movement in Unsaturated Systems." In *Handbook of Soil Science*, Sumner, M.E., ed. Boca Raton: CRC Press LLC, 2000.

Schaefer, C.E., Arands, R.R., van der Sloot, H.A., Kosson, D.S. Modeling of the Gaseous Diffusion Coefficient through Unsaturated Soil Systems. Journal of Contaminant Hydrology, 29:1–21, 1997.

Selker, J.S., Keller, C.K., McCord, J.T., *Vadose Zone Processes*. Boca Raton: Lewis Publishers, 1999.

Shearer, R.C., Letey, J., Farmer, W.J., Klute, A. Lindane Diffusion in Soil. Soil Science Society of America Proceedings, 37:189–193, 1973.

Shimamura, K. Gas Diffusion Through Compacted Sands. Soil Science, 153(4):274–279, 1992.

Sleep, B.E. Modeling Transient Organic Vapor Transport in Porous Media with the Dusty Gas Model. Advances in Water Resources, 22(3):247–256, 1998.

Taylor, S.A. Oxygen Diffusion in Porous Media as a Measure of Soil Aeration. Soil Science Society of America Proceedings, 14:55–58, 1949.

Weeks, E.P., Earp, D.E., Thompson, G.M. Use of Atmospheric Fluorocarbons F-11 and F-12 to Determine the Diffusion Parameters of the Unsaturated Zone in the Southern High Plains of Texas. Water Resources Research, 18(5):1365–1378, 1982.

CHAPTER 8

GAS INJECTION AND FINGERING IN POROUS MEDIA

MUHAMMAD SAHIMI[1], M. REZA RASAEI[1,2] AND MANOUCHEHR HAGHIGHI[1,2]

[1] *Department of Chemical Engineering, University of Southern California, Los Angeles, California, 90089–1211, USA*

[2] *Institute of Petroleum Engineering and Department of Chemical Engineering, Faculty of Engineering, The University of Tehran, Tehran, 11365–4563, Iran*

8.1 INTRODUCTION

On average, two-thirds of the original oil in any reservoir remains unrecovered, even after water injection into the reservoir. The same is true if a low-pressure gas is injected into a reservoir which is largely immiscible with the oil. The oil is trapped in the reservoir due to the capillary forces and the interfacial tension, and remains entrapped regardless of how much water or low-pressure gas is injected into the reservoir. It forms either a discontinuous phase in the swept zone, or a continuous phase in the unswept zone of the reservoir.

One way of enhancing the recovery of oil is by injecting into a reservoir a fluid which is miscible with the oil in place. In principle, the injected fluid should drastically reduce the capillary and interfacial forces. If that happens, then one may recover a large fraction of the trapped oil. For example, if one injects a gas (or a gas mixture) into an oil reservoir and if, at the temperature and pressure of the reservoir, the gas is in a critical or near critical state, it will mix in large proportions (if not completely) with the oil in place. Under these conditions, one will have an efficient miscible displacement process. For example, the critical temperature of CO_2 is only about 31°C and, therefore, it can be an ideal agent for miscible displacement of oil. In fact, field-scale CO_2 injection has been successfully carried out in the United States. It is such gas injection processes and the associated phenomena in oil reservoirs that are of prime interest to us in the present chapter.

If we inject a gas into a reservoir which is completely or partially saturated with oil, and if the gas and the oil-in-place mix in all proportions, the gas and the oil are said

C. Ho and S. Webb (eds.), Gas Transport in Porous Media, 133–168.
© 2006 *Springer.*

to be *first-contact miscible*. Light and intermediate-molecular weight hydrocarbons, such as methane, propane and butane, are first-contact miscible agents. However, the injected gas and the oil may form two different phases, that is, they may not be first-contact miscible, but mass transfer between the two phases and long-time contact between the fluids may achieve miscibility. This is usually called *multiple-contact* or *dynamic* miscibility. In the petroleum industry, the gases or fluids that achieve either first-contact or dynamic miscibility are usually called *miscible solvents*. In this chapter we restrict ourselves mostly to first-contact miscible displacements, as modeling multiple-contact miscibility involves computation of the phase equilibria diagrams for the mixture of the reservoir fluids and the injected gas, a subject which is beyond the scope of our discussions.

A typical miscible injection, or flooding, is carried out by injecting into the reservoir a limited volume or slug of the solvent. The solvent slug may be displaced miscibly by an appropriate drive fluid (miscible slug process), or immiscibly by, for example, water. The latter process leaves a residual solvent saturation in the reservoir, and causes dilution and fingering of the drive fluid in the solvent, and fingering of the solvent in the oil, all of which reduce the overall effectiveness of the process. Such miscible displacements have received considerable attention since the early 1950s. Over 100 studies were undertaken in the 1950s and early 1960s to investigate the feasibility and economics of miscible displacement processes, and in particular gas injection, as an effective tool for enhanced oil recovery (EOR).

The early studies indicated that natural gas, flue gas, nitrogen at high pressure, and enriched hydrocarbon gases achieve miscibility with the reservoir oil. The high injection pressure and the specific composition that an enriched gas must have limit the number of prospective oil reservoirs in which a miscible flood can be used. Slug of fluids containing oil, water, surfactants, and co-solvents (e.g., alcohols) in various compositions – known as micellar polymer solutions – have also been found to be efficient miscible displacement agents, although they are not used commonly.

Perhaps the best advantage of N_2 flooding is that it can potentially be used anywhere in the world, if it can be cheaply extracted from the air, where other injection fluids are either not available or the cost of their delivery to the oil reservoir is prohibitive. On the other hand, since light hydrocarbons contain considerable combustion energies, their use as EOR agents is generally limited to remote locations, which makes the cost of delivery of the oil to commercial markets too high. In the United States, the main emphasis has been on miscible CO_2 flooding since it offers two main advantages over other gases. One is that CO_2 requires a relatively low operating pressure to attain miscibility with reservoir fluids. The second advantage is that CO_2 is relatively inexpensive as it has no value as a fuel, and is available in large amounts from natural deposits or as a waste product of industrial processes.

However, even if the gas that one wishes to use as a miscible displacement agent is economically available, its use is not without problems. Gases are normally less viscous than typical crude oil. The viscosity contrast between the oil and the injected gas, together with the phenomenon of *gravity segregation* (see below), render the miscible displacement much less efficient than desired. In addition, in most miscible displacement processes involving a gas the required temperature and pressure for miscibility of the oil and the displacement agent are often high enough that they limit the number of prospective reservoirs. For example, medium to heavy hydrocarbons become miscible with oil only at high temperatures and/or pressures (see, e.g., Sahimi et al., 1985; Sahimi and Taylor, 1991). Another negative aspect of a miscible displacement process is its cost. It may happen that a miscible displacement is more efficient in terms of the amount of the recovered oil than an immiscible injection, but the total cost of the miscible displacement (including the cost of transporting the displacing gas to the oil field from other locations) is so high that makes it economically unattractive. Moreover, flue gas and nitrogen have only limited application as agents of a miscible displacement in deep and high pressure reservoirs. For these reasons, EOR processes based on gas injection have not been as common as immiscible displacement processes, such as water flooding, although they are used relatively extensively.

8.2 FACTORS THAT AFFECT THE EFFICIENCY OF MISCIBLE DISPLACEMENTS

Miscible gas injection as an EOR process is influenced by several factors. Although these factors are well-known, there is still some disagreement on the extent of the influence of each individual factor. For example, while *laboratory* experiments indicate that the viscosity contrast between the oil and the displacing gas has a strong effect on the efficiency of the miscible displacements, the same may not be true at the *field scale* which is dominated by large-scale heterogeneities, such as the spatial distributions of the permeability and porosity and the presence of fractures and/or faults. In what follows, we describe the effect of some of the most important factors on the efficiency of a miscible displacement by a gas (or any other fluid, for that matter).

8.2.1 Mobility and Mobility Ratio

The mobility λ_i of a fluid i is defined as the ratio of the effective permeability K_i of the porous medium, experienced by fluid i, and the fluid's viscosity μ_i, $\lambda_i = K_i/\mu_i$. When one fluid displaces another, the mobility ratio M, defined as the ratio of the mobilities of the displacing and displaced fluids, is one of the most important influencing factors in any displacement process. Normally, M is not constant because mixing of the displacing gas and the oil changes the effective viscosities of the two fluids. In addition, the viscosity of the mixed zone also depends on concentrations of the displacing and displaced fluids. In many cases the viscosity μ_m of the mixed zone is estimated from the following empirical law due to Koval (1963)

$$\mu_m = \left(\frac{C_s}{\mu_s^{1/4}} + \frac{1 - C_s}{\mu_o^{1/4}} \right) \tag{8.1}$$

where C_s is the solvent concentration, and μ_s and μ_o are the viscosities of the solvent and oil, respectively. Sometimes, instead of C_s, the solvent volume fraction is used in Eq. (8.1). If, in addition to the solvent, another fluid, such as water, is also injected into the porous medium, as is often done in order to reduce the mobility of the solvent, then defining an effective value of M can become problematic. The mobility ratio when a solvent displaces oil at irreducible water saturation (the saturation at which the water phase becomes disconnected), with negligible mixing of the solvent and oil, is simply the ratio of oil and solvent viscosities, $M = \mu_o/\mu_s$ (assuming that the porous medium is homogeneous, though microscopically disordered, so that the permeabilities K_o and K_s experienced by the two fluids are essentially equal), and is always greater than unity. When $M > 1$, the mobility ratio is said to be *unfavorable*, since it may lead to formation of fingers which reduce the efficiency of a miscible displacement (see below). Defining an effective mobility ratio is more complicated and uncertain when mobile water is present.

In many miscible displacement processes, one must deal with more than one displacing front. For example, in tertiary oil recovery (which is carried out when, for example, water flooding is no longer effective) there are usually more than one displacing fronts. The problem of defining an effective M is even more complex in such situations, and no completely satisfactory method has been developed yet to address it. In such processes, motion of any particular front is affected not only by the mobility ratio across that front (i.e., the mobility ratio of the two fluids on the two sides of the front), but also by the mobilities of the other regions behind and ahead of the front and by their relative sizes.

8.2.2 Diffusion and Dispersion

Under certain conditions, mixing of the displacing and displaced fluids by diffusion can be important. For example, during both secondary and tertiary displacement of a reservoir's oil by CO_2, the development of multiple-contact miscibility strongly controls the ultimate recovery efficiency. At the microscopic scale, molecular diffusion is the mechanism by which molecular mixing of CO_2 and oil occurs. It is at this scale that the usual assumption of rapid local equilibrium is made, and is used for numerical simulation of CO_2 flooding. Similarly, during flooding of rich gases that contain hydrocarbons of intermediate molecular weight, miscibility with the in-place oil is developed by a multiple-contact condensing mechanism, during which diffusive mass transfer plays an important role. Moreover, a significant oil saturation may exist in the dead-end pores, or be traped by water films in a water-wet porous medium during a miscible displacement. Such isolated oil remains largely unrecoverable unless the gas injected into the porous medium can efficiently traverse the surrounding water barriers to contact and swell the trapped oil. The injected gas also penetrates the oil by molecular diffusion which in turn inhibits viscous fingering (since diffusion helps the fingers join together), delays premature gas breakthrough and, therefore, increases the oil production rate.

If two miscible fluids, with an initially sharp front separating them, are put in contact, their subsequent mixing caused by molecular diffusion is described by the diffusion equation:

$$\frac{\partial G_i}{\partial t} = -D_{io} A \frac{\partial C_i}{\partial x} \tag{8.2}$$

where G_i is the amount (in moles) of fluid i that has diffused across the front at time t, D_{io} is the effective diffusion coefficient of fluid i in the porous medium, A is the cross-sectional area for diffusion, and C_i is the molar concentration of i at position x at time t. The diffusion coefficient D_{io} depends, in principle, on the mixture's composition, but in reservoir simulations an average diffusion coefficient at 50% solvent concentration usually yields adequate representation of the diffusive mixing.

Many experimental methods have been developed for measurement of the effective diffusion coefficient D_{io} involving oil in porous media (see, for example, Reamer and Sage, 1958; Gavalas et al., 1968; Schmidt et al., 1982; Renner, 1988; Nguyen and Farouq-Ali, 1995). Experimental data for the effective diffusivity are still necessary because, despite several decades of research, no accurate theoretical method for estimating the effective diffusivities of mixtures in porous media is yet available (Sahimi, 1993a, 1995, 2003). Unfortunately, even the experimental measurements are generally difficult and very time consuming. Most conventional methods require composition analysis which is tedious and expensive (Moulu, 1989). Simpler methods of measuring the effective diffusion coefficients for gas-oil mixtures, which use PVT cells with no compositional analysis, have also been proposed (Riazi, 1996; Zhang et al., 2000).

Dispersion is mixing of two miscible fluids flowing in a system, such as a porous medium. Therefore, unlike diffusion, the flow velocity field plays an important role in this type of mixing process. Similar to diffusion, mixing by a dispersion process can decrease the viscosity and density contrasts between the displacing gas and the displaced fluids, which in most cases is very useful to the displacement process.

Two major mechanisms of dispersive mixing are small- and large-scale variations of fluid velocities (or, equivalently, the permeabilities), and molecular diffusion, both of which help mixing of the two miscible fluids. If, for example, a first-contact miscible solvent is injected into a linear packed-bed column to displace oil from it, the effluent concentration profile of the solvent will have an S-shape, which is the result of mixing of the solvent and oil in the packed-bed. Because of this, a transition zone of solvent/oil mixtures separates a zone of 100% solvent from one which is pure oil. This mixing, which is in the direction of the macroscopic flow, is called *longitudinal dispersion*.

Dispersion can also occur in the direction(s) perpendicular to the direction of macroscopic flow; this is referred to as the *transverse dispersion*, which occurs when, for example, a solvent is injected into a stratified porous medium which consists of layers of different permeabilities parallel to the macroscopic flow. In this case, the solvent in the more permeable layer moves faster, and mixes with the oil not only by longitudinal dispersion in the direction of flow, but also transverse (perpendicular) to the direction of flow in the less permeable layers. Dispersion processes in porous media are usually modeled based on the convective-diffusion (CD) equation:

$$\frac{\partial C}{\partial t} + v.\nabla C = D_L \frac{\partial^2 C}{\partial x^2} + D_T \nabla_T^2 C \tag{8.3}$$

where v is the macroscopic mean velocity (the macroscopic flow is assumed to be in the x direction), C is the mean concentration of the solute, and ∇_T^2 is the Laplacian in the transverse directions. Thus, the basic idea is to model the dispersion process as anisotropic diffusional spreading of the concentration, with the effective diffusivities being the longitudinal dispersion coefficient D_L and the transverse dispersion coefficient D_T.

The longitudinal dispersion coefficient D_L is usually larger than the transverse dispersion coefficient D_T by a factor that, depending on the morphology of the porous medium, ranges anywhere from 5 to 24. This large difference between D_L and D_T implies that more dispersive mixing takes place in the direction of the macroscopic flow. Since pore space heterogeneities strongly affect D_L and D_T, the implication is that the heterogeneities also affect miscible displacements. Unlike the effective diffusivities, the dispersion coefficients D_L and D_T depend on the mean flow velocity. In principle, the relation between the dispersion coefficients and the mean flow velocity depends on the value of the Peclet number Pe, defined as, $\mathrm{Pe} = \lambda v / D_m$ where λ is a characteristic length scale of the porous medium, and D_m is the molecular diffusivity. The relation between the dispersion coefficients and Pe can be non-linear (see, e.g., Sahimi, 1993a, 1995, for comprehensive discussions), but in numerical simulations one usually assumes that

$$D_L = \alpha_L v, \quad D_T = \alpha_T v \tag{8.4}$$

where α_L and α_T are, respectively, the longitudinal and transverse *dispersivities*. Roughly speaking, if the length scale of the observations or measurements is larger than the dispersivities, then the CD equation is applicable to describing the dispersion process (and, thus, miscible displacements). Dispersion is called *anomalous* if it cannot be described by the CD equation (Sahimi, 1993b).

Use of the dispersivities permits a phenomenological description of transport through porous media. However, a fundamental understanding of the phenomena is obtained only if the dispersivities can be related to the basic physical properties of porous media, such as their porosity, hydraulic conductivity and/or pore size distribution. To develop such relations, two approaches have been developed in the past. One approach constructs empirical correlations between the measured dispersivities and important morphological parameters of the porous medium. For example, Harleman and Rumer (1963) found, based on the analysis of their experimental data, that the longitudinal dispersivity is proportional to the square root of the mean hydraulic conductivity of the porous medium. However, attempts to verify this relationship under field conditions (Taylor et al., 1987), using the results of tracer tests, were not successful. The second approach is theoretical, using various models of heterogeneous porous media. Numerous attempts have been made in this direction (Sahimi, 1993a, 1995).

It is important to realize that values of the dispersion coefficients measured in the laboratory are usually *smaller* than the corresponding field-scale values, which presumably explains why the empirical relations between the dispersivities and the hydraulic conductivity, obtained based on laboratory experiments, fail under field conditions. Thus, one must devise an appropriate method for scale-up of the dispersion coefficients from the laboratory to the field scales (Johns et al., 2000a,b). Mixing by dispersion in a field-scale porous medium is likely to be much greater than in a laboratory-scale sample of the same porous medium, since large-scale variations in the permeability and porosity of the field-scale porous medium strongly increase the mixing process at that scale.

Mobility ratio and gravity also affect dispersion. If $M > 1$, viscous instability develops (see below) in which case the displacement is no longer a simple process. However, if $M < 1$, the usual dispersion mechanisms discussed above are operative. Moreover, since no instability develops for $M < 1$, the effect of pore space heterogeneities is also suppressed. On the other hand, if in a miscible displacement a less dense fluid displaces a denser fluid, the density difference suppresses the effect of dispersive mixing.

In some situations, longitudinal dispersion affects a miscible displacement more strongly than the transverse dispersion, and vice versa. For example, if large fingers of the displacing fluids develop, which is often the case when $M > 1$, or when there are large-scale variations in the permeabilities, there would be a large surface area on the sides of the fingers which allows for significant transverse dispersion to occur. This can help join the fingers, stabilize the displacement, and increase its efficiency. By contrast, longitudinal dispersion can only take place at the tip of the fingers and, therefore, its effect is much weaker than that of transverse dispersion. For this reason, models that ignore transverse dispersion are usually not completely adequate for describing a miscible displacement.

8.2.3 Anisotropy of Porous Media

As mentioned above, dispersion processes are sensitive to the distribution of the heterogeneities of a porous medium, including its stratification, which in turn affects the performance of miscible displacements. This is particularly true when the displacing agent is a gas. It is clear that the displacing gas preferentially chooses the strata with higher permeabilities. As a result, large amounts of the fluid (oil) to be displaced can be left behind in the strata with low permeabilities. If we attempt to displace this fluid by injecting larger amounts of gas into the low permeability strata, some of the gas will inevitably enter the high permeability strata and do essentially nothing, as such strata have already been swept by the previously-injected gas and, therefore, additional gas injection would not be very efficient. The effect of stratification is even stronger when the mobility ratio $M > 1$. Another phenomenon that affects miscible displacements in stratified media is the crossflow of displacing and displaced fluids between the strata. Depending on the direction of the displacement process, crossflows can help or hinder the efficiency of the process.

8.3 THE PHENOMENON OF FINGERING

If the injected gas and the fluids that it displaces in the porous medium are first-contact miscible, and if the mobility ratio $M < 1$, then the displacement process is very simple and efficient. The displaced fluids move ahead of the displacing fluid, and the displacement front is stable and, aside from small perturbations, is relatively smooth. There is also a mixed zone between regions of pure displacing and displaced fluids. However, in practice a miscible displacement process is not so simple. Since typically $M > 1$, the front is unstable and many fingers of the mixture of the gas and the displaced fluid develop, leaving behind large amounts of oil. The formation of the fingers, which have very irregular shapes, reduces strongly the efficiency of the miscible displacements. Figure 8.1 shows the effect of the mobility ratio M on the formation and shape of the fingers. This phenomenon is usually referred to as *viscous fingering*. The terminology is an accurate representation of the phenomenon in macroscopically-homogeneous porous media, for which the mobility ratio reduces to the ratio of the viscosities of the displaced and displacing fluids. Under field conditions, fingering may be dominated by the distribution of the heterogeneities of the porous formation and, therefore, one should simply refer to the phenomenon as fingering.

Due to its significance, fingering has been studied for a long time. Several older reviews discussed this subject, among which are those of Wooding and Morel-Seytoux (1976), Homsy (1987), and Sahimi (1993a, 1995). Many of the initial studies of fingering, in flow of both miscible and immiscible fluids, were carried out in a Hele-Shaw cell which is an essentially two-dimensional system confined between two flat plates of length L that are separated by a small distance b. Flow in this cell can

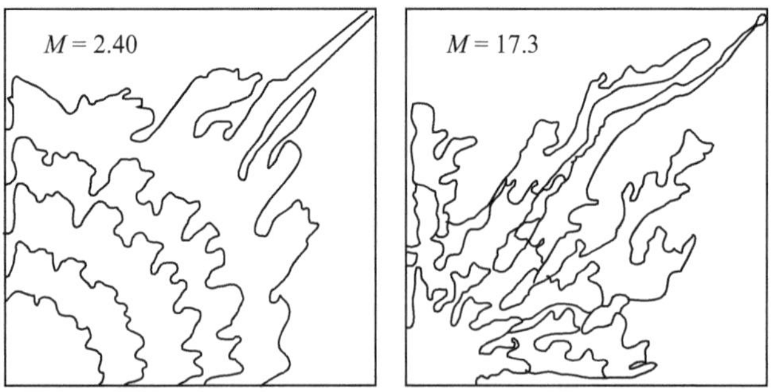

Figure 8.1. Displacement fronts for two values of the mobility ratio M. The injection point is the lower left corner (after Habermann, 1960)

be rectilinear, that is, the displacing fluid is injected into the system at one face of the cell and produced at the opposite face, or it can be radial in which the fluid is injected into the system at its center, and the system acts as a cylinder of a large radius and small height. As long as dispersive mixing is absent, the analogy between formation of fingers in Hele-Shaw cells and a two-dimensional porous medium is valid, which explains why the study of miscible displacements in Hele-Shaw cells has been popular. If dispersion is present, the analogy breaks down because transverse dispersion is always present in a porous medium, whereas a Hele-Shaw cell with its thin gap between the two plates cannot support significant transverse dispersion in the third direction.

There have also been many experimental studies of fingering phenomena, both in miscible and immiscible displacements. Some of these include those of Blackwell et al. (1959), Benham and Olson (1963), Slobod and Thomas (1963), Greenkorn et al. (1965), Kyle and Perrine (1965), Perkins et al.(1965), Mahaffey et al. (1966), Perkins and Johnston (1969), and more recently, those of Paterson (1981, 1983, 1985), Paterson et al. (1982), Chen and Wilkinson (1985), Lenormand et al. (1988), and Bacri et al. (1991). Most of these experiments were also accompanied by a linear stability analysis (see Section 8.9).

8.4 FACTORS AFFECTING GAS FINGERING

Let us point out that, in addition to the various factors that affect fingering, which will be described below, three types of interesting phenomena occur during fingering which are worth describing here. The first one is *tip splitting* in which the tip of a finger becomes unstable and splits into two branches that compete with each other for further growth. The second phenomenon is *spreading*, which usually occurs in a porous medium when the flow velocity is not large. Under this condition, transverse dispersion causes *lateral* spreading of the fingers which helps them to join and, therefore, increase the efficiency of the displacement process. However, due to the spreading phenomenon the tips of the fingers can become unstable since their typical breadth exceeds the cutoff length scale for stability (see Section 8.9), which is also set by transverse dispersion. Thus, tip splitting occurs only if the Peclet number Pe exceeds a critical value which depends on the mobility ratio M, the density difference between the displacing and displaced fluids, the flow velocity, and the effective permeability of the porous medium. Finally, one may have *shielding* during fingering, which is when one finger spreads out and grows much faster than other fingers and, therefore, shields them. Figure 8.2 shows the phenomena of tip splitting and finger shielding.

Generally speaking, the fingers' patterns are influenced by the fluids' characteristics, including the possible non-monotonicity in their viscosity profile (Manickam and Homsy, 1993), and the heterogeneity of the porous medium manifested by the variance and the correlation structure of the permeability field (Waggoner et al., 1992; Araktingi and Orr, 1993; Sorbie et al., 1994). Fingering is mitigated by small scale dispersion (at the pore and laboratory scales) and by large-scale heterogeneity effects

Figure 8.2. Tip splitting (top) and shielding (bottom) in viscous fingering

(large-scale permeability variations and channeling). In what follows, we describe the effect of some these factors on fingering phenomena.

8.4.1 Displacement Rate

Compared to immiscible displacements, miscible displacements are much less sensitive to the displacement rate. In immiscible displacements at high flow rates smaller and more numerous fingers are formed than at low rates. In contrast, fingers that emerge during miscible displacements are only mildly sensitive to the displacement rate. This is due to the fundamental role that dispersive mixing plays which help the smaller fingers to merge.

The mild sensitivity of the fingers' patterns during unstable miscible displacements to the displacement rate has practical implications. To control the stability of an immiscible displacement it is often enough to control the displacement rate. This is clearly not the case with miscible displacements. While it is possible to obtain a stable displacement by using a rate less than a critical rate, in most practical situations this would imply using a rate which would not be economical.

8.4.2 Heterogeneity Characteristics

As mentioned earlier, one factor that plays a fundamental role in finger formation is the heterogeneity of the porous medium. Once a finger starts to grown, its subsequent growth is closely linked to its interaction with the heterogeneity and, in particular, to the spatial variations of the porosity and permeability. Permeability variations have been found to play an important role in finger initiation and growth (Stalkup, 1983; Moissis et al., 1987). The spatial variation of the permeability is usually described by two parameters. One is the coefficient of permeability variation C_k defined by

$$C_k = \frac{\sigma_k}{k_m} \tag{8.5}$$

where σ_k is the standard deviations of the permeability distribution, and k_m is its mean. The second characteristic quantity is a permeability correlation length ξ_k, which is the length scale over which the permeabilities are correlated.

Moissis et al. (1987) found that the permeability distribution near the inflow end of their two-dimensional model determines the number of the fingers, their initial locations, and the relative growth rates. The locations where the fingers form are controlled by the maxima in the permeability distribution near the inflow end. The initial number of the fingers depends strongly on the correlation length of the permeability distribution. Highly-correlated porous media have fewer maxima in their permeability distribution and, consequently, fewer fingers form in such porous media. As mentioned above, some of the initial fingers grow faster than the rest and eventually dominate the displacement. Due to the shielding effect, the long fingers suppress the growth of small fingers. The smaller fingers may merge later on with the larger ones, leaving upswept areas that may be fairly extensive. This process results in a number of large fingers which grow quite independently of each other, at least until the breakthrough, and are referred to as the *active fingers*. The number of the initial fingers, that of the active fingers, and the time of their formation all depend on C_k, ξ_k, and the viscosity ratio.

The number of the active fingers is a decreasing function of the permeability correlation length ξ_k since, as discussed above, the initial number of the fingers is smaller for larger ξ_k. Moreover, the effect of downstream permeability distribution (see below) is more significant for large ξ_k, resulting in more merger of the fingers, which reduces their number. For a given value of C_k, the growth rate of the fingers increases with ξ_k up to a limiting value. For large values of ξ_k, the permeability variation tends to generate fingers of large wavelengths which grow relatively slowly. However, once these fingers grow beyond the zone in which their growth is approximately described by the linear stability analysis (see Section 8.9), it is easier for the displacing gas to develop flow channels which accelerates the subsequent growth rate of the fingers.

The number of active fingers is also a decreasing function of the coefficient of permeability variations C_k. As C_k increases, the difference in permeability between high- and low-permeability regions increases. In the initial stages of the displacement, this difference favors the growth of the longest fingers, which tend to grow in the high-permeability regions. Thus, for highly heterogeneous porous media, the longest fingers grow much faster than the rest and dominate the displacement relatively early, shielding the smaller fingers. Later on during the displacement, the large difference in permeabilities from region to region facilitates merging of the fingers. The net result of these two mechanisms is the reduction of the number of active fingers. Since almost all the displacing fluid flows through one or at most a few fingers, breakthrough occurs early and, therefore, the sweep efficiency is poor.

The effect of downstream permeability variations on the finger formation and growth is negligible for random (uncorrelated) porous media, but it becomes increasingly more important as the correlation length ξ_k increases. This effect is more pronounced for small viscosity ratios. The cause of this behavior can be traced to the scale of the permeability variations. For a finger to be significantly affected by the permeability variations, the length scale over which the variations are significant must be of the same order of magnitude as the finger's width or larger. In porous media with a correlated permeability distribution, the existence of high- and low-permeability regions causes flow channeling and, thus, affects finger growth mainly by causing additional merging of the fingers. For uncorrelated porous medium, on the other hand, the fingers' width is always larger than the length scale over which the permeability variations are significant.

8.4.3 Viscosity Ratio

As discussed above, one of the main characteristics of finger growth is that the large initial number of fingers is reduced by their merging and formation of a smaller number of active fingers. The effect of the viscosity ratio on this process is important: At low viscosity ratios, there are no fingers that grow dramatically faster than the rest. Consequently, finger merging and growth suppression occur to a much lesser extent, resulting in a larger number of active fingers with growth rates that do not vary much. At high viscosity ratios, on the other hand, the active fingers start to outgrow the others earlier, and more finger merging and suppression of growth occur. This results in a smaller number of long active fingers. Note also that, as the viscosity ratio increases the fingers become more unstable, their shapes become more irregular, and exhibit some tip splitting.

8.4.4 Dispersion

Strong longitudinal dispersion helps formation of a thick (more diffused) displacement front, hence making the fingers less susceptible to flow disturbances that are caused by small-scale heterogeneities. Moreover, since the front is more diffused, the waves that correspond to small values of the concentration grow faster and, therefore, the breakthrough occurs earlier. While it may appear that the fingers should grow faster when longitudinal dispersion is larger, it is in fact difficult to define finger lengths or growth rates in this case. The effect of transverse dispersion on the fingers was already described. We will come back to the effect of dispersion later in this chapter, when we discuss quantitative modeling of fingers.

8.4.5 Aspect Ratio and Boundary Conditions

We define the aspect ratio $\Re_0$ of a porous medium as the ratio of its length and width, $\Re_0 = L_x/L_y$. If $\Re_0$ increases (narrower porous media, somewhat like a slim tube commonly used in laboratory studies of CO_2 injection), the initial fingers are closer to one another and, consequently, their interaction is stronger. This results in suppression of growth of the smaller fingers and their merging at the initial stages of the displacement, hence yielding, even at early times, a small number of active fingers. At very large aspect ratios, only one active finger will form.

The effect of the geometry of a porous medium on fingering phenomena has received considerable attention (see, e.g., Zimmerman and Homsy, 1991; Waggoner et al., 1992; Sorbie et al., 1994). Typically, simulations of unstable miscible displacements at field scales are carried out in geometries that have an aspect ratio of about 3 (Christie, 1989). Waggoner et al. (1992) simulated displacements at conditions of transverse (or vertical) equilibrium. This limit is reached when the generalized aspect ratio, $\Re = (L_x/L_y)(k_v/k_h)^{1/2}$, is large, where k_h and k_v are the horizontal and vertical permeabilities. Sorbie et al. (1994) studied the sensitivity of the displacement patterns to this parameter in heterogeneous reservoirs, and showed that it significantly affects the delineation in the parameter space of the various displacement regimes (fingering, dispersion, and channeling).

In another study, Yang and Yortsos (1995, 1996) provided an asymptotic description of the displacements in porous media, including formation of the fingering, in the limits that the parameter $\Re$ is large or small. The case of large $\Re$ corresponds to conditions of transverse equilibrium. This regime is reached in long and narrow isotropic reservoirs, in those in which the permeability transverse to the applied pressure gradient exceeds significantly the permeability parallel to it, and in slim tubes. It is a regime in which intense transverse mixing occurs. Small $\Re$ corresponds to the opposite regime of zero transverse mixing and is better known as the Dykstra-Parsons approximation (Dykstra and Parsons, 1950).

In parallel, Yang (1995) reported on the sensitivity of viscous fingering to $\Re$ by means of high-resolution simulations. He reported that for uncorrelated, weakly heterogeneous porous media at conditions near transverse equilibria, most of the viscous fingering ultimately occurs near the lateral, no-flow boundaries of the system. More specifically, he found that narrow, single fingers originate at these boundaries and propagate faster than the fingers in the interior of the domain, until a small permeability value was randomly encountered, at which point the fingers turned inwards. The intensity of this effect was found to depend on the viscosity ratio, and on the heterogeneity parameter. Yang and Yortsos (1998, 2002) showed that this effect is not a numerical artifact, but arises as a consequence of the slip boundary condition implied by the use of Darcy's law along no-flow boundaries. They found that the origin of the boundary effect is the vanishing of the transverse, but not of the streamwise, velocity component at the no-flow boundary. When $\Re$ is small (for example, when $k_v < k_h$), transverse mixing is minimal everywhere (including the boundary), and so is the boundary effect. By contrast, at large $\Re$ (for example, when $k_v > k_h$), transverse mixing is intense everywhere, except at the no-flow boundary. Therefore, the growth of all the fingers, except the one at the boundary, is mitigated.

8.5 GRAVITY SEGREGATION

An important factor that influences vertical sweeps in miscible displacements is the gravity. Solvents are usually less dense than either oil or brine, and drive gases, such as hydrocarbons or flue gas, are even less dense. Because of the density differences, solvents and drive gases may segregate and override the other reservoir fluids, which in turn decrease the vertical sweep in horizontal floods. On the other hand, gravity can be used to advantage in dipping reservoirs to improve the sweep and the displacement efficiency.

To study the effect of gravity on miscible displacements, two dimensionless numbers are introduced:

$$\eta = (\rho_o - \rho_s)/\rho_s \tag{8.6}$$

$$N_G = \frac{gk(\rho_o - \rho_s)}{q\mu_o} \tag{8.7}$$

where N_G is the gravity number. If N_G is small, gravity is unimportant and viscous fingering dominates the flooding behavior (see above). As N_G increases, viscous fingering can still occur but gravity influences the growth rates of the individual fingers. If the solvent is lighter than oil, it tends to flow upwards. Hence, more solvent enters the fingers in the upper part of the porous medium, resulting in faster growth of these fingers, while the growth of the rest of the fingers is somewhat suppressed, partly due to gravity drainage and partly because of the shielding effect described earlier. This phenomenon results in early breakthrough and reduced sweep efficiency. If, however, the solvent is heavier than oil, it is conceivable that, for a range of N_G, gravity might improve the sweep efficiency by delaying the breakthrough of the fastest-growing finger.

If N_G is large enough, a gravity tongue is formed and grows at the top of the porous medium. Moreover, viscous fingering may occurs near the tongue, while fingering in the rest of the medium will be suppressed. Under this condition, both gravity override and viscous fingering are important and affect the displacement. For very large values of N_G, gravity override completely dominates the displacement, suppressing any fingers that may form at early times. In this case, the gravity tongue breaks through very early, and the rate of oil recovery will be very low. Therefore, it should be clear that as N_G increases, the breakthrough time decreases, while the rate of oil recovery after the breakthrough also decreases (Moissis et al., 1989).

The effect of gravity is more pronounced at high viscosity ratios, because the ratio of gravitational and viscous forces is inversely proportional to the viscosity of the fluid currently present in the porous medium. This ratio is equal to N_G only at time $t = 0$. As the displacement proceeds and the lower viscosity fluid (the displacing gas) enters the porous medium at a constant rate, the ratio of gravitational and viscous forces increases and, therefore, gravity becomes progressively more important.

8.6 AVERAGED CONTINUUM MODELS OF MISCIBLE DISPLACEMENTS

One approach to describing miscible displacements and fingering phenomena in a porous medium is based on averaged continuum equations. This method can describe any instability that is smooth on the length scale over which of the continuum description is applicable. In the strict absence of dispersion (i.e., the limit that the Peclet number Pe $\rightarrow \infty$, where Pe is based on D_L or D_T and not D_m), fingers form at all length scales, with growth rates that increase with decreasing length scale. This means that a length scale is reached at which a continuum description is no longer appropriate, and one must develop a pore-scale model. Such models will be described below. In this case, the initial-value problem that describes the phenomenon is ill-posed, but one can seek solutions that contain discontinuities. Since dispersion is absent in this case, there will be a step jump in the viscosity profiles (from the gas or the displacing fluid to the displaced fluid). As a result, the pressure obeys the Laplace equation, and the pressure and fluid fluxes are continuous across the front separating the displaced and displacing fluids. One may have all types of singularities in the solution, with different non-uniformities appearing in different boundary-value problems.

Two popular one-dimensional and semi-empirical continuum models of miscible displacements are those due to Koval (1963) and Todd and Longstaff (1972). Koval recognized that the central feature of the physics of viscous fingers is linear growth of the fingers' length with time. Thus, to ensure that this is a feature of his model, Koval cast the problem as a hyperbolic transport equation similar to the more familiar Buckley-Leverett equation of two-phase flow. In Koval's model the displacing fluid is assumed to travel at a constant, characteristic velocity v. Koval made an analogy between miscible and immiscible displacements. For an immiscible displacement of oil by water, the Buckley-Leverett equation, when gravity and the capillary pressure are negligible, relates f_w, the fractional flow of water, to the permeabilities k_w and k_o of the water and oil phases. Koval argued that permeability to either oil or the displacing fluid can be expressed as the total permeability, k, multiplied by the average saturation of each fluid. Thus, if viscous fingering is the dominant phenomenon, one can write

$$f_s = \frac{1}{1 + \frac{1}{H}\frac{\mu_{es}}{\mu_{eo}}\frac{1-S}{S}} \tag{8.8}$$

where μ_{es} and μ_{eo} are the *effective* viscosities of the solvent and oil (the displaced fluid), respectively, and H is called the *heterogeneity index* that characterizes the inhomogeneity of a porous medium. To correlate H with some measurable quantity, Koval defined a homogeneous porous medium as one in which the oil recovery, after 1 pore volume of the solvent has been injected into the medium, is 99%. Thus, for a homogeneous porous medium, $H = 1$, while any other porous medium for which the recovery is less than 99% is characterized by $H > 1$. Empirically, H and the recovery appear to be linearly related. Based on experimental data, Koval (1963) also suggested the following expression,

$$\frac{\mu_{eo}}{\mu_{es}} = \left[0.78 + 0.22\left(\frac{\mu_o}{\mu_s}\right)^{1/4}\right]^4 \tag{8.9}$$

which is similar to Eq. (8.1).

The predictions of Koval's model are in good agreement with the experimental data of Blackwell et al. (1959). Despite its success, Koval's model does suffer from two fundamental shortcomings. One is the empirical nature of Eq. (8.9). The model's second shortcoming is the inadequacy of H for describing the heterogeneity of a field-scale porous medium which is characterized by large-scale variations in the permeabilities and long-range correlations in their distribution.

The second model is due to Todd and Longstaff (1972). In their model the average concentration $\overline{C}_s$ of the solvent is described by

$$\frac{\partial \overline{C}_s}{\partial t_D} + \frac{\partial \bar{f}_s}{\partial x_D} = 0 \tag{8.10}$$

where $\bar{f}_s$ represents the average of f_s and is a function of $\overline{C}_s$, and x_D and t_D are dimensionless distance and time, respectively. This equation is, of course, a limiting case of a convective-diffusion equation in which dispersion has been neglected. Todd and Longstaff assumed that

$$\mu_{eo} = \mu_o^{1-\varsigma} \mu_m^{\varsigma} \tag{8.11}$$

with a similar expression for μ_{es},where μ_m is given by Eq. (8.1) in which $\overline{C}_s$, instead of, C_s is used, and $\varsigma \cong 2/3$. $\bar{f}_s$ is given by

$$\bar{f}_s = \frac{\overline{C}_s}{\overline{C}_s + M_e^{-1}\left(1 - \overline{C}_s\right)} \tag{8.12}$$

One finds that the average concentration $\overline{C}_s$ moves with a speed $d\bar{f}_s/d\overline{C}_s$. Therefore, the leading edge of the finger, where $\overline{C}_s = 0$, moves at a speed $d\bar{f}_s/d\overline{C}_s = \mu_{eo}/\mu_{es}$, whereas the trailing edge, where $\overline{C}_s = 0$, moves at a speed $d\bar{f}_s/d\overline{C}_s = \mu_{es}/\mu_{eo}$. Despite some success, the model of Todd and Longstaff (1972) suffers from the same shortcomings as those of Koval's model. However, because of their simplicity, and despite their shortcomings, these two models have been used heavily in the petroleum industry.

More recent models, which are more or less based on the same type of averaged continuum equations, are those of Vossoughi et al. (1984), Newley (1987), Fayers (1988), Odeh and Cohen (1989), and Fayers et al. (1990). Although these models may give an adequate fit to the production/effluent data in an (either one- or two-dimensional) experiment, they often lead to quite different predictions of the pressure field during the unstable displacement process. In the evaluation of the averaged models of viscous fingering, their performance in the two-dimensional model, where viscous instability occurs, has been considered (Newley, 1987; Fayers et al., 1990). To distinguish between such models experimentally, it is necessary to measure the pressure field directly during unstable miscible displacements, an extremely difficult task to achieve for either two-dimensional linear or other geometries (see, however, Sorbie et al., 1995). Alternatively, the computed pressure distribution may be used as "experimental" data for assessing the predictions of averaged equations. However, such calculations should also be validated by comparison with experimental results.

8.7 NUMERICAL SIMULATION OF MISCIBLE DISPLACEMENTS

Numerical simulations of unstable miscible displacements have been carried out for a long time. The main difficulty in this problem lies in the correct reproduction of the wide range of the relevant length and time scales that typically characterize these phenomena. At the smallest length scale, the size of the smallest grid block is determined by the action of physical diffusion or dispersion, while the reservoir linear size, or the distance between the wells, determines the large scales.

In a pioneering work, Peaceman and Rachford (1962) developed a finite-difference algorithm for computing an unstable rectilinear miscible displacement. Due to the low order of their numerical algorithm, a fingering instability did not develop on its own, but had to be triggered artificially by imposing small random spatial variations in the permeability. Over the four decades following this early work, a host of novel numerical approaches for the simulation of miscible displacements has been introduced and tested on problems of varying degrees of difficulty. In what follows, we briefly discuss some of these methods.

8.7.1 Finite-Element Methods

For a comprehensive discussion of the efforts up until 1983, see the review by Russell and Wheeler (1983). Douglas et al. (1984) presented the results of computer simulations of unstable miscible displacements for one quarter of the five-spot geometry. They used a self-adaptive finite-element (FE) method; see also Bell et al. (1985). Darlow et al. (1984) described the so-called mixed FE methods, which offer certain computational advantages, in particular for strongly heterogeneous porous media, by solving for pressure and velocity *simultaneously*. Ewing et al. (1984) analyzed a modified method of characteristics for handling the governing equation for the concentration.

Ewing et al. (1989) performed more detailed FE simulations of miscible displacements in anisotropic and heterogeneous porous media in one quarter of five-spot geometry, using computational grids of up to size 100×100. They typically observed dominance of the viscous fingering instability over the permeability-related effects. In general, the relative importance of these effects is expected to depend on the viscosity ratio and the degree of heterogeneity of the porous medium. Comprehensive reviews of the above methods, as well as related work (Ewing and Wang, 1994) are provided by Ewing et al. (1994), along with a discussion of these methods in light of recent developments in computer architectures; see also Arbogast et al. (1996).

8.7.2 Finite-Difference Methods

Conventional finite-difference (FD) discretizations have also been used to obtain improved physical understanding of miscible displacements; see, for example, Christie and Bond (1987), Christie (1989), Bradtvedt et al. (1992), Fayers et al. (1992), and Christie et al. (1993), as well as Sorbie et al. (1995) and Zhang et al. (1996). The overall accuracy of these methods are typically of second order, but often they still suffer from significant amount of numerical dispersion. In terms of formal accuracy, an important seminal work was that of Leventhal (1980), who applied the fourth-order Operator Compact Implicit (OCI) family of FD schemes (see, e.g., Berger et al., 1980; Morton, 1996) to one-dimensional, two-phase immiscible waterflood problems, and demonstrated a significant reduction of the adverse effect of numerical diffusion. Tchelepi and Orr (1993a,b) minimized the amount of numerical diffusion by employing a particle tracking technique (Araktingi and Orr, 1990; see below), although the solution of the accompanying pressure equation still resulted in some artificial smoothing.

8.7.3 Streamline Method

In this approach, the flow problem is decoupled into a set of 1D problems, solved along streamlines, which reduces the simulation time and suppresses the numerical dispersion. The method has also been used as a scale-up technique. The obvious advantages of the streamline method have increased its wide application and fast commercialization. However, compared with the conventional FD simulations, the streamline-based simulations usually account only for relatively simple physics. These simulators are limited to production scenarios where the effect of gravity can be neglected. Given that compositional simulations are much more time consuming, it is of great interest to speed up the compositional simulators using a streamline method.

Thiele et al. (1995) proposed a compositional simulation approach to investigate compositional displacements in two-dimensional heterogeneous reservoirs. Thiele et al. (1997) combined a three-dimensional streamline method with a one-dimensional two-phase numerical solver and developed a streamline-based three-dimensional compositional reservoir simulator. Jessen and Orr (2002) developed a three-dimensional compositional reservoir simulator based on the analytical mapping approach. By mapping the analytical solution to one-dimensional two-phase multicomponent gas injection along each streamline, the simulator can be orders of magnitude faster and completely free of numerical dispersion. A limitation of the analytical mapping method is, however, that it is only applicable to uniform initial condition.

More recently, Yan et al. (2004) developed a three-phase compositional streamline simulator by integrating a 1D numerical solver. The numerical solver is optimized for calculating the efficiency, and various descriptions of phase equilibrium between hydrocarbon and water, as well as the gravity effect, are included by the operator-splitting method. Operator splitting (Batycky et al., 1996; Bradtvedt et al., 1996;

Batycky, 1997) relies on the consistency of treating the convective flux independently of gravity flux within a given time step of the simulation. For small time steps, the operator-splitting approximation is fairly accurate, whereas large time steps may lead to significant errors. Jessen and Orr (2004) used the operator-splitting technique to develop a compositional simulator suitable for displacement processes with significant gravity segregation.

8.7.4 Spectral Methods

To overcome the limitation in the accuracy of many of the above grid-based methods, significantly more accurate spectral methods have also been developed (see, e.g., Tan and Homsy, 1988; Zimmerman and Homsy, 1992a,b; Rogerson and Meiburg, 1993a,b). These methods possess superior accuracy for smooth flows, that is, for flows without discontinuities and singularities (Gottlieb and Orszag, 1977), and avoid problems associated with numerical diffusion. In addition, the spectral simulations are carried out for the governing equations that are formulated in terms of the vorticity and the stream functions. Since one no longer needs to solve for the pressure distribution, such formulation of the governing equations often leads to higher computational efficiency. Moreover, this formulation offers the advantage that it satisfies the conservation of mass identically. Thus, these methods allow for detailed investigation of mobility- and gravity-driven fingering processes in rectilinear displacements at relatively low levels of physical dispersion.

8.8 STOCHASTIC MODELS OF MISCIBLE DISPLACEMENTS

In addition to the above methods, several models of miscible displacement processes in porous media, for both the laboratory and field-scale porous media, have been developed in which probabilistic concepts have been utilized. Many of such models (DeGregoria, 1985; Paterson, 1984; Sherwood and Nittman, 1986; Siddiqui and Sahimi, 1990a,b) are applicable when the effect of dispersion can be neglected. We describe briefly two stochastic models that do take into account the effect of dispersion.

The first model that we describe is due to King and Scher (1987, 1990). Consider first the case of miscible displacements *without* dispersion. For point injection of fluids the governing equations are

$$\frac{\partial C}{\partial t} + u \cdot \nabla C = \delta^2(x) \tag{8.13}$$

$$u = \frac{v}{\hat{Q}} = \nabla \times (\psi \hat{z}) \tag{8.14}$$

where $\hat{Q}$ is the injection rate (volume per unit thickness per unit time), ψ is the stream function, and $\delta^2(x)$ is the two-dimensional Dirac delta function. The injected volume of fluid provides a natural time variable

$$\tau = \frac{\int_0^t \hat{Q}(t')dt'}{\psi} \tag{8.15}$$

In the absence of dispersion the solution is simple: A concentration bank $C = 1$ (i.e., the pure fluid) displacing $C = 0$. In general, however, the concentrations need not form a bank, since dispersion intervenes and develops a mixed zone. To develop a probabilistic model that takes this effect into account, King and Scher (1987, 1990) interpreted $(\partial C/\partial\tau)d^2x$ as a *two-dimensional* probability density function for concentration evolution. According to Eqs. (8.13) and (8.14) we can write

$$\frac{\partial C}{\partial \tau}d^2x = -\frac{\partial C}{\partial\xi_1\partial\xi_2}d\xi_1 d\xi_2 = -dCd\psi \tag{8.16}$$

where ξ_1 and ξ_2 are local tangential and normal coordinates on the front. Equation (8.16) can now be given a probabilistic interpretation: The probability of concentration evolution at x [i.e., $(\partial C/\partial\tau)d^2x$] is the product of the probability $d\psi$ of fluid flow through x and the probability dC of a concentration gradient moving through x. For a given *realization*, one samples the cumulants C and ψ, that is, determines the flux contour $C = r_1$ and the streamline $\psi = r_2$, and calculates their intersection in the plane, which is also the point at which concentration is modified, where r_1 and r_2 are two random numbers uniformly distributed in (0,1).

In such a simulation, one must employ a probabilistic interpretation of the FD version of Eq. (8.13). We integrate this equation over a rectangular spatial region A_{ij} and time interval $\Delta\tau$ to obtain ΔC, the change in the average concentration C_{ij}. This is given by

$$\delta C_{ij} = -\frac{\Delta\tau}{\Delta x\Delta y}\oint_{\partial A_{ij}} \overline{C}d\psi \tag{8.17}$$

Obviously, if $\delta C_{ij}/\Delta C$ is properly normalized, then it can be interpreted as the growth probability at site ij (i.e., as the probability that the displacing fluid and the front between the two fluids advance). One now has to fix ΔC. According to Eq. (8.17), the growth probability is non-zero only when the boundary integral overlaps the edge of the cluster of the displacing fluid mixed with the displaced fluid. This method has the advantage that finite values of the viscosity ratio can be used in the simulations.

We can now add the effect of dispersion. Consider first the static case, $\mathrm{v} = 0$. The evolution equation is simply the diffusion equation, $\partial C/\partial T - D_L\nabla^2 C = 0$, where $T = D_L t$ (King and Scher assumed that, $D_L = D_T$, which is, however, not justified). In discrete form,

$$\delta C_{ij} = \frac{\Delta T}{\Delta x\Delta y}\oint_{\partial A_{ij}} \hat{n}.\nabla\overline{C}dl \tag{8.18}$$

which should be compared with Eq. (8.17). If the time step ΔT satisfies $\Delta T/(\Delta x \Delta y) \leq \Delta C$, then

$$\frac{\delta C_{ij}}{\Delta C} = -\sum_{faces} \oint_{\partial A_{ij}} \left(\frac{\Delta T}{\Delta x \Delta y \Delta C}\right)(-\hat{n}.\nabla \overline{C})\, dl \tag{8.19}$$

For the full problem (convection and dispersion) we can split the time evolution as

$$\phi \frac{\partial C}{\partial t} = \hat{Q}\frac{\partial C}{\partial \tau} + \phi D_L \frac{\partial C}{\partial T} \tag{8.20}$$

In this equation $\partial/\partial t$ and $\partial/\partial T$ represent separate convection and dispersion processes. Convective evolution is described by Eq. (8.18), while dispersion evolution is represented by Eq. (8.20). Thus, we have a sequential finger evolution in which convection initiates the growth, which is then modified or moderated by dispersion. If we fix the time interval ∂t, then, $\partial \tau = (\hat{Q}/\psi)\delta t$, and $\partial T = D_L \delta t$, which then define the Peclet number, $\mathrm{Pe} = \delta\tau/\delta T$. In practice, $\delta\tau$ is set by, $\delta\tau = \Delta\tau = \Delta x \Delta y \Delta C$, implying that $\Delta T = \Delta\tau/\mathrm{Pe}$. In most of practical cases, $\mathrm{Pe} > 1$, and Eq. (8.20) is properly normalized. However, if $\mathrm{Pe} < 1$, δT is subdivided into n_D intervals to obtain $\Delta T = (\delta\tau/\mathrm{Pe})n_D$, where $n_D > \mathrm{Pe}^{-1}$. King and Scher (1990) simulated miscible displacements, with dispersion effect included, for various values of the viscosity ratio and obtained reasonable agreement with the experimental data.

The second probabilistic model is due to Araktingi and Orr (1990). In their model the porous medium is represented by a three- or two-dimensional grid of cubic (square) blocks. At the beginning of each time step the pressure field is calculated, given the distribution of the permeability and the current distribution of fluid viscosities. Tracer particles that carry a finite concentration of displacing fluid are injected into the system and are moved with velocities based on the pressure field. The velocities are calculated at the midpoint between grid nodes. Velocities for particles that are not on such nodes are obtained by linear interpolation. After moving the particles by convection to their current positions, the effect of dispersion is simulated by random perturbations of particle positions in the longitudinal and transverse directions. Since for diffusive dispersion the standard deviations of the particles' motion are given by, $\sigma_x = \sqrt{2D_L t}$ and $\sigma_y = \sqrt{2D_T t}$, the distribution of the particles about a mean position can be simulated by multiplying these standard deviations by a number between -6 and $+6$. This number is obtained by adding a sequence of 12 random numbers, distributed normally with a zero mean and unit standard deviation, to -6. The values $+6$ and -6 are used because, on a practical basis, the probability of a particle moving beyond 6 standard deviations on either side of the mean is less than 1%. After the particles arrive at their new position, the current time step is determined. To avoid having particles travel a distance greater than a grid block, the time step is chosen to allow movement equal to half of a grid block length (or width), traveled at the highest existing velocity. The new pressure field is calculated, and a new position for each particle is determined. This procedure is repeated many times.

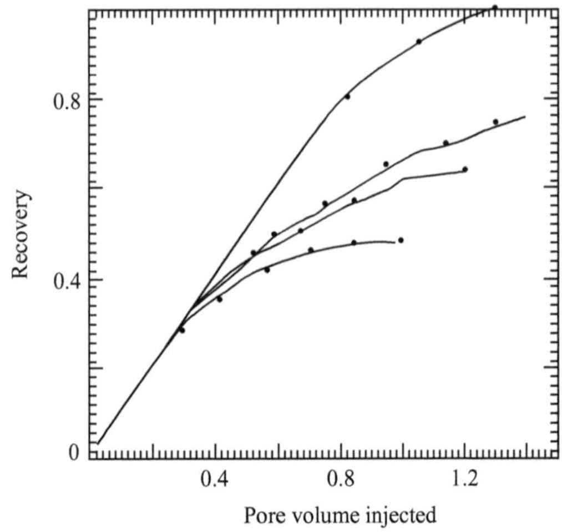

Figure 8.3. Comparison of the simulation results of Araktingi and Orr (1990) (curves) with the experimental data (symbols) of Blackwell et al. (1959). The results are, from top to bottom, for $M = 5$, 86, 150, and 375

Araktingi and Orr (1990) compared their results with the experimental data of Blackwell et al. (1959). Figure 8.3 compares their results with the data; the agreement is very good. Although this model suffers from fluctuations due to the limited number of particles used in the simulation, the ensemble-averaged properties over several realizations compare well with the data of Blackwell et al. (1959). Generalization of this model to irregular computational grids was achieved by Ebrahimi and Sahimi (2004).

8.9 STABILITY ANALYSIS OF MISCIBLE DISPLACEMENTS

Theoretically, miscible displacements are very efficient EOR processes. Because capillary forces, which are usually responsible for oil entrapment, are absent, the displacement can potentially be 100% efficient. In practice, however, complete recovery is usually not realized due to the instability phenomena mentioned above, which leads to macroscopic fingering of the solvent.

Over the past many decades, many stability analyses of miscible displacements have been carried out (see, e.g., Perrine, 1963; Dumore, 1964, Schowalter, 1965; Heller, 1966; Gardner and Ypma, 1982; Lee et al., 1984; Peters et al., 1984). Among the unsolved problems is the question of what causes the disturbances that propagate as the fingers. It is widely believed that reservoir heterogeneities are the source of the disturbances. However, the presence of a porous space is not necessary for the development of the fingers. Indeed, pioneering studies of viscous fingering did not even use a porous medium. Instead, they used the Hele-Shaw cell. It may seem that the uniformity of a Hele-Shaw cell should preclude any cell-related source of instability. This is actually seen in the early time radial behavior at the inlet of a radial Hele-Shaw Cell (Kopf-Sill and Homsy, 1988; Meiburg and Homsy, 1988). At the same time, explanation of fingering in Hele-Shaw cells is typically based on the assumption of the presence of a perturbation in the system. Moreover, some studies (Christie and Bond, 1987; Fanchi and Christianson, 1989; Fanchi, 1990) suggest that non-linear dynamics of a miscible displacement may provide an alternative source for the disturbances in porous media.

Kelkar and Gupta (1988) reported that they were unable to initiate a finger without introducing some type of perturbation, such as a permeability variation or a concentration perturbation. As described above, Araktingi and Orr (1990) introduced randomness in their model. These works illustrate the approach taken by many researchers in the field: Identify a parameter in the system with values that exhibit some degree of spatial randomness. The randomness then becomes the source of the perturbation. A problem with this approach is its dependence on the *ad hoc* incorporation of a probabilistically-distributed variable.

We now provide a more quantitative discussion of stability analysis of miscible displacements. The main goals of this discussion are to illustrate, (i) how a stability analysis is actually carried out, and (ii) how far such an analysis can take one.

For simplicity, we assume that the dispersion coefficients D_L and D_T are related to the flow velocities $v_x$and v_y through the following relations (Sahimi, 1995):

$$D_L = D_m + \alpha_L \left| v \right| + \frac{(\alpha_L - \alpha_T)v_x^2}{\left| v \right|} \tag{8.21}$$

$$D_T = D_m + \alpha_T \left| v \right| + \frac{(\alpha_L - \alpha_T)v_y^2}{\left| v \right|} \tag{8.22}$$

For convenience, we follow the notation in Yortsos and Zeybek (1988). The governing equations are the CD and the continuity equations, coupled with Darcy's law:

$$\phi\frac{\partial C}{\partial t} + v_x\frac{\partial C}{\partial x} + v_y\frac{\partial C}{\partial y} = \frac{\partial}{\partial x}\left(D_L\frac{\partial C}{\partial x}\right) + \frac{\partial}{\partial y}\left(D_T\frac{\partial C}{\partial y}\right) \tag{8.23}$$

$$\frac{\partial v_x}{\partial x} + \frac{\partial v_y}{\partial y} = 0 \tag{8.24}$$

$$v_x = -\frac{k}{\mu}\frac{\partial P}{\partial x}, \quad v_y = -\frac{k}{\mu}\frac{\partial P}{\partial y} \tag{8.25}$$

Implicit in the above continuum description is the assumption that the local Peclet number vl/D_m is small.

The base-state solution (i.e., the solution to unperturbed or mean quantities), corresponding to a constant injection rate in a rectilinear flow geometry, is given by the well-known diffusive profile

$$\overline{C} = \frac{1}{2}\mathrm{erfc}\left[\frac{\zeta}{2\sqrt{t}}\right] \tag{8.26}$$

$$\frac{\partial \overline{P}}{\partial \zeta} = -\frac{1}{\lambda(\overline{C})} \tag{8.27}$$

where $\zeta = (x - \upsilon t)/L$ is a moving coordinate, λ denotes a normalized mobility, erfc denotes the complementary error function, and all the lengths are scaled with D_{L0}/υ, where D_{L0} is the base-state dispersion coefficient. Writing $C = \overline{C} + C'$, $\upsilon = \bar{\upsilon} + \upsilon'$, and, $P = \overline{P} + P'$, and using normal modes for concentration and flow rate, respectively

$$(C', \upsilon_x') = (\sum, \Phi) \exp(\omega t + i\alpha y) \tag{8.28}$$

the following equations are obtained

$$\sum_{\zeta\zeta} -(\varepsilon\alpha^2 + \omega) \sum = \Phi C_\zeta - L_c(\Phi C_\zeta)\zeta \tag{8.29}$$

$$\lambda(\lambda^{-1}\Phi_\zeta)_\zeta - \alpha^2\Phi = -\alpha^2 R \sum \tag{8.30}$$

where the concentration dependence of the mobility is taken to be,

$$\lambda(C) = \exp(RC) \tag{8.31}$$

with $R = lnM$. The subscripts denote derivatives with respect to the variables. Two important quantites,

$$\varepsilon = \frac{D_m + \alpha_T \upsilon}{D_m + \alpha_L \upsilon} \tag{8.32}$$

$$L_c = \frac{\alpha_L \upsilon}{D_m + \alpha_L \upsilon} \tag{8.33}$$

appear in Eq. (8.29). One, ε, is a measure of flow-induced anisotropic dispersion and is characteristic of porous media, while the other, L_c, is a measure of the contribution of longitudinal dispersion to total dispersion.

Based on our discussion in the previous sections, one expects that the onset of instability and related features should be dictated by the sharpest mobility contrast, namely, those associated with a step concentration profile, which also allow for an analytical solution given by (Yortsos and Zeybek, 1988),

$$\alpha R \left[1 + L_c \gamma_0 \tanh\left(\frac{R}{2}\right) \right] = 2\gamma_0(\alpha + \gamma_0) \tag{8.34}$$

where

$$\gamma_0 = \sqrt{\varepsilon\alpha^2 + \omega} > 0 \tag{8.35}$$

In general, the solution of (8.29) leads to parabolic-like profiles, examples of which are shown in Figure 8.4. In the limit $L_c = 0$ and for an unfavorable mobility ratio

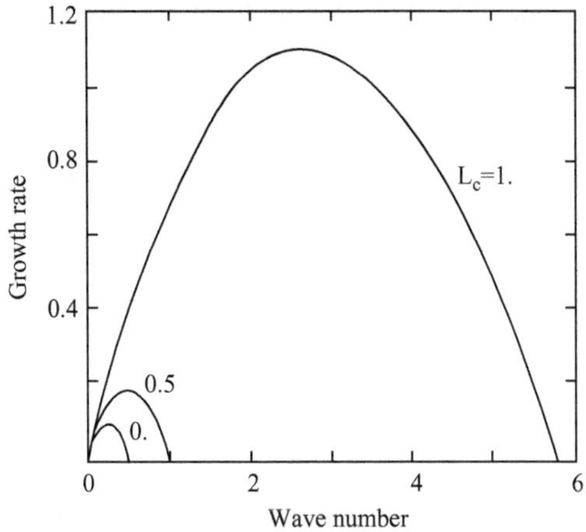

Figure 8.4. Step-profile results for the growth rate of fingers (after Yortsos and Zeybek, 1988)

($R = lnM > 0$), large wavelengths are unstable, while a strong stabilization due to transverse dispersion is exerted at smaller wavelengths. A cutoff wave number can be identified

$$\alpha_c = \frac{R}{2(\varepsilon + \sqrt{\varepsilon})} \tag{8.36}$$

As expected, α_c increases with increasingly unfavorable mobility, and with an increase in the ratio of longitudinal and transverse dispersion ($1/\varepsilon$). However, the limits of the continuum description should be kept in mind. The size of the most unstable disturbance scales with the characteristic length, which for large enough flow rates becomes equal to the dispersivity α_L, which is normally a multiple of the typical pore size (or the length scale of the heterogeneities). It is apparent that a possible conflict may develop between the above result and the continuum description, precluding meaningful predictions over scales of the order of the microscale.

While the limit $L_c = 0$ leads to expected results, a distinct sensitivity develops for $L_c \neq 0$ (see Figure 8.4). This effect is present only due to the velocity dependence of the dispersion coefficients, which can be best quantified in terms of the following combination

$$B_c \frac{RL_c}{2} \tanh\left(\frac{R}{2}\right) - \sqrt{\varepsilon} - 1 \tag{8.37}$$

The following results may then be shown (Yortsos and Zeybek, 1988):

(i) When $B_c < 0$ (which is *always* the case if $L_c = 0$), at small enough viscosity ratio and for $L_c \neq 0$, the cutoff wave number is finite:

$$\alpha_c = \frac{R}{2\sqrt{\varepsilon}}(-B_c) \tag{8.38}$$

although it increases as L_c or B_c does.

(ii) On the other hand, when $B_c > 0$, a finite cutoff does not exist, with the rate of growth increasing indefinitely at large α as

$$\omega \approx B_c(B_c + 2\sqrt{\varepsilon})\alpha^2 > 0 \tag{8.39}$$

Clearly, such is the case for a sufficiently high (but finite) mobility contrasts, so long as $L_c \neq 0$, as shown in Figure 8.4. This unexpected and rather remarkable result is obtained on the basis of a step base state, which is subject to a singular behavior in the large (as well as in the small) α region. To better understand the proper dependence, Yortsos and Zeybek (1988) attempted a more rigorous asymptotic analysis valid for base states near a step profile, namely, for

$$\overline{C} = \frac{1}{2}\text{erfc}(c\xi) \tag{8.40}$$

where $c \gg 1$. Their results showed that the step profile prediction, inequality (8.39), is invalid at large α when $B_c > 0$ and, in fact, a cutoff wave number does exist. However, the latter was found to increase monotonically and without bound as c increases, namely, as the profile becomes step-like, provided that $B_c > 0$.Thus, the essential prediction that qualitatively different instability behavior is obtained by changing the sign of B_c, remains intact. Most of the above results were confirmed by the experimental study of Bacri et al. (1991).

The implications of these results are straightforward. Because of the dependence of the dispersion coefficients on the flow velocity, an essential feature of hydrodynamic dispersion in porous media, and for mobility ratios that exceed a critical value dictated by the given process conditions, a miscible displacement is predicted to be unstable *at all wavelengths*.

Under such conditions, there is no finite preferred mode and, in fact, the above continuum description is ill-posed and breaks down. This remarkable result raises serious doubts about our ability to describe the conditions for the onset of instability in miscible displacements. Recall that this result is obtained if we make several hypotheses, including the validity of a continuum description, and with the dispersion process formulated by a CD equation and the dispersion coefficients represented by Eqs. (8.21) and (8.22). If these predictions are to be taken seriously, the breakdown of the continuum hypothesis beyond a finite M calls for an alternative description.

8.10 MAIN CONSIDERATIONS IN MISCIBLE DISPLACEMENT PROJECTS

Because of their huge cost, field-scale gas-flood projects require careful technical design. The basic design parameters include the geology of the reservoir; its porosity and permeability distributions, and temperature and pressure; the relative permeability curves characterizing two-phase flow in the reservoir, the amount of residual oil, and the viscosity and *minimum miscibility pressure* of the oil-in-place.

The minimum miscibility pressure is typically determined phenomenologically by measuring displacement of crude oil by CO_2 (or another injection fluid) in a long capillary tube (slim tube) at a series of successively higher pressures. If we plot the amount of displaced oil versus pressure, the resulting curve usually has a break at about 95% recovery, the pressure at which is taken as the minimum miscibility pressure. However, we must keep in mind that many crudes contain asphaltenes that precipitate and do not dissolve, even after a series of theoretical multiple contacts between the crude and propagation of the mixture of the injection fluid and the crude's non-asphaltenic components.

In what follows we describe some of the issues that are most important to gas injection processes.

8.10.1 Reservoir Characterization and Management

In virtually all gas injection projects, the most critical decisions are made only after lengthy computer simulations that attempt to optimize the amounts of the fluid injected, the injection and production rates, and other operation variables. Thus, utilizing a realistic model of the reservoir is very important and, in fact, the accuracy of any predictive simulation technique is limited directly by the accuracy with which the reservoir can be described. Despite the extensive sets of experimental data that are used as the input parameters, the sophistication of the simulators, the size of the computers on which the simulations are carried out, and the large number of cases that are simulated, the simulator's predictions may still be subject to very large uncertainties. Therefore, a much less expensive pilot flood is usually carried out first, and the reservoir's model is tuned by changes in the values of the input parameters in order to make the simulator's output fit the pilot data.

For the field-scale projects that have been carried out, calculated optimal CO_2 injection volumes have ranged from 20 to 50% of the hydrocarbon pore volume. The predicted CO_2 utilization factors range from 5 to 15 Mcf CO_2/bbl of recovered oil. The projected ultimate oil recoveries range from 5 to 30% of the original oil-in-place. These numbers represent only the consensus of the current expectations. Significant revisions of many of these estimates may be required after a large amount of actual full-scale production data become available.

8.10.2 Mobility Control

As discussed above, the control of unfavorable mobility ratios is recognized as a major technological problem of gas-flood EOR, as such mobility ratios produce the fingering phenomena which are a principal reason for the failure of many of the liquefied petroleum gas floods of the 1950s and early 1960s (Craig, 1970). These problems occur because the injection gases have very small viscosities at the temperatures and pressures at which they are used, hence producing an unfavorable mobility ratio. The problems associated with the mobility are greatly aggravated, and their theoretical and experimental study greatly complicated by the facts that, (1) the reservoir rock is highly heterogeneous at every length scale, from sub-millimeter to kilometer, and (2) the number of fluid phases is often three rather than two.

To control the mobility, many different methods have been used, such as water-alternating-gas (WAG), gas-soluble viscofiers, gelled and cross-linked polymers, and surfactants and foams (Smith, 1988). We describe one of these, namely, the WAG process.

8.10.3 Miscible Water-Alternative-Gas Process

This method is the only gas-flood mobility control technique which is used regularly in field applications. Its chief virtues are familiarity to the petroleum engineers, as it consists of the familiar water flooding, alternated with gas flooding; simplicity, having timing and the ratio of water to gas as its only design parameters, and its low cost. It is used mostly to maintain pressure while the solvent supply is interrupted, or to stretch out injection costs by substituting water for the more expensive solvent.

Miscible WAG injection has been implemented successfully in a number of fields around the world (Christensen et al. 1998a,b). In principle, it combines the benefits of miscible gas injection and water flooding by injecting the two fluids either simultaneously or alternatively. Miscible gas injection has excellent microscopic sweep efficiency, but poor macroscopic sweep efficiency due to fingering and gravity override. Furthermore, it is expensive to implement. The optimal ratio for simultaneous WAG injection in a relatively homogeneous reservoir can be estimated by matching the advance rates of the water-oil and solvent-oil displacement fronts. Water flooding, on the other hand, is cheaper and less vulnerable to gravity segregation and frontal instabilities. However, the residual oil saturations after water flooding are relatively high.

Johns et al. (2003) studied optimization of the WAG processes for enriched gas floods, particularly as a primary recovery method. Al-Shuraiqi et al. (2003) carried out laboratory investigations of first-contact miscible WAG to study the effects of WAG ratio and flow rate on recovery efficiency. Their experiments (using beadpacks) indicated that, (1) recovery from WAG may vary with rate, even when gravity is neglected, and (2) water-solvent and water-oil relative permeabilities may not be the same if they are first-contact miscible. The influence of the rate on WAG recovery appears to be due to the rate-dependency of water-oil and water-solvent permeabilities.

8.10.4 Relative Permeabilities

Enhanced oil recovery processes that are carried out under near miscible conditions require relative permeability data to calculate the flow behavior of the low interfacial tension (IFT) fluids. In systems where the displacing and displaced fluids are miscible and the flow velocity is low enough that the process is controlled by diffusion, the fractional flow versus saturation curve is a 45° line, and is referred to as miscible relative permeability function. The term *near-miscible relative permeability functions* is used to denote the curves found in the region between the immiscible limit and miscible limits. If the IFT is finite and there is no mass transfer, the relative permeabilities of the phases are, as is well-known, more complicated.

It is not yet completely clear how near miscibility changes the relative permeabilities, and which parameters are controlling this change (Blom, 1999). Some investigators (Ameafule and Handy, 1982; Harbert, 1983; Henderson et al., 1996) found that the relative permeability to the non-wetting phase is affected more strongly, while others (Asar and Handy, 1988; Schechter and Haynes, 1992) reported the same for the wetting phase. For example, Cinar and Orr (2004) recently reported a 10-fold increase of the non-wetting phase relative permeabilities against a 100-fold decrease in the IFT. Others did not find any effect of IFT at all (Delclaud et al., 1987; Kalaydjian et al., 1996). Equally contradictory are the reports on the effect of the flow velocity on the near-miscible relative permeability. Some investigators (Fulcher et al., 1985; Schechter, 1988) reported no effect, whereas others (Boom et al., 1995; Ali et al., 1997; Mott et al., 1999; Saevareid et al., 1999; Whitson et al., 1999; Henderson et al., 2000) observed an increase with the velocity.

There appear to be two conflicting views on the mechanism that controls the change in the relative permeability. Bardon and Longeron (1980), Jerauld (1996), and Blom et al. (2000) argued that a low IFT affects the relative permeabilities through the capillary number, the ratio of the viscous and capillary forces. Most of these authors suggest, however, that there is a threshold IFT below which the capillary-number dependence becomes important. On the other hand, Haniff and Ali (1990), Morel et al. (1992), Hartman and Cullick (1994), and Munkerud (1995) interpreted their relative permeability data in terms of the IFT alone. In two of these cases, the interpretation was done in view of the fact that a transition from partial wetting to complete wetting, as predicted by Cahn (1977), may affect the mobility of both phases. The influence of such a wetting transition cannot be described in terms of the capillary number, because it is directly induced by a change in the interfacial tension between the near-miscible phases. Blom and Hagoort (1998) reviewed the progress in resolving some of these issues.

8.10.5 Upscaling

In a miscible gas injection, the effects of channeling and fingering, which occur due to the heterogeneity and adverse mobility ratio between the injected gas and oil, must be properly accounted for in order to obtain an accurate estimate of the displacement efficiency. However, the resolution of the required computational grid is extremely high. This fact, coupled with the high number of components and complex phase behavior, render fine-scale simulations of miscible processes prohibitively time consuming. For this reason, most field-scale simulations of miscible displacements are not carried out on high-resolution grids. Therefore, one must have a proper scheme for scale-up of the high-resolution grid to coarser grids that can be used in the computations.

Approaches to modeling miscible/near miscible displacements can be broadly divided into two groups. In one group are multi-contact miscible processes, for which compositional simulations are required. First-contact miscible processes are in the second group, for which the limited compositional formulations – described earlier as averaged continuum models – may be preferable, due to their computational efficiency. A key issue with any upscaling procedure is how well the coarsened (upscaled) model replicates important aspects of the fine-scale flow behavior, such as total injection and production rate, average pressure or saturation throughout the reservoir, and breakthrough times of injected flows. Additional issues are the degree of coarsening achievable by a given method, the level of robustness of the coarse scale model (that is, its applicability to models with different global boundary conditions or well locations), and whether or not the method introduces modifications to the form of the governing equations. Reviewing all aspects of upscaling methods would require a full review by itself. The interested reader is referred to Ebrahimi and Sahimi (2002, 2004), where some of these issues are discussed in detail, and a new and efficient upscaling method for miscible displacements is proposed.

REFERENCES

Ali, J.K., McGualey, P.J., and Wilson C.J., The effects of high velocity flow and PVT changes near the wellbore on condensate well performance. SPE 38923 (1997).

Al-Shuraiqi, H.S., Muggeridge A.H., and Grattoni, C.A., Laboratory investigations of first contact miscible WAG displacement: the effects of WAG ratio and flow rate. SPE 84894 (2003).

Ameafule, J.O., and Handy, L.L., The effect of interfacial tensions on relative oil/water permeabilities of consolidated porous media. SPE 9783 (1982).

Araktingi, U.G., and Orr, F.M., Viscous fingering, gravity segregation, and reservoir heterogeneity in miscible displacements in vertical cross sections. SPE 20176 (1990).

Araktingi, U.G., and Orr, F.M., Viscous fingering in heterogeneous porous media. SPE 18095 (1993).

Arbogast, T., Bryant, S., Dawson, C., Saaf, F., Wang, C., and Wheeler, M.F., Computational methods for multiphase flow and reactive transport problems arising in subsurface conatinant remediation. *J. Comp. Appl. Math.* 74, 19 (1996).

Asar, H., and Handy, L.L., Influence of interfacial tension on gas/oil relative permeability in a gas-condensate system. SPE 11740 (1988).

Bacri, J.C., Salin, D., and Wouméni, R., *Phys. Rev. Lett.* 67, 2005 (1991).

Bardon, C., and Longeron, D.G., Influence of very low interfacial tensions on relative permeability. SPE 7609 (1980).

Batycky, R.P., A Three-Dimensional Two-Phase Field scale Streamline Simulator. Ph.D. dissertation, Stanford University, Stanford, California (1997).

Batycky, R.P., Blunt, M.J., and Thiele, M.R., A 3D fine-scale streamline simulator with gravity and changing well conditions. SPE 36726 (1996).

Bell, J.B., Shubin, M.J., and Wheeler, M.F., Analysis of a new method for computing the flow of miscible fluids in a porous medium. *SIAM J. Numer. Anal.* 22, 1041 (1985).

Benham, A.L., and Olson R.W., A model study of viscous fingering. SPE 513 (1963).

Berger, A.E., Solomon, J.M., Ciment, M., Leventhal, S.H., and Weiberg, B.C., Generalized OCI schemes for boundary layer problems. *Math. Comp.* 37, 79 (1980).

Blackwell, R.J., Rayne, J.R., and Terry, W.M., Factors influencing the efficiency of miscible displacements. *Trans. AIME* 216, 1 (1959).

Blom, S.M.P., Relative Permeability to Near-Miscible Fluids. Ph.D. dissertation, Delft University of Technology, The Netherlands (June 1999).

Blom, S.M.P., and Hagoort, J., How to include the capillary number in gas condensate relative permeability functions? SPE 49268 (1998).

Blom, S.M.P., Hagoort, J., and Soetekouw, D.P.N., Relative permeability at near critical conditions. SPE 38935 (2000).

Boom, W., Wit, K., Schulte, A.M., Oedai, S., Zeelenberg, J.P.W., and Maas, J.G., Experimental evidence for improved condensate mobility at near-wellbore flow conditions. SPE 30766 (1995).

Bratvedt, F., Bratvedt, K., Buchholz, C.F., Holden, L., Holden, H., and Risebro, N.H., A new front-tracking method for reservoir simulation. SPE 19805 (1992).

Bradtvedt, F., Gimse, T., and Tegnander, C., Streamline computation for porous media flow including gravity. *Transport in Porous Media* 25, 1, 63 (1996).

Cahn, J.W., Critical point wetting. *J. Chem. Phys.* 66, 3667 (1977).

Chen, J.D., and Wilkinson, D., Pore-scale viscous fingering in porous media. *Phys. Rev. Lett.* 55, 1892 (1985).

Christensen, J.R., Stenby, E.H., and Skauge, A., Review of WAG field experience. SPE paper 39883, Tulsa, OK (1998a).

Christensen, J.R., Stenby, E.H., and Skauge, A., Compositional and relative permeability effects on near-miscible WAG. SPE paper 39627, Tulsa, OK (1998b).

Christie, M.A., High-resolution simulation of unstable flow in porous media. SPE 16005 (1989).

Christie, M.A., and Bond, D.J., Detailed simulation of unstable processes in miscible flooding. SPE/DOE 14896 (1987).

Christie, M.A., Muggeridge, A.H., and Barley, J.J., 3D simulation of viscous fingering and WAG schemes. SPE 21238 (1993).

Cinar, Y., and Orr, Jr., F.M., Measurement of three phase relative permeability with IFT variation. SPE 89419 (2004).

Craig, F.F., A current appraisal of field miscible slug projects. JPT 22, 529 (SPE 2418). (1970).

Darlow, B.L., Ewing, M.F., and Wheeler, M.F., Mixed finite element method for miscible displacement problems in porous media. SPE 10501 (1984).

DeGregoria, A.J., A Predictive Monte Carlo simulation of two-fluid flow through porous media at finite mobility ratio. *Phys. Fluids* 28, 2933 (1985).

Delclaud, J., Rochon, J., and Nectoux, A., Investigation of gas/oil relative permeabilities: high-permeability oil reservoir application. SPE 16966 (1987).

Douglas, J., Wheeler, M.F., Darlow, B.L., and Kendall, R.P., Self adaptive finite element simulation of miscible displacement in porous media. *Comp. Meth. Appl. Mech. Eng.* 47, 131 (1984).

Dumore, J.M., Stability considerations in downward miscible displacements. SPE 961, 358 (1964).

Dykstra, H., and Parsons, R.L., The prediction of oil recovery by waterflooding. In, *Secondary Recovery of Oil in the United States*, 2nd ed. (American Petroleum Inst., New York City, 1950), p. 160.

Ebrahimi, F., and Sahimi, M., Multiresolution wavelet coarsening and analysis of transport in heterogeneous media. *Physica A* 316, 160 (2002).

Ebrahimi, F., and Sahimi, M., Multiresolution wavelet scale up of unstable miscible displacements in flow through heterogeneous porous media. *Transport in Porous Media* 51, 1, 57–102 (2004).

Ewing, R.E., Russell, T.F., and Wheeler, M.F., Convergence analysis of an approximation of miscible displacement in porous media by mixed finite elements and a modified method of characteristics. *Comp. Meth. Appl. Mech. Eng.* 47, 73 (1984).

Ewing, R.E., Russell, T.F., and Young, L.C., An isotropic coarse-grid dispersion model of heterogeneity and viscous fingering in five-spot miscible displacement that match experiments and fine grid simulations. SPE 18441 (1989).

Ewing, R.E., and Wang, H., Eulerian-Lagrangian localized adjoint methods for variable coefficient advective-diffusive-reactive equations in groundwater contaminant transport. In: *Advances in Optimization and Numerical Analysis* 275, edited by S. Gomez and J.P. Hennart (Kluwer, Netherlands, 1994), p. 185.

Fanchi, J.R., Chaos: a source of miscible viscous fingering instabilities. SPE 21587 (1990).

Fanchi, J.R., and Christianson, R.L., Applicability of fractals to the description of viscous fingering. SPE 19782 (1989).

Fayers, F.J., An approximate model with physically interpretable parameters for representing miscible viscous fingering. SPE 13166 (1988).

Fayers, F.J., Blunt, M.J., and Christie, M.A., Accurate calibration of empirical viscous fingering models. In, *Proceedings of 2nd European Conference on the Mathematics of Oil Recovery* (SPE, Paris, 1990), 45.

Fayers, F.J., Blunt, M.J., and Christie, M.A., Comparisons of empirical viscous fingering models and their calibration for heterogeneous problems. SPE 22148 (1992).

Fulcher, R.A., Ertekin, T., and Stahl, C.D., Effect of capillary number and its constituents on two-phase relative permeability curves. JPT 249 (SPE 12170) (1985).

Gardner, J.W., and Ypma, J.G.J., An investigation of phase behavior-macroscopic bypassing interaction in CO_2 flooding. SPE paper 10686, SPE Enhanced Oil Recovery Symposium, Tulsa, OK (1982).

Gavalas, G.R., Reamer, H.H., and Sage, B.H., Diffusion coefficients in hydrocarbon systems. *Ind. Eng. Chem. Fund.* 7, 306 (1968).

Gottlieb, D., and Orszag, S.A., *Numerical Analysis of Spectral Methods: Theory and Applications* (SIAM, Philadelphia, 1977).

Greenkorn, R.A., Johnson, C.R., and Haring, R.E., Miscible displacement in a controlled natural system. JPT 5, 329 (SPE 1232) (1965).

Habermann, B., The efficiencies of miscible displacement as a function ofmobility ratio. *Trans. AIME* 219, 264 (1960).

Haniff, M.S., and Ali, J.K., Relative permeability and low tension fluid flow in gas condensate systems. SPE 20917 (1990).

Harbert, L.W., Low interfacial tension relative permeability. SPE 12171 (1983).

Harleman, D.R.F., and Rumer, R.R., Longitudinal and lateral dispersion in an isentropic porous medium. *J. Fluid Mech.* 16, 385 (1963).

Hartman, K.J., and Cullick, A.S., Oil recovery by gas displacement at low interfacial tension. *J. Pet. Sci. Eng.* 10, 197 (1994).

Heller, J.P., Onset of Instability Patterns between Miscible Fluids in Porous Media. *J. Appl. Phys.* 37, 1566 (1966).

Henderson, G.D., Danesh, A., Tehrani, D.H., and Peden, J.M., The effect of velocity and interfacial tension on the relative permeability of gas condensate fluids in the well bore region. In, *8th European Symposium on Improved Oil Recovery* (SPE, Vienna, Austria, 1995).

Henderson, G.D., Danesh, A., Tehrani, D.H., Al-Shaldi, S., and Peden, J.M., Measurement and correlation of gas condensate relative permeability by the steady state method. SPE 31065 (1996).

Henderson, G.D., Danesh, A., Al-Kharusi, B., and Tehrani, D., Generating reliable gas condensate relative permeability data used to develop a correlation with capillary number. *J. Pet. Sci. Eng.* 25, 79 (2000).

Homsy, G.M., Viscous fingering in porous media. *Ann. Rev. Fluid Mech.* 19, 271 (1987).

Jerauld, G.R., General three-phase relative permeability model for Prudhoe Bay. SPE 36178 (1996).

Jessen, K., and Orr, F.M., Compositional streamline simulation. SPE 77379 (2002).

Jessen, K., and Orr F.M., Gravity segregation and compositional streamline simulation. SPE 89448 (2004).

Johns, R.T., Bermudez, L., and Parakh, H., WAG optimization for gas floods above MME. SPE 84366 (2003).

Johns, R.T., Sah, P., and Subramanian, S., Effect of gas enrichment above the MME on oil recovery in enriched-gas floods. SPE 56826 (2000a).

Johns, R.T., Sah, P., and Solano, R., Effect of dispersion on local displacement efficiency for multicomponent enriched-gas floods above the MME. SPE 64725 (2000b).

Kalaydjian, F.J.M., Bourbiaux, B.J., and, Lombard, J.M., Predicting gas condensate reservoir performance: how flow parameters are altered when approaching production wells. SPE 36715 (1996).

Kelkar, M.G., and Gupta, S.P., A numerical study of viscous instabilities: Effect of controlling parameters and scaling considerations. SPE 18094 (1988).

King, M.J., and Scher, H., A probability approach to multi-phase, multi-component fluid flow in porous media. *Phys. Rev. A* 35, 929 (1987).

King, M.J., and Scher, H., Geometric dispersion and unstable flow in porous media. *Phys. Rev. A* 41, 874 (1990).

Kopf-Sill, A.R., and Homsy, G.M., Nonlinear unstable viscous fingers in Hele-Shaw flows. I. Experimental. *Phys. Fluids* 31, 242 (1988).

Koval, E.J., A method for predicting the performance of unstable miscible displacement in heterogeneous media. *Trans. AIME* 228, 145 (1963).

Kyle, C.R., and Perrine, R.L., Experimental studies of miscible displacement instability. SPE 1113 (1965).

Lee, S.T., Gary Li, K.M., and Culham, W.E., Stability analysis of miscible displacement processes. SPE/DOE 12631 (1984).

Lenormand, R., Toubol, E., and Zarcone, C. *J. Fluid Mech.* 189, 165 (1988).

Leventhal, S.H., The operator compact implicit method for reservoir simulation. SPE 7678 (1980).

Mahaffey, J.L., Rutherford, W.M., and Mathews, C.S., Sweep efficiency by miscible displacement in a five-spot. SPE 1233 (1966).

Manickam, O., and Homsy, G.M., Stability of miscible displacements in porous media with nonmonotonic viscosity profiles. *Phys. Fluids A* 5, 1356 (1993).

Meiburg, E., and Homsy, G.M., Nonlinear unstable viscous fingers in Hele-Shaw flows. II. numerical simulation. *Phys. Fluids* 31, 429 (1988).

Moissis, D.E., Miller, C.A., and Wheeler, M.F., A parametric study of viscose fingering, Numerical simulation in oil recovery. Institute for Mathematics and its Applications Series in Applied Mathematic (Springer, Berlin, 1987), p. 227.

Moissis, D.E., Miller C.A., and Wheeler, M. F., Simulation of miscible fingering using a modified method of characteristics: effects of gravity and heterogeneity. SPE 18440 (1989).

Morel, D.C., Lomer, J.F., Morineau, Y.M., and Putz, A.G., Mobility of hydrocarbon liquids in gas condensate reservoirs: interpretation of depletion laboratory experiments. SPE 24939 (1992).

Morton K.W., *Numerical Solution of Convective-Diffusion Problems*, (Chapman and Hall, New York, 1996).

Mott, R, Cable, A., and Spearing, M., A new method of measuring relative permeabilities for calculating gas condensate well deliverability. SPE 56484 (1999).

Moulu, J.C., Solution-gas drive: experiments and simulation. *J. Pet. Sci. Eng.* 2, 379 (1989).

Munkerud, P.K., The effects of interfacial tension and spreading on relative permeability in gas condensate systems. Proceedings of 8th European IOR Symposium (1995).

Newley, T.M.J., Comparison of empirical models for unstable miscible displacement. Proceedings of the European Symposium on EOR, Hamburg, Germany (1987).

Nguyen, T.A., Farouq-Ali, S.M., Role of diffusion and gravity segregationin oil recovery by the immiscible carbon dioxide wag process. In, UNITAR International Conference on Heavy Crude and Tar Sands, Houston, Texas, February 1995, p. 393.

Odeh, A.S., and Cohen, M.F., A technique for improving the simulation of linear, miscible hydrocarbon displacement in the event of viscous fingering. *Energy Sources* 11, 9 (1989).

Paterson, L., Radial fingering in a Hele-Shaw cell. *J. Fluid Mech.* 113, 513 (1981).

Paterson, L., *Int. J. Hydrogen Energy* 8, 53 (1983).

Paterson L., Diffusion-limited aggregation and two-fluid displacements in porous media. *Phys. Rev. Lett.* 52, 1621 (1984).

Paterson, L., Fingering with iscible fluids in a Hele-Shaw cell. *Phys. Fluids* 28, 26 (1985).

Paterson, L., Hornof, V., and Neale, G.H., A consolidated porous medium for visualization of unstable displacements. *Powder Technol.* 265 (1982).

Peaceman, D.W., and Rachford, H., Numerical calculation of multidimensional miscible displacement. SPE 471 (1962).

Perkins, T.K., and Johnston, O.C., A study of immiscible fingering in linear models. SPE 2230 (1969).

Perkins, T.K., Johnston, O.C., and Hoffman, N.R., Mechanics of viscous fingering in miscible systems. SPE 1229 (1965).

Perrine, R.L., A unified theory for stable and unstable miscible displacement. SPE 543 (1963).

Peters, E.J., Broman, W.H., and Broman, J.A., A stability for miscible displacement. SPE 13167 (1984).

Reamer, H.H., Sage, B.H., Diffusion coefficients in hydrocarbon systems. Methane in the liquid phase of the methane-propane system. *Ind. Eng. Chem.* 3, 54 (1958).

Renner, T.A., Measurement and correlation of diffusion coefficients for CO and rich-gas applications. SPE 15391 (1988).

Riazi, M.R., A new method for experimental measurement of diffusion coefficients in reservoir fluids. *J. Pet. Sci. Eng.* 14 (1996).

Rogerson, A.M., and Meiburg, E., Shear stabilization of miscible displacement process in porous media. *Phys. Fluids A* 5, 1344, (1993a).

Rogerson, A.M., and Meiburg, E., Numerical simulation of miscible displacement processes in porous media flows under gravity. *Phys. Fluids A* 5 (1993b).

Russell, T.F., and Wheeler, M.F., Finite element and finite difference methods for continuous flows in porous media, In, *Mathematics of Reservoir Simulation*, edited by R.E. Ewing, (SIMA, Philadelphia, 1983), p. 35.

Saevareid, A., Whitson, C.H., and Fevang, O., An engineering approach to measuring gas condensate relative permeabilities, SCA 9930, International Symposium of the Society of Core Analysis, Golden, Colorado (1999).

Sahimi, M., 1993a, Flow phenomena in rocks: From continuum models, to fractals, percolation, cellular automata, and simulated annealing. *Rev. Mod. Phys.* 65, 1393 (1993a).

Sahimi, M., Fractal and superdiffusive transport and hydrodynamic dispersion in heterogeneous porous media. Transport in Porous Media 13, 3 (1993b).

Sahimi, M., *Flow and Transport in Porous Media and Fractured Rock*, (VCH, Weinheim, Germany, 1995).

Sahimi, M., *Heterogeneous Materials*, Volume I (springer, New York, 2003).

Sahimi, M., Davis, H.T., and Scriven, L.E., Thermodynamic modeling of phase and tension behavior of carbon dioxide-hydrocarbon systems. SPE 10268 (1985).

Sahimi, M., and Taylor, B.N., Surface tension of binary liquid-vapor mixtures: A comparison of mean-field and scaling theories. *J. Chem. Phys.* 95, 6749 (1991).

Schechter, D.S., Immiscible flow behaviour in porous media. Ph.D. dissertation, University of Bristol, United Kingdom (1988).

Schechter, D.S., and Haynes, J.M., Relative permeabilities of a near-critical binary fluid. *Transport in Porous Media* 9, 241 (1992).

Schmidt, T., Leshchyshyn, T.H., and Puttagunta, V.R., Diffusion of carbon dioxide into athabasca bitumen. 33rd Annual Technical Meeting of the Petroleum Society of CIM, Calgary, Canada, Paper No. 82–33–100 (1982).

Schowalter, W.R., Stability criteria for miscible misplacement of fluid from a porous medium. *AIChE J.* 11, 99 (1965).

Sherwood, J.D., and Nittmann, J., Gradient-governed growth: the effect of viscosity ratio on stochastoic simulation of Saffman-Taylor instability. *J. Physique* 47, 125 (1986).

Siddiqui, H., and Sahimi, M., Computer simulation of miscible displacement processes in disordered porous media. *Chem. Eng. Sci.* 45, 163 (1990a).

Siddiqui, H., and Sahimi, M., A statistical model for simulating miscible viscous fingers in porous media and other growth phenomena. *J. Phys. A* 23, L497 (1990b).

Slobod, R.L., and Thomas, R.A., Effect of transverse diffusion on fingering in miscible displacement. SPE 464 (1963).

Smith D.H., ed., *Surfactant-Based Mobility Control: Progress in Miscible Flood Enhanced Oil Recovery*, (American Chemical Society, Washington, DC, 1988).

Sorbie, K.S., Feghi, F., Pickup, G.E., Ringrose, P.S., and Jensen, J.L., Flow regimes in miscible displacements in heterogeneous correlated random fields. SPE Advanced Technology Series (April 1994), 78.

Sorbie, K.S., Zhang, H.R., and Tsibuklis, N.B., Linear viscous fingering: new experimental results, direct simulation and the evaluation of averaged models. *Chem. Eng. Sci.* 50, 601 (1995).

Stalkup, F.I., Miscible displacement. In, *Henry L. Doherty Memorial*, (Society of Petroleum Engineers, Dallas, 1983).

Tan, C.T., and Homsy, G.M., Simulation of nonlinear viscous fingering in miscible displacement. *Phys. Fluids* 31, 1330 (1988).

Taylor, S.R., Moltyaner, G.L., Howard, K.W.F., and Killey, R.W.D., A compression of the field and laboratory methods for determining contaminant flow parameters. *Ground Water* 25, 321 (1987).

Tchelepi, H.A., and Orr, F.M., Dispersion, permeability heterogeneity, and viscous fingering: acoustic experimental observations and particle-tracking simulations. *Phys. Fluids A* 5, 1558 (1993a).

Tchelepi, H.A., and Orr F.M., The Interaction of viscous fingering, permeability heterogeneity, and gravity segregation in the 3D. SPE 25235 (1993b).

Thiele, M.R., Blunt, M.J., and Orr, F.M., Modelling flow in heterogeneous media using streamtubes-II. compositional displacements. *In-situ* 19, 367 (1995).

Thiele, M.R., Batycky, R.P., and Blunt, M.J., A streamline-based 3D field-scale compositional reservoir simulator. SPE 38889 (1997).

Todd, M.R., and Longstaff W.J., The development, testing and application of a numerical simulator for predicting miscible flood performance. *Trans. AIME* 253, 874 (1972).

Vossoughi, S., Smith, J.E., Green, D.W., and Willhite, G.P., A new method to simulate the effects of viscous fingering on miscible displacement processes in porous media. *Trans. AIME* 277, 56 (1984).

Waggoner, J.R., Castillo, J.L., and Lake, L.W., Simulation of EOR processes in stochastically generated permeable media. *SPE Formation Eval.* 5, 173 (June 1992).

Whitson, C.H., Fevang, O., and Saevareid, A., Gas condensate relative permeability for well conditions. SPE 56476 (1999).

Wooding, R.A., and Morle-Seytoux, H.J., Multiphase fluid flow through porous media. *Ann. Rev. Fluid Mech.* 8, 233 (1976).

Yan, W., Michelsen, M.L., Stenby, E.H., Berenblyum, R.A., and Shaprio, A.A., Three-phase compositional streamline simulation and its application to WAG. SPE 89440 (2004).

Yang, Z.M., Application of transverse flow equilibrium in miscible displacements. Ph.D. dissertation, University of Southern California, Los Angeles (1995).

Yang, Z.M., and Yortsos, Y.C., A study of the asymptotic regimes of displacements in porous media flows. In, Proceedings of Mathematical Modelling of Flow Through Porous Media (World Scientific, Singapore, 1995), p. 358.

Yang, Z.M., and Yortsos, Y.C., Asymptotic regimes in miscible displacements in random porous media. SPE 35456 (1996).

Yang, Z.M., and Yortsos, Y.C., Effect of no-flow boundaries on viscous fingering in porous media of large aspect ratio. SPE 51257 (1998).

Yang, Z.M., and Yortsos, Y.C., Asymptotic regimes in unstable miscible displacements in random porous media. *Adv. Water Resour.* 25, 885 (2002).

Yortsos, Y.C., and Zeybek, M., *Phys. Fluids* 31, 3511 (1988).

Zhang, H.R., Sorbie, K.S., and Tsibuklis, N.B., Viscous fingering in five-spot experimental porous media: new experimental results and numerical simulation. *Chem. Eng. Sci.* 52, 37 (1996).

Zhang, Y.P., Hyndman, C.L., and Maini, B.B., Measurement of gas diffusivity in heavy oils. *J. Pet. Sci. Eng.* 25, 37 (2000).

Zimmerman, W.B. and Homsy, G.M., Nonlinear viscous fingering in miscible displacement with anisotropic dispersion. *Phys. Fluids A* 3, 1859 (1991).

Zimmerman, W.B. and Homsy, G.M., Viscous fingering in miscible displacements: unification of effects of viscous contrast, anisotropic dispersion and velocity dependence of dispersion on nonlinear viscous propagation. *Phys. Fluids A* 4, 1099 (1992a).

Zimmerman, W.B. and Homsy, G.M., Three-dimensional viscous fingering: a numerical study. *Phys. Fluids A* 4, 1901, (1992b).

CHAPTER 9

UNSTABLE AND FINGERING GAS FLOW IN FRACTURES

PETER PERSOFF
Lawrence Berkeley National Laboratory, Berkeley, CA, USA

In this chapter, two causes of unstable and fingering gas flow in fractures will be considered: capillary instability in horizontal fractures and gravitational instability in vertical or inclined fractures.

Several physical situations may induce gas flow in unsaturated fractures. Any chemical, biochemical, or radiolytic reaction that produces or consumes gas increases or decreases the pressure. Gas may also be released from solution when the water pressure is reduced. Any injection or extraction of gas through wells induces a gas pressure gradient; changes in barometric pressures are also felt below the ground surface. Instability results when a gas flow path does not remain continuously open, but is intermittently blocked by water. This can occur if there is sufficient water, or sufficient water flow, in the fracture to block the gas flow path. In this discussion, we consider air and water as the nonwetting and wetting fluids, respectively; some of these observations will also be applicable to oil and water or air and oil.

9.1 PARTIALLY SATURATED FRACTURES ABOVE THE WATER TABLE

Above the water table, water is held in fractures at asperities (points of contact) and wherever capillary forces are greater than the gravitational force tending to drain water from the pores. That is, pores of radius smaller than r are saturated at distance d above the water table, where r and d are related by

$$r = 2\sigma/(p_g - p_l) = 2\sigma/\rho g d \tag{9.1}$$

where p is pressure, subscripts g and l are gas and liquid, g is the acceleration of gravity and σ and ρ are the surface tension and density of water. The capillary pressure is defined as $p_{cap} = (p_g - p_l)$; and $(p_l - p_g)$ is also called the moisture tension. Fractures

C. Ho and S. Webb (eds.), Gas Transport in Porous Media, 169–178.
© 2006 *Springer.*

do not actually have pores in the same sense as a porous medium such as sandstone, but the term "pore size" will refer to the aperture at any location in the fracture. Similarly as for porous media, hysteresis in the capillary pressure curve may also cause water to be "trapped" in pores larger than r given by Eq. (9.1) when the water table is falling. Water, being the wetting fluid, occupies the smallest pores. Given a pore-size distribution, one can calculate the saturation of the fracture at any distance above the water table. Because the smallest pores contribute the least to the fracture conductivity, if the fracture liquid saturation is low, (e.g., 10%), gas flow is scarcely impeded by the presence of water, and the permeability to gas is almost as great in the partially saturated fracture as in the dry fracture (i.e., relative permeability to gas is close to 1). At low liquid saturation, gas flow will normally be stable. Because flow velocities in fractures may be much greater than in porous media, inertial effects may be significant (i.e., non-darcy flow).

At smaller values of d (closer to the water table), larger pores are saturated and gas permeability is reduced significantly; that is, gas relative permeability is less than 1. Experiments to measure gas and liquid relative permeabilities in fractures at a range of moisture tension and water saturation were conducted by Persoff and Pruess (1995). Those experiments were conducted in horizontally-mounted transparent replicas of natural fractures, to permit flow visualization and avoid gravitational effects. Gas flow was unstable at low capillary pressures (at high water saturations) and this was attributed to intermittent blocking of gas flow paths by water, as discussed below.

9.2 UNSTABLE GAS FLOW IN AN UNSATURATED FRACTURE

Persoff and Pruess (1995) conducted experiments to measure relative permeabilities to gas and liquid flow in 7.6-cm square, horizontally mounted, transparent replicas of natural rock fractures and in a natural rock fracture sample. The hydraulic apertures (aperture of a parallel plate fracture with the same absolute permeability) of these samples ranged from 9 to 22 μm. An aperture map was produced by comparing digitized images of a fracture replica filled with either plain water or dyed water. Light attenuation by Beer's law was used to calculate the aperture, a technique similar to that used by Detwiler et al. (1999), calibrated with measurements on a "stair-step" fracture of known apertures.

In these experiments gas and liquid were injected simultaneously at one edge of the sample and removed from the opposite edge. Gas was injected directly to the edge of the fracture and liquid through a porous plate; this allowed gas to be injected at a greater pressure than liquid without invading the liquid injection line (an adaptation of the Hassler method for measuring relative permeability). Liquid was injected at constant rate, and gas was injected at either constant mass flow rate or constant pressure. Inlet and outlet pressures were measured, and boundary conditions were adjusted to achieve equal capillary pressures (defined as $p_{\mathrm{cap}} = p_{\mathrm{g}} - p_l$) at the inlet and outlet.

All experiments started with the fracture saturated with liquid, and liquid flowing. When gas was introduced at constant pressure at the inlet, gas fingers penetrated into the sample. Gas pressure was increased stepwise and the gas fingers grew (although not with each pressure increase), until a single continuous flow path was established through the sample from inlet to outlet, generally following a path of maximum aperture. If the map of apertures were interpreted as a topographic contour map, the initial gas flow path would be like walking along a ridge line. In some experiments more than one gas flow path was established.

In experiments where gas was injected at a controlled flow rate, when the gas:liquid volumetric flow-rate ratio was less than about 20:1, flow was not always stable, resulting in discontinuous and cycling gas flow as described below. Pressures were averaged over many cycles to calculate capillary pressure and relative permeability values. Then the gas flow rate was increased stepwise while liquid flow rate was decreased stepwise. In this manner relative permeabilities to both phases were measured through several steady states of decreasing liquid saturation and increasing capillary pressure. An example plot of the resulting relative permeability against capillary pressure is shown in Figure 9.1. Error bars are the standard deviation of measurements made over 24 hours. The larger error bars at low capillary pressure result from the pressure cycling that accompanied unstable gas flow; full details are presented in Persoff and Pruess (1995).

In the present discussion of unstable gas flow, it is the measurements in liquid-dominated conditions (i.e., low capillary pressures, and low gas:liquid flow ratios) that are of interest. Under these conditions it was impossible to achieve steady flow conditions. When both phases were injected at controlled flow rates, the measured gas pressures were seen to cycle repeatedly. Examples of this behavior are shown in

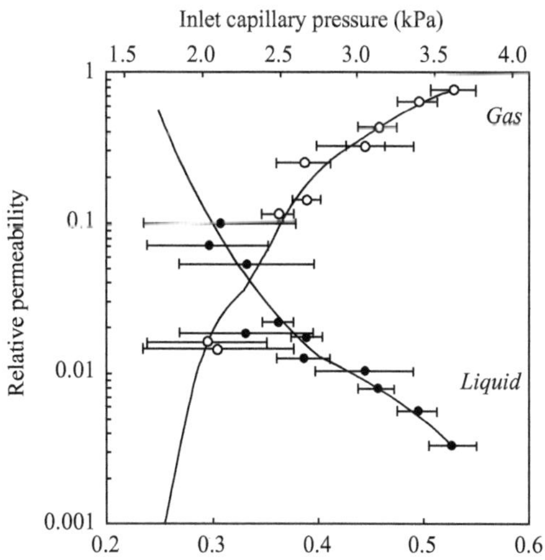

Figure 9.1. Relative permeability in a fracture replica (8.5 μm hydraulic aperture, Dixie Valley replica, Expt D)

Figure 9.2 through Figure 9.4. The use of transparent replicas allowed visualization which showed that this was due to intermittent blockage of gas flow paths at pore throats.

Capillary theory suggests that for gas-liquid flow in a fracture, the larger pore spaces will be occupied by gas (the nonwetting phase), and the smaller ones by liquid; as long as the boundary pressures do not change, the pore occupancy should be static. During two-phase flow in liquid-dominated conditions, pore occupancy was indeed generally static in almost the entire fracture, but frequently certain critical pores switched between gas and liquid occupancy, especially in liquid-dominated conditions. In a typical liquid-dominated flow condition, one or more distinct gas flow paths were evident. Part of the gas flow path was always occupied by gas, but several stretches of the gas flow path were intermittently occupied by liquid. This

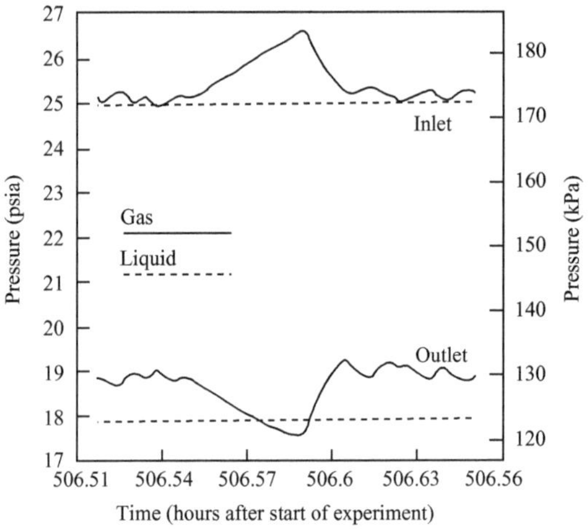

Figure 9.2. Example of "throat-clearing" event that recurred approximately every 20 minutes. (8.5 μm hydraulic aperture, Stripa replica, Expt A, gas:liquid volume flow ratio was 9.5)

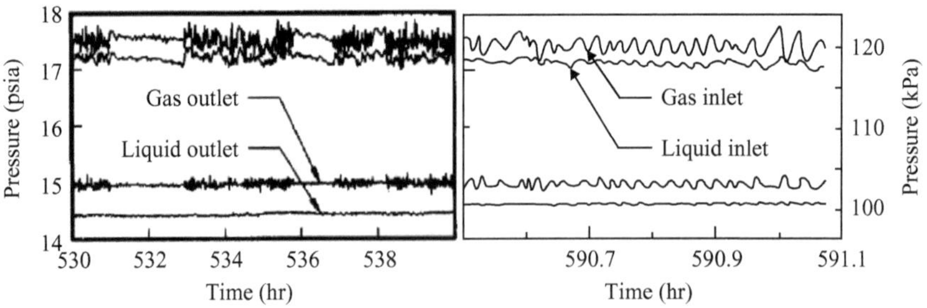

Figure 9.3. Examples of repeated blocking and unblocking events, alternating with periods of steady flow (21.7 μm hydraulic aperture, Stripa rock, Expt C. gas:liquid mass flow ratio was 0.025)

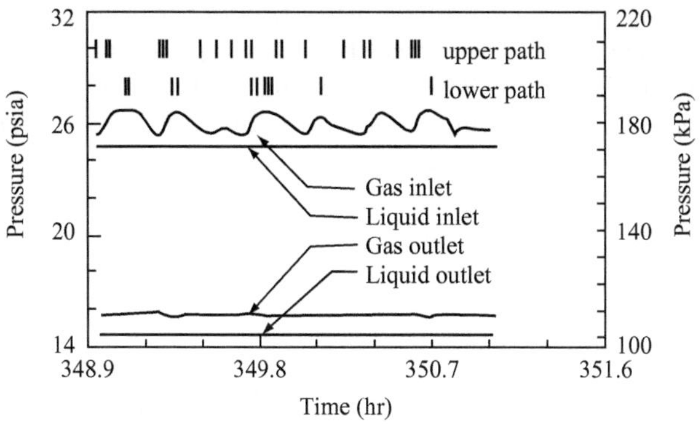

Figure 9.4. Examples of repeated blocking and unblocking events, on two gas flow paths (17.8 μm hydraulic aperture, Stripa replica, Expt B)

temporary invasion of water where it is "not allowed" caused the observed pressure excursions.

In Experiment A, the first gas flow path to be formed was near one edge of the fracture. When the gas:liquid flow rate ratio was low, steady pressures could not be reached; rather the system was disturbed by regular "hiccups" as shown in Figure 9.2 and Figure 9.3. These events were caused by periodic invasion and blockage of the gas flow path by liquid. A temporary blockage of the gas flow path caused the upstream gas pressure to increase, and the downstream gas pressure to decrease, as shown in Figure 9.3. Visual observation showed that all gas flowed through a critical path (choke point) near one side edge of the fracture. A dead-end gas-filled pore upstream of the choke point became larger, acting as a surge chamber, until the pressure drop across the blocked throat became great enough for gas to displace water from the throat, and flow was restored, with several bubble trains flowing toward the exit. The repeated reinvasion of gas filled pores by liquid is similar to the "snap-off" events in foam flow described by Ransohoff and Radke (1988), except that the liquid films between bubbles in these experiments collapsed instantly because they were not stabilized by surfactant. As the gas flow rate was increased and the liquid flow rate decreased, less water was available to invade gas flow paths and the throat-clearing events became less frequent and eventually ceased altogether.

In Experiment C in a natural rock fracture, the outlet liquid pressure was steady, but the outlet gas pressure and both inlet pressures cycled irregularly. As an example, pressures measured during two-phase flow with a gas:liquid volumetric flow ratio of 14.4 are shown in Figure 9.4. Generally, the pressures cycled irregularly around average values, but at times the pressure cycling stopped; and at others, it became very regular. As the gas:liquid ratio increased, the size of the pressure excursions decreased relative to the pressure drop across the fracture, as did their frequency. When the gas:liquid mass flow rate ratio was greater than 1, such pressure excursions were essentially eliminated.

To further examine the tendency of the system for persistent pressure cycling, we reviewed videotapes from Experiment B. A videotape record was available for the period shown in Figure 9.4. When viewing the videotape, no motion can be seen when gas and liquid are flowing in their own flow paths. Motion can only be seen when some portion of the fracture changes from gas-occupied to liquid-occupied or the reverse. Such phase-occupancy-change events (POC events) appeared to occur instantly and generally lasted only a few seconds. Typically, a slug of water would emerge from the liquid-occupied area and invade a gas flow path, move quickly along the gas flow path for one or two centimeters, and then disappear back into the liquid-occupied area. Afterward, there would often be a slight change in the overall pattern of phase occupancy in the fracture replica. By tracing the location of such motion on the monitor, it became clear that all such events occurred along two paths through the fracture; comparison with the aperture map produced from light-attenuation measurements showed, as expected, that they were in the regions of greatest aperture. Often, an event along one part of the flow path would be followed within a minute or so by an event farther along the same path. The occurrences of POC events are noted in Figure 9.4. POC events along the "lower" flow path were associated with maxima in gas inlet pressure and inlet capillary pressure; this indicates unblocking of a gas flow path.

The smallest pore along a gas flow path is the one most likely to be invaded and blocked by water. In these experiments gas was injected at constant flow rate, so when this happened the gas pressure increased upstream of the point of blockage until the pressure was sufficient to displace the water. But once gas broke through, the gas pressure (determined by the outlet pressure and flow rate) was insufficient to keep water out of the pore. This explains the instability for gas injected at constant flow rate. This blockage was only observed for gas:liquid flow rate ratios less than about 20:1. At greater gas:liquid flow rate ratios, the gas flow path is wide enough that it can contract without being shut off entirely. Another way to interpret these phenomena is to suppose that both phases could flow continuously in stable flow paths, but this would require a particular arrangement of gas and liquid pore occupancy. But if this arrangement does not minimize the gas-liquid interfacial area, then surface tension forces will tend to destabilize it. This may explain why in Experiment B, prolonged periods without pressure cycling were observed.

In Experiment D, gas was injected at constant pressure. Pressure cycling was therefore not observed, but changes in pore occupancy still occurred; at the upstream sides of regions of large aperture, where only gas would be "allowed," liquid would accumulate. Generally liquid would accumulate along the upstream edge of such a region, and occasionally jump across. This appeared to be an important mechanism of liquid transport.

Thus it appears that under certain conditions one phase or the other must be intermittently blocked by the other. Gas flow in a fracture is more susceptible to this kind of instability than in porous media because (unless the adjoining matrix is significantly permeable to gas) the two-dimensionality of the medium prevents gas from finding an alternative flow path; also, a fracture is highly permeable and can deliver water to block and reblock the critical pore.

Unstable gas flow in wells as observed, for example, at the Wellenberg site in Switzerland (Finsterle, 1994), may be an example of this type of multi-phase flow phenomena in which a non-wetting phase cannot maintain sufficient pressure to keep a flow path continuously open. Thunvik and Braester (1990) performed numerical simulations of two-phase flow in networks of capillary tubes. They showed that a constant gas pressure boundary condition in a fracture network could result in a cyclic gas flow through the network, as a result of some fractures alternating between water and gas occupancy. Recently, Faybishenko (2002) calculated diagnostic parameters of deterministic chaos from time-series data from the experiments of Persoff and Pruess (1995) and of Su et al. (1999).

9.3 BUBBLE FLOW IN FRACTURES BELOW THE WATER TABLE

Below the water table, fractures are by definition fully saturated, so that relative permeability to gas is zero. The presence of gas bubbles below the water table is by definition unstable; also the presence of gas makes the fracture at least locally unsaturated. In these cases gas exists not in continuous flow paths but as discrete bubbles. In vertical or inclined fractures, buoyancy is a significant force driving gas flow upward.

Water in a dry vertical or inclined fracture is gravitationally unstable and must flow down; if the rate of water delivery to the fracture is less than the conductivity of the fracture, the water will flow in fingers. This process has been studied for example by Su et al. (1999). The analogous process for gas occurs when gas is injected or formed in a fracture below the water table. This case is of interest in studying the loss of gas from underground storage caverns (Kostakis, 1998); also bubbles of gas may be formed below the water table by chemical reaction or exsolution of dissolved gas in response to pressure drop. The rise of bubbles in a saturated vertical fracture is analogous to the downward fingering of water in a dry fracture in that both events occur in response to gravitational instability. The processes differ, however, as shown in Table 9.1, because of different wetting character of the fluid relative to the fracture wall.

Natural gas can be stored in excavated caverns in hard rock. To prevent leakage of gas through fractures, water pressure in fractures surrounding the cavern is maintained greater than the gas pressure inside the cavern; this is called hydrodynamic containment, or a water curtain. (i.e., one does not depend on the air-entry value of the fracture to prevent gas from entering the fracture.) Simulating this situation, Kostakis (1998) investigated bubble initiation and flow in vertical and inclined water-filled fractures experimentally and with numerical simulations. A transparent replica of a natural rough fracture was filled with water and connected to a gas-filled reservoir at the bottom. Initially slugs of gas flowed rapidly through the fracture. Water pressure was increased stepwise to reduce the gas flow to a stream of bubbles, then individual

Table 9.1. Difference between gravity-driven flow of wetting and non-wetting fluids

Feature	Liquid fingering downward	Gas bubble rising	Factors affecting
Force causing fluid to enter fracture	Gravity aided by capillarity	Gravity opposed by capillarity. Air entry pressure must be exceeded	Surface tension opposing gas entry must be exceeded. Gas enters at the largest pore space
Shape of moving fluid – infinite source at fracture edge	Long fingers form	String of bubbles, size limited by water flow to cause snap-off	
Shape of moving fluid – finite source at fracture edge	Liquid remains as a film on walls and slug may be completely consumed in this way. Water will not penetrate indefinitely deep unless replenished from above	Finite number of gas bubbles, but will keep rising	Gas invades as a finger but bubble forms in a "snap-off" process. Bubble size depends on properties of liquid ρ, σ, and μ. Snap off occurs sooner (bubbles smaller) if σ or ρ is larger, μ is smaller
Fluid contact with fracture walls	Water wets fracture walls	Gas does not wet fracture wall, film of liquid remains between gas and wall	
Effect of constriction	Constriction restricts liquid flow but does not block penetration	Constriction opposes bubble rise and may cause blockage (Jamin effect)	

bubbles and finally to shut off (contain) the gas flow. Gas flow rates were estimated from bubble sizes and velocities as recorded on videotape.

The following features were observed: Gas flow was not contained until the increasing ΔP (water pressure minus gas pressure) was approximately 2.5 kPa. For ΔP below this value, bubbles always formed in the largest pore along the bottom edge of the fracture, which was approximately 1.5 mm. Observations from an actual gas-storage cavern showed that ΔP of approximately 25 kPa was needed to prevent gas leakage, suggesting that a fracture with a larger entry pore intersects the cavern.

In porous media, gas will not generally displace liquid from a pore until the gas pressure exceeds the liquid pressure by the "air-entry pressure," which is determined by the pore diameter and liquid surface tension. The data of Kostakis (1998), however, show, in effect, a negative air-entry pressure. This suggests that the invasion of gas into the fracture was determined not only by capillary forces but also, probably, gravity.

Gas never formed as a continuous finger but always as a series of bubbles; this shape minimizes the bubble surface area. Bubbles were always larger in dimension than the aperture, and were therefore "squashed flat" between the fracture walls. Bubbles were not elongated but took a distorted-circle shape determined by the aperture field through which they flowed. Largest bubbles were formed when the fracture was oriented 30° from horizontal, and smallest when it was 90° from horizontal, that is, vertical. This suggests, in accordance with snap-off theory, that the bubble size is controlled by the rate at which water can flow around the gas finger to snap off the bubble (i.e., water flows more slowly when the angle is small). Bubbles rose generally vertically, favoring larger pores, similar to the observation of Persoff and Pruess (1995). Bubbles rose more rapidly in the vertical than inclined fracture, and larger bubbles rose more rapidly, both explained by greater buoyancy. The flow is unstable in the sense that it is discontinuous.

Kostakis (1998) developed a mathematical model of bubble generation and flow in a fracture. Using an equation for the flow rate through an orifice at the point of bubble generation, the bubble size (radius of equivalent sphere) was first solved by a balance of all forces acting on the bubble (expressing acceleration as a virtual mass force), and then the gas momentum conservation equation was solved numerically, tracking the bubble through a 2-D grid representing the fracture. Resulting predictions were compared favorably with data from Kostakis' and other experiments.

9.4 SUMMARY

Gas flow in fractures is unstable when there is sufficient water to block the gas flow path. Below the water table, this causes gas to flow as bubbles. Above the water table gas flow is unstable when the gas pressure is insufficient to overcome surface tension forces that tend to draw water into the smallest pore in a gas flow path. Unstable gas flow may cycle regularly, and if boundary conditions remain steady, may be treated as constant in the average. Because of the greater permeability of fractures compared to porous media, velocities may be great enough to cause non-darcy flow, in which case pipe-flow models may be appropriate.

REFERENCES

Detwiler, R.L.; S.E. Pringle; and R.J. Glass . Measurement of fracture aperture fields using transmitted light: An evaluation of measurement errors and their influence on simulations of flow and transport through a single fracture , *Water Resour. Res.*, **35**(9), 2605–2617, 1999.

Faybishenko, B. Chaotic dynamics in flow through unsaturated fractured media. *Advances in Water Resources*, **25**, 793–816, 2002.

Finsterle, S. Inverse Modeling of Test SB4-VM2/216.7 at Wellenberg. Lawrence Berkeley Laboratory Report LBL-35454, 1994.

Kostakis E. Investigations on the movement of gas bubbles in a water-filled rock fracture. Ph.D. thesis, Imperial College of Science, Technology, and Medicine, London, 1998.

Persoff, P. and K. Pruess. Two-phase flow visualization and relative permeability measurement in natural rough-walled rock fractures, *Water Resour. Res.*, **31**(5), 1175–1186, 1995.

Ransohoff, T.C., and C.J. Radke Mechanisms of foam generation in glass-bead packs, SPE Reservoir Eng. **3**, 573–585, 1988.

Su, G.W., J.T. Geller, K. Pruess, and F. Wen, Experimental studies of water seepage and intermittent flow in unsaturated, rough-walled fractures, *Water Resour. Res.*,**35**(4), 1019–1037, 1999.

Thunvik, R. and C. Braester Gas migration in discrete fracture networks, *Water Resour. Res.*, **26** (10), 2425–2434, 1990.

CHAPTER 10

NATURAL CONVECTION GAS TRANSPORT IN POROUS MEDIA

K. KHANAFER AND K. VAFAI
Department of Mechanical Engineering, University of California, Riverside, CA 92521, USA

10.1 INTRODUCTION

Natural convection is the study of heat transport processes by fluid motion carrying energy with it as a result of the temperature difference between the fluid and the solid. Convection heat transfer consists of two main mechanisms due to both diffusion (random molecular motion) and bulk motion of the fluid.

Convection heat transfer can be classified according to the nature for the flow as forced convection and free convection. Forced convection investigates the heat transfer between a moving fluid and a solid surface. The fluid flow is caused by an external means such as a fan, a pump, or atmospheric winds. Thus the flow has a nonzero streaming motion in the farfield away from the solid surface. There are various types of forced convection such as duct flows and bodies immersed in a uniform stream.

In contrast, a free convection flow field is a self-sustained flow driven by the presence of a temperature difference, which is opposed to a forced convection where external means are used to cause the flow. Natural convection fluid motion is due solely to buoyancy force caused by the density differences as a result of the temperature difference. This force is a strong function of the temperature difference between the solid and the fluid. As such the buoyancy force will induce a flow current due to the gravitational field and the variation in the density field. Convection above a hot surface occurs because hot air expands, becomes less dense, and rises. In general, natural convection heat transfer is usually smaller compared to a forced convection heat transfer.

Natural convection can be divided in two main branches namely external and internal natural convection. External natural convection may occur along different geometries such as free convection along vertical walls, inclined walls, and horizontal

C. Ho and S. Webb (eds.), Gas Transport in Porous Media, 179–200.
© 2006 *Springer.*

walls. In addition, natural convection may take place around horizontal and vertical cylinders as well as around spheres. Natural convection may take place around other immersed bodies such as cubes and spheroids.

Unlike the external natural convection boundary layer that is caused by the heat transfer interaction between a single surface and a large reservoir, internal natural convection in enclosures is a result of the complex interaction between the fluid and all the surfaces that confine it.

In convection heat transfer, Grashof number replaces the Reynolds number in the convection correlation equations. Grashof number is defined as the ratio between the buoyancy force and the viscous force. In free convection, buoyancy driven flow dominates the flow inertia and therefore the Nusselt number is a function of Grashof and Prandtl numbers. However, Reynolds number is significant if there is an external flow. In many situations, it is advantageous to combine the Grashof number and the Prandtl number to define a new parameter called Rayleigh number ($\mathrm{Ra} = \mathrm{Gr}\,\mathrm{Pr}$). The significance of the Rayleigh number is to characterize the laminar to turbulent transition of a natural convection boundary layer flow.

The boundary conditions for convection heat transfer problems include no slip boundary conditions on the solid surfaces and either temperature or heat flux on the walls. In forced convection, the free stream farfield velocity is assigned along with the ambient temperature while temperature or heat flux may be applied on the solid surfaces. For the case of natural convection, either temperature or heat flux can also be applied on the solid walls while usually stagnant, isothermal infinite reservoir boundary conditions are applied for the far field conditions.

Fundamental studies associated with natural convection gas transport in porous media have increased substantially over the past decades because of the importance of porous media in diverse technological and industrial applications such as migration of pollutants, storage of nuclear waste, oil recovery enhancement, thermal insulation, electronics cooling and packed bed chemical reactors. Natural convection gas transport in porous media has been widely studied and well documented in the literature both experimentally and numerically. Major developments have been made in modeling natural convection gas transport in porous media including several important physical aspects. Some of the studies have used what is now commonly known as Brinkman-Forchheimer-extended Darcy or the generalized model. Significant advances have been made in developing the momentum equation that governs the fluid flow in porous media starting from Darcy's law to the generalized model. Darcy's law revealed proportionality between the velocity and the applied pressure difference for low speed flow in an unbounded porous medium. As such Darcy's law does not account for inertial effects or no-slip condition at the wall. To account for the solid boundary, Brinkman's equation which also known as Brinkman's extension of Darcy's law was developed. Brinkman's equation incorporated two viscous terms. The first is the typical Darcy term and the second is similar to the Laplacian term. Darcy's law is linear in the Darcy velocity, which holds for a sufficiently small velocity. At higher velocities, inertial effects become appreciable causing an increase in the form drag. Forchheimer's equation was developed as an extension to the Darcy's law to account for a quadratic drag. A generalized model for the fluid flow through a porous media was developed during the past couple of decades which accounts for the inertial and boundary effects, and the quadratic drag. These effects are incorporated by using this general flow model known as the Brinkman-Forchheimer-extended Darcy model.

A comprehensive literature review on natural convection gas transport in porous media is conducted targeting different categories of interest. These categories are

1. Natural Convection from a Horizontal Surface Embedded in a Porous Medium
2. Natural Convection from a Vertical Surface
3. Natural Convection from an Inclined Surface
4. Natural Convection in Vertical Channels
5. Natural Convection around Cylinders
6. Natural Convection around Spheres
7. Natural Convection in Enclosures
8. Double-Diffusive Natural Convection within a Porous Medium.

10.2 NATURAL CONVECTION FROM A HORIZONTAL SURFACE

Convection heat transfer from surfaces embedded in porous media has received substantial amount of attention because of applications in geophysical and energy-related engineering problems. Comprehensive reviews were presented by Cheng (1985) and Nield and Bejan (1992).

The problem of steady free convection in a porous medium adjacent to a horizontal impermeable heated surface, with variable wall temperature was investigated by Chang and Cheng (1983) using the method of matched asymptotic expansions. The effects of fluid entrainment, streamwise heat conduction and upward-drift induced friction were taken into consideration in the second and third-order theory for which similarity solutions were obtained.

The free convective boundary layer above a near-horizontal heated flat surface bounding a saturated porous medium was studied by Rees and Riley (1985). Two configurations were considered: one where the component of the buoyancy force along the surface assisted the flow, the other opposed the flow. Series solutions were developed: one valid near the leading edge, where the flow was driven along by an induced pressure gradient, and for the favorable case only, an asymptotic solution, where the flow was driven along by the direct action of buoyancy forces.

Boundary-layer analysis was performed by Pop and Gorla (1991) for free convection flow over a hot horizontal surface embedded in a porous medium saturated with a gas with variable properties. The variable gas properties were accounted for via the assumption that thermal conductivity and dynamic viscosity were proportional to temperature. A similarity solution was shown to exist for the case of constant surface temperature. Numerical results for the stream function, horizontal velocity, and temperature profiles within the boundary layer as well as for the mass of entrained gas, surface slip velocity, and heat transfer rate at different values of the wall-temperature parameter were presented. Asymptotic solutions for large heating were also available to support the numerical work.

Lesnic et al. (1995) presented a detailed analysis of the conjugate free convection from a horizontal flat plate of finite length embedded in a porous medium. The boundary-layer approximation was considered in this study. The problem was reduced to a pair of coupled parabolic partial differential equations for the temperature and the stream function and full numerical and asymptotic solutions were obtained for a wide range of coordinate values along the plate. The results for the temperature and velocity profiles on the plate and in the convective fluid were presented. A comparison between the finite-difference solution and the small and large series expansion solutions illustrated that the full numerical solution is accurate.

The effect of inertia on free convection from a horizontal surface embedded in a porous medium was considered by Rees (1996). In this investigation, the author considered the case where the inertia effects were sufficiently large that the leading-order boundary layer theory is modified; this was equivalent to reconsidering flow sufficiently near the leading edge that the induced velocities were large enough for inertia effects to arise, but sufficiently far from the leading edge that the boundary layer approximation remained valid. The resulting nonsimilar boundary layer equations were solved using the Keller box method. Near the leading edge inertial effects were found to dominate, but Darcy flow was re-established further downstream.

The free convection in a porous medium adjacent to a horizontal surface subject to spatial wall temperature distribution was analyzed by Lesnic and Pop (1998). In the

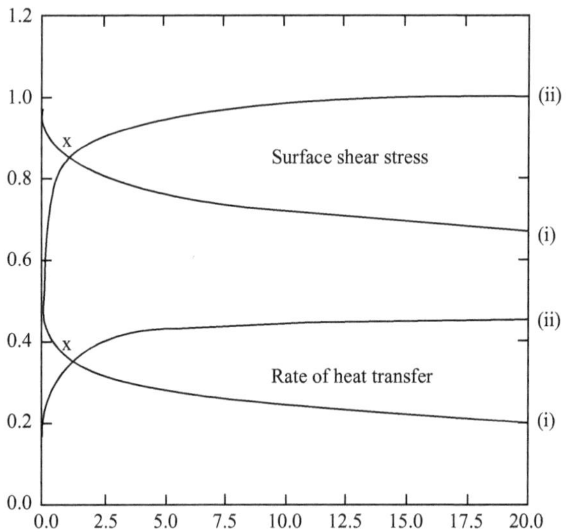

Figure 10.1. Surface shear stress and rate of heat transfer given by (i) Eqs. (51a) and (ii) Eqs. (51b). Also shown are the corresponding results obtained from the small-x asymptotic analysis (from Rees and Vafai (1999))

steady case, when the wall temperature was a quadratic function along the surface, it was shown by that a steady free convective boundary-layer flow solution existed if and only if the quadratic wall temperature was a convex function. This was also confirmed for the unsteady case for a concave parabolic wall temperature prescription. Rees and Vafai (1999) investigated boundary and inertial effects for natural convection from a horizontal plate embedded in a porous medium. The governing equations were studied using both numerical and asymptotic methods.

10.3 NATURAL CONVECTION FROM A VERTICAL SURFACE

Buoyancy-induced flow along a vertical plate embedded in a porous medium is a fundamental problem. This problem has received considerable attention due to possible applications in industries involving heat exchanger design, petroleum production, filtration, chemical catalytic reactors, and nuclear waste repositories.

The problem of natural convection about a heated impermeable surface embedded in fluid-saturated porous media was investigated by Cheng and Minkowycz (1977). They obtained similarity solutions, which were based on Darcy's law and boundary layer approximations. To extend the range of applicability of the boundary layer analysis to relatively lower modified Rayleigh numbers, Cheng and Hsu (1984) and Joshi and Gebhart (1984) examined higher order effects such as the entrainment from the edge of the boundary layer, the axial heat conduction, and the normal pressure gradients using the method of matched asymptotic expansions. The Brinkman model was used for the theoretical study of boundary effects for a natural convection in a porous medium adjacent to a semi-infinite vertical plate with a power law variation of wall temperature by Hsu and Cheng (1985).

The heat-transfer rate from an isothermal vertical plate placed next to saturated high-permeability porous media was studied experimentally and analytically by Kaviany and Mittal (1987). The medium was polyurethane foam saturated with air. An integral method was applied in predicting the heat-transfer rate by including non-Darcy effects, which was expected to be significant at high permeabilities and high Rayleigh numbers. The agreement between experimental and predicted results was good except when the permeability was relatively low and the Peclet number was very small. Although some of the permeabilities encountered were large due to experimental constraints, Rayleigh numbers were not very high, resulting in insignificant non-Darcian effects.

Hong et al. (1987) examined analytically the effects of non-Darcian and nonuniform permeability conditions on natural convection from a vertical plate embedded within a porous medium. The non-Darcian effects, which include the no-slip and inertia effects, decrease the flow and heat transfer rate, while the nonhomogeneity effect enhances the heat transfer. For packed spheres, in particular, the nonhomogeneity in permeability due to the packing of spheres near the solid wall results in a strong flow-channeling effect that significantly increases the heat transfer. The effect of transverse thermal dispersion was also examined.

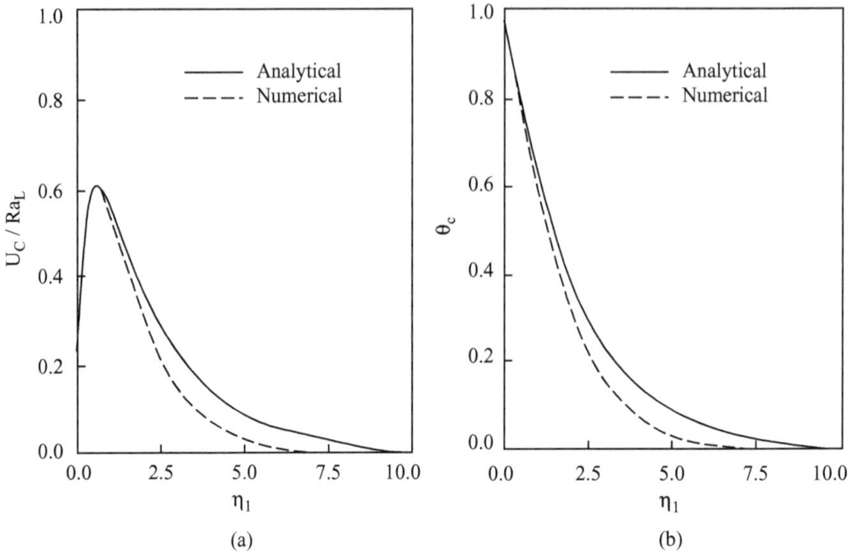

Figure 10.2. (a) Comparison between the analytical and numerical velocity distribution for Case TI, $\Gamma_T x^{-1/2} = 0.3$ and (b) Comparison between the analytical and numerical temperature distribution for Case TI (from Kim and Vafai (1989))

Buoyancy-driven fluid flow and heat transfer about a vertical plate embedded in a porous medium were analyzed in depth by Kim and Vafai (1989) for constant wall temperature and constant wall heat flux cases as depicted in Fig. 10.2. The method of matched asymptotic expansions was used to obtain comprehensive analytical solutions for both the velocity and the temperature fields as well as Nusselt numbers.

Heat transfer through a vertical skin surface covered with perpendicular hair strands of uniform density was investigated numerically by Lage and Bejan (1991). The heat transfer rate was the result of (1) direct heat transfer to the air that makes contact with the skin and (2) the heat conducted by each strand away from the skin. The hair strand and its surrounding air were not in local thermal equilibrium. It was found that hair strands have the desirable effect of slowing the air that sweeps the vertical surface and the undesirable effect of acting as fins, thereby augmenting the overall heat transfer rate. Two distinct air flow models were considered: Darcy model and the Forchheimer-Brinkman extended Darcy model. The overall heat transfer charts reported illustrated that the heat transfer rate can greatly exceed the estimate based on the traditional homogeneous porous medium model. By means of numerical examples, the Darcy model was shown to be adequate for modeling air flow through mammal hair.

The effects of uniform suction and injection on the linear stability theory of laminar natural convection boundary layer flow along a vertical flat plate at a constant temperature and embedded in a porous medium was studied by Watanabe et al. (1996).

The nonsimilar boundary layer equations were solved numerically employing a very efficient finite difference scheme in combination with an iterative method for solving the resulting ordinary differential equations. The temporal neutral stability theory for wavelike disturbances of Tollmien-Schlichting type is then presented for the velocity and temperature functions. The corresponding eigenvalue problem for the disturbance amplitude functions is also solved numerically using a very accurate method. Results are presented graphically for both the basic and disturbance velocity and temperature profiles for some values of the modified local Grashof number G and suction or injection parameter X. The Prandtl number Pr is taken to be 0.73 (air) throughout this paper. The results show clearly the important role that the suction or injection parameter X may have on the base and disturbed flow characteristics. Chamkha (1997a) analyzed free convection flow of an electrically conducting fluid along a vertical plate embedded in a thermally stratified porous medium in the presence of a uniform normal magnetic field. In his study Chamkha (1997a) considered non-Darcian effects, Hartmann and Hall effects of magnetohydrodynamics, as well as thermal stratification effects.

The transient free convection boundary-layer flow of a viscous and incompressible fluid adjacent to a semi-infinite vertical flat plate is investigated for Prandtl number of unity by Harris et al. (1998). An analytical solution was presented which is valid at small values of τ. The existence of two solutions was demonstrated which is physically acceptable. Hossain et al. (1999) analyzed the effect of radiation on the natural convective flow of an optically dense incompressible fluid along a uniformly heated vertical plate with uniform suction. The governing nonsimilar boundary-layer equations were analyzed using (i) series solution for small values of ξ (a scaled streamwise coordinate); (ii) asymptotic solution for large ξ; and (iii) a full numerical solution. The solutions were expressed in terms of the local shear stress and local rate of heat transfer. The effects of varying the Prandtl number ($1 \leq Pr \leq 5$), the radiation parameter, and the surface temperature parameter were determined.

Watanabe et al. (1999) conducted a theoretical study for the stability characteristics of the laminar free convection boundary layer flow along a vertical porous (permeable) flat plate subject to a constant heat flux. The disturbance equations were solved numerically on the basis of the linear stability theory for a wide range of values of the modified Grashof number, G, and some values of the suction or injection parameter X using air as the fluid medium. These solutions indicate the important role of the parameters G and X on the flow and heat transfer characteristics.

The effect of thermal radiation on the natural convection flow along a uniformly heated vertical plate embedded in porous medium with variable viscosity and uniform suction velocity was investigated numerically by Hossain et al. (2001). The fluid considered in this study was considered to be optically dense. The governing equations were analyzed using a variety of methods: (i) a series solution for small values of a scaled streamwise coordinate; (ii) an asymptotic solution for large values of the scaled streamwise coordinate; and (iii) a full numerical solution using the Keller box method. The solutions are expressed in terms of the local shear stress and the local heat transfer rate. The working fluid was taken to have a Prandtl number of unity.

Jian et al. (2001) analyzed numerically natural convection flow from a vertical flat plate with a surface temperature oscillation. Prandtl number was assumed to be unity. In the steady case, numerical results for the Grashof numbers 0–625 were obtained using an iterative approach and the results for small Grashof numbers were validated using a perturbation method. For larger values of the Grashof numbers, an unsteady numerical scheme was constructed and the results obtained at large times were compared with steady state solutions.

10.4 NATURAL CONVECTION FROM AN INCLINED SURFACE

A number of analytical and numerical studies on the natural convection flow along an inclined surface embedded in a saturated porous medium have been carried out by a number of investigators due to its applications in geothermal energy, insulation systems, and heat storage in aquifers. Jang and Chang (1988) have analyzed the free convection boundary layer flow over an inclined surface embedded in a saturated porous medium retaining both the streamwise and normal components of the buoyancy force in the momentum equation. The analysis was valid for the wide range of inclination of the surface ranging from zero to close to 90 degrees from the horizontal.

Transient natural convection for an inclined flat plate embedded in a porous medium was presented by Zeghmati et al. (1991). The problem was treated by considering two separate regions that is, the boundary layer and the capillary-porous plate-for which a specific differential system of equations was developed. The two systems were linked with the wall heat and mass balances from which the local and average

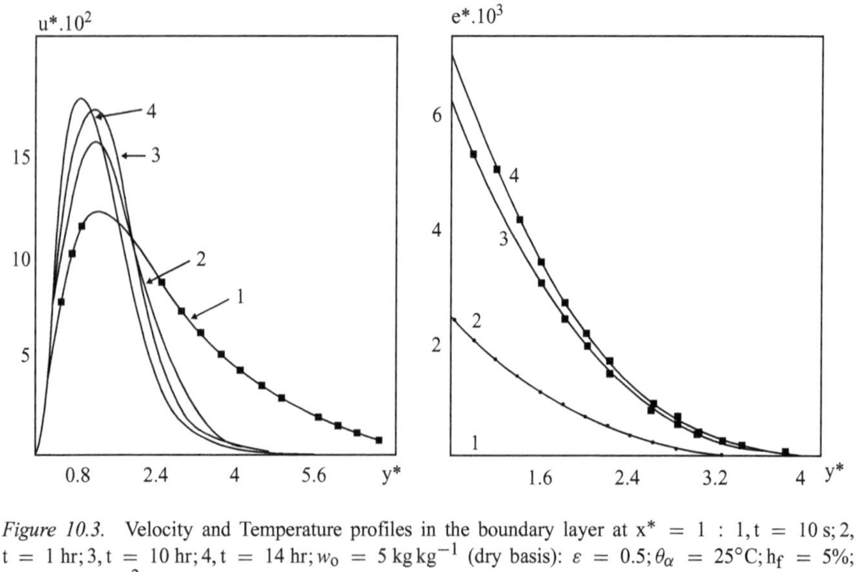

Figure 10.3. Velocity and Temperature profiles in the boundary layer at $x^* = 1 : 1, t = 10$ s; 2, $t = 1$ hr; 3, $t = 10$ hr; 4, $t = 14$ hr; $w_o = 5\ \mathrm{kg\,kg^{-1}}$ (dry basis): $\varepsilon = 0.5; \theta_\alpha = 25°C; h_f = 5\%$; $Q = 500\ \mathrm{W\,m^{-2}}; \alpha = 40°$ (from Zeghmati et al. (1991))

Nusselt and Sherwood numbers are deduced. For some particular cases, quantitative comparisons with previous works reported in the literature were reported and the agreement between theoretical results and experimental data was found to be satisfactory.

A mathematical model governing free convection boundary-layer flow over an isothermal inclined plate embedded in a thermally stratified porous medium in the presence of a non-uniform transverse magnetic field was developed by Chamkha (1997b). The resulting equations of that study account for non-Darcian boundary and inertial effects of the porous medium and allow for variable ambient temperature. In addition, the applied magnetic field was assumed to be an increasing function of the distance along the plate. Prandtl number of unity was assumed in that study. The results showed that as the Hartmann number increases, both the skin-friction coefficient and the local Nusselt number are increased. The radiation effect of an optically dense viscous incompressible fluid along a heated inclined flat surface maintained at a uniform temperature and embedded in porous media was studied numerically by Hossain and Pop (1997). The results of that study showed that as the buoyancy parameter increases, the local Nusselt number increases.

10.5 NATURAL CONVECTION IN A VERTICAL CHANNEL

Analytical solutions for fully developed MHD natural convection flow in open-ended vertical porous channels were presented by Al-Nimr and Hader (1999) using a Darcy model. Four fundamental boundary conditions have been investigated and the corresponding fundamental solutions were obtained. Expressions for the flow and heat transfer parameters were given in terms of the Darcy number, Prandtl number and other parameters.

Transient natural convection between two vertical walls filled with a porous material having variable porosity was analyzed numerically by Paul et al. (2001) in which inertial and variable porosity effects were investigated. The Brinkman Forchheimer-extended Darcy model was used to simulate the momentum transfer in the porous domain. A numerical scheme for the transient solution of the governing equations was devised to investigate the channeling effect on the velocity field due to variable porosity of the porous matrix. The finer grid points taken near the walls revealed that the influence of variable porosity is significant only near the heated wall. It was found that the Darcy model predicts an enhancement in the air velocity near the heated wall compared to the general flow model. Formation of the boundary layer near the heated wall was the result of the Brinkman term and its effect was strongly dependent on the diameter of the spherical beads.

10.6 NATURAL CONVECTION ABOUT CYLINDERS

The local nonsimilarity solution method was applied by Minkowycz and Sparrow (1974) to solve for natural convection on a vertical cylinder for conditions where there were large deviations from the flat plate results. To assess and insure the accuracy

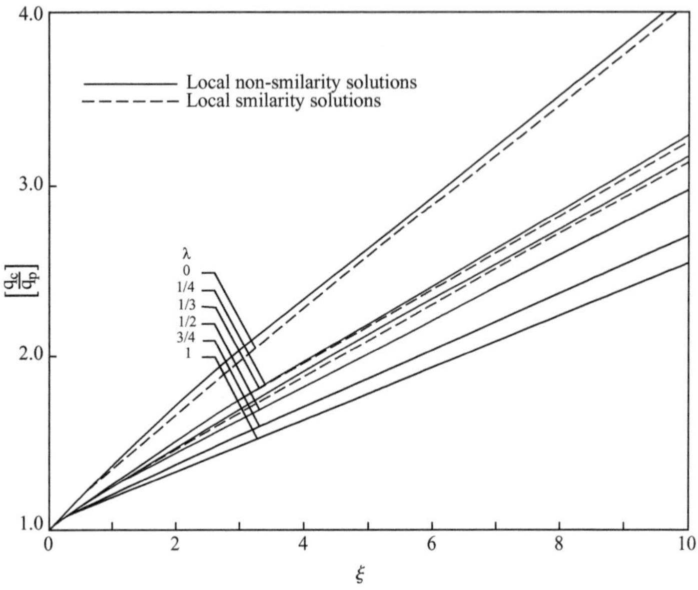

Figure 10.4. Local heat transfer ratio for selected values of λ (from Minkowycz and Cheng (1976))

of the results, solutions were obtained for three levels of truncation of the governing equations. The solutions were carried out for Pr $= 0.733$ and for a range of cases extending from small deviations from the flat plate results to a factor-of-four deviation between the local heat fluxes for the cylinder and the flat plate. The results provided by the local similarity solutions were found to be quite accurate over this range. Comparisons were made with available results for local and surface-integrated heat transfer, and solution methods were identified, which appeared promising for other applications. A presentation of representative temperature and velocity profiles showed only small deviations between the local similarity solutions and those for the third level truncation.

Natural convection about a vertical cylinder with a power-law temperature and embedded in a saturated porous medium was studied by Minkowycz and Cheng (1976). Within the framework of boundary-layer approximations, an exact solution was obtained for the special case where surface temperature varies linearly with x. For a given value of the power-law exponent, the results show that the ratio of local surface heat flux of a cylinder to that of a flat plate was nearly a linear function of a curvature parameter.

The steady free convection boundary layer flow of non-Darcy fluid along an isothermal vertical cylinder embedded in a saturated porous medium using the Ergun model was studied by Kumari et al. (1986). The partial differential equations governing the flow were solved numerically using an implicit finite-difference scheme. The results showed that the heat transfer was strongly affected by the modified Grashof number, which characterizes the non-Darcy fluid, and the curvature parameter. The natural convection from a heated circular cylinder into an unbounded porous medium was investigated for the full range of Rayleigh numbers by Ingham and Pop (1987). At small Rayleigh numbers a qualitative solution was obtained and at large Rayleigh numbers the second-order boundary-layer solution was found that took into account the first-order plume solution.

The effect of conduction-radiation on natural convection flow of an optically dense viscous incompressible fluid along an isothermal cylinder of elliptic cross section was investigated by Hossain et al. (1998). The boundary layer equations governing the flow were shown to be nonsimilar. Full numerical solutions of the governing equations were obtained using the implicit finite difference method. The solutions were expressed in terms of the Nusselt number against the eccentric angle alpha in the range $(0, \pi)$. The working fluid was assumed to have a unit value for the Prandtl number, Pr. It was found that the rate of heat transfer from the slender body is higher than the corresponding one for the blunt body and that the radiation field further enhances this increase.

The flow and heat transfer by transient natural convection in a vertical cylinder filled with air saturated porous medium was studied by Slimi et al. (1998). The cylinder is opened at both ends and heated with a constant wall heat flux density. That study was carried out using the Forchheimer-extended Darcy flow model and a two-temperature model. The results of that study provided the validity of the Darcy flow model and the thermal boundary layer approximations. In a related study, Hossain et al. (1999) considered non-Darcy convection heat and mass transfer along a vertical permeable cylinder embedded in a porous medium for Prandtl number of unity.

10.7 NATURAL CONVECTION ABOUT SPHERES

Natural convection flow and heat transfer due to the presence of a heated sphere embedded in a porous medium has motivated many researchers owing to its wide-ranging applications in a number of fields such as chemical engineering, thermal insulation systems, and nuclear waste management. Yamamoto (1974) was the first to consider the natural convection problem around a sphere in a porous medium. Asymptotic solutions for small Rayleigh numbers were obtained for steady natural convection from a constant surface temperature sphere.

Sano and Okihara (1994) analyzed natural convection around a sphere embedded in a porous medium for small Rayleigh numbers. The sphere was suddenly heated and, subsequently, maintained at a constant heat flux over the surface. Asymptotic solutions were obtained for the transient and steady-state temperature distribution around the sphere. The effects of viscous dissipation, Joule heating and heat source/sink on non-Darcy MHD natural convection flow over an isoflux permeable sphere in a porous medium are numerically analyzed by Yih (2000). The governing equations are transformed into the nonsimilar boundary layer equations and solved by the Keller box method. Numerical results for the wall shear stress and the Nusselt number are presented for the dimensionless coordinate along the surface, the Forchheimer number,

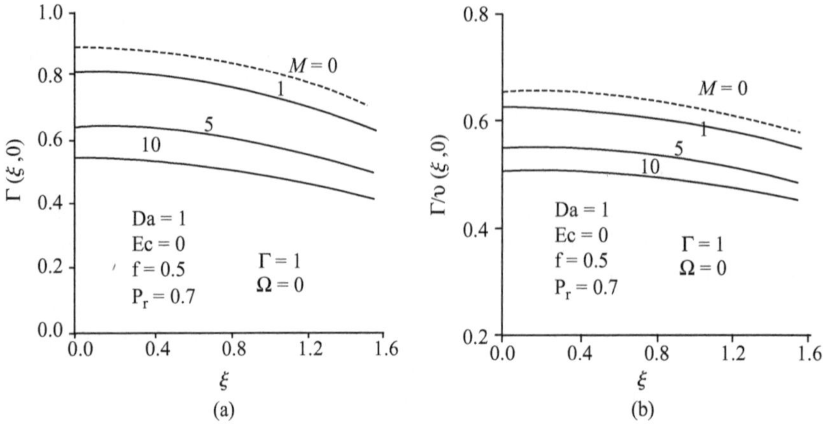

Figure 10.5. (a) $f''(\xi, 0)$ for various values of M; (b) $1/\theta(\xi, 0)$ for various values of M (from Yih (2000))

the magnetic parameter, the Darcy number Da, the Prandtl number Pr $= 0.7$, the Eckert number Ec, the heat source/sink parameter, and the suction/blowing parameter.

10.8 NATURAL CONVECTION IN ENCLOSURES

Natural convection in a porous medium has several important geophysical and engineering applications. These include regenerative heat exchangers containing porous materials, solar power collection, petroleum reservoir, building insulation, heat storage beds, nuclear waste repository, gain storage, and underground water contamination. Several investigators have presented analytical, numerical and experimental results for natural convection within a fluid saturated porous medium.

Measurements of the heat transfer by natural convection across vertical and inclined air layers are reported by Elsherbiny et al. (1982). The air layer is bounded by flat isothermal plates at different temperatures and around the edges by a perfectly conducting boundary. Measurements are reported with aspect ratio 5–110, covering a portion of the range of practical interest for windows, solar collectors, etc. A range of Rayleigh numbers from 10^2 to 10^7 were covered. The measurements enable the role of the aspect ratio to be defined clearly.

Tong and Gerner (1986) studied the effect a vertical partition has on steady-state natural convection in air-filled rectangular enclosures. A finite-difference scheme was used to solve the governing equations. Computed Nusselt numbers were presented as a function of the governing parameters. It was found that placing a partition midway between the vertical walls of an enclosure produces the greatest reduction in heat transfer and often compares favorably with insulating the entire enclosure with a porous material. Natural convection in rectangular enclosures partially filled with a porous medium was analyzed by Tong and Subramanian (1986). The two regions were separated by an impermeable wall and the vertical and horizontal boundaries were considered to be isothermal and adiabatic, respectively. The flow in the porous region was modeled by a modified Darcy's law where Brinkman's extension was incorporated to allow the non-slip condition to be satisfied. A finite-difference scheme was used to numerically solve the field equations in the two regions.

Beckermann et al. (1986) performed a numerical study of non-Darcian natural convection in a vertical enclosure filled with a porous medium. The flow was modeled using the Brinkman-Forchheimer-extended Darcy equations. The governing equations were solved using the SIMPLER algorithm and good agreement with previously reported numerical and experimental results was found. They demonstrated the importance of non-Darcian effects. For high Darcy numbers ($Da > 10^{-4}$), both boundary and inertial effects were of the same order of magnitude and had to be used simultaneously. In addition, Forchheimer's extension had to be included for $Pr \leq 1.0$ for all Darcy numbers. Finally, Nusselt number correlations were presented for three different ranges of the Darcy number covering a wide range of governing parameters.

Vafai and Sarkar (1987) analyzed the condensation and phase change processes in an enclosure partially filled with a porous insulation. The effect of variations in the porous insulation thickness on the moisture, relative humidity, temperature, and condensation rate fields was investigated. The problem was modeled as a transient, multiphase flow in a composite slab consisting of a porous portion followed by an air gap with impermeable, adiabatic horizontal boundaries and permeable vertical boundaries. The thickness of the porous insulation was varied between 60 and 100% of the overall thickness of the enclosure. For some typical conditions in a building insulation, it was found that the condensation rate and the resultant liquid accumulation did not increase significantly as the thickness of the insulation was decreased in the aforementioned range. Condensation and phase change processes in an enclosure, which was completely filled with a porous medium, was investigated earlier by Vafai and Sarkar (1986).

The applicability of the Boussinesq approximation was investigated for natural convection in a fluid-saturated porous cavity with vertical walls maintained at different temperatures and insulated horizontal walls by Peirotti et al. (1987). Numerical calculations were performed for water and air for a wide range of Rayleigh numbers and aspect ratios. Flow and temperature fields and heat-transfer rates, obtained through the evaluation of a model that includes temperature-dependent properties, were presented. The authors concluded that under certain circumstances the Nusselt number evaluated through the Boussinesq approximation differs substantially from the Nusselt number obtained with this model.

A numerical study of heat and mass transfer with phase change in porous materials was performed by Vafai and Tien (1989). The problem was modeled by a system of transient inter-coupled equations governing the two-dimensional multiphase transport process in porous media. The solution algorithm allowed full simulation without any significant simplifications. The variations and the interrelationships between the temperature vapor density, condensation rate, liquid content and the fluid velocity fields were demonstrated and discussed in detail. The aspect ratio of the porous matrix

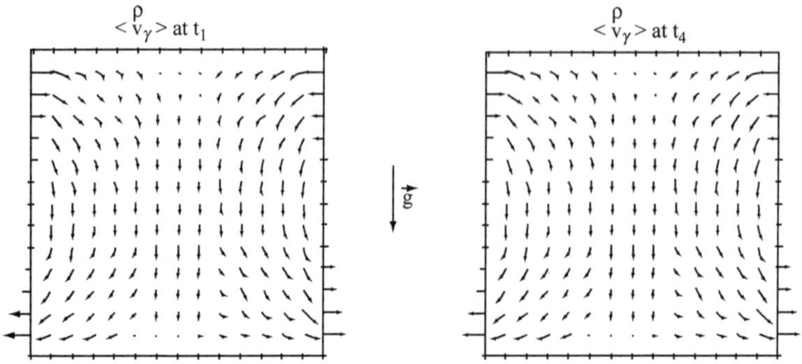

Figure 10.6. Gas phase velocity plots for $A = 1$, $\omega_h = 1.0$, at two different times: $t_1 = 0.0005$: $t_4 = 0.01$ (from Vafai and Tien (1989))

did not have a significant effect on the Nusselt number results. The authors showed that the one-dimensional model was not valid when the boundaries of the porous matrix were subjected to a small or zero pressure difference and that the constant pressure simplifications would generate significant errors under some circumstances. The results of this study were applicable to a class of problems on heat and mass transport with phase change through a porous medium.

Later on, the same authors analyzed the infiltration and thermal stratification effects on heat and mass transfer in porous insulation (Tien and Vafai (1990)). The vertical boundaries of the porous system were partially permeable for simulating holes or cracks in walls. Hydrostatic pressure variations were considered on the vertical boundaries and a set of realistic boundary conditions was imposed on the system under consideration. The transient inter-coupled equations governing the complicated transport process along with the convective boundary conditions were solved by an efficient numerical scheme. The dependence of the Nusselt number and the field variables on several important parameters was investigated systematically. The results showed that the infiltration had a major effect on the overall heat transfer even for small pressure gradients across the insulation slab. Furthermore, the opening locations had a pronounced effect on the heat transfer rate across the slab and the corresponding condensation rate.

Steady natural convection heat transfer in a two dimensional cavity filled with a uniform heat generating, saturated porous medium was studied by Du and Bilgen (1992) for a wide range of Rayleigh number and aspect ratios. Various heat transfer modes were identified in that study depending on the Rayleigh number and the aspect ratio range.

The thermal impact of the convective inertia term was investigated numerically considering Benard convection in a fluid-saturated porous medium by Lage (1992). The results of that study showed that the inertial term included in the general momentum equations had little effect on the calculation of the overall heat transfer. Later on, the same author (1993) performed an analytical study to identify the range of influence of each term of the general equation for flow within a fluid saturated porous medium.

Natural convection in a shallow laterally heated air-filled cavity was investigated numerically by Wright et al. (1995) using a primitive variable formulation of the governing equations. Solutions were obtained for the entire flow field using a coupled solver combined with an FAS non-linear multigrid convergence accelerator. While care was taken to use a high-order, bounded discretization scheme for convective transport, the overall stability and efficiency of the approach was enhanced through the use of a defect correction. This combination of features enabled solutions to be found for very fine grids. Results for the flow in the end region of such cavities were compared qualitatively with the predictions of asymptotic theory for large aspect ratio, A (length: height), up to values of A higher than previously reported ($A = 100$); the particular case of $A = 20$ was considered for Rayleigh numbers in the range of 10^3 to 10^8. Finally it was demonstrated how such solutions can be further enhanced by using locally refined grids in the end regions.

The theory of thermally driven convection of dry air in a porous medium was reviewed by Stauffer and Auer (1997). Because of the differences in the physical properties of air and water, initiation of convection required the product of thermal gradient and permeability to be thousands of times greater for air than for water. Finite amplitude analysis of the problem for Ra $<$ 300 revealed that (1) at low thermal gradient, Ra vs. Nu curves are nearly the same for air and water; (2) the slope of the Ra vs. Nu curve matches well with experimental data reported by others for water; (3) time to reach steady state decreases approximately as the square root of Nusselt number.

Multiple steady-state solutions of natural convection in an inclined enclosure with a fluid layer and a heat-generating porous bed were investigated numerically using the finite volume method by Chen and Lin (1997). The conservation equations for the porous layer were based on a general flow model, which includes boundary and inertial effects. The flow in the fluid layer was modeled by Navier-Stokes equations. The method of pseudo arc-length continuation was adapted in studying the effects of tilt angle on flow pattern and heat transfer. It was found that, based on the title angle, there exists two groups of solutions with quite different flow pattern and heat transfer behavior. The effects of aspect ratio on flow pattern and heat transfer have also been studied. Prandtl number of 0.7 is assumed in that study.

Natural convection heat transfer in a fluid saturated variable porosity medium was investigated by Nithiarasu et al. (1997a). A generalized non-Darcian porous medium for natural convection was developed taking into account linear and non-linear matrix drag components as well as the inertial and viscous forces within the fluid. A Prandtl number of unity was assumed in that study. The authors established that the thickness of the porous layer and the nature of variation in porosity substantially affect the natural convection flow pattern as well as the heat transfer features.

The problem of unsteady, laminar, two-dimensional hydromagnetic natural convection heat transfer in an inclined square enclosure filled with a fluid-saturated porous medium in the presence of a transverse magnetic field and fluid heat generation effects was studied numerically by Khanafer and Chamkha (1998). The walls of the enclosure were maintained at constant temperature. The flow in the porous region was modeled using the Brinkman-extended Darcy's law to account for the no-slip conditions at the walls. The control volume method was used to solve the governing balance equations for different values of the Darcy number, Hartmann number, and the inclination angle. The obtained numerical results were presented graphically in terms of streamlines and isotherms as well as velocity and temperature profiles at midsections of the cavity to illustrate interesting features of the solution.

A pressure-velocity solution for natural convection for fluid saturated heat generating porous medium in a square enclosure was analyzed using the finite element method by Das and Sahoo (1999). The numerical solutions obtained for a wide range of fluid Rayleigh number, Darcy number, and heat generating number for a Prandtl number of unity. It was observed that the peak temperature occurs at the top central part and weaker velocity prevails near the vertical walls of the enclosure due to the heat generation. The modified Rayleigh number used by earlier investigators cannot explain explicitly the effect of heat generation parameter on natural convection within an enclosure having differentially heated vertical walls. As expected, increasing Darcy number, results in a higher peak velocity. Experimental investigations are performed for natural convective heat transfer in a composed rectangular-parallelogrammic enclosure by Hwang and Kim (1999). The experiments covered a range of modified Grashof numbers utilizing various inclination angles. The air-filled enclosure consists of vertical heat source and sink walls, and adiabatic top and bottom walls. The effects of a guide vane installed within the composed enclosure and the dimensionless channel depth were studied. Experimental results were given as plots of Nusselt number vs. the inclined angles and dimensionless channel depth.

Thick layers of highly permeable loose-fill insulation are used to thermally insulate attics in modern building technology. Since the use of highly permeable insulation leads to an increase in air movement not only in the porous layer (insulation) but also in the fluid layer (attics), it is interest to determine the influence of natural and forced convection on thermal properties of the insulating porous medium. In related to this type of application, a numerical investigation of natural convection in horizontal porous media heated from below was analyzed both numerically and experimentally by Shankar and Hagentoft (2000). The combined effect of air flow within the open portion of the cavity as well as the porous layer was investigated. The results obtained from the numerical computations were compared with experimental findings.

As an application to the nuclear waste management, Webb (2001) provided a method that allows one to compute an equivalent permeability using Darcy's law based on open space flow physics. It was done for a square enclosure with and without a porous media fill. Webb and Hickox (2001) applied the equivalent permeability analysis to horizontal concentric cylinders and compared the results to a CFD calculation for horizontal concentric cylinders.

10.9 DOUBLE-DIFFUSIVE NATURAL CONVECTION WITHIN A POROUS MEDIUM

Double-diffusive natural convection in a fluid-saturated porous medium has received considerable attention in a variety of applications such as food processing, geophysical systems, contaminant transport in groundwater, and grain storage. Karimi et al. (1997) conducted a numerical analysis of double-diffusive convection in a square cavity filled with a porous medium for various pertinent controlling parameters. In double-diffusive heat transfer convection, the flow is driven by a combined buoyancy effect due to both temperature and concentration variations. The Prandtl number was taken to be unity in this study. Several different pertinent flow models for porous media, such as Darcy flow, Forchheimer's extension, Brinkman's extension, and the generalized flow were considered. The coupled equations were solved using a finite volume approach with a projection algorithm for the momentum equation. Non-Darcian effects were analyzed through investigating the average heat and mass transfer rates. The study consisted of a global analysis for each model and comparison between them when Darcy number varies. The influence of the Lewis number on the inertial and boundary effects was established and it was shown that the inertial and boundary effects have a profound effect on the double-diffusive convection.

Double-diffusive natural convective flow for air within a rectangular enclosure was studied using a generalized porous medium approach by Nithiarasu et al. (1996). The results were validated with the help of theoretical heat transfer results available for various porous medium flow models and also with experimental results

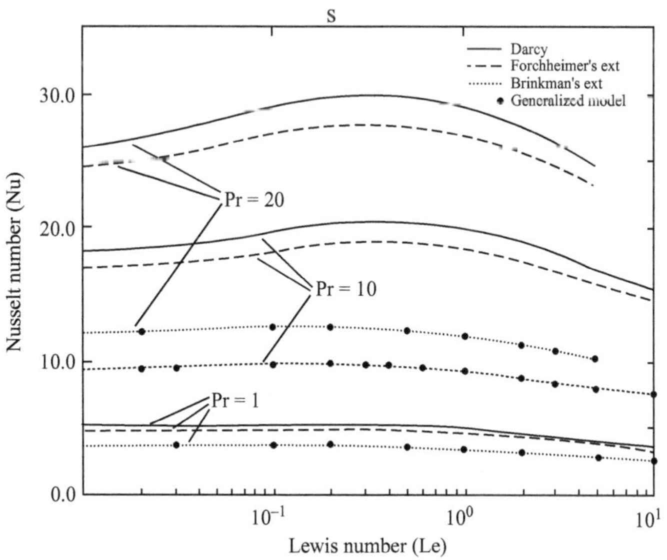

Figure 10.7. Variation of Nusselt number as a function of Lewis and Prandtl numbers for $Gr_T = Gr_C = 10^5$, $Da = 10^{-3}$, and $\Lambda = 2.34$ (from Karimi et al. (1997))

for double-diffusive convection in a fluid-filled rectangular cavity. The generalized model covered the entire range from Darcy flow to free fluid flow. Numerical predictions by the model indicate that the flow pattern as well as heat and mass transfer is profoundly influenced by the buoyancy ratio. Also non-Darcy effects on flow, heat, and mass transfer become significant when the Rayleigh or Darcy numbers are large. The Sherwood and Nusselt numbers were found to be sensitive to bed porosity variation in the non-Darcy regime.

Later on, the same authors (1997b) analyzed double-diffusive natural convection in fluid saturated porous medium using a generalized porous medium model for a Prandtl number of $Pr = 1.0$. One of the vertical walls of the porous cavity considered was subject to convective heat and mass transfer conditions. It was shown that heat and mass transfer rates were sensitive to the applied mass transfer coefficient in both the Darcy and non-Darcy flow regimes. It was also observed that the Sherwood number approaches a constant value as the solutal Biot number increases.

The problem of coupled heat and mass transfer by natural convection from a vertical, semi-infinite flat plate embedded in an air saturated porous medium in the presence of an external magnetic field and internal heat generation or absorption was formulated by Chamkha and Khaled (2000). The plate surface was maintained at constant temperature or constant heat flux and was permeable to allow for possible fluid wall suction or blowing. Useful correlations containing various physical parameters for both isothermal and isoflux walls were reported in that study.

REFERENCES

Horizontal Surface

J.D. Chang and P. Cheng, Matched asymptotic expansions for free convection about an impermeable horizontal surface in a porous medium, Int. J. Heat Mass Transfer, vol. 26, pp. 163–174 (1983).

P. Cheng, Geothermal heat transfer, in W.M. Rohsenow, J.P. Hartnett, and E.N. Ganic (eds.), Handbook of Heat Transfer Applications, 2nd ed., Chap. 11, McGraw-Hill, NY (1985).

D. Lesnic, D.B. Ingham and I. Pop. Conjugate free convection from a horizontal surface in a porous medium. Zeitschrift fur Angewandte Mathematik und Mechanik, vol. 75, pp. 715–722 (1995).

D. Lesnic and I. Pop, Free convection in a porous medium adjacent to horizontal surfaces. Zeitschrift fur Angewandte Mathematik und Mechanik, vol. 78, pp. 197–205 (1998).

D.A. Nield and A. Bejan, Convection in Porous Media, Springer-Verlag, NY (1992).

I. Pop and R.S.R. Gorla, Horizontal boundary-layer natural convection in a porous medium saturated with a gas, Transport In Porous Media, vol. 6, pp. 159–171 (1991).

D.A. Rees and D.S. Riley, Free convection above a near horizontal semi-infinite heated surface embedded in a saturated porous medium, Int. J. Heat Mass Transfer, vol. 28, pp. 183–190 (1985).

D.A.S. Rees, The effect of inertia on free convection from a horizontal surface embedded in a porous medium, Int. J. Heat Mass Transfer, vol. 39, pp. 3425–3430 (1996).

D.A.S. Rees and K. Vafai, Darcy-Brinkman free convection from a heated horizontal surface, Numerical Heat Transfer, Part A, vol. 35, pp. 191–204 (1999).

Vertical Surface

A. Chamkha, MHD-free convection from a vertical plate embedded in a thermally stratified porous medium with Hall effects, Appl. Math. Modeling, vol. 21, pp. 603–609 (1997a).

P. Cheng and W.J. Minkowycz, Free convection about a vertical flat plate embedded in a porous medium with application to a heat transfer from a dike, J. Geophys. Res., vol. 82, pp. 2040–2044 (1977).

P. Cheng and C.T. Hsu, Higher-order approximations for Darcian free convective flow about a semi-infinite vertical flat plate, J. Heat Transfer, vol. 106, pp. 143–151 (1984).

S.D. Harris, L. Elliott, D.B. Ingham and I. Pop, Transient free convection flow past a vertical flat plate subject to a sudden change in surface temperature. Int. J. Heat Mass Transfer, vol. 41, pp. 357–372 (1998).

J.T. Hong, Y. Yamada and C.L. Tien, The effects of non-Darcian and nonuniform porosity on vertical-plate natural convection in porous media. J. Heat Transfer, vol. 109, pp. 356–362 (1987).

M.A. Hossain, M. Alim and D.A.S. Rees, The effect of radiation on free convection from a porous vertical plate. Int. J. Heat Mass Transfer, vol. 42, pp. 181–191 (1999).

M.A. Hossain, K. Khanafer and K. Vafai, The effect of radiation on free convection flow of fluid with variable viscosity from a porous vertical plate. Int. J. Thermal Sciences, vol. 40, pp. 115–124 (2001).

C.T. Hsu and P. Cheng, The Brinkman model for natural convection about a semi-infinite vertical flat plate in a porous medium, Int. J. Heat Mass Transfer, vol. 28, pp. 683–697 (1985).

Jian Li, D.B. Ingham and I. Pop, Natural convection from a vertical flat plate with a surface temperature oscillation. Int. J. Heat Mass Transfer, vol. 44, pp. 2311–2322 (2001).

Y. Joshi and B. Gebhart, Vertical natural convection flows in porous media: calculations of improved accuracy, Int. J. Heat Mass Transfer, vol. 27, pp. 69–75 (1984).

M. Kaviany and M. Mittal, Natural convection heat transfer from a vertical plate to high permeability porous media: an experiment and an approximate solution. J. Heat Mass Transfer, vol. 30, pp. 967–977 (1987).

S.J. Kim and K. Vafai, Analysis of natural convection about a vertical plate embedded in a porous medium, Int. J. Heat Mass Transfer, vol. 32, pp. 665–677 (1989).

J.L. Lage and A. Bejan, Natural convection from a vertical surface covered with hair, Int. J. Heat Fluid Flow, vol. 12, pp. 46–53 (1991).

T. Watanabe, H. Taniguchi and I. Pop, Instability of laminar natural convection boundary layer flow on a vertical porous flat plate. Transactions of the ASME. Journal of Applied Mechanics, vol. 63, pp. 404–410 (1996).

T. Watanabe, H. Taniguchi, M. Kumagai and I. Pop, Stability of free convection with uniform suction or injection from a vertical flat plate subjected to a constant wall heat flux. Acta Mechanica, vol. 136, pp. 143–53 (1999).

Inclined Surface

A. Chamkha, Hydromagnetic natural convection from an isothermal inclined surface adjacent to a thermally stratified porous medium, Int. J. Engng. Sci., vol. 35, pp. 975–986 (1997b).

M.A. Hossain and I. Pop, Radiation effect on Darcy free convection flow along an inclined surface placed in porous media, Heat and Mass Transfer, vol. 32, pp. 223–227 (1997).

J.Y. Jang and W.J. Chang, Buoyancy-induced inclined boundary layer flow in saturated porous medium, Comp. Method Appl. Mech. Engng., vol. 68, pp. 333–344 (1988).

B. Zeghmati, M. Daguenet and G. Lepalec, Study of transient laminar free convection over an inclined wet flat plate, Int. J. Heat Mass Transfer, vol. 34, pp. 899–909 (1991).

Channel

M.A. Al-Nimr and M.A. Hader, MHD free convection flow in open-ended vertical porous channels, Chemical Engineering Science, vol. 54, pp. 1883–1889 (1999).

T. Paul, A.K. Singh, and A.K. Mishra,Transient natural convection between two vertical walls filled with a porous material having variable porosity, Mathematical Engineering In Industry, vol. 8, pp. 177–185 (2001).

Cylinder

M.A. Hossain, M.A. Alim and D.A.S. Rees, Effect of thermal radiation on natural convection over cylinders of elliptic cross section, Acta Mechanica, vol. 129, pp. 177–186 (1998).
M.A. Hossain, K. Vafai and M. Khalil, Khanafer, Non-Darcy natural convection heat and mass transfer along a vertical permeable cylinder embedded in a porous medium. Int. J. Thermal Sciences, vol. 38, pp. 854–862 (1999).
D.B. Ingham and I. Pop, Natural convection about a heated horizontal cylinder in a porous medium, J. Fluid Mechanics, vol. 184, pp. 157–181 (1987).
M. Kumari, I. Pop and G. Nath, Non-Darcy natural convection on a vertical cylinder in a saturated porous medium. Waerme und Stoffuebertragung, vol. 20, pp. 33–37 (1986).
W.J. Minkowycz and E.M. Sparrow, Local nonsimilar solutions for natural convection on a vertical cylinder, J. Heat Transfer, vol. 96, pp. 178–183 (1974).
W.J. Minkowycz and Ping Cheng, Free convection about a vertical cylinder embedded in a porous medium, Int. J. Heat Mass Transfer, vol. 19, pp. 805–813 (1976).
Slimi K., Ben Nasrallah S. and Fohr J.-P., Transient natural convection in a vertical cylinder opened at the extremities and filled with a fluid saturated porous medium: validity of Darcy flow model and thermal boundary layer approximations Int. J. Heat Mass Transfer, vol. 41, pp. 1113–1125 (1998).

Sphere

T. Sano and R. Okihara, Natural convection around a sphere immersed in a porous medium at small Rayleigh numbers. Fluid Dynamics Research, vol. 13, pp. 39–44 (1994).
K. Yamamoto, Natural convection about a heated sphere in a porous medium, J. Physical Society of Japan, vol. 37, pp. 1164–1166 (1974).
K.A. Yih. Viscous and Joule heating effects on non-Darcy MHD natural convection flow over a permeable sphere in porous media with internal heat generation. Int. Comm. Heat Mass Transfer, vol. 27, pp. 591–600 (2000).

Enclosure

C. Beckermann, R. Viskanta and S. Ramadhyani, A numerical study of non-Darcian natural convection in a vertical enclosure filled with a porous medium. Numer. Heat Transfer, vol. 10, pp. 557–570 (1986).
Y.-H. Chen and H.-T. Lin, Natural convection in an inclined enclosure with a fluid layer and a heat-generating porous bed. Heat and Mass Transfer, vol. 33, pp. 247–255 (1997).
S. Das, and R.K. Sahoo. Effect of Darcy, fluid Rayleigh and heat generation parameters on natural convection in a porous square enclosure: a Brinkman-extended Darcy model. Int. Comm. Heat Mass Transfer, vol. 26, pp. 569–578 (1999).
Z.-G. Du and E. Bilgen, Natural convection in vertical cavities with internal heat generating porous media, Heat and Mass Transfer, vol. 27, pp. 149–155 (1992).
S.M. ElSherbiny, G.D. Raithby and K.G.T. Hollands, Heat transfer by natural convection across vertical and inclined air layers. J. Heat Transfer, vol. 104, pp. 96–102 (1982).
In-Ju Hwang and Y.J. Kim, An experimental study of natural convection in the composed rectangular-parallelogrammic enclosure. Hwang, In-Ju; Kim, Youn J. Proceedings of Renewable and Advanced Energy Systems for the 21st Century, Maui, HI, United States, pp. 271–275 (1999).
K.M. Khanafer and A.J. Chamkha, Hydromagnetic natural convection from an inclined porous square enclosure with heat generation. Numer. Heat Transfer, Part A, vol. 33, pp. 891–910 (1998).

J.L. Lage, Effect of the convective inertia term on Benard convection in a porous medium, Numer. Heat Transfer, vol. 22, pp. 469–485 (1992).

J.L. Lage, Natural convection within a porous medium cavity: predicting tools for flow regime and heat transfer. Int. Comm. Heat Mass Transfer, vol. 20, pp. 501–513 (1993).

P. Nithiarasu, K.N. Seetharamu and T. Sundararajan, Natural convective heat transfer in a fluid saturated variable porosity medium, Int. J. Heat Transfer, vol. 40, pp. 3955–3967 (1997a).

M.B. Peirotti, M.D. Giavedoni, M.D., and J.A. Deiber. Natural convective heat transfer in a rectangular porous cavity with variable fluid properties - validity of the Boussinesq approximation. Int. J. Heat Mass Transfer, vol. 3, pp. 2571–2581(1987).

V. Shankar and C.-E. Hagentoft, Numerical investigation of natural convection in horizontal porous media heated from below – comparisons with experiments, Journal of Thermal Envelope and Building Science, vol. 23, pp. 318–338 (2000).

P.H. Stauffer, L.H. Auer, N.D. Rosenberg, Compressible gas in porous media: A finite amplitude analysis of natural convection, vol. 40, pp. 1585–1589 (1997).

H.C. Tien and K. Vafai, Pressure stratification effects on multiphase transport across a vertical slot porous insulation. Transactions of the ASME. Journal of Heat Transfer, vol. 112, pp. 1023–1031(1990).

T. Tong and F.M. Gerner, Natural convection in partitioned air-filled rectangular enclosures, Int. Comm. Heat Mass Transfer, vol. 13, pp. 99–108 (1986).

T.W. Tong and E. Subramanian, Natural Convection in Rectangular Enclosures Partially Filled with a Porous Medium, Int. J. Heat and Fluid Flow, vol. 7, pp. 3–10 (1986).

K. Vafai and S. Sarkar, Condensation effects in a fibrous insulation slab, ASME Journal of Heat Transfer, vol. 108, pp. 667–675 (1986).

K. Vafai and S. Sarkar, Heat and mass transfer in partial enclosures. Journal of Thermophysics and Heat Transfer, vol. 1, pp. 253–259 (1987).

K. Vafai and H.C. Tien, A numerical investigation of phase change effects in porous materials. Int. J. Heat Mass Transfer, vol. 32, pp. 1261–1277 (1989).

S.W. Webb, Modeling of open space natural convection in porous media codes, Proceedings of the 9th International High-Level Radioactive Waste Management Conference, April 29–May 3, Las Vegas, NV (2001).

S.W. Webb and C.E. Hickox Jr., Assessment of the equivalent permeability relationship for open space natural convection, Proceedings of the 9th International High-Level Radioactive Waste Management Conference (IHLRWM), April 29–May 3, Las Vegas, NV (2001).

N.G. Wright, P.H. Gaskell and P.A. Sleigh, Natural convection in a shallow laterally heated air field cavity, Comm. Numer. Methods Engineering, vol. 11, pp. 937–950 (1995).

Double-Diffusive Natural Convection within a Porous Medium

A. Chamkha and A. Khaled, Hydromagnetic combined heat and mass transfer by natural convection from a permeable surface embedded in a fluid-saturated porous medium, Int. J. Numer. Heat Fluid Flow, vol. 10, pp. 455–476 (2000).

M. Karimi-Fard, M.C. Charrier-Mojtabi and K. Vafai, Non-Darcian effects on double-diffusive convection within a porous medium, Numerical Heat Transfer, Part A, vol. 31, pp. 837–852 (1997).

P. Nithiarasu, K.N. Seetharamu and T. Sundararajan, Double-diffusive natural convection in an enclosure filled with fluid-saturated porous medium: a generalized non-Darcy approach. Numerical Heat Transfer, Part A (Applications), vol. 30, pp. 413–426 (1996).

P. Nithiarasu, T. Sundararajan and K.N. Seetharamu, Double-diffusive natural convection in a fluid saturated porous cavity with a freely convecting wall. Int. Comm. Heat Mass Transfer, vol. 24, pp. 1121–1130 (1997b).

CHAPTER 11

SCALING ISSUES IN POROUS AND FRACTURED MEDIA

VINCENT C. TIDWELL
Sandia National Laboratories, P. O. Box 5800, MS 0735, Albuquerque, NM 87185, USA

The continuum hypothesis of rational mechanics forms the basis of most subsurface flow and transport models. In this approach, the time-and-space dependence of state variables is expressed in the form of differential balance equations formulated on the principles of mass, momentum, and energy conservation. To achieve tractable solutions for the resulting balance equations, simplifying assumptions are used that typically introduce constitutive properties into the flow/transport equations. These properties (e.g., permeability, dispersivity) account for the integrated effects of heterogeneities and physical processes that occur at scales much smaller than the desired scale of analysis. Constitutive properties are related not to a discrete point within the porous media but to a control volume or sample support and are assumed to vary smoothly enough in time and space so that the resulting balance equations can be solved by standard analytical/numerical methods of differential equations.

Because of technological and computational constraints, it is rarely possible to measure constitutive properties at the desired scale of analysis. For this reason, some averaging or scaling model is required to transfer information from the scale of measurement to the desired scale of analysis. If the averaging process of the particular property under study were known, the problem would be alleviated. For example, the average porosity of a volume is simply the arithmetic average of the porosities of all the samples that constitute it. The simple arithmetic averaging process holds true for additive variables such as porosity and ore grade. Unfortunately, many constitutive properties (e.g., permeability) are not additive; that is, the scaling process not only depends on the volume fraction present but other factors as well. These factors include, but are not limited to, the heterogeneous characteristics (i.e., length scales, variance, spatial patterns) of the medium (e.g., Gelhar and Axness, 1983; Dagan, 1984; Fogg, 1986), the nature (e.g., linear vs. convergent flow) of the flow field (e.g., Desbarats, 1992a; Indelman and Abramovich, 1994; Tidwell et al., 1999), and scale

C. Ho and S. Webb (eds.), Gas Transport in Porous Media, 201–212.
© 2006 *Springer.*

dependent flow/transport processes (e.g., hydrodynamic dispersion is superceded by macrodispersion at larger length scales).

Below we briefly consider the role of scaling in the modeling of fluid flow and mass transport in porous and fractured media. We begin by reviewing some basic approaches to modeling material property scaling. We then present examples from the laboratory and field demonstrating scaling effects. The focus of this section is on the scaling of permeability and dispersivity.

11.1 SCALING THEORIES

Simple averaging rules (arithmetic, geometric, and harmonic) represent the point of entry for classical scaling theory. These averaging rules are only valid for a narrow range of aquifer/reservoir conditions. Specifically, an infinitely stratified formation comprised of individual layers with constant permeability that is subject to steady, linear flow. Where flow is oriented parallel to the stratification the effective permeability is given by the arithmetic mean k_a,

$$k_\mathrm{a} = \sum_{i=1}^{n} \frac{k_i d_i}{d} \tag{11.1}$$

where n is the sample set size, d is the total thickness of the system, and k_i and, d_i are the permeability and thickness of each discrete bed, respectively. When flow is oriented normal to stratification the harmonic mean k_h, given by

$$k_\mathrm{h} = \frac{d}{\sum_{i=1}^{n} d_i \Big/ k_i} \tag{11.2}$$

represents the appropriate averaging rule. These two effective permeabilities are derived by solving the governing flow equation (Darcy's Law) for each layer in parallel and in series, respectively. Where the formation lacks spatial correlation, Warren and Price (1961) found the geometric mean to provide a good estimate of the effective permeability. The geometric mean k_g is given by

$$k_\mathrm{g} = \exp\left[1/n \sum_{i=1}^{n} \ln(k_i)\right] \tag{11.3}$$

It follows that, $k_\mathrm{a} > k_\mathrm{g} > k_\mathrm{h}$.

Unfortunately, most natural systems lie somewhere in-between these extremes as do the appropriate scaling rules. The power law average provides a convenient means of handling this wide range of behavior. The power law average k_p is given by:

$$k_\mathrm{p} = \left[1/n \sum_{i=1}^{n} k_i^{\omega}\right]^{1/\omega} \tag{11.4}$$

where ω is the power coefficient. With this approach the selection of an appropriate scaling rule is predicated on finding the proper power coefficient. Where flow is oriented parallel or normal to an infinitely stratified system ω is simply set to 1 or -1 to yield an arithmetic or harmonic average, respectively. Alternatively, the geometric mean is defined as $\omega = 0$, and thus Equation (11.4) must undergo a limited expansion as currently defined. In general, empirical methods are used to determine ω. Journel et al. (1986) and Deutsch (1989) provide algorithms for estimating ω for bimodal sand-shale sequences, while Desbarats (1992b) provides direction in the case of log-normally distributed data.

Stochastic methods (e.g., Bakr et al., 1978; Gelhar, 1993) have been developed that allow direct calculation of effective properties based on the heterogeneous characteristics of the aquifer/reservoir. Application of these methods requires an unbounded domain, and uniform flow (i.e., the extent of the domain and the characteristic scale of the flow nonuniformity are much larger than the correlation length scale of the medium). The permeability distribution must also be a weakly stationary and ergodic random variable with a relatively small variance (generally Var[lnk] $<$ 1, however under certain circumstances this limit can be exceeded). In this approach, the aquifer/reservoir is viewed as an ensemble of homogenous, isotropic blocks whose spatial distribution is fully characterized by its first two moments. Using a small perturbation, first-order approximation of the governing stochastic differential equation, an expression for the effective permeability has been obtained. Gutjahr et al. (1978), extending the earlier work of Matheron (1967), found the effective permeability of a heterogeneous, isotropic medium to be:

$$k_{\text{eff}} = k_{\text{g}}[1 - \sigma_y^2/2] \tag{11.5}$$

in one dimension,

$$k_{\text{eff}} = k_{\text{g}} \tag{11.6}$$

in two dimensions, and

$$k_{\text{eff}} = k_{\text{g}}\left[1 + \frac{\sigma_y^2}{6}\right] \tag{11.7}$$

in three dimensions, where k_{g} and σ_y^2 are the geometric mean and variance of the natural log permeability distribution, respectively. An interesting result of this work is the dependence of k_{eff} on the dimensionality of the flow domain. Gelhar and Axness (1983) extended this work to a 3-D statistically anisotropic medium. In this case, the effective permeability is expressed as:

$$k_{ii} = k_{\text{g}}[1 + \sigma_y^2(0.5 - g_{ii})] \tag{11.8}$$

where g_{ii} is a geometric factor accounting for the degree of anisotropy and orientation of flow relative to the principal permeability axes and the subscript ii designates the tensor components. For the case of infinite stratification and flow parallel or normal to the bedding, Equation (11.8) reduces to that of an arithmetic and harmonic mean, respectively. Indelman and Abramovich (1994) demonstrated g_{ii} to also be a function of the shape of the covariance function.

Scaling has also been approached by purely numerical means. In this case (Ababou et al., 1989; Bachu and Cuthiell, 1990; Desbarats, 1987; Kossack et al., 1990) the effective permeability of a spatially heterogeneous permeability field is determined by numerically solving the steady-state flow equation for a prescribed domain. Realizations of the heterogeneous permeability field are generated by statistical methods, based on a variety of principles ranging from sedimentological process models to geostatistics (e.g., Koltermann and Gorelick, 1996). In formulating the flow model, care must be taken to assign boundary conditions that reflect conditions the computational element will experience within the larger flow domain (Lasseter et al., 1986; Gomez-Hernandez and Journel, 1994). The statistical moments of the upscaled permeability field can then be extracted though Monte Carlo simulation or from a single realization provided that the ergodicity requirements are satisfied. Use of numerical methods in applied scaling problems is limited due to the excessive, sometimes insurmountable computational requirements.

In general, each of the aforementioned theories assume the porous media exhibits a discrete hierarchy of scales (i.e., a finite correlation length scale exists). However, not all porous media behave in such manner, but rather exhibit a continuous hierarchy of scales or continuous evolving heterogeneity (i.e., infinite length scale). A convenient means of modeling evolving heterogeneities is with fractals. In fact, a number of researchers have found geologic materials to display fractal characteristics. Fractal behavior was noted for the case of sandstone porosity (Katz and Thompson, 1985), soil properties (Burrough, 1983), fracture networks (Barton and Larson, 1985), and reservoir porosity (Hewett, 1986). Upon examining transmissivity and permeability data measured over scales of 10 cm to 45 km, Neuman (1994) argued that the data scale according to a power-law semivariogram (i.e., infinite length scale). Using the stochastic framework of Gelhar and the apparent power-law scaling, Neuman offers a model for the effective permeability. The model predicts that the effective isotropic permeability will decrease with increasing sample support (i.e., sample volume) in one-dimensional media, increase in three-dimensional media, and show no systematic variation within two-dimensional media. The concept of multifractal scaling of hydraulic conductivity distributions has also been developed to deal with data sets where conductivity variations are more heterogeneous at smaller scales than at larger scales (Liu and Molz, 1997).

Similar approaches have been used to predict effective transport parameters such as the dispersion coefficient or macrodispersivity. The macrodispersivity tensor is a measure of the influence hydraulic conductivity imposes on large-scale solute mixing. Calculations based on stochastic theory and assuming a finite correlation length scale predict that the longitudinal macrodispersivity, A_{ij} increases directly with the variance of log hydraulic conductivity in the isotropic case (Gelhar and Axness, 1983)

$$A_{11} = \frac{\sigma_y^2 \lambda}{\gamma^2} \qquad (11.9)$$

where λ is the correlation length scale and $\gamma = q_1 / K_g J_1$ (where q_1 and J_1 are the flux and average gradient in the longitudinal direction, respectively). In stratified media, the transverse mixing process is highly anisotropic; that is, the horizontal transverse macrodispersivity in the plane of bedding is much larger than vertical transverse macrodispersivity associated with a direction perpendicular to that of bedding. Assuming finite correlation length scales, Dagan (1988) shows that the longitudinal macrodispersivity grows with travel time to an asymptotic value that is independent of the anisotropy ratio. These travel distances are on the order of tens of horizontal log-conductivity correlation length scales to reach asymptotic mixing conditions. One limitation of these and similar theories is the assumption of mild heterogeneity (i.e., $\sigma_y^2 < 1$). For more heterogeneous materials, Neuman and Zhang (1990) employed quasi-linear theory to describe the evolution of the dispersion process.

Fractal concepts have also been used to explain the scaling of basic transport processes like diffusion (Spoval et al., 1985) and macrodispersivity (Wheatcraft and Tyler, 1988). For porous media exhibiting evolving scales of heterogeneity, classical Fickian dispersive and diffusive concepts fail; that is the macrodispersivity continues to grow with scale rather than reaching an asymptotic limit. Wheatcraft and Tyler (1988) developed a Lagrangian model for dispersion in a set of fractal streamtubes. They found the dispersivity is proportional to the straight-line travel distance raised to a power of $2D - 1$ where D is the fractal dimension. This evolving scale dependent behavior is seen when one plots dispersivity versus experimental scale (e.g., Gelhar et al., 1992). Assuming a hierarchy of fractal media and quasi-linear theory, Neuman (1990) offers a theoretical basis for interpreting this long recognized behavior.

11.2 RESULTS FROM LABORATORY AND FIELD TESTS DOCUMENTING SCALING EFFECTS

Field and laboratory studies have been performed to gain a better understanding of scaling processes. In general, these studies involve the measurement of some property over a range of different scales. However, the manner in which the scaling studies are conducted can vary significantly in terms of the measurement techniques employed, the range in sample support (i.e., sample volume) investigated, and the geologic media interrogated. Laboratory tests have the advantage that greater control over the measurement process can be maintained, while field studies allow a larger range of scales to be investigated.

One obvious approach to investigating scaling processes is to dissect a heterogeneous medium and then reconstruct its effective properties from its component parts. Laboratory-based studies conducted by Henriette et al. (1989) reconstructed the effective permeability of two $15 \times 15 \times 50$ cm blocks of rock, one a sandstone and the other a limestone. The reconstruction was based first on permeability measurements made on $15 \times 15 \times 15$ cm intermediate-scale blocks cut from the former and then 300 core samples cut from them. Scaling was manifest as trends in the statistical properties of the measured permeability values. Specifically, the sample variance of the permeability decreased with increasing sample volume. This is a common finding reflecting the fact that larger sample volumes integrate over more heterogeneity. The investigators also found the sample mean of the permeability decreased slightly with increasing sample volume. This is consistent with theory that suggests the effective permeability of a medium with short-range spatial correlation, as was the case here, approaches that of the geometric mean of the smaller scale measurements (e.g., Gelhar and Axness, 1983).

An alternative approach to scaling investigations is to collect samples from the field that are subsequently analyzed in the laboratory. For example, Parker and Albrecht (1987) acquired soil cores of three different volumes (92, 471, and 1770 cm^3) taken from two different soil layers along closely spaced transects. Saturated hydraulic conductivities and solute dispersivities were then measured in the laboratory. As expected, the variance of the natural-log conductivity and dispersivity decreased with increasing core volume. However, in this case the mean permeability was found to increase with increasing sample volume, which was inconsistent with the short-range spatial correlation and linear flow imposed in the test. Upon closer inspection the investigators concluded that the integrity of the smaller core samples had been compromised during collection. This highlights but one of the difficulties with conducting scaling experiments. That is, care must be taken to avoid introducing bias into the experiment due to changes in sample integrity, sample density, or measurement precision/accuracy for different scales of measurement.

In an effort to avoid such biasing, Tidwell and Wilson (1997, 1999a, 1999b, 2000) employed a consistent measurement device and sampling strategy to acquire permeability data over a range of different scales. Specifically, they used a computer automated minipermeameter test system with six different size tip seals, each providing approximately an order of magnitude larger sample volume than the next smaller. Over 150,000 permeability values were collected from three, meter-scale blocks of rock; including, two cross-bedded sandstones and one volcanic tuff. Characterization of each block face involved high-resolution mapping of the heterogeneous permeability field with each of five different size tip seals, plus the collection of a single large tip seal measurement designed to interrogate most of the sampling domain. These exhaustive data sets, measured under consistent experimental conditions yielded empirical evidence of permeability scaling (e.g., Figure 11.1). Specifically, as the sample support increased the sample variance decreased, the semivariogram range increased linearly, while the small-scale (i.e., smaller than the tip seal) spatial structure was preferentially filtered from the permeability maps and semivariograms. Although all three-rock samples exhibit similar qualitative scaling trends, distinct differences were also noted. These differences were most evident in the quantitative characteristics of the aforementioned trends and in the scaling of the mean permeability. These differences in part can be explained on the basis of the spatial characteristics of the three rock samples and the divergent flow geometry imposed by the minipermeameter tip seal.

Beyond the work of Tidwell and Wilson, there is a growing body of evidence demonstrating the role of flow geometry on scaling processes. One example is the

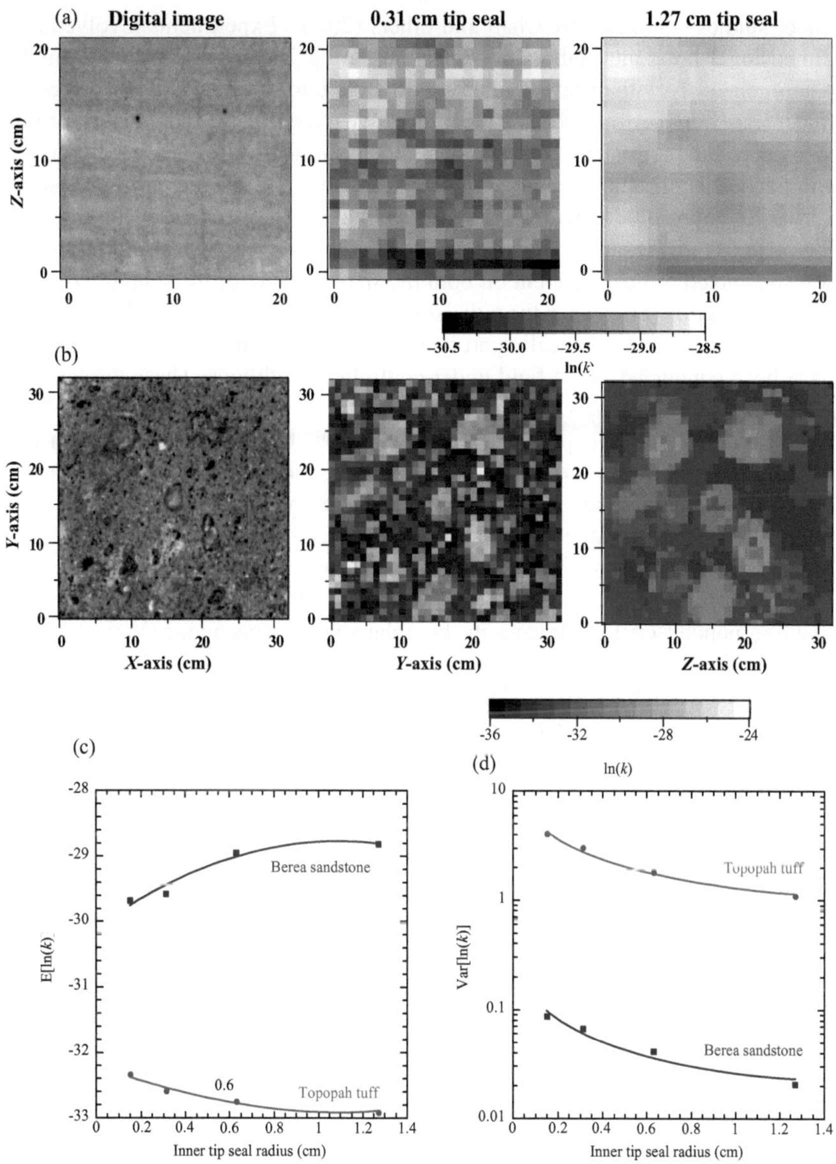

Figure 11.1. Permeability scaling measured on a block of Berea Sandstone and Topopah Spring Tuff. (a) Digital image of the Berea Sandstone sample and corresponding natural log permeability fields measured with two different size tip seals. (b) Same as (a) but for a block of Topopah Spring Tuff. (c) Scaling of the mean of the natural log permeability as measured on the two rock samples shown above (note that permeability maps for two of the tip seals are not shown). (d) Scaling of the variance of the natural log permeability

tracer studies conducted by Chao and others (2000). Experiments involved a two-dimensional horizontal laboratory tank measuring 244 by 122 by 6.35 cm filled heterogeneously with different silica sands. Both uniform flow and convergent flow tracer tests were performed using a conservative tracer. Tests involving uniform flow yielded results consistent with classical stochastic theory that is the dispersivities approached a constant value at large displacements. The convergent flow tests gave very different results; specifically the dispersivities exhibited sustained scale dependence across a wide range of displacements. For the convergent tests, the dispersivity was also noted to be dependent on both the spatial extent of the source term and its point of injection (even when the radial distance from the pumping well was constant).

To investigate transport scaling processes over a larger range of scales, experiments have been conducted in the field under controlled conditions. These tests generally included detailed descriptions of the subsurface geology followed by a natural and/or forced gradient tracer experiment. The advection, dispersion and reaction of the contaminant plume are then monitored over time through the sampling of a dense network of boreholes. Efforts are then made to interpret the measured tracer behavior with that predicted by stochastic transport models.

One of the best-known field tests was conducted at the Borden Site in Ontario, Canada. Natural gradient experiments were conducted using a variety of conservative and nonconservative tracers in this relatively homogeneous (Var[lnk] = 0.29) sand aquifer (Freyberg, 1986; Sudicky, 1986). In much the same manner, field tests were conducted in Cape Cod, Massachusetts (Garabedian, 1987; Hess et al., 1991). Natural-gradient tracer experiments conducted in this gravel aquifer extended over a travel distance of about 275m. Additionally, Moltyaner and Killey (1988a, 1988b) describe a 40 m long tracer test conducted in a fluvial sand aquifer at Chalk River, Ontario, termed the Twin Lakes site.

In each of these tests, macrodispersivities calculated from the field data were compared with that predicted by the stochastic theory of Gelhar and Axness (1983) subject to the spatial distributions of conductivity characterizing each aquifer. For each of the field tests, favorable comparisons were found. As predicted by theory, longitudinal spreading dominated the evolution of the tracer plume with the transverse dispersion playing a relatively weak role, particularly in the vertical direction.

More recently several natural-gradient tracer tests were conducted at the Columbus Air Force Base in Mississippi, commonly known as the MADE (Macrodispersion Experiment) site. The MADE site is associated with a shallow alluvial aquifer that is at least one order of magnitude more heterogeneous than the aquifers noted above. Owing to the considerable heterogeneity in the measured hydraulic conductivity distribution the classical Fickian advection-dispersion model had difficulty reproducing the measured tracer behavior (Zheng and Jiao, 1998; Eggleston and Rojstaczer, 1998). Part of the problem was believed to be due to transport along preferential flow pathways. Similar studies of tracer dispersion and diffusion in fractured dolomite required higher-order models to explain mass-transfer between the fractures and host matrix (Haggerty et al., 2001).

REFERENCES

Ababou, R., D. McLaughlin, L.W. Gelhar, and A.F.G. Thompson, Numerical simulation of three-dimensional saturated flow in randomly heterogeneous porous media, *Transport in Porous Media,* 4, 549–565, 1989.

Bachu, S. and D. Cuthiell, Effects of core-scale heterogeneity on steady state and transient fluid flow in porous media: Numerical analysis, *Water Resour. Res.*, 26(5), 863–874, 1990.

Bakr, A.A., L.W. Gelhar, A.L. Gutjahr, and J.R. MacMillan, Stochastic analyses of spatial variability in subsurface flows, 1, Comparison of one- and three-dimensional flows, *Water Resour. Res.*,14(2), 263–272, 1978.

Barton, C.C. and E. Larson, Fractal geometry of two-dimensional fracture networks at Yucca Mountain, Southwest Nevada, paper presented at the *International Symposium on Fundamentals of Rock Joints*, Swedish National Group of the International Society of Rock Mechanics, Bjorkliden, Sweden, September, 1985.

Burrough, P.A., Multiscale sources of spatial variation in soil. I. The application of fractal concepts to nested levels of soil variation, *Journal of Soil Science*, 34, 577–597, 1983.

Chao, H.C., H. Rajaram, and T. Illangasekare, Intermediate-scale experiments and numerical simulations of transport under radial flow in a two-dimensional heterogeneous porous medium, *Water Resour. Res.*, 36(10), 2869–2884, 2000.

Dagan, G., Time-dependent macrodispersion of solute transport in anisotropic heterogeneous aquifers, *Water Resour. Res.,* 24(9), 1491–1500, 1988.

Dagan, G., Solute transport in heterogeneous porous formations, *J. Fluid Mech.,* 145, 151–177, 1984.

Desbarats, A.J., Spatial averaging of transmissivity in heterogeneous fields with flow toward a well, *Water Resour. Res.*, 28(3), 757–767, 1992a.

Desbarats, A.J., Spatial averaging of hydraulic conductivity in three-dimensional heterogeneous porous media, *Math. Geol.*, 24(3), 249–267, 1992b.

Desbarats, A.J., Numerical estimation of effective permeability in sand-shale formations, *Water Resour. Res.*, 23(2), 273–286, 1987.

Deutsch, C.V., Calculating effective absolute permeability in sandstone/shale sequences, *SPE Form. Eval.*, 4(3), 343–348, 1989.

Eggleston, J. and S. Rojstaczer, Identification of large-scale hydraulic conductivity trends and the influence of trends on contaminant transport, *Water Resour. Res.,* 34(9), 2155–2168, 1998.

Fogg, G.E., Groundwater flow and sand body interconnectedness in a thick, multiple-aquifer system, *Water Resour. Res.*, 22(5), 679–694, 1986.

Freyberg, D.L., A natural gradient experiment on solute transport in a sand aquifer, 2, Spatial moments and the advection and dispersion of nonreactive tracers, *Water Resour. Res.*, 22(13), 2031–2046, 1986.

Garabedian, S.P., Large-scale dispersive transport in aquifers: Field experiments and reactive transport theory, Ph.D. thesis, Dept. Civil Eng., Massachusetts Inst. Techn., Cambridge, Massachusetts, 290 pp., 1987.

Gelhar, L.W., and C.L. Axness, Three-dimensional stochastic analysis of macrodispersion in aquifers, *Water Resour. Res.*, 19(1), 161–180, 1983.

Gelhar, L.W., C. Welty, and K.R. Rehfeldt, A critical review of data on field-scale dispersion in aquifers, *Water Resour. Res.,* 28(7), 1955–1974, 1992.

Gelhar, L.W., *Stochastic Subsurface Hydrology*, Prentice Hall, Englewood Cliffs, New Jersey, p. 390, 1993.

Gomez-Hernandez, J.J. and A.G. Journel, Stochastic characterization of grid-block permeabilities, *Stanford Center for Reservoir Forecasting, Reprt 7,* Stanford, CA, p. 21 1994.

Gutjahr, A.L., L.W. Gelhar, A.A. Bakr, and J.R. Macmillan, Stochastic analysis of spatial variability in subsurface flows, 2. Evaluation and application, *Water Resour. Res.*, 14(5), 953–960, 1978.

Haggerty, R., S.W. Fleming, L.C. Meigs, and S.A. McKenna, Tracer tests in a fractured dolomite 2. Analysis of mass transfer in single-well injection-withdrawal tests, *Water Resour. Res.*, 37(5), 1129–1143, 2001.

Henriette, A., C.G. Jacquin, and P.M. Adler, The effective permeability of heterogeneous porous media, *PCH PhysicoChemical Hydrodynamics*, 11(1), 61–80, 1989.

Hess, K.M., S.H. Wolf, M.A. Celia, and S.P. Garabedian, Macrodispersion and spatial variability of hydraulic conductivity in a sand and gravel aquifer, Cape Cod, Massachusetts, *EPA/600/M-91/005, Environmental Research Brief*, Environmental Protection Agency, Ada, OK, 1991.

Hewett, T.A., Fractal distributions of reservoir heterogeneity and their influence on fluid transport, *paper SPE 15386 presented at the 1986 Annual Technical Conference and Exhibition*, New Orleans, La., 5–8 Oct, 1986.

Indelman, P. and B. Abramovich, Nonlocal properties of nonuniform averaged flows in heterogeneous media, *Water Resour. Res.*, 30(12), 3385–3393, 1994.

Journel, A.G., C.V. Deutsch , and A.J. Desbarats, Power Averaging for Block Effective Permeability, SPE 15128, in *Proceeding of the 56th Annual California Regional Meeting of the Society of Petroleum Engineers,* Oakland, 329–334, 1986.

Katz, A.J. and A.H. Thompson, Fractal sandstone pores: Implications for conductivity and pore formation, *Physical Review Letters*, 54, 1325–1328,1985.

Koltermann C.E. and S.M. Gorelick, Heterogeneity in sedimentary deposits: A review of structure-imitating, process-imitating, and descriptive approaches,*Water Resour. Res.*, 32(9), 2617–2658, 1996.

Kossack, C.A., J.O. Aasen, and S.T. Opdal, Scaling up heterogeneities with pseudofunctions,*SPEFE*, 226–232, 1990.

Lasseter, T.J., J.R. Waggoner, and L.W. Lake, Reservoir heterogeneities and their influence on ultimate recovery, in *Reservoir Characterization*, ed. by L.W. Lake and H.B. Carroll, Jr., Academic Press, New York, 545–560, 1986.

Liu, H.H. and F.J. Molz, Multifractal analyses of hydraulic conductivity distributions, *Water Resour. Res.*,33(11), 2483–2488, 1997.

Matheron, G., *Elements pour une theorie des milieux poreux*, 166 pp., Maisson et Cie, Paris, 1967.

Moltyaner, G.L. and R.W.D. Killey, Twin Lake tracer tests: Longitudinal dispersion, *Water Resour. Res.*, 24(10),1613–1627,1988a.

Moltyaner, G.L. and R.W.D. Killey, Twin Lake tracer tests: Transverse dispersion, *Water Resour. Res.*, 24(10),1628–1637,1988b.

Neuman, S. P., Generalized scaling of permeabilities: Validation and effect of support scale, *Geophys. Res. Lett.*, 21(5), 349–353, 1994.

Neuman, S.P., Universal scaling of hydraulic conductivities and dispersivities in geologic media, *Water Resour. Res.*, 26(8), 1749–1758, 1990.

Neuman, S.P. and Y.-K. Zhang, A quasi-linear theory of non-Fickian and Fickian subsurface dispersion, 1, Theoretical analysis with application to isotropic media, *Water Resour. Res.,* 26(5), 887–902, 1990.

Parker, J.C., and K.A. Albrecht, Sample volume effects on solute transport predictions, *Water Resour. Res.*, 23(12), 2293–2301, 1987.

Spoval, B., M. Rosso, and J.F. Bouyet, The Fractal nature of a Diffusing front and the relation to percolation, *Journal of Physics Letters*, 46, L149-L156, 1985.

Sudicky, E.A., A natural gradient experiment on solute transport in a sand aquifer: Spatial variability of hydraulic conductivity and its role in the dispersion process, *Water Resour. Res.*, 22(13), 2069–2081, 1986.

Tidwell, V.C. and J.L. Wilson, Laboratory method for investigating permeability upscaling, *Water Resour. Res.,* 33(7), 1607–1616, 1997.

Tidwell, V.C. and J.L. Wilson, Permeability upscaling measured on a block of Berea Sandstone: Results and interpretation, *Math. Geol.*, 31(7), 749–769, 1999a.

Tidwell, V.C., and J.L. Wilson, Upscaling experiments conducted on a block of volcanic tuff: Results for a bimodal permeability distribution, *Water Resour. Res.,* 35(11), 3375–3387, 1999b.

Tidwell, V.C. and J.L. Wilson, Heterogeneity, permeability patterns, and permeability upscaling: Physical characterization of a block of Massillon Sandstone exhibiting nested scales of heterogeneity, *SPE Reservoir Eval. & Eng.*,3(4), 283–291, 2000.

Tidwell, V.C., A.L. Gutjahr, and J.L. Wilson, What does an instrument measure? Empirical spatial weighting functions calculated from permeability data sets measured on multiple sample supports, *Water Resour. Res.*, 35(1), 43–54, 1999.

Warren, J.E. and H.S. Price, Flow in heterogeneous porous media, *SPE J.*, 1, 153–169, 1961.

Wheatcraft, S.W. and S.W. Tyler, An explanation of scale-dependent dispersivity in heterogeneous aquifers using concepts of fractal geometry, *Water Resour. Res.*, 24(4), 566–578, 1988.

Zheng, C. and J.J. Jiao, Numerical simulation of tracer tests in a heterogeneous aquifer, *J. Environ, Eng.*, 124(6), 510–516, 1998.

CHAPTER 12

NUMERICAL CODES FOR CONTINUUM MODELING OF GAS TRANSPORT IN POROUS MEDIA

KARSTEN PRUESS

Lawrence Berkeley National Laboratory, Berkeley, CA 94720, USA

12.1 INTRODUCTION

A variety of techniques and approaches are available for quantitative mathematical modeling of gas transport in permeable media. Analytical solutions can be obtained for suitably simplified problems. Numerical simulation approaches are more versatile. They can accurately represent the fundamental physical processes, as well as honoring non-ideal flow and transport properties, general boundary and initial conditions, and heterogeneities of the medium. Most numerical approaches currently in use are based on "macroscale continuum concepts," such as Darcy's law for advection, Fick's law for diffusion, or the "dusty gas model" (DGM). Alternatives such as numerical solution of the Navier-Stokes equation or lattice gas techniques are employed for fundamental studies of gas behavior in individual pore channels. The approach selected for any particular modeling study depends upon the nature of the flow system, the space and time scales of interest, and on the objectives of the study.

12.2 CONTINUUM MODELING

Here we review mathematical models and numerical solution approaches that are currently being used in areas such as oil and gas recovery and storage, geothermal systems, vadose zone studies, and industrial processing. The mathematical formulation given below makes specific reference to the TOUGH2 numerical simulation program (Pruess et al., 1999), although similar models and methods are employed in other commonly used codes. In some cases gas is the only fluid filling the pore space, while usually additional phases are present (aqueous and non-aqueous), that may profoundly affect flow and transport behavior in the gas phase.

C. Ho and S. Webb (eds.), Gas Transport in Porous Media, 213–220.
© 2006 *All rights reserved.*

12.2.1 Advective-Diffusive Model

We begin with the formulation of an "advective-diffusive" model (ADM), in which advective and diffusive contributions to gas transport are evaluated separately, using phenomenological flux laws, and then added.

Advective gas flux is commonly described with a multiphase version of Darcy's law,

$$\mathbf{F}_g = \rho_g \mathbf{u}_g = -k \frac{k_{rg} \rho_g}{\mu_g} (\nabla \mathbf{P}_g - \rho_g \mathbf{g}) \tag{12.1}$$

where $\mathbf{u}_g$ is the volume flux, often referred to as "Darcy velocity." Whereas for liquids absolute permeability k is a material constant, this parameter becomes dependent on pressure for gases, according to the Klinkenberg (1941) relationship

$$k = k_\infty \left(1 + \frac{b}{P_g}\right) \tag{12.2}$$

The physical effect that is causing an apparent increase in permeability for low gas pressures is "slip flow:" at low pressures, the mean free path of gas molecules is no longer negligibly small in comparison to pore sizes, so that the assumption of advective velocity at the pore walls approaching zero is no longer valid.

Eq. (12.1) gives the total flux of gas phase. The flux for an individual mass component κ is obtained by multiplying with the mass fraction of that component in the gas phase, $\mathbf{F}_g^\kappa = X_g^\kappa \mathbf{F}_g$. Under general multi-phase conditions, an expression like Eq. (12.1) is also written for advective transport in the liquid phase.

Diffusive transport in the gas phase is modeled as Fickian in the ADM; that is, diffusive flux is proportional to the concentration gradient, usually expressed as mass fractions or mole fractions. In the TOUGH2 code, the diffusive flux of component κ in the gas phase is written as

$$\mathbf{j}_g^\kappa = -\phi\, \tau_0\, \tau_g\, \rho_g\, D_g^\kappa \nabla \mathrm{X}_g^\kappa \tag{12.3}$$

Here ϕ is porosity, $\tau_0 \tau_g$ is the tortuosity which includes a porous medium dependent factor τ_0 and a coefficient that depends on gas phase saturation S_g, $\tau_g = \tau_g(S_g)$, ρ_g is gas density, D_g^κ is the diffusion coefficient of component κ in bulk gas, and X_g^κ is the mass fraction of component κ in the gas. For a binary system with equal diffusion coefficients, Eq. (12.3) yields a net mass flux of zero, so that total mass flux in the gas phase is given just by the advective term, Eq. (12.1). Formulations similar to Eq. (12.3) are employed in other continuum codes (see Table 12.1).

In some cases the mole fraction gradient rather than the mass fraction gradient is used as driving force, giving equimolar counterdiffusion in a binary system with equal diffusion coefficients (Bird et al., 1960). The TETRAD code (Vinsome and Shook,

Table 12.1. Treatment of gas diffusion in continuum codes

Code	Comments	Diffusive flux formulation	Vapor	Reference
TOUGH2	Multiphase, multicomponent flows	Eq. (12.3)	Optional enhancement factor	Pruess et al., 1999
NUFT	Multiphase, multicomponent flows	Eq. (12.3)	Treated like NCG	Nitao, 1998
PORFLOW	Multiphase, multicomponent flows	Eq. (12.3)	Treated like NCG	ACRI, 2003
STOMP	Multiphase, multicomponent flows	similar to Eq. (12.3) but using mole fraction	Treated like NCG	White and Oostrom, 2000
TETRAD	Multiphase, multicomponent flows	Similar to Eq. (12.3) with two choices, (1) using mole fraction gradient (2) using $\nabla(\rho x)$ as driving force	Treated like NCG	Vinsome and Shook, 1993
FEHM	Multiphase, multicomponent flows	Eq. (12.3)	No vapor diffusion	Zyvoloski et al., 1999
UNSAT-H	Unsaturated zone moisture transport in liquid and gas phases	Vapor diffusion only	Enhanced vapor diffusion after Philip and de Vries (1957)	Fayer, 2000

1993) also provides an option to use the gradient of molar density as driving force, which will give equal-volume counterdiffusion in a binary system with equal diffusion coefficients. Some codes include provisions for enhancement of vapor diffusion as compared to diffusion of non-condensible gases, to represent effects of pore-level condensation-vaporization phenomena (Philip and de Vries, 1957; Jury, 1973; Cass et al., 1984).

The coupling between gas and liquid phase transport involves mass fractions for advection and mass fraction gradients for diffusion, leading to profound differences in the space-discretized treatment of the two processes. Advective transport must be treated with total variation diminishing (TVD) discretization schemes, in order to avoid spurious oscillations and unphysical behavior. The TOUGH2 code employs the simplest TVD scheme possible, namely, full upstream weighting (upwinding) of fluid mobilities and compositions. Advective transport is then calculated by evaluating upstream-weighted flux expressions separately for gas and liquid phases, and adding them.

New issues arise in the space-discretized treatment of diffusion under multiphase conditions. Let us consider a finite difference formulation for total diffusive flux at the interface between two finite difference grid blocks n and m.

$$\left(f^{\kappa}\right)_{nm} = -\left(\Sigma_g^{\kappa}\right)_{nm} \frac{\left(X_g^{\kappa}\right)_m - \left(X_g^{\kappa}\right)_n}{D_{nm}} - \left(\Sigma_l^{\kappa}\right)_{nm} \frac{\left(X_l^{\kappa}\right)_m - \left(X_l^{\kappa}\right)_n}{D_{nm}} \quad (12.4)$$

Here we have introduced the as yet unknown diffusive strength coefficients $(\Sigma_g^{\kappa})_{nm}$ and $(\Sigma_l^{\kappa})_{nm}$ at the interface, which must be expressed in terms of the strength coefficients in the participating grid blocks. Invoking conservation of diffusive flux across the interface between two grid blocks leads in the usual way to the requirement of harmonic weighting of the diffusive strength coefficients. However, such weighting may in general not be applied separately to the diffusive fluxes in gas and liquid phases, because these may be strongly coupled by phase partitioning effects. This can be seen by considering the extreme case of diffusion of a water-soluble and volatile compound from a grid block in single-phase gas conditions to an adjacent grid block which is in single-phase liquid conditions. Harmonic weighting applied separately to liquid and gas diffusive fluxes would result in either of them being zero, because for each phase effective diffusivity is zero on one side of the interface. Thus total diffusive flux would vanish in this case, which is unphysical. In reality, tracer would diffuse through the gas phase to the gas-liquid interface, would establish a certain mass fraction in the aqueous phase by dissolution, and would then proceed to diffuse away from the interface through the aqueous phase. Similar arguments can be made in the less extreme situation where liquid saturation changes from a large to a small value rather than from 1 to 0, as may be the case in the capillary fringe, during infiltration events, or at fracture-matrix interfaces in variably saturated media.

A consistent space-discretized treatment of diffusion in multiphase conditions can be achieved by introducing gas and liquid phase mass fractions $(X_g^{\kappa})_i$ and $(X_l^{\kappa})_i$ at the interface *i* between grid blocks n and m as additional unknowns. Diffusive fluxes to and from the interface can be calculated by using finite difference expressions such as $\left((X_g^{\kappa})_i - (X_g^{\kappa})_n\right)/D_{ni}$ for mass fraction gradients, and applying the effective diffusive strength coefficients in the individual grid blocks. Two additional constraints are available to determine the unknown gas and liquid phase mass fractions at the interface, namely, (1) the solubility relation $\left(C^{\kappa} = X_l^{\kappa}/X_g^{\kappa}\right)$, and (2) the conservation of total diffusive flux at the interface.

12.2.2 Dusty Gas Model

The Dusty Gas Model represents the porous medium as a collection of giant spherical molecules (dust particles) held in space by external force. The movement of gas molecules in the spaces between dust particles is described by the kinetic theory of gases. See Solcova and Schneider (this book) for more details.

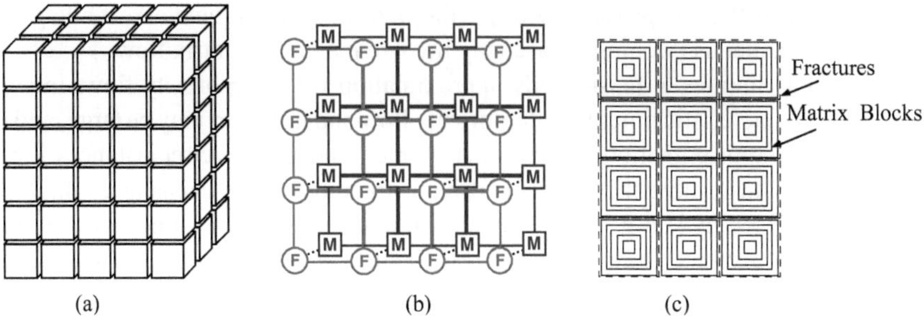

Figure 12.1. Illustration of concepts used for modeling of multiphase flow in fractured rocks: (a) double-porosity model (DPM; after Warren and Root, 1963); (b) dual permeability model (DKM), with global flow in both fracture (F) and matrix continua (M); (c) MINC sub-gridding for resolution of gradients in the matrix blocks (after Pruess and Narasimhan, 1985).

12.2.3 Fractured Media

In many cases of interest the media through which gas transport occurs are fractured. Conceptually, the simplest approach for transport modeling involves an explicit representation of fractures, which are described as more-or-less planar regions of large permeability and porosity, with "small" spatial extent perpendicular to the fracture plane. Such explicit representation of fractures is feasible only for flow systems with a small number of fractures. For flow systems with ubiquitous interconnected fractures, their explicit representation is neither practically possible nor desirable, and continuum representations are used instead. Such representations always entail volume averaging on some scale. For gas transport this has a firm basis in the fundamental physical processes which tend to be diffusive in nature (described by parabolic or nearly-parabolic PDEs). Commonly used modeling approaches for fractured media include double-porosity (DPM), dual permability (DKM), and multiple interacting continua (MINC); see Figure 12.1.

The DPM considers that global flow occurs only through the network of interconnected fractures, while fractures and matrix rock of generally low permeability may exchange fluid, solutes, and heat locally. The fracture system is characterized and modeled with customary porous medium parameters. At each point (or grid node) of the system, two sets of thermodynamic parameters are defined to characterize the state of the flow system: one set involves an average over the fractures, the other an average over the matrix rock. "Interporosity flow" between the fracture and matrix continua is assumed to be proportional to the difference in the average values of the intensive quantities driving flow and transport (such as pressures, mass fractions, or temperatures).

For flow systems in which both the fracture and matrix continua contribute to global flow, the DPM is generalized to allow global fracture-fracture as well as matrix-matrix flow, in addition to fracture-matrix exchange (Figure 12.1b). This type of approach is applicable for multiphase (or unsaturated) flow in fractured-porous media, where global flow of the gas phase may occur through the fracture network, while global flow of the aqueous (wetting) phase may involve matrix-to-matrix flow.

The DPM implies a quasi-steady approximation for interporosity flow. In some cases the characteristic times for interporosity flow and transport may be large, and the quasi-steady approximation becomes inaccurate. Under these circumstances it may be necessary to resolve the gradients driving fracture-matrix exchange, which can be accomplished by sub-gridding of matrix blocks into a series of continua defined according to distance from the matrix block surface, the so-called MINC-concept (Figure 12.1c).

12.2.4 Example: Diffusive Effects in the Migration of Volatile Tracers in Fractured Media

Tracer testing has been extensively applied in fractured geothermal reservoirs as a means for mapping fast preferential pathways and characterizing reservoir processes that involve boiling and condensation phenomena. Diffusive propagation of tracers tends to be slow in comparison to advective transport, so that its contribution to transport over interwell distances is negligible. However, diffusion can be a very important process for fracture-matrix exchange, and as such can have a strong influence on tracer breakthrough curves (BTCs). As fluids (gas or liquid) are advecting through a fracture, tracer will diffuse into the rock matrix, causing a delay in tracer breakthrough and reduced peak concentrations (Figure 12.2). If tracer is injected as a slug, the concentration gradient between fractures and matrix will reverse after peak concentrations have passed, and tracer that previously entered the rock matrix will diffuse back out into the fractures, giving rise to long tails in the BTCs.

Pruess (2002) examined the behavior of volatile tracers during water injection into depleted vapor zones of a geothermal reservoir. In this case, injected water is subject to vigorous vaporization. Much of the tracer will partition into the gas (vapor) phase, and the exchange between fractures and matrix will involve diffusion in both gas and liquid phases. Figure 12.3 shows simulated breakthrough curves for two different phase-partitioning tracers with and without inclusion of diffusive effects. Diffusion

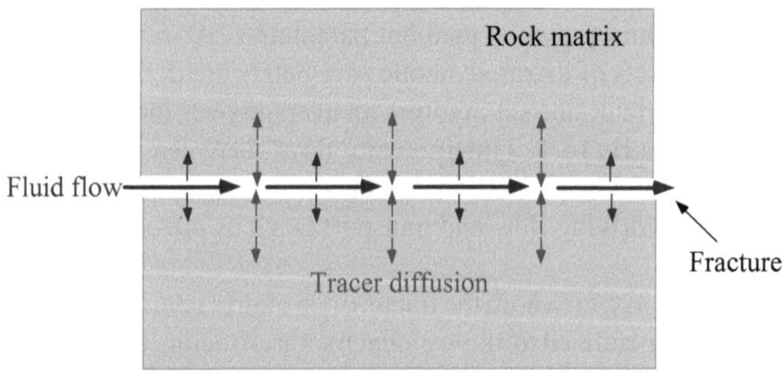

Figure 12.2. Schematic of tracer behavior in a fractured-porous medium

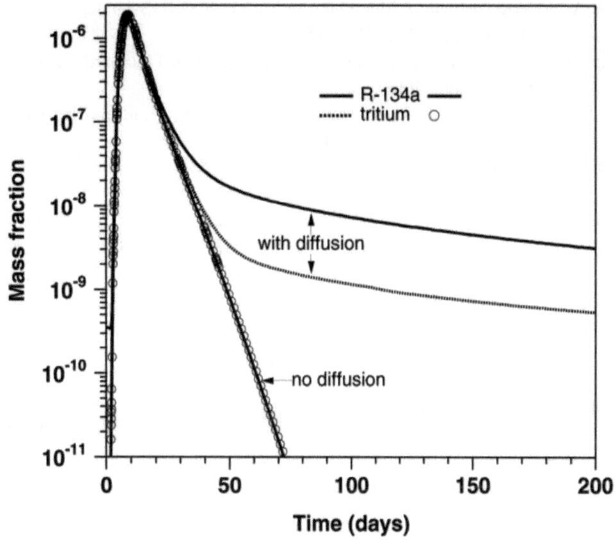

Figure 12.3. Simulated tracer breakthrough curves in a fractured vapor-dominated geothermal reservoir

gives rise to long tails in the BTCs which may permit an estimate of fracture-matrix interface area, a crucial parameter for determining the rate at which heat can be transferred from the matrix rocks to the fluids flowing in the fractures. BTCs with long tails similar to Figure 12.3 were observed during tracer testing at The Geysers geothermal field (Beall et al., 1994). Further discussion of coupled gas and liquid diffusion effects during tracer migration in two-phase geothermal reservoirs is given in Pruess (2002) and Shan and Pruess (2003).

REFERENCES

ACRI (Analytical and Computational Research, Inc.). User's Manual, Version 5.0, Rev: 5, Bel Air, CA, 2003.

Beall, J.J., M.C. Adams and P.N. Hirtz. R-13 Tracing of Injection in The Geysers, *Transactions,* Geothermal Resources Council, Vol. 18, pp. 151–159, 1994.

Bird, R., W.E. Stewart and E.N. Lightfoot. *Transport Phenomena*, John Wiley & Sons, New York, NY, 1960.

Cass, A., G.S. Campbell and T.L. Jones. Enhancement of Thermal Water Vapor Diffusion in Soil. *Soil Sci. Soc. Am. J.,* Vol. 48, no. 1, pp. 25–32, 1984.

Fayer, M.J. UNSAT-H Version 3.0: Unsaturated Soil Water and Heat Flow Model, Theory, User Manual, and Examples, Pacific Northwest National Laboratory, Report PNNL-13249, Richland, WA, June 2000.

Jury, W. A. Simultaneous Transport of Heat and Moisture Through a Medium Sand, Ph.D. Thesis, University of Wisconsin, 1973.

Klinkenberg, L.J. The Permeability of Porous Media to Liquids and Gases, *API Drilling and Production Practice*, pp. 200–213, 1941.

Nitao, J. Reference Manual for the NUFT Flow and Transport Code, Version 2.0, Lawrence Livermore National Laboratory, report UCRL-MA-130651, Livermore, California, 1998.

Philip, J.R. and D.A. de Vries. Moisture Movement in Porous Materials Under Temperature Gradients, *EOS Trans. AGU*, Vol. 38, no. 2, pp. 222–232, 1957.

Pruess, K. Numerical Simulation of Multiphase Tracer Transport in Fractured Geothermal Reservoirs, *Geothermics*, Vol. 31, pp. 475–499, 2002.

Pruess, K. and T.N. Narasimhan. A Practical Method for Modeling Fluid and Heat Flow in Fractured Porous Media, *Soc. Pet. Eng. J.*, 25 (1), 14–26, February 1985.

Pruess, K., C. Oldenburg and G. Moridis. TOUGH2 User's Guide, Version 2.0, Lawrence Berkeley National Laboratory Report LBNL-43134, Berkeley, CA, November 1999.

Shan, C. and K. Pruess. Numerical Simulation of Noble Gases as Natural Tracers for Injection Returns and Reservoir Processes in Vapor-Dominated Systems, Proceedings, 28th Workshop on Geothermal Reservoir Engineering, report SGP-TR-173, Stanford University, Stanford, CA, January 2003.

Vinsome, P.K.W. and G.M. Shook. Multipurpose Simulation, *J. Petroleum Science and Engineering*, 9 (1), March, 1993.

Warren, J.E. and P.J. Root. The Behavior of Naturally Fractured Reservoirs, *Soc. Pet. Eng. J., Transactions, AIME, 228*, 245–255, September 1963.

White, M.D. and M. Oostrom. STOMP, Subsurface Transport over Multiple Phases, Version 2.0 Theory Guide, Pacific Northwest National Laboratory, Report PNNL-12030, Richland, WA, March 2000.

Zyvoloski, G.A., B.A. Robinson, Z.V. Dash and L.L. Trease. Models and Methods Summary for the FEHM Application, Los Alamos National Laboratory Report SC-194, Los Alamos, NM, 1999.

CHAPTER 13

LATTICE BOLTZMANN METHOD FOR CALCULATING FLUID FLOW AND DISPERSION IN POROUS AND FRACTURED MEDIA

HARLAN W. STOCKMAN
Sandia National Laboratories, P.O.Box 5800, Albuquerque, NM, 87185, USA

13.1 INTRODUCTION

The lattice Boltzmann (LB) method is a numerical technique suited for modeling flow of fluids, particularly those obeying the Navier-Stokes and advection-dispersion equations (Rothman and Zaleski 1997; Martys and Chen, 1996). The strength of the method lies in its ability to calculate transport in the presence of geometrically complex solid boundaries and initial conditions. Added effects may include chemical reaction, dissolution and precipitation, immiscibility, and buoyancy.

In the past ten years, the LB method has been applied increasingly to problems involving flow through porous and fractured rocks and soils (Stockman et al., 1997; Stockman et al., 1998; Zhang et al., 2002; Pan et al., 2004; Sukop and Or, 2004, and references therein). While the technique can be applied to most fluids, there are some special considerations for LB models of gases, and in particular, for assemblages of many pores, or macroscopic fractures. These special considerations, along with illustrative examples and benchmarks, are the subject of this chapter. The focus of the chapter is on pore-scale models, where fluid behavior is approximated by the Navier-Stokes equations. These pore-scale models are used to obtain permeability and dispersion coefficients applicable to a larger scale, and the inherent and practical limitations of the method are illustrated with examples.

13.2 METHOD

13.2.1 Basic Method

In the LB method, physical space is broken up into a set of nodes, usually on a Cartesian grid; the set of these nodes is called the automaton. The term *lu* is used for

C. Ho and S. Webb (eds.), Gas Transport in Porous Media, 221–242.
© 2006 *Springer.*

lattice unit, the fundamental unit of distance in the automaton (the distance between adjacent nodes is one *lu*); *ts* refers to one time step in the automaton. Mass is allowed to move about the grid, but via a fixed set of n velocity vectors $\vec{e}_i$ (n is typically 15 or 19 for 3D LB). At each node, there is set of n floating-point (non-integral) particle distribution functions f_i, corresponding to each ith velocity. The mass density at each node is $\rho = \sum_{i=0}^{n-1} f_i$. The evolution of the particle distribution function f_i with time t,is calculated according to

$$f_i(\vec{r} + \vec{e}_i, t + 1) = f_i(\vec{r}, t) - \omega \cdot (f_i(\vec{r}, t) - f_i^{eq}(\vec{r}, t)) \tag{13.1}$$

where $\vec{r}$ is a node position (x, y, z) in the lattice, f_i^{eq} is the Boltzmann equilibrium distribution, and ω is the collision operator, which determines the viscosity and diffusion coefficients. With a carefully chosen form for f_i^{eq}, the $\rho(x, y, z)$ will obey the incompressible Navier-Stokes equation for fluid flow. There are several ways to add momentum "kicks" with each step, so the fluid will accelerate, perhaps reaching a viscous equilibrium. In this chapter, it is assumed that all LB calculations use some variant of the BGK (Bhatnagar-Gross-Krook) method, where the collision operator is simply a constant $1/\tau$ (Qian et al., 1992). It is important to understand that LB does not presume to be a model of molecular behavior; while one refers to distribution functions, and there are allusions to translations of particles, the distributions are merely convenient artifices for tracking the movement of mass and momentum.

In typical LB computer programs, a translation or streaming function performs the left side of Eq. (13.1), and a collision function performs the right side. Thus, Eq. (13.1) is really carried out in two steps:

1. At each node, an equilibrium distribution f_i^{eq} is calculated for each vector direction. The distribution of f_i at that node is then replaced with a new distribution $f_i - (f_i - f_i^{eq})/\tau$.
2. The new f_i are then translated, or "streamed," along the vector directions to the adjacent nodes. This process is handled by the computer in such a way that none of the new values are accidentally overwritten.

A velocity field is calculated at each x, y, z site as

$$\vec{u} = (u_x, u_y, u_z) = \sum_{i=0}^{n-1} f_i \cdot \vec{e}_i \tag{13.2}$$

$$u^2 = \vec{u} \cdot \vec{u} \quad \text{(a scalar)} \tag{13.3}$$

In this chapter, we use the 19-vector D3Q19 lattice (Qian et al., 1992). This lattice has one "rest" particle with a velocity vector $e_0 = (0, 0, 0)$, six velocity vectors along the Cartesian axes (permutations of $(\pm 1, 0, 0)$) and twelve vectors that are permutations of $(\pm 1, \pm 1, 0)$. For this lattice, the equilibrium distribution is calculated as (Martys and Chen, 1996):

$$f_0^{\text{eq}} = \frac{\rho}{3}\left[1 - \frac{3}{2}u^2\right] \tag{13.4}$$

$$f_i^{\text{eq}} = q_i\,\rho\left[1 + 3\vec{e}_i \cdot \vec{u} + \frac{3}{2}((3\vec{e}_i \cdot \vec{u})^2 - u^2)\right] \tag{13.5}$$

where $q_i = 1/18$ for the six vectors along the Cartesian axes, and $q_i = 1/36$ for the remaining vectors. This distribution, when used in Eq. (13.1), will give back a very good approximation of incompressible Navier-Stokes behavior for $\vec{u}$.

13.2.2 Dispersion

To model dispersion of dilute tracers, the simplest approach is to use one lattice for a "carrier fluid" that obeys the Navier-Stokes equation, and calculate $\vec{u}$ for that carrier fluid only. A separate lattice exists for each tracer, and changes in the tracer distributions are propagated according to Eqs. (13.1), (13.2), (13.4) and (13.5), using the $\vec{u}$ determined from the carrier.

Unlike the Navier-Stokes equations, the advection-dispersion equation (which governs tracer movement) is linear in $\vec{u}$. Flekkøy et al. (1995) and Noble (1996) reasoned that the tracer equilibrium distributions should be linear in $\vec{u}$ as well. Thus for a tracer s, the equilibrium distribution is of the form:

$$f_{s,i}^{\text{eq}} = A + B \cdot (\vec{e}_{s,i} \cdot \vec{u}) \tag{13.6}$$

and the A and B are fixed by the requirement that solute is conserved ($\rho_s = \sum f_{s,i}^{\text{eq}}$), and the requirement that the solute flux at equilibrium is due entirely to advection ($\sum f_{s,i}^{\text{eq}}\vec{e}_{s,i} = \rho_s \cdot \vec{u}$). Compared to Eq. (13.5), Eq. (13.6) requires fewer floating point operations to evaluate the equilibrium distribution, hence is faster to compute.

The real bottleneck in the LB calculation, is the time taken to read all the vectors f_i from memory (for both carrier and tracers), and to write them back after they have been updated. Thus a great deal of memory traffic can be avoided by using lattices with fewer vectors for the tracers. In 2D, only four Cartesian vectors are needed, and in 3D, the six Cartesian vectors will suffice. Wolf-Gladrow (1994) suggested low-vector lattices for modeling diffusion, and Noble (1996) performed a Chapman-Enskog expansion for Cartesian lattices, recovering the advection-dispersion equation and estimating the error. The present author compared the 6-vector and 19-vector methods for tracer problems in extreme conditions and found the agreement was extremely close. (It must be emphasized that low-vector lattices, and Eq. (13.6), are used *only* for the dilute tracers, not the carrier.)

In many dispersion problems, the flow field $\vec{u}$ is at steady state, and need not be recalculated at each timestep. Furthermore, periodic boundary conditions on $\vec{u}$ are often appropriate, even when tracer dispersion is not periodic, as in the SC dispersion problem described in Section 13.3.2. For such problems, one can use a 19-vector carrier fluid in a single repeat unit, $N_x \times N_y \times N_z\, lu^3$, then save the equilibrium

$\vec{u}(x, y, z)$. Typically, the flow reaches steady-state at time $> L^2/\nu$, where ν is the kinematic viscosity of the carrier fluid, and L is a characteristic length scale perpendicular to the flow direction (e.g., the channel width). Then the memory used for the 19-vector carrier fluid can be reclaimed, and reallocated for additional repeat units of solid geometry (and the lattice used for the tracers) in the x direction. The velocity field at x, y, z is then $\vec{u}((x \text{ MOD } N_x), y, z)$. The lattice Boltzmann updates for the tracers are still carried out at each step, but rather than recalculate the velocity at each point, the saved velocity field is used. We refer to these additional repeat units as "clones."

As mentioned, the viscosity of the carrier fluid, and the diffusion coefficients of the tracers in the fluid, are entirely dependent on the collision parameter τ. In the 3D model above,

$$\nu = \frac{\left(\tau_0 - \frac{1}{2}\right)}{3} \quad \text{and} \quad D_{ms} = \frac{\left(\tau_s - \frac{1}{2}\right)}{3} \tag{13.7}$$

where D_{ms} is the molecular diffusion coefficient of tracer s, in units of lu^2/ts.

In one important way, the LB method is better suited to model dispersion in gas transport, than dispersion in liquids such as water. It is difficult to set up a dispersion calculation wherein the τ_0 of the bulk fluid, or carrier, is vastly different than the τ_s of the tracers, either for reasons of numerical stability or convenient scaling. In water, the $\nu \sim 0.01$ cm^2/s, but $D_m \sim 10^{-5}$ cm^2/s, so there is a three-order of magnitude difference that must be spanned. But in gases, kinetic theory leads us to expect the kinematic viscosity and diffusion coefficients (for similar molecules) will be approximately the same. Thus ν for air is $\sim$0.14 cm^2/s at 18 C and 1 atm, and D_m for CO_2, O_2 and H_2O in air are approximately $\sim$0.14, 0.18 and 0.24 cm^2/s, respectively, at the same temperature and pressure. Hence, one is more likely to find a suitable ν/D_m for gases.

13.2.3 Boundaries

The equations and calculation method above describe the update of the free fluid. The processing at calculation boundaries and solid walls require special attention, to ensure mathematical "closure;" i.e., that the total mass of each component is conserved in the absence of intentional sinks. The calculation domain is finite in size; thus during the translation stage, the distributions that point out of the domain would be lost, unless a special effort were made to preserve that mass. A simple way to solve this dilemma is to use "wrap" boundaries, so that a vector leaving the right side of the automaton at x_{max} is reinjected at 0. Similarly, a vector that leaves the left side of the automaton at $x = 0$, is reinjected at x_{max}, the last site in the x-direction. This wrap is performed automatically in the translation step, and applies to the y- and z-directions as well.

One must also decide how to treat solids during the collision step. The basic problem is this: before a collision step, the preceding translation step moves vectors from free fluid to solid sites, and these vectors point into the solid. Yet there may be no vectors pointing out of the wall into the fluid, since there is no fluid inside the solid before translation. A simple solution is to invert the directions of all vectors on the wall (for both the carrier fluid and each tracer); this approach is called the "bounce-back" boundary, and it assures that there is a no-slip velocity boundary near the wall (Chen et al., 1996). Section 13.4 will discuss the consequences of bounce-back in more detail.

If a solid boundary is placed at the ends of the automaton, it automatically supercedes the wrap condition, because the solid wall deflects all outward-pointing vectors back into the automaton.

Self Study Example:
We will simulate dispersion in a long duct, with square cross-section, by the method described in Section 13.2.2. The initial duct contains 24 nodes in the x-direction, and 26 in each of the y- and z-directions, and is open in the x-direction, with the other 4 sides closed off with solid walls, as shown below:

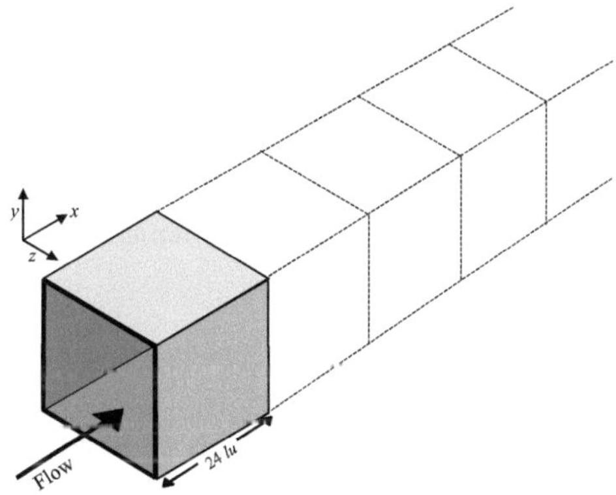

A x-direction body force is placed on the fluid, causing the fluid to accelerate along the x-axis. The viscosity ν is 0.01 lu^2/ts, and the average speed is limited to 0.01 lu/ts. The initial calculation begins with one 19-vector carrier fluid, to calculate the fluid velocity field. After 60000 ts, the memory from the flow calculation is reclaimed, the flow field is "frozen in," and the channel is cloned 47 additional times in the x-direction. The resultant $(48 \times 24) \times 26 \times 26$ channel is used for a dispersion study with a single 6-vector tracer. A slug of tracer is injected into the left end of the channel, and is followed as it disperses in the flow; this part of the calculation proceeds for additional 60000 ts.

(1a) Are 60000 ts adequate to reach viscous equilibrium?

(1b) On your computer, the LB program processes a single 19-vector component at a rate of 9 million site-updates per second (MUPS), and processes a single 6-vector component (in dispersion-only mode) at 35 MUPS. How many seconds will be needed to complete the first (19-vector) and last (6-vector) parts of the calculation?

(1c) The calculation is performed in single precision, and each vector is represented by a 32-bit (4 byte) floating-point number. How much memory is required for the first (19-vector) and second (6-vector) parts of the calculation?

13.2.4 Efficient Algorithms

The LB method can be quite calculation-intensive, and is simply not efficient for most simple geometries. However, many of the problems experienced by newcomers to the technique probably stem from coding inefficiencies. The calculations reported herein have benefited from many speed optimizations in the source code. Some of these optimizations, such as disambiguation of pointers, have been discussed previously (Stockman et al., 1998). Since that publication, the code has been modified to use SIMD (single-instruction, multiple data) programming, along with explicit prefetch; these improvements resulted in ~3-fold improvement in performance. The compiler used for producing the executable code can make a great deal of difference. For example, we found a factor 3.5 speedup in going from the most efficient version of code compiled under gcc (the public domain compiler available through the GNU project), to code compiled with Intel's version 5.01 C. We found that the choice of operating systems (OS; e.g., Linux vs. Microsoft Windows) made no difference on the program speed, so long as the same version compiler was used on each OS. Currently, the code is limited by the speed with which it can read and write vectors to main memory, and performance is greatest on desktop computers with high-speed, multiple-channel memory. It is notable that we have seen little improvement in the code speed in 3.5 years, simply because the speed of memory access has not improved greatly in that time.

13.3 BENCHMARKS AND ACCURACY

The following benchmarks illustrate the bounds of accuracy of the method, and also provide background for the discussions in Section 13.5. Where necessary for clarification, *lu* and *ts* units will be converted into centimeters and seconds; however, most comparisons are made in dimensionless parameters. Of particular interest are the Reynolds number (Re), Peclet number (Pe), and Schmidt number (Sc), defined as:

$$\mathrm{Re} = U \cdot \frac{L}{\nu}, \qquad \mathrm{Pe} = U \cdot \frac{L}{D_{\mathrm{m}}}, \text{ and } \mathrm{Sc} = \frac{\mathrm{Pe}}{\mathrm{Re}} \tag{13.8}$$

where U is a characteristic speed and L is a characteristic length scale, defined for the particular problem at hand. Usually L is the width or a particle or a pore.

(2) The D_{m} for the self-study problem is 0.01 lu^2/ts. What are the Re, Pe, and Sc for the channel calculation in Part 1?

13.3.1 Permeability of an Array of Spheres

The first test is calculation of permeability in a non-trivial geometry; in this case, in an infinite array of spheres in a simple cubic (SC) geometry (Zick and Homsy, 1982).

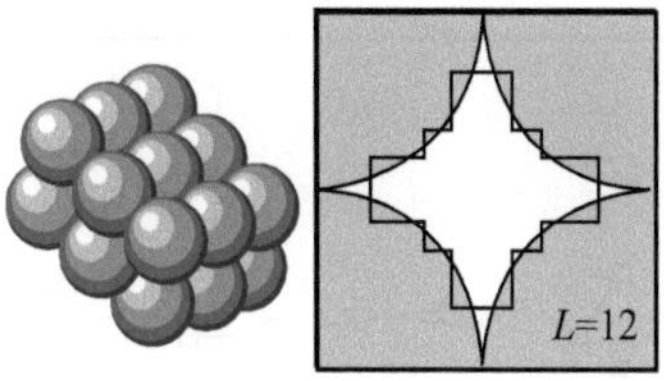

Figure 13.1. Left: part of an array of simple cubic (SC) spheres. Right: the ideal shape of the pore space between spheres (thin smooth curves) and the actual pixelation (black solid line) for a calculation with $L = 2R = 12$ lattice units

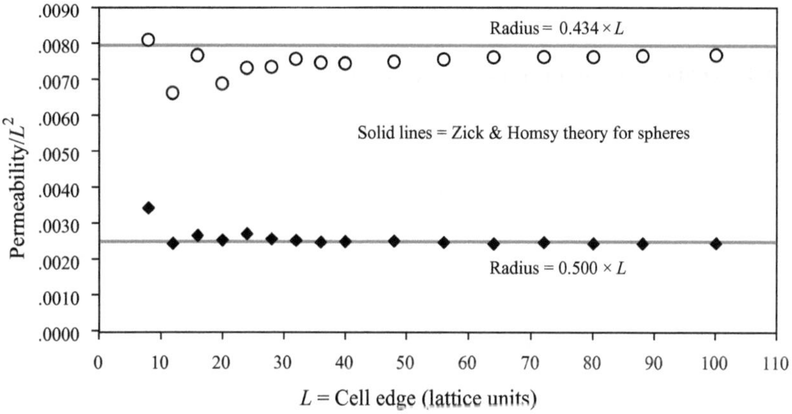

Figure 13.2. LB estimates of permeability for infinite SC array of spheres (Figure 13.1) as symbols, and Zick and Homsy (1982) theoretical values as solid horizontal lines

A portion of the geometry is shown in Figure 13.1. The LB simulation uses wrap boundaries on the faces of the automaton, so only one sphere need be considered in the calculation, as the wrap boundaries effectively produce an infinite array of spheres in the x, y, and z-directions. Figure 13.1 also shows a potential problem; in the relatively small LB simulations, the narrow channel between adjacent spheres is "pixelated" by the limited number of nodes ($12 \times 12 \times 12$ in this case), and is not well-resolved.

Figure 13.2 shows the LB calculated permeability, plotted against the separation L between the spheres (in *lu* on a side), for the case where the sphere radius $R = 0.5 \cdot L$ (touching) and the case $R = 0.434 \cdot L$ (slightly separated spheres), and $\mathrm{Re} \sim 0$. The Zick and Homsy values are shown as solid lines; the agreement is fairly good, and the measured permeabilities converge on the "true" values as the number of nodes in the simulation increases. Figure 13.3 shows the correlation of the porosity with the permeability, for the LB simulations with $R = 0.5 \cdot L$; the target porosity is 0.476. The deviations from the target porosity result from the pixelation of the LB model, and are reflected in the errors in estimated permeability.

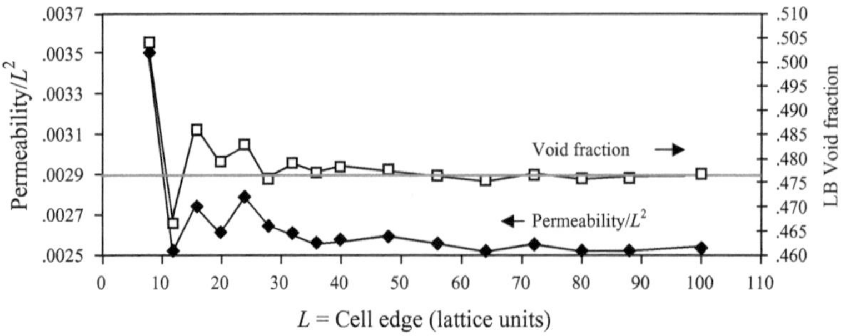

Figure 13.3. Correlation of void fraction in LB (ideal value = 0.476) with estimated permeability

13.3.2 Dispersion in an Array of Spheres

Dispersion coefficients can be estimated for the same SC array. This example also illustrates the "cloning" process described above, and is a necessary precursor to the interpretations in Section 13.5.1; it also shows the significant differences in modeling dispersion in gases, vs. liquids. The technique for measuring dispersion is the method of moments; that is, a slug of solute is injected into a flow field, and the spread of the solute 2nd moment is measured as it disperses downstream, as illustrated in Figure 13.4.

The dispersion coefficient D^* is then defined by:

$$D^* = \frac{1}{2}\,\mathrm{d}m_2/\mathrm{d}t \tag{13.9}$$

where m_2 is the second moment, or variance of the solute distribution, projected onto the x-axis (more detail about the method of moments can be found in Stockman, 1997). For the SC Geometry, $\mathrm{Pe} \equiv L \cdot U/(D_\mathrm{m} \cdot (1-\varepsilon))$, and $\mathrm{Re} \equiv L \cdot U/(\nu \cdot (1-\varepsilon))$ and where ε is the porosity and U is the average Darcy flow speed.

Figure 13.4 shows the LB-calculated dispersion patterns for the tracer for two cases. The upper figure shows dispersion for a case of $\mathrm{Sc} = 0.88$, which matches the Sc of argon in air; that gas mixture was used for an experimental investigation of gaseous dispersion in an SC array (Gunn and Pryce, 1969). The bottom of the figure shows the pattern for a hypothetical case with the same Pe, but the Re lowered so $\mathrm{Sc} = 32 = \mathrm{Pe}/\mathrm{Re} = \nu/D_\mathrm{m}$. The latter case does not represent any real tracer gas, but shows the hypothetical dispersion in near-Stokesian flow, with little inertia. At these early stages in the calculation, the lower Sc case appears to show greater dispersion. Figure 13.5 shows the flow patterns for the $\mathrm{Sc} = 0.88$ case ($\mathrm{Re} = 50$), with those for the $\mathrm{Sc} = 32$ case ($\mathrm{Re} = 4$). The flow patterns have two significant differences; the higher Re flow has (1) recirculation zones, and (2) greater maximum fluid speed in the channels between the spheres, at the same Darcy speed. Either difference may help account for the enhanced dispersion.

Figure 13.6 compares the measured D^*/D_m for LB calculations, with a set of experiments by Gunn and Pryce (1969), and with two other numerical estimates.

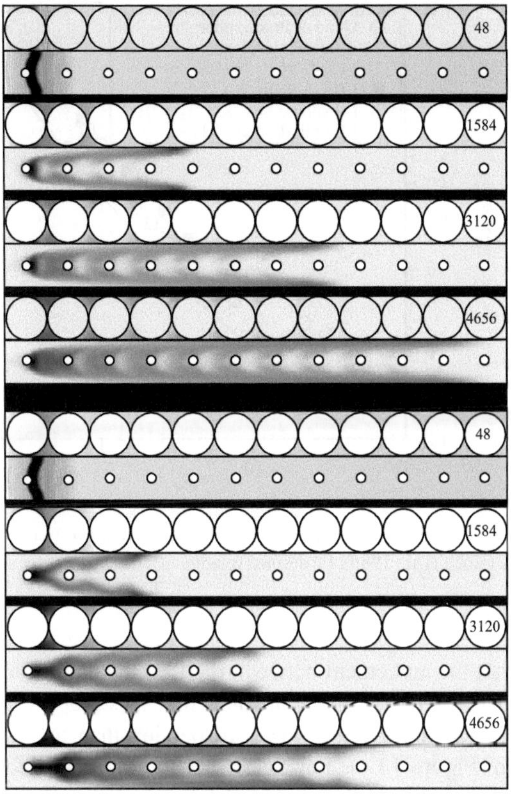

Figure 13.4. Solute patterns in the early stages of the LB calculations, for Re/Pe = 0.88 (top) and Re/Pe = 32 (bottom); both are at Pe = 132. The step number is indicated on the upper right of each frame. Each frame shows cuts at depths = 18 and 0 *lu* (out of a total depth of 36 *lu*). The solute is indicated in black

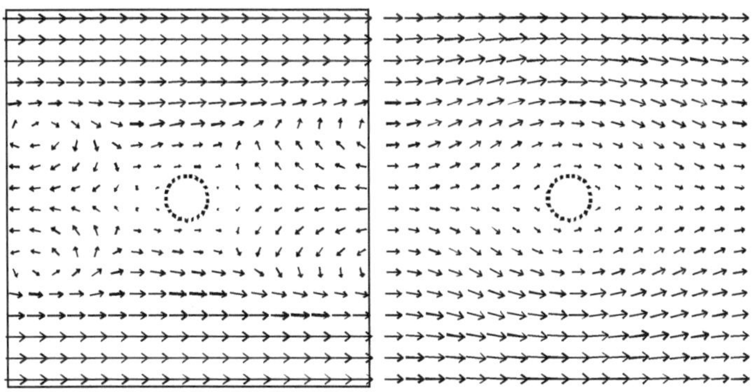

Figure 13.5. Vector flow fields around one sphere, for Re = 150 (left) and Re = 4 (right). The $x - y$ cut is at depth where the two spheres just touch. Vectors are 4th-root scaled to emphasize smaller vectors (i.e., a vector that appears half as long as another, is really 1/16 as long)

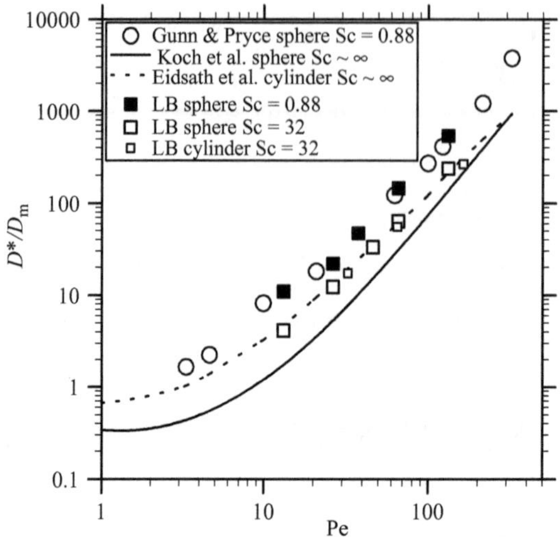

Figure 13.6. Comparison of experimental results (Gunn and Pryce, 1969), LB calculations (this study) and Stokesian theory (Koch et al., 1989). Pe defined relative to particle diameter and averaged Darcy flow speed

First, one sees that the agreement between the LB calculations and the experiments (both at Sc = 0.88) is rather good. Second, one sees that the LB calculations for the Sc = 32 produce significantly lower dispersion than for Sc = 0.88 (as might be expected from Figure 13.4). For Sc = 32, the inertial effects should be much smaller; but it is surprising to see that the deviation at even Pe ~ 15, corresponding to Re ~ 17 and 0.5 (for Sc = 0.88 and 32 respectively). Third, Figure 13.6 also contains a comparison of LB calculations of dispersion in an array of cylinders, with the 2D numerical results of Eidsath et al. (1983) for the same geometry. The latter comparison is included to show that the LB technique is capable of generating agreement with other numerical techniques, in non-trivial geometries, in case the near-agreement of the experimental results is not deemed sufficient. And fourth, the figure also plots the "Stokesian" theory of Koch et al. (1989) for a 3D SC array. The latter involved a Stokesian flow field (i.e., no inertial effects), but also included several other simplifications; e.g., effective diffusion within the spheres. In any case, it is clear that assuming a Stokesian model – widely used for dispersion in liquids – may not be appropriate for modeling dispersion in gases.

13.3.3 Unsteady Flow

Unsteady LB can be used to model high frequency wind-oscillation pumping of near surface soils (Neeper, 2001), mixing at fractures junctions, and transport in periodic dispersion experiments. Reynolds et al. (2000) used LB for flow, combined with a particle-tracking method for diffusion, to study oscillatory dispersion of

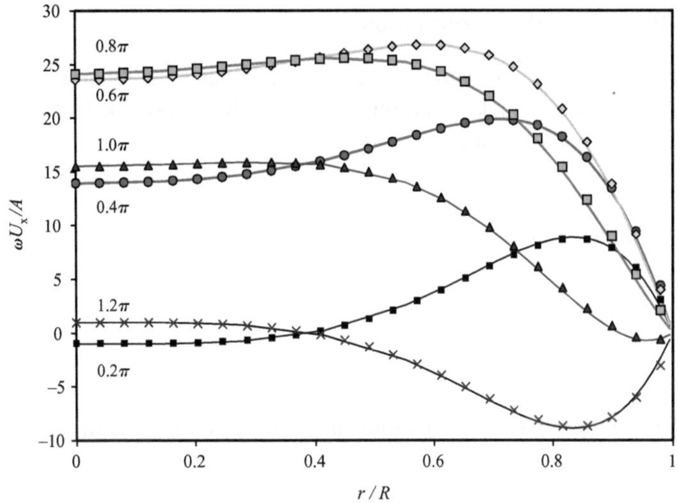

Figure 13.7. Comparison of theoretical normalized flow speed (unbroken lines) with LB calculations (symbols) at different times (where the fractions of π indicate the time offsets in fractions of one period). The abscissa is the radial position, $R = 25$ *lu*, and A is the axial pressure gradient

gases in closest-packed porous beds. However, models of deeper barometric pumping (e.g., Nilson et al., 1991) are generally out-of-reach for pore-scale LB, since the required time and length-scales are simply too large.

The following benchmark shows the ability of the model to handle oscillatory flow; in Section 13.4 we will use this example to illustrate scaling. We reproduce Lambossy's (1952) analytical solution for oscillatory flow in a cylindrical tube at a "Womersly" number Wo $= 7.237 - R \cdot (\omega/\nu)^{1/2}, \omega = 2\pi/T$, where T is the oscillation period. A similar 2D LB benchmark was recently published by Cosgrove et al. (2003), and a 2D and 3D LB comparison was published by Artoli et al. (2002).

Figure 13.7 shows the comparison of the theoretical Lambossy solution, and an LB calculation done for this study, for a cylinder with radius $= 25$ *lu*. No special effort was made to prevent the pixelation (stair-stepping) of the walls. The agreement is quite good, with the LB capturing the antegrade flow that continues after a reversal of the forcing (there is no "normalization" in this figure; the good fit is despite the fact that the $R = 25$ channel was pixelated).

13.4 FUNDAMENTAL AND PRACTICAL LIMITATIONS

There are several limitations of the LB method that give it insufficient flexibility for all arbitrary problems, or make the method far too computation-intensive. Below we explore the sources of some of these limitations, and use some illustrations to show the constraints on calculation time.

13.4.1 Bounce-Back Interpretation

During the development of lattice gas automata, it was noted (Cornubert, 1991) that the traditional bounce-back condition did not put the zero-velocity boundary on the solid nodes, but rather, at a position roughly halfway between a solid node and the adjacent fluid node. The accuracy of the reinterpreted bounce-back is a function of τ (thus the viscosity; Chen et al., 1996), but the nature of the variation is not made clear in most publications.

Figure 13.8 shows the position of the solid-fluid boundary in a small 2-D model, 8 columns wide by 8 rows high. The "nodes" of the model are shown as black circles. A solid at node (x, y) is considered a square "block" extending from $x-0.5$ to $x+0.5$, and from $y-0.5$ to $y+0.5$; solids are represented in the figure as gray blocks. This interpretation assigns a constant volume to each solid block, hence it is very easy to account for volume changes when (e.g.) solids dissolve or precipitate. In addition, when a geometry is supplied as a "voxel map" (volumetric pixels) from an X-ray tomographic scan, this interpretation allows a 1:1 map from voxel volume to LB volume.

There are several problems with the bounce-back method, as regards estimation of permeability and dispersion. Some of the LB fluid is resident on the solid nodes. Presumably, the fluid resident on the nodes must be considered in an account of applied body force, and it is not straightforward to apply forcing at the solid site. However, tests performed by the author suggest that forcing on the solid can be ignored, and forcing may be applied only at the fluid sites. Some tracers are resident on the nodes, behind the 1/2 boundary implied by the reinterpreted bounce-back; thus the effective width of a channel for dispersion may be slightly greater than the width for flow. The effective lateral diffusion coefficient of tracer at the solid nodes is debatable; with

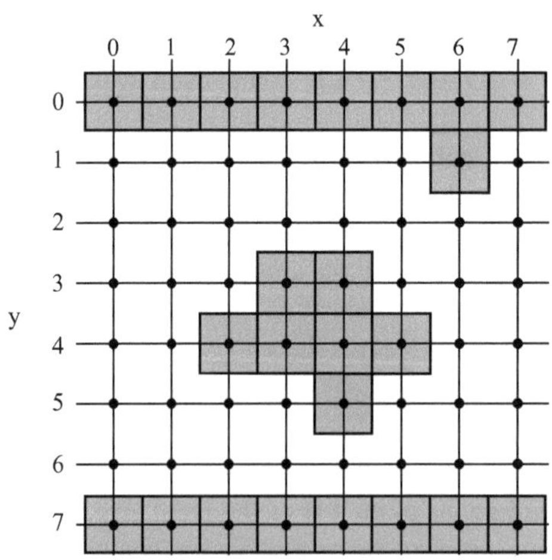

Figure 13.8. The black dots represent the node positions in the LB automaton; the gray blocks show the solid volume associated with each node in the "reinterpreted bounce-back" approach

bounce-back, this fluid does not experience the free-fluid collision step, and appears to have a lateral diffusion coefficient of zero. Again, tests suggest the overall error (for measured vs. theoretical dispersion coefficients) is small, even for narrow channels (Stockman et al., 1998). Obviously, the effect of the fluid at the interface decreases as the system is scaled to use more nodes to represent the same solid.

13.4.2 Error in Permeability for Small Channels

Figure 13.9 shows the ratio of measured to true permeability for flow between two flat plates, as a function of the width of the plate separation W (measured in nodes or lattice units, *lu*), for LB calculations with the bounce-back condition. The "measured" permeability is taken as $\mathrm{U_{meas}} \cdot \nu \cdot \rho/(\mathrm{d}P/\mathrm{d}x)_{\mathrm{meas}}$ (with $\rho \equiv 1$), while the true permeability is taken as $W^2/8$, where ν is the kinematic viscosity, U_{meas} and $(\mathrm{d}P/\mathrm{d}x)_{\mathrm{meas}}$ are the measured average flow speed and pressure gradient. The upper curve is for $\tau = 0.998$ ($\nu = 0.166\ lu^2/ts$), and the lower curve is for $\tau = 0.530$ ($\nu = 0.01\ lu^2/ts$). Clearly, there is not a monotonic increase in accuracy from high to low τ. However, with both τ choices, there is a rapid increase in accuracy, so that by $W = 10$, the error is only 1 to 2%.

13.4.3 Instability at Irregular Surfaces

With the bounce-back method, instabilities in tracer distributions may arise at irregular solid walls. Under steady-state flow at higher Pe (ranging from 25 to 500, depending on the geometry), anomalous concentrations form adjacent to protruding solids. Abnormally high concentrations form on the windward side of solids, and lows form

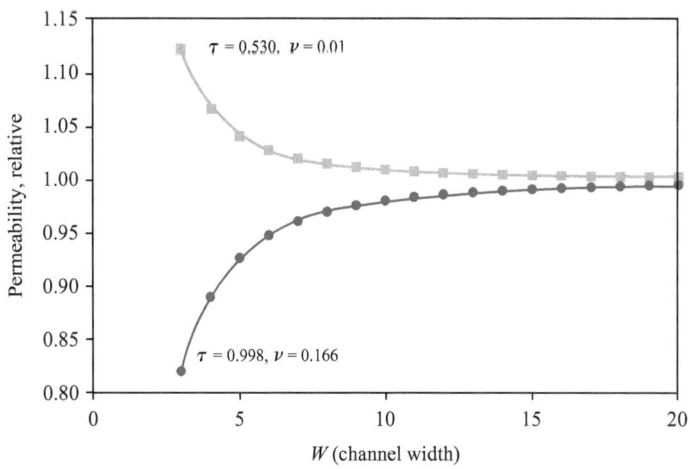

Figure 13.9. LB calculation of permeability for flow between two parallel plates, against the separation of the plates in *lu*, for two different viscosities

on the leeward side. A single node may have both a very high concentration on one side, and a very low on the other, resulting in very asymmetrical non-equilibrium distributions in the adjacent nodes. At some point, an oscillatory instability sweeps across the entire automaton, producing nonsensical concentrations.

Two conditions seem to stabilize systems against this form of instability, without causing perceptible changes in the measurement of quantities such as the dispersion coefficient. The first employs a function that sweeps through the automaton, searching for boundary solid nodes with asymmetrical concentration distributions. Any nodes which have vectors with an $(f_{max}/f_{min}) >$ constant, have the concentration on the node equally redistributed among the vectors. Typically, there are few such nodes; constant is currently set to 1.5.

The second stability-enhancing condition is oscillatory flow, when it is appropriate. Reversal of flow also reverses the build-up of anomalous concentrations, as the "leeward" and "windward" directions are reversed from time to time.

13.4.4 Limits Imposed by Time and Space Scaling

The LB method is limited by the Courant-Friedrichs-Lewy (CFL) criterion. Methods have been proposed for overcoming CFL criterion (Nourgaliev et al., 2003); but to date, all these methods have been of limited application, and may have severe drawbacks (e.g., calculational intensity). In the classic LB, $c \cdot ts/lu$ = CFL ~ 1, where c is the speed of sound in the automaton, and is typically $\sim$1. Thus the LB method is expected to be marginally stable, and becomes unstable as the $\tau \rightarrow 0.5$ (Sterling et al., 1996).

The average fluid speed in the spaces of the LB automaton cannot exceed c. However, experience shows us that the practical upper limit is more like $U = 0.02\ lu/ts$ (where U is the average flow speed). The wide variation in pore diameter ensures that local fluid speeds may be close to 0.1 lu/ts, even when the average speed in just 0.01. This restriction puts a severe limit on the applicability of the method.

Similarly, the τ cannot be varied arbitrarily. To avoid significant error in the reinterpreted bounce-back (Chen et al., 1996), $\tau < 1$, hence (from Eq. 13.7) we must have ν and $D_m < 0.1667$. From experience, we know that we must have ν and $D_m > 10^{-6}$ for numerical stability.

Given these limits, there is an inexorable tie between time and length scales. One often finds that LB is quite capable of modeling a phenomenon of interest; but only for seconds. Thus LB has typically been used to model pore-scale phenomena that are expected to reach steady-state in a short time. Before deciding to apply LB to any problem, it is wise to calculate the number of steps and physical nodes required.

For example, suppose one wished to model a process that involved transfer between a minor condensed phase and flowing air, in a matrix of sand. To be concrete, we assume sand with 0.1 cm diameter grains and $\sim$40% porosity. Such a process might be the transfer of CO_2 between a water film and air, the evaporation of an organic phase in remediation, or the sorption of a tracer gas on the solid grains. We take a pore flow speed $U \sim 2$ cm/s as an example, implying Re $\sim$ Pe $\sim$ 10 (the length scale is taken as the grain diameter). Such flow would be fast enough to justify using a Navier-Stokes solver. We pick 10 *lu* for the average grain size; from Figure 13.2, it can be seen that this choice of grain size may still leave substantial (10–20%) permeability errors. If near-spherical, the average grain will then have a volume of $\sim 524\ lu^3$. If we aim for ~40% porosity, and we choose a modest LB model with $100 \times 50 \times 50$ nodes (the longest axis in the direction of flow), the simulation will contain ~ 500 grains. The long distance (100 *lu*) will correspond to just 1 cm. At the maximum average LB speed of $0.02\ lu/ts$, it would take 5000 *ts* for air to travel the length of the automaton, which would correspond to 0.5 s of real time; that is, 10000 *ts* correspond to 1 s. To model 1 minute of the process, $6 \cdot 10^5\ ts$ would be required. Using the fast dispersion-only modes, a personal computer with fast memory (e.g., 6 GByte/s transfer while streaming) would update at $\sim 4 \cdot 10^7$ sites/s, so it would take $100 \times 50 \times 50$ nodes $\times\ 6 \cdot 10^5\ ts/(4 \cdot 10^7$ node-$ts/s) = 3750$ seconds of computer clock time, or about 1 hour – to model just one minute of the real process – and then, over a flow path of just 1 cm. If we opted instead for an average grain diameter of 20 *lu*, but still wanted ~ 500 grains in the model, the clock time would increase by a factor of 16.

Now suppose one wishes to measure the dispersion coefficient in a similar porous medium, as a function of Pe ~ Re, with $1 \leq \text{Pe} \leq 100$, by injecting a slug of solute and measuring the second moment of the tracer (Stockman, 1997) as it disperses downstream along the x-axis. We begin by assuming that the $100 \times 50 \times 50$ segment is sufficient to define a representative volume; however, we expect that this volume must be cloned in the x-direction, as described in Section 13.2.2. First, continuity of flow must be provided at the boundaries of the cloned units; the easiest approach is to mirror the representative volume along the x-axis, so our 100 x-direction *lu* might really contain a $50 \times 50 \times 50$ volume and its mirror image. Then how many of these cloned units would we need, to measure D^* with reasonable confidence?

Typically, at least one characteristic diffusion time $\tau_D = L^2/D_m$, is needed to reach a steady-state, linear growth of the second moment, where L is a characteristic length scale. We take $L = 10\ lu =$ the grain diameter. To achieve $\text{Pe} = U \cdot L/D_m = 100$, with a maximum U of $0.02\ lu/ts$, D_m must be $2 \cdot 10^{-3}$, and thus $\tau_D = 5 \cdot 10^4\ ts$. To use the second moment method, the slug must stay entirely within the solid domain – it can't fall off the edge. Thus the automaton must be at least as long as the distance the slug would travel in τ_D, or $\sim U \cdot \tau_D = 0.02 \cdot 5 \cdot 10^4 = 10^3\ lu$. Thus the 100-long geometry must be cloned 10 times in the x-direction. The number of cells in the total cloned automaton would be $50 \times 50 \times 10^3 = 2.5 \cdot 10^6$. In the code described by Stockman (1999), ~25 bytes per node are required in diffusion-only mode. Thus the simulation would require 62.5 Mbytes. Again assuming the update rate (that is, the real computer calculation rate) of $\sim 4 \cdot 10^7$ sites/s, the $\text{Pe} = 100$ run would take $(2.5 \cdot 10^6 \text{sites}) \times (5 \cdot 10^4\ ts)/(4 \cdot 10^7 \text{ site} \cdot ts/s) = 3125\ s$, or less than an hour. However, note that the calculation time would go as Pe^2, so a substantially higher Pe might be impractical. At Pe ~ 500, the calculation would take nearly a day.

> (**3**) Are the *ts* allotted, and the length of the cloned channel in question 1 of the example, sufficient to reach a dispersive steady-state?

Finally, consider what the oscillatory flow simulation in Figure 13.7 represents, in terms of both real gaseous flow through a pore or fracture, and the automaton needed to carry out the calculation. The results in Figure 13.7 were carried out with an $8 \times 53 \times 53$ lu^3 automaton, the radius of the cylinder (parallel to the x-axis) was $R = 25$ *lu* (nodes on the x-axis could have been 1, but the vectorization built into the program kept the simulation to $x = 8$). The simulation was carried out to match the $Wo = 7.237$ used in Lambossy's original benchmark (Lambossy, 1952). This Wo was achieved by setting the period of oscillation to 750 *ts*, with a $\nu = 0.100$; the corresponding $\mathrm{Re} = 4.47$, defined relative to the diameter and the maximum flow speed. This simulation was taken through 400 oscillation periods ($3 \cdot 10^5$ *ts* total), to ensure a steady state was reached ($3 \cdot 10^5$ *ts*), and took ~2 hours of computer time.

Suppose then that the simulation in Figure 13.7 were to represent gaseous oscillations, first at the diameter of a large pore (1 mm), then at the diameter of a macropore in soil (1 cm). The period of oscillation, $T = \pi(d/Wo)^2/(2\nu)$, where d is the diameter, and $\nu = 0.14$ cm^2/s, would be 0.00214 s for the 1 mm pore, and 0.214 sec for the 1 cm pore. Obviously, these periods are probably unrealistic for surface wind driven phenomena; but the power two dependence on Wo means that by merely dropping Wo to 2, the period corresponds to ~3 seconds for the 1 cm macropore. However, for modeling dispersion, a greater period of oscillation means that a physically longer simulation is required (both in *ts* and in axial *lu*), to keep a slug of solute from running off the end of the automaton. Thus pore-scale LB is not practical for modeling phenomena that have longer periods, such as deep barometric pumping (Nilson et al., 1991).

13.5 REPRESENTATIVE VOLUME

When one uses pore-scale LB to estimate permeability or dispersion coefficients in porous media, it is necessary to determine a representative volume with confidence. For reasons of computational efficiency, the volume should contain as few nodes as practical; yet the more nodes, and the better the resolution of pores (i.e., the higher the number of nodes/cm), the more likely the calculation is to be accurate.

13.5.1 Porous Filter Cake

To illustrate the choice of representative volume with a practical example, consider the determination of permeability in a porous filter cake, as might be formed during minerals processing. The geometry for this example is a digital $256 \times 256 \times 100$ voxel, X-ray microtomagraphic image (provided by C.L. Lin; see Lin and Miller, 2004 for methods).

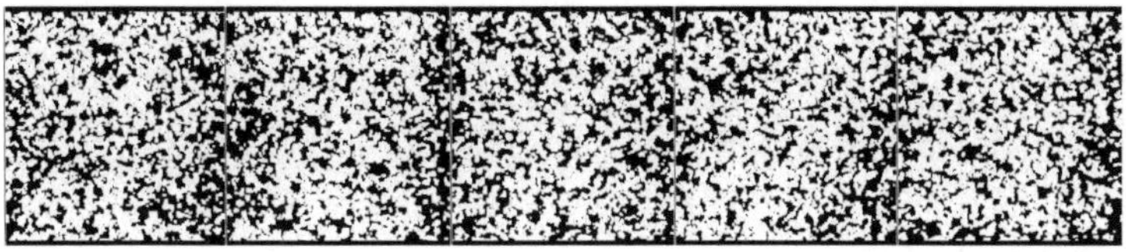

Figure 13.10. Five cuts through the porous filter cake, at different depths into the sample.

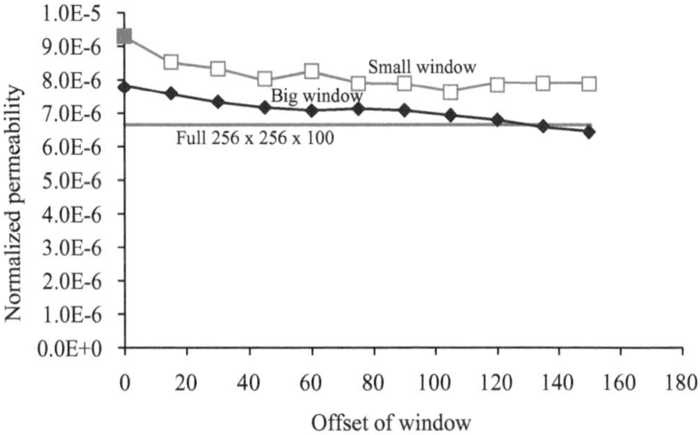

Figure 13.11. Measured LB permeability, for 50-node-wide and 100-node-wide windows moved left-to-right across the 256 × 256 × 100 geometry

Figure 13.10 shows five cuts through a 256 × 256 × 100 geometry. The particles have apparently not compacted evenly, and there are variations in the density of each. To test the effects of these density variations, we take successive subsamples of the full geometry, marching from left to right across the geometry, and determine the LB permeability on each slice, using windows of two sizes.

Figure 13.11 shows the variation of the LB-measured permeability (Re $\sim$ 0), for subsamples within a 50-wide pixel window ("small window") and a 100-window, as the windows are moved from left to right across the 256-pixel-wide sample. The horizontal line marked "full" represents the 256 × 256 × 100 pixel LB calculation. Clearly, both windows yield a slight underestimation in the aggregate permeability (but never by more than a factor of 1.39), and the error varies with position across the sample; but the net variation for the larger window is much smaller. Actually, the distribution of higher speed "jets" among the pores in quite inhomogeneous; in a few channels, the fluid moves up to 20 times faster than the net Darcy flow speed, so the small amount of variation is actually rather surprising.

To estimate the dispersion coefficient in this same sample, a 1500-node long porous medium was constructed by allowing the flow to reach steady state, then the solids were cloned in the x-direction, as described in Section 13.2.2. A solute slug was then released at one end of the automaton, and evolution the 2nd moment of the slug

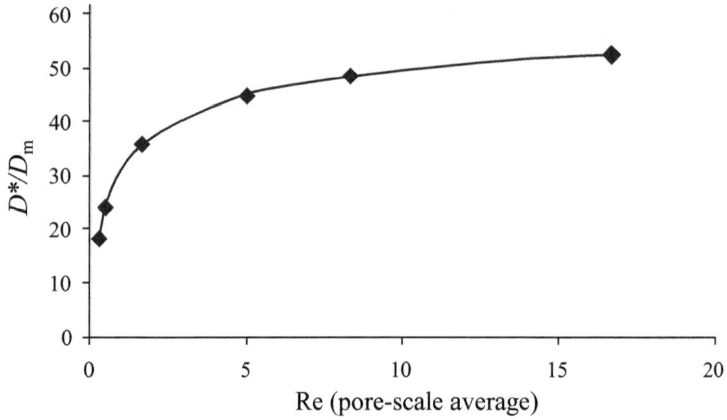

Figure 13.12. The LB measured dispersion coefficient D^*, as a friction of the speed of increasing inertial effects (Re), for a fixed Pe = 5

was used to calculate the dispersion coefficient (per Section 13.3.2). Figure 13.12 shows the variation of measured dispersion coefficient, as a function of Re (defined relative to the average diameter of the grains in the aggregate). One would not expect to see this type of plot with a real gaseous system, as the ratio of $D_{\text{m(tracer)}}/\nu_{\text{(bulkfluid)}}$ would be fixed. Here, however, this numerical exercise allows us to see the surprising variation of D^*, particularly at modest Re. The source of this variation is probably the inhomogeneous distribution of the flow, with a few channels through the automaton having speeds much greater than the Darcy speed; in effect, the local Re (in a fast-moving channel) may exceed 20, even for average Re $\sim$ 1.

13.5.2 Fracture Flow

The selection of representative volume also affects modeling of flow in fractures, even synthetic fractures that are intentionally kept simple. Consider the dispersion of a water-borne solute between a flat and an irregular plate; while this is clearly not a "vapor" application, it does illustrate some limits of the method at high Pe. The experimental setup is described in detail by Detwiler et al. (2000). In brief: two 14.8 by 30.2 cm sheets of textured glass plate ("shower glass") were placed together, with an average separation (aperture) of 0.0221 cm. The full plate images were captured on a digital camera in a 958 $\times$ 1958 pixel image. The pattern on the glass was small-scale, so the average correlation length (λ) of the aperture (defined in Detwiler et al.) was only $\sim$0.044 cm. A steady-state flow of water was induced between the transparent plates, then a slug of solute was introduced and the dispersion of solute was determined by the method of moments. Initially, the slug of solute was $\sim$1.5 to 2 cm wide, and experienced little transverse dispersion on its path through between the plates; hence the actual geometry sampled by the slug was actually $\sim$2 cm $\times$ 28 cm. In the original paper, the experiment was modeled by approximating the flow field with the Reynolds equation, with particle tracking to determine the dispersion in that flow field. The aperture field between the two plates was determined by filling the gap with a constant concentration of dye.

For the LB simulation of the Detwiler et al. experiment, a small subset of the geometry was chosen, mirrored, and replicated in the x-direction after a steady-state flow field was achieved. Then a slug source was introduced at one end of the automaton, allowed to disperse downstream, and the method of moments was used to determine the D^*. Two 1.54 cm by 1.54 cm subsamples of the real system, corresponding to 100×100 pixels, were used for the LB simulation. This LB size was picked to contain 35λ, since Detwiler et al. reasoned that at least 20λ was needed to overcome ergodic effects. This subsample also approximated the width of the initial solute pulse in the real experiments. Since only the aperture field was known (not the distribution of porosity along the axis perpendicular to the plates), LB runs were performed with the aperture symmetrically disposed between the plates, and all on one side (Figure 13.13 inset).

There were several difficult constraints on the LB model, which deserve mention as illustrations of limits. First, the experiment was designed to allow reliable determination of the solute distribution, and to be suitable for Reynolds equation modeling; for those reasons, the slopes on the textured surface were gentle. Unfortunately, the LB method for this study used uniform gridding in the x-, y-, and z-directions, so nodes

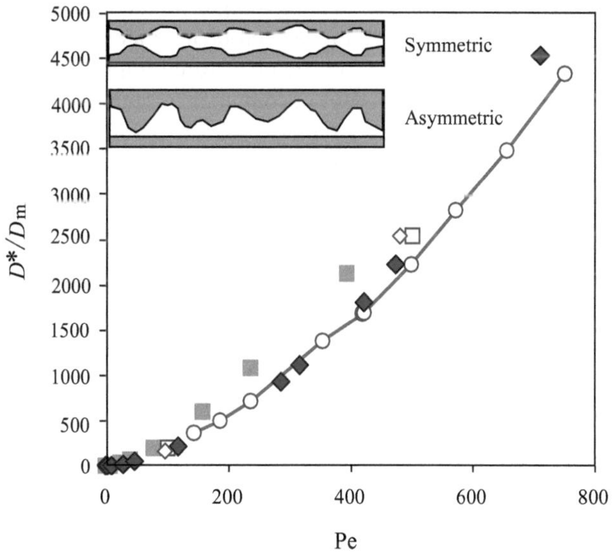

Figure 13.13. Comparison of Detwiler et al. (2000) experimental results (circles connected by line segments) and LB calculations. The diamonds are symmetric LB calculations (see inset, upper left), and the squares are for asymmetric calculations. The filled diamonds and squares are for LB subsample 1, and the open squares and diamonds (two each) are for subsample 2

were not distributed efficiently, and the 100×100 pixel image had to be stretched into a $400 \times 400 \times 20$ LB automaton to achieve the correct aspect ratio. The geometry was then reflected in the flow direction, to assure continuity of flow fields. Finally, after the automaton reached a steady-state flow, the geometry was cloned up to 16 times in the flow direction (to provide a long path for method-of-moments dispersion, as described in Section 13.2), and a slug of solute introduced. Second, the irregular surfaces induced the instability described in Section 13.4.3, which limited the maximum Pe to $\sim$700. (This study was completed before the technique described in Section 13.4.3 was available.)

Figure 13.13 compares the LB estimated dispersion coefficient, with the experimental dye dispersion measurements. Despite all the limitations, the LB technique gave reasonable agreement. The two subsamples give significantly different results, but the differences are comparable to the differences created by other assumptions – e.g., the assumption of asymmetric vs. symmetric disposition of the apertures, which is an inherent uncertainty in the Reynolds equation modeling used by Detwiler et al. (2000).

13.6 CONCLUSIONS

Lattice Boltzmann modeling provides a powerful technique for investigating gas permeability and dispersion in complex geometries. Results averaged over many pore scales can be used to estimate macroscopic transport properties. However, the method is not a panacea, and for many geologic problems, pore-scale modeling is far too calculation-intensive. Significant problems include the need to define a representative volume; the fixed CWL number, often implying that only hours of physical time can be modeled with days of computer computation; and the restriction to "modest" Pe and Re (<1000 in a typical problem). Nonetheless, LB can model many coupled phenomena that are difficult to approach with traditional methods, such as finite difference and finite element techniques.

Answers to Self-Study Examples:

(1a) Yes. From Section 13.2.2, viscous equilibrium requires $\sim L^2/\nu = (26-2)^2/0.01 = 57600$. (The right and left, and top and bottom sites of the channel are solid, so the channel is 24 lu wide.)

(1b) First part: (sites $= 26 \times 26 \times 24) \times (60000\ ts)/(9 \cdot 10^6$ ts $\cdot$ sites/s$) = 108$ s.

Second part: (sites $= 26 \times 26 \times 24 \times 48) \times (60000\ ts)/(35 \cdot 10^6$ ts $\cdot$ sites/s$) =$ 1335 s.

(1c) First part: memory to store vectors: (sites $= 26 \times 26 \times 24) \cdot$ (19 vectors) $\times$ (4 bytes/vector) $= 1233024$ bytes $= 1.18$ MB. There is a smaller amount of memory required to store pointers, $26 \cdot 26 \cdot 24$ bytes required to store solid site information, and (sites $= 26 \times 26 \times 24) \times$ (3 speed components) $\times$ (4 bytes/component) $= 0.185$ MB required to store the x, y, z components of the flow velocity at each site.

Second part: memory to store vectors: (sites = 26 × 26 × 24 × 48) × (6 vectors) × (4 bytes/vector) = 1233024 bytes= 17.8 MB. The memory to store the flow speed is the same as in (1b).

(2) From Eq. (13.8), Pe = Re = 24 = (L = 24 *lu*) × (U = 0.01 *lu/ts*)/(ν or $D_m = 0.01\ lu^2/ts$); Sc = 1.

(3) From Section 13.4.4, a dispersive steady-state is reached after a slug travels for a characteristic diffusion time $\sim L^2/D_m = 24^2\ lu^2/0.01\ (lu^2/ts) = 57600\ ts$. The channel must be long enough to contain the entire slug during this time, else part of the slug will "fall off" the right end of the channel, and will wrap back if there is no concentration boundary at the channel end. While the average speed across the channel is 0.01 *lu/ts*, the solute in the middle of the channel will travel faster. In a round pipe, the middle fluid travels twice the average speed; this speed would be reached by the solute only at very high Pe, but to be on the safe side, we assume this factor 2. Thus the distance traveled by the leading edge of the solute, in the characteristic diffusion time, is (57600 *ts*)× (2 × 0.01 *lu/ts*)=1152 *lu*. The cloned channel is 48 × 24 = 1152 *lu* long, so the channel length is adequate.

REFERENCES

Artoli, A.M., Hoekstra, A.G. and Sloot, P.M.A. (2002) 3D pulsatile flow with the lattice Boltzmann BGK method, *Internat. J. Modern Phys. C* 13(8), 1119–34.

Chen, S.J., Martínez, D. and Mei, R. (1996) On boundary conditions in lattice Boltzmann methods, *Phys. Fluids* 8, 2527–2535.

Cornubert, R., d'Humières, D. and Levermore, D. (1991) A Knudsen layer theory for lattice gases, *Physica D* 47, 241–259.

Cosgrove, J.A., Buick, J.M., Tonge, S.J., Munro, C.G., Greated, C.E. and Campbell, D.M. (2003) Application of the lattice Boltzmann method to transition in oscillatory channel flow, *J. Phys. A. Math. Gen.* 36, 2609–2620.

Detwiler, R.L., Rajaram, H. and Glass, R.J. (2000) Solute transport in variable-aperture fractures: An investigation of the relative importance of Taylor dispersion and macrodispersion , *Water Resources Res.* 36, 1611–1625.

Eidsath, A., Carbonell, R.G., Whitaker, S. and Herrmann, L.R. (1983) Dispersion in Pulsed Systems – III Comparison between Theory and Experiments for Packed Beds, *Chem. Engng. Sci.* 38, 1803–1816.

Flekkøy, E.G., Oxaal, U., Feder, T. and Jøssang, T (1995) Hydrodynamic dispersion at stagnation points: simulations and experiments, *Phys. Rev.* E 52, 4952–4962.

Gunn, D.J. and Pryce, C. (1969) Dispersion in Packed Beds, *Trans. Inst. Chem. Engrs.* 47, T341-T350.

Koch, D.L., Cox, R.G., Brenner, H. and Brady, J.F. (1989) The effect of order on dispersion in porous media, *J. Fluid Mech.* 200, 173–188.

Lambossy, P (1952) Oscillations forcées d'un liquide incompressible et visqueaux dans un tube rigide et horizontal. Calcul de la force frottement, *Helv. Phys. Acta* XXV, 371–386.

Lin, C.L. and Miller, J.D. (2004) Pore structure analysis of particle beds for fluid transport simulation during filtration, *Inter. J. Mineral Process* 73(2–4), 281–294.

Martys, N. and Chen, H. (1996) Simulation of multicomponent fluids in complex three-dimensional geometries by the lattice Boltzmann method, *Phys. Rev. E* 53, 743–750.

Neeper, D.H. (2001) A model of oscillatory transport in granular soils, with application to barometric pumping and earth tides, *J. Contam. Hydrology* 48, 237–252.

Nilson, R.H., Peterson, E.W., Lie, K.H., Burkhard, N.R. and Hearst, J.R. (1991) Atmospheric pumping: a mechanism causing vertical transport of contaminated gases through fractured permeable media, *J. Geophys. Res.* 96(B13), 21933–21948.

Noble, D.R. (1996) Lattice Boltzmann Study of the Interstitial Hydrodynamics and Dispersion in Steady Inertial Flows in Large Randomly Packed Beds. Ph.D. thesis, University of Illinois Urbana-Champaign.

Nourgaliev, R.R., Dinh, T.N., Theofanous, T.G. and Joseph, D. (2003) The lattice Boltzmann equation method: theoretical interpretation, numerics and implications, *International Journal of Multiphase Flow* 29, 117–169. (CFL, methods for dissociating grids and getting CFL < 1).

Pan, C., Hilpert, M. and Miller, C.T. (2004) Lattice-Boltzmann simulation of two-phase flow in porous media, *Water Resources Res.* 40(1), W01501 14pp.

Qian, Y.H., d'Humières, D. and Lallemand, P. (1992) Lattice BGK models for Navier-Stokes equation, *Europhys. Lett.* 7(6 BIS), 479–484.

Reynolds, A.M., Reavell, S.V. and Harral, B.B. (2000) Flow and dispersion through a close-packed fixed bed of spheres, *Phys. Rev. E* 62(3), 3632–3639.

Rothman, D.H. and Zaleski, S. (1997) *Lattice-Gas Cellular Automata: Simple Models of Complex Hydrodynamics.* Cambridge University Press, Cambridge, 297p.

Sterling, J.D. and Chen, S.Y. (1996) Stability analysis of Lattice Boltzmann methods, *J. Computat. Phys.* 165(1), 288–306.

Stockman, H.W. (1997) A lattice-gas study of retardation and dispersion in fractures: Assessment of errors from desorption kinetics and buoyancy, *Water Resources Res.* 33 (8), 1823–1832.

Stockman, H.W., Li, C. and Wilson, J.L. (1997) A lattice-gas and lattice Boltzmann study of mixing at continuous fracture junctions: importance of boundary conditions, *Geophys. Res. Lett.* 24(12), 1515–1518.

Stockman, H.W., Glass, R.J., Cooper, C. and Rajaram, H. (1998) Accuracy and computational efficiency in 3D dispersion via lattice-Boltzmann: models for dispersion in rough fractures and double-diffusive fingering, *International Journal of Modern Physics* C 9(8), 1545–1557.

Stockman, H.W. (1999) *A 3D Lattice Boltzmann Code for Modeling Flow and Multi-Component Dispersion*. SAND 99–0162, Sandia National Laboratories, Albuquerque, NM. 165 pp.

Sukop, M.C. and Or, D. (2004) Lattice Boltzmann method for modeling liquid-vapor interface configurations in porous media, *Water Resources Res.* 40(1), W01509 11pp.

Wolf-Gladrow, D. (1994) A lattice Boltzmann equation for diffusion, *J. Stat. Phys.* 79, 1023–1032.

Zhang, X., Bengough, A.G. and Deeks, L.K. (2002) A novel three-dimensional lattice Boltzmann model for solute transport in variably saturated porous media, *Water Resources Res.* 38(9), 6–1 to 6–10.

Zick, A.A. and Homsy, G.M. (1982) Stokes Flow through Periodic Arrays of Spheres, *J. Fluid Mech.* 115, 13–26.

PART 2

MEASUREMENT AND MONITORING

CHAPTER 14

EXPERIMENTAL DETERMINATION OF TRANSPORT PARAMETERS

OLGA ŠOLCOVÁ AND PETR SCHNEIDER

Institute of Chemical Process Fundamentals, Czech Academy of Science, Rozvojová 135, 165 02 Praha 6, Czech Republic

Any prediction or simulation of gas transport in porous solids is based on mass balances. The balance equations incorporate inevitably constitutive equations, which relate the intensity of mass flux to mass flux driving forces. Constitutive equations comprise information of three kinds:

1. Properties of components of the gas mixture: bulk diffusion coefficients of all pairs of components of the gas mixture, component viscosities and their mean speeds. Such properties are, usually, readily available.
2. The physical laws, which describe the gas transport, with idealized units of the porous medium. Usually, cylindrical capillaries are utilized because of the availability and simplicity of the laws.
3. Characteristics of the porous medium. Since the diverse complicated nature of the porous medium is usually unknown it must be modeled. Modeling of the porous structure is based on units from paragraph 2 and accounts for pore-size distribution and other factors not apprehended by pore units. Today two basic models are available for description of combined (diffusion and permeation) transport of multicomponent gas mixtures: the Mean Transport-Pore Model (MTPM) (Fott et al., 1983; Schneider and Gelbin, 1984) and the Dusty Gas Model (DGM) (Jackson, 1977; Mason and Malinauskas, 1983).

Mean Transport-Pore Model (MTPM) assumes that the decisive part of the gas transport takes place in transport-pores that are visualized as cylindrical capillaries with radii distributed around the mean value $\langle r \rangle$ (first model parameter). The width of this distribution is characterized by the mean value of the squared transport-pore radii, $\langle r^2 \rangle$ (second model parameter). The third model parameter is the ratio of porosity, ε_t, and tortuosity of transport-pores, q_t, $\psi = \varepsilon_{t/q_t}$ (Schneider, 1978; Novák et al., 1988).

C. Ho and S. Webb (eds.), Gas Transport in Porous Media, 245–272.
© 2006 *Springer.*

Dusty Gas Model (DGM) visualizes the porous medium as a collection of giant spherical molecules (dust particles) kept in space by external force. The movement of gas molecules in the spaces between dust particles is described by the kinetic theory of gases. Formally, two of MTPM transport parameters, $\langle r \rangle$ and $\langle r^2 \rangle$, can be used also in DGM. The third DGM transport parameter, B_0, characterizes the viscous (Poiseuille) gas flow in pores and can be, formally, replaced by $\langle r^2 \rangle \psi/8$.

14.1 CONSTITUTIVE EQUATIONS

The net molar flux density of component i in a n-component gas mixture per unit *total* cross-section of the porous solid, N_i, due to the combined influence of composition gradients and total pressure gradient in a porous solids is given as the sum of the permeation molar flux density of component i, N_i^p, and the diffusion molar flux density of this component, N_i^d, (Mason et al., 1967):

$$N_i = N_i^p + N_i^d \quad i = 1, \ldots, n \tag{14.1}$$

For the net *mixture* molar flux density, N, it follows

$$N = N^p + N^d \quad i = 1, \ldots, n \tag{14.2}$$

where N^d is the mixture molar diffusion flux density ($N^d = \sum_{i=1}^{n} N_i^d$) and N^p is the mixture molar permeation flux density ($N^p = \sum_{i=1}^{n} N_i^p$).

In MTPM and DGM the steady-state *isothermal diffusion transport* in multicomponent gas mixtures, for cylindrical pores with diameter $\langle r \rangle$ and the transition region (i.e., when the pore diameter, 2 $\langle r \rangle$, is comparable with the mean free-path length of gas molecules, λ; $2\langle r \rangle \approx \lambda$), is described by the modified Maxwell-Stefan equation (Rothfeld, 1963):

$$\frac{N_i^d}{D_i^k} + \sum_{\substack{j=1 \\ j \neq i}}^{n} \frac{y_j N_i^d - y_i N_j^d}{D_{ij}^m} = (df)_i \quad i = 1, \ldots, n \tag{14.3}$$

(df) is the driving force, y_i the mole fraction of component i, D_{ij}^m is the effective diffusion coefficient of the pair i–j in the bulk diffusion region:

$$D_{ij}^m = \psi \mathcal{D}_{ij}^m \tag{14.4}$$

with $\mathcal{D}_{ij}^m$ the binary bulk diffusion coefficient and D_i^k, is the effective Knudsen diffusion coefficient of component i:

$$D_i^k = \psi \langle r \rangle K_i \tag{14.5}$$

with the Knudsen coefficient K_i

$$K_i = \frac{2}{3}\sqrt{\frac{8R_gT}{\pi M_i}} \tag{14.6}$$

The driving force term $(df)_i$ differs for MTPM and DGM:

$$\text{MTPM} \quad (df)_i = -c_T\frac{\mathrm{d}y_i}{\mathrm{d}x} \tag{14.7}$$

$$\text{DGM} \quad (df)_i = -\frac{\mathrm{d}\,(c_Ty_i)}{\mathrm{d}x_i} \tag{14.8}$$

c_T is the total molar concentration of the gas mixture and x is the length coordinate in the transport direction. For pure (isobaric) diffusion, where $\mathrm{d}c_T/\mathrm{d}x = 0$, both driving forces (14.7), (14.8) are identical: $-\mathrm{d}(c_Ty_i)/\mathrm{d}x = -c_Tdy_i/dx$. The (rather small) difference starts to appear in combined diffusion and permeation cases.

By summing modified Maxwell-Stefan isobaric diffusion equations (14.3) for all gas mixture components the generalized Graham's law appears

$$\sum_{i=1}^{n} N_i^d\sqrt{M_i} = 0 \tag{14.9}$$

which is the condition that must be fulfilled in order to have pure diffusion mass transport.

The *MTPM permeation molar flux density* of gas mixture component i, N_i^P, in porous solids is described by the Darcy equation:

$$N_i^P = -y_iB_i\frac{\mathrm{d}c_T}{\mathrm{d}x} \quad i = 1,\ldots,n \tag{14.10}$$

B_i is the effective permeability coefficient of mixture component i (Schneider, 1978):

$$B_i = \langle r\rangle\psi K_i\frac{\omega\nu_i + Kn_i}{1 + Kn_i} + \frac{\langle r^2\rangle\psi p}{8\eta} \quad i = 1,\ldots,n \tag{14.11}$$

which includes the MTPM transport parameters, ψ, $\langle r\rangle$, $\langle r^2\rangle$. The numerical coefficient ω depends on the details of the wall-slip description ($\omega = 0.9, \pi/4, 3\pi/16$, etc., see, Schneider, 1978); ν_i is the square root of the relative molecular weight of the gas mixture component i:

$$\nu_i = \sqrt{M_i/\sum_{j=1}^{nj} y_jM_j} \tag{14.12}$$

η is the gas mixture viscosity and Kn_i is the Knudsen number of component i ($Kn_i \equiv \lambda_i/2\langle r\rangle$) based on mean free-path length of component i in the gas mixture.

The *DGM permeation molar flux density* of component i, N_i^p, is described by Darcy law

$$N_i^p = -y_i B \frac{\mathrm{d}c_T}{\mathrm{d}x} \quad i = 1, \ldots, n \tag{14.13}$$

with identical effective permeability coefficients, B, for all gas mixture components:

$$B = B_o p/\eta \quad i = 1, \ldots, n \tag{14.14}$$

B_o is the third DGM transport parameter that can be, formally, replaced by $\langle r^2 \rangle \psi/8$.

Description of the *combined diffusion and permeation gas transport* is obtained by applying Eq. (14.1) to diffusion and permeation constitutive Equations (14.3) and (14.10) (for MTPM) or Equations (14.3) and (14.13) (for DGM). In both cases the combined transport can be expressed in a vector form

$$-\frac{\mathrm{d}\mathbf{c}}{\mathrm{d}x} = \mathbf{FN} \tag{14.15}$$

with the $(n * 1)$ vector of component molar concentrations, $\mathbf{c} = [c_1, c_2, \ldots, c_n]^T$, $(n * 1)$ vector of combined molar flux densities, $\mathbf{N} = [N_1, N_2, \ldots, N_n]^T$ and a $(n * n)$ matrix $\mathbf{F} = \{f_{ij}\}$ with elements

$$f_{ij} = -\frac{c_i}{c_T D_{ij}^m} + \frac{c_i \alpha_i}{D_i^k} \quad i \neq j; \quad i = 1, \ldots, n; j = 1, \ldots, n \tag{14.16}$$

$$f_{ii} = \frac{1}{D_i^k} + \frac{1}{c_T} \sum_{\substack{j=1 \\ j \neq i}}^{n} \frac{c_j}{D_{ij}^m} + \frac{c_i \alpha_i}{D_i^k} \quad i = 1, \ldots, n \tag{14.17}$$

Parameter α_i is different for MTPM and DGM. For MTPM it is defined as:

$$\alpha_i = \frac{1 - \frac{B_i}{D_i^k} + \frac{1}{c_T} \sum\limits_{\substack{j=1 \\ j \neq i}}^{n} \frac{c_j (B_j - B_i)}{D_{ij}^m}}{\sum\limits_{j=1}^{n} \frac{c_j B_j}{D_j^k}} \quad i = 1, \ldots, n \tag{14.18}$$

For DGM:

$$\alpha_i = -\frac{B/D_k^i}{1 + B \sum\limits_{j=1}^{n} c_j / D_j^k} \quad i = 1, \ldots, n \tag{14.19}$$

14.2 TRANSPORT PARAMETERS

The unknown real pore structure makes an *a priori* determination of transport characteristics unfeasible. The pore structure characteristics relevant to transport in pores have to be determined experimentally. Two approaches are used in this respect: (i) textural analysis of the porous solid and (ii) evaluation of simple transport processes taking place in the porous solid in question.

The advantage of *textural analysis of the porous solid* derives from the wealth of available experimental methods and evaluation procedures (physical adsorption of gases, high-pressure mercury porosimetry, etc.). These methods are frequently used, but they are far from being the best choice; intrusion of a liquid metal into pores or multilayer physical adsorption and condensation for example, of nitrogen at 77 K are governed by completely different laws than gas transport (Schneider and Gelbin, 1984).

The relevance of *evaluation of transport parameters from simple transport processes which take place in the porous solid in question* stems from the possibility to use the same pore-structure model both for evaluation of transport parameters and for description of the process in question. It is a good idea to use a mass transfer process, which is similar to the gas transport process under consideration. It is of advantage to choose for determination of transport parameters a (simple) process that can be easily followed at near-laboratory conditions, which does not require sophisticated instrumentation.

Several choices can be made:

(a) Pure counter-current diffusion of binary gas mixtures under steady-state conditions.
(b) Binary diffusion under dynamic conditions.
(c) Dynamic or steady-state permeation of individual gases.
(d) Combined diffusion and permeation gas transport.

At the same time it is good choice to use inert (i.e., nonadsorbale) gases; this eliminates transport of adsorbed gas along the surface of pores (surface diffusion) the nature of which is not very well understood.

14.3 STEADY-STATE BINARY COUNTER-CURRENT DIFFUSION

14.3.1 Wicke-Kallenbach Cell

The classic Wicke-Kallenbach cell (Figure 14.1) (Wicke and Kallenbach, 1941) consists of upper and lower compartment and a metallic disc with cylindrical holes sandwiched between both compartments. A porous pellet is forced into undersized rubber tube and the pellet-tubing assemblies are then forced into holes of the metallic disc. The absence of gaps between porous pellets and rubber tubes as well as between rubber tubes and metallic surfaces of disc holes can be verified by replacing porous pellets by identically sized metallic cylinders. One gas flows steadily through one cell compartment and another gas through the other cell compartment; both chambers are

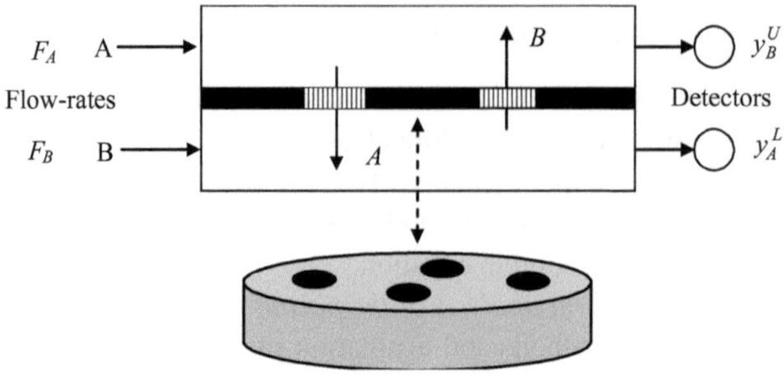

Figure 14.1. Wicke-Kallenbach diffusion cell. Porous pellets (hatched) – mounted in cylindrical holes of the impermeable cell partition (black). F_A,F_B flow-rates of gases A and B; y_B^U mole fraction of B in the upper outlet gas stream; y_A^L mole fraction of A in the lower outlet gas stream. Detail of metallic disc with holes for porous pellets is shown

kept at precisely the same pressure. The outlet gas streams from both compartments are analyzed for the content of the gas from the opposite compartment (mole fractions y_B^U, y_A^L). Gases in both cell compartments should be ideally mixed, that is, composition in the compartment should be the same as at the compartment outlets. From mass balance of both gases it is possible to determine the diffusion flux densities through the porous solids.

$$N_A^d = c_T F_B y_A^U / S \quad N_B^d = c_T F_A y_B^L / S \tag{14.20}$$

Here F_A and F_B are the volumetric flow-rates of gases A and B entering both cell compartments, c_T is the total molar concentration and S is the total cross-section of porous pellets of length L placed in the impermeable disc. For binary counter-current diffusion the following constitutive equation follows from the modified Stefan-Maxwell equation:

$$N_A^d = \left(\frac{1}{D_A^k} + \frac{1-\alpha y_A}{D_{AB}^m}\right)^{-1} \left(-c_T \frac{dy_A}{dx}\right) \quad \text{with} \quad \alpha_{AB} = 1 - \sqrt{\frac{M_A}{M_B}} \tag{14.21}$$

Integration of this ODE with boundary conditions, that is, mole fractions of gases A and B in both compartments $y_B^U, y_B^L, y_A^L, y_A^U$, yields

$$N_A^d = \frac{c_T}{\alpha_{AB} L} D_{AB}^m \ln \frac{1 - \alpha y_A^L + (D_{AB}^m / D_A^k)}{1 - \alpha y_A^U + (D_{AB}^m / D_A^k)} \tag{14.22}$$

If diffusion flux densities, N_A^d, are experimentally determined for different gas pairs A–B or/and for different total pressure or/and different temperature, then by comparison of experimental molar diffusion flux densities with Eq. (14.22) it is possible to obtain the corresponding bulk- and Knudsen-effective diffusion coefficients. These coefficients are connected with transport parameter via Eqs. (14.4) and (14.5).

The disadvantage of this cell is: (a) the necessity to monitor composition of the two outlet gas streams, which requires analytical instruments and (b) the strict requirement of equal pressure in both cell compartments (even small pressure differences between compartments can cause large errors due to the intrusion of permeation flow).

14.3.2 Graham Cell

The two-component version of the general Graham's law (Eq. (14.9)) reads

$$\frac{N_A^d}{N_B^d} = -\sqrt{\frac{M_B}{M_A}} \tag{14.23}$$

where M_A and M_B are molecular weights of A and B. Thus, in binary countercurrent diffusion the diffusion fluxes are not equimolar as is often assumed. The minus sign in Eq. (14.23) reflects the opposite directions of molar diffusion flux densities. For gases with different molecular weight this ratio is far from unity. Thus, the net diffusion flux, $N^d = N_A^d + N_B^d$, is nonzero. This offers a simple way for determination of flux densities of individual gases simply by following the net diffusion flux, N^d.

The diffusion cell based on Graham's law (Figure 14.2) (Valuš and Schneider, 1981, 1985) employs the easily determinable nonzero net molar diffusion flux densities, N^d (Eqs. (14.1) or (14.2)). The cell design is similar as in the Wicke-Kallenbach

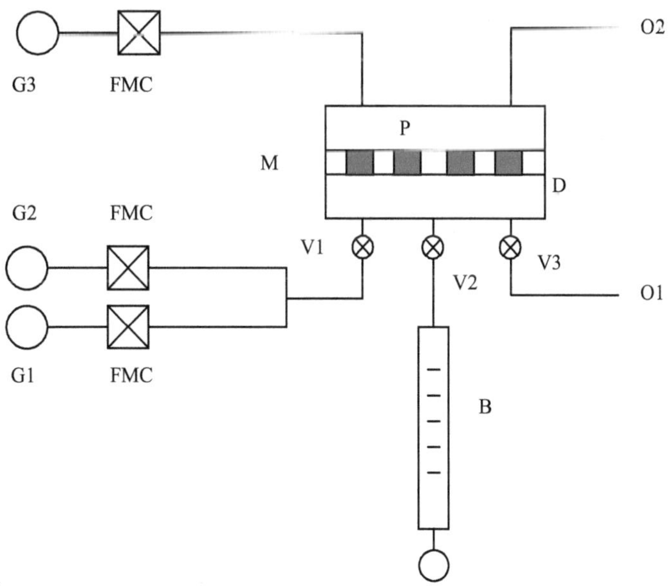

Figure 14.2. Graham diffusion cell. G1, G2, G3 pressure cylinders with gases 1, 2, 3; FMC flow-meter-controllers; O1, O2 gas outlets; D diffusion cell, M metallic disc, P porous pellets, B optical digital flowmeter; V1, V2, V3 valves

cell. Different gases A and B flow steadily through the both compartments until the steady state is established. Inlet flow-rates of gases A and B should be high enough to guarantee that gases flowing from both compartments are nearly pure (i.e., $y_A^U = y_B^L = 1$ and $y_A^L = y_B^U = 0$). After the gas inlet and outlet in one of the cell compartment is stopped, the net volumetric diffusion flux is determined. For this purpose the rate of movement of a soap-film, $V(t)$, in a optical digital flowmeter connected on one side to the isolated diffusion cell compartment and on the other side to atmosphere, is followed. The obtained dependence, $V(t)$, is extrapolated to the moment of the compartment closure ($t = 0$) and the net molar flux density, $N^d = N_A^d + N_B^d$, determined from the initial slope of $V(t)$, corresponds to the steady state (for illustration see Figure 14.3).

$$N^d = \left(\frac{\mathrm{d}V}{\mathrm{d}t}\bigg|_{t=0} \right) \frac{c_T}{S} \tag{14.24}$$

Thus, it is not necessary to monitor the composition of both gas streams at the outlets of cell compartments. This removes the first disadvantage of the Wicke-Kallenbach cell. The second disadvantage of the Wicke-Kallenbach cell (viz. equality of pressure in both compartments) is removed because both compartments are open to atmosphere and, hence, no special pressure regulation is required. Disadvantage of this cell stems from the fact, that the more different the molecular weights of both gases the higher the net molar diffusion flux density, N^d, and then the more accurate is its determination. However, gases with high molecular weights tend to adsorb, which can cause (unwanted) surface diffusion to occur. This restricts the choice of suitable gas pairs to nonadsorbable inerts (e.g., H_2, He, N_2, Ar). The volume of experimental information can be increased by replacing one of the gases (say A) by a binary mixture

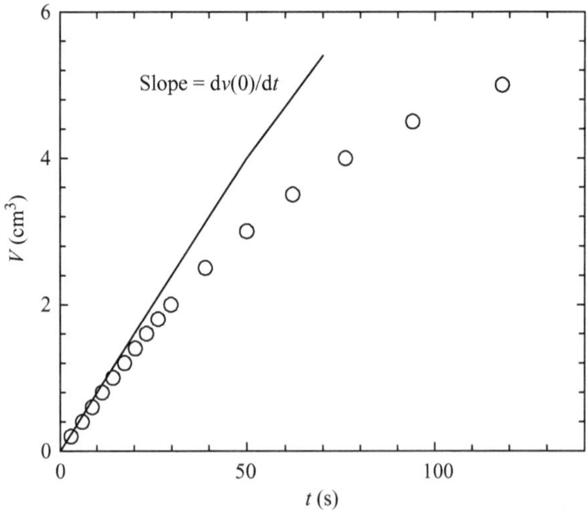

Figure 14.3. Time dependence of the volume swept by the soap-film, $V(t)$

with gas B (gas system $(A+B)$ vs. B) or with another gas C (gas system $(A+C)$ vs. B) in order to increase the confidence of evaluated transport parameters. Equations corresponding to the use of more than two gases can be found elsewhere (Šolcová et al., 2001).

With boundary conditions

$$y_A^U = y_B^L = 1 \quad y_A^L = y_B^U = 0$$

the integration of differential Equation (14.21) with the Graham's law (14.23) on application of Eqs. (14.4) and (14.5) yields

$$N^d = N_A^d + N_B^d = \frac{c_T}{L}\psi \mathcal{D}_{AB}^m \ln \frac{1 + \left(\mathcal{D}_{AB}^m / \langle r\rangle K_A\right)}{1 - \alpha + \left(\mathcal{D}_{AB}^m / \langle r\rangle K_A\right)} \tag{14.25}$$

If net diffusion flux densities, N^d, are experimentally determined for different gas pairs $A-B$, then by comparison with Eq. (14.25) it is possible to obtain the corresponding bulk- and Knudsen-effective diffusion coefficients. These coefficients are connected with transport parameter via Eqs. (14.4) and (14.5). Figure 14.4 shows 95% confidence regions of transport parameters for several commercial and laboratory-prepared porous catalyst pellets evaluated from Eq. (14.25) with three-component gas systems (($A+B$) vs. B and ($A+C$) vs. B). The optimum parameter sets are marked as circles. It is interesting that the shape of the confidence region is intimately connected with region in which the diffusion process takes place. There are two limiting diffusion mechanisms: the Knudsen diffusion, which appears only in narrow pores and the bulk diffusion mechanisms, which takes place in wide pores.

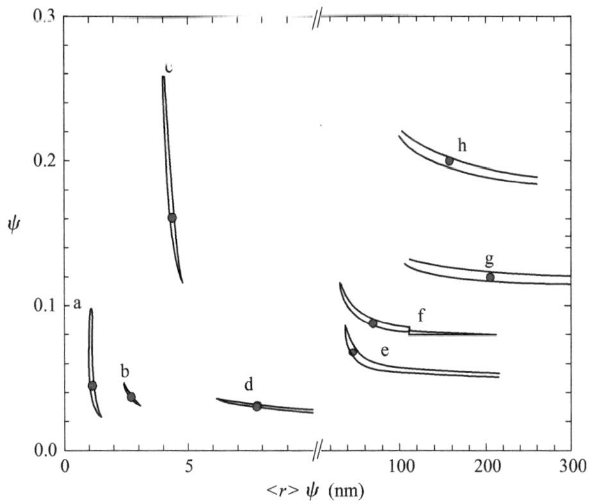

Figure 14.4. 95% confidence regions of transport parameters for porous catalyst samples a–g

In the transition diffusion region both mechanisms play a role and the contribution of each mechanism can be seen from the shape of the confidence region. When the confidence region in the plane ψ vs. $\langle r\rangle\psi$ is nearly vertical (e.g., samples a, b, c) the influence of ψ is quite weak; hence, the contribution of bulk diffusion is small. When the shape of the confidence region is nearly horizontal (samples d, g, h) the influence of $\langle r\rangle\psi$ is weak. The minor role of Knudsen diffusion mechanism and prevailing role of bulk diffusion is demonstrated. Shapes of confidence regions for samples in which diffusion in the transition region operates are combinations of the above limiting shapes; thus, the contribution of each diffusion mechanism can be qualitatively judged from the confidence region shape (samples e, f).

14.4 DYNAMIC BINARY DIFFUSION

14.4.1 Gas Chromatography (Dynamic Diffusion)

One of the well-established techniques (Suzuki and Smith, 1975; Haynes, 1988), the chromatographic method, permits diffusion study under dynamic condition. The simplicity of the equipment as well as the speed of data acquisition are the main reasons for the frequent use of these techniques.

Effective diffusivities are evaluated from response (chromatographic) curves of a columns packed with the tested porous particles. The carrier-gas (C) flows through the packed column at constant rate and a pulse of another gas (tracer – T) is injected into the carrier stream at the column inlet. At the column outlet the tracer concentration is monitored by a suitable detector and the recorded outlet response peak is then analysed. Several processes occur during the passage of the tracer band through the column: besides convection and axial dispersion, transport of tracer through the laminar film around the packing particles takes place, followed by diffusion in the pore structure and possibly by adsorption (for adsorbable tracers) on the internal surface of the porous packing.

Porous particles can be packed in the column in two ways (i) a wide bed packed with particles; the column to particle diameter ratio should be large enough, say about 1:20, to satisfy conditions for axially dispersed plug-flow, or, (ii) particles are packed one by one into a column with diameter that is only 10–20% larger than the particle diameter. This arrangement is known as Single Pellet String Column (SPSC) (Scott et al., 1974) and it is used usually for spherical or cylindrical porous particles.

14.4.2 SPSC and Convolution

In SPSC arrangement (see Figure 14.5) high linear carrier-gas velocities – which suppress the mass transfer resistance of the laminar film around particles and tracer peak broadening due to axial dispersion – can be easily attained. This is significant when inert (nonadsorbable) gases are used as tracers, which move through the column

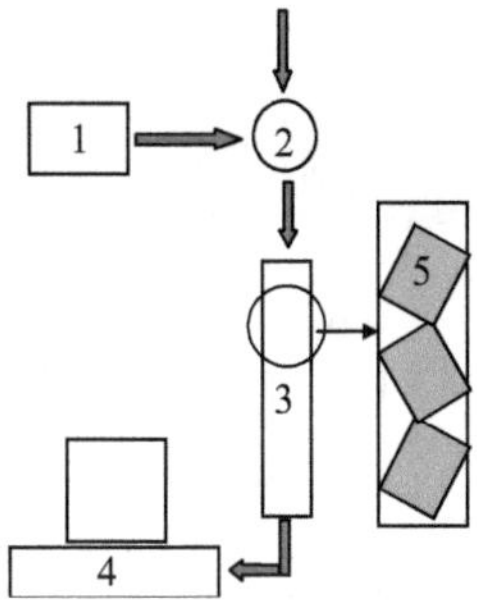

Figure 14.5. Chromatographic set-up in SPSC arrangement; 1 carrier gas; 2 tracer sampling valve; 3 column with pellets; 4 tracer detector + A/D conversion + data logger; 5 detail of the packed column

with the high carrier-gas velocity. The use of inert tracers prevents their adsorption and the possible surface diffusion, which obscures the effective diffusion coefficients (Schneider and Smith, 1968a) and transport characteristics. Another advantage of the SPSC arrangement is the averaging of obtained transport characteristics over many pellets present in the column. In addition: SPSC guarantees low carrier gas consumption.

The analysis of outlet peaks is based on the model of processes in the column. Today the Kubín-Kučera model (Kubín, 1965; Kučera, 1965), which accounts for all the above-mentioned processes, as long as they can be described by linear (differential) equations, is used nearly exclusively. Several possibilities exist for obtaining rate parameters of intracolumn processes (axial dispersion coefficient, external mass transfer coefficient, effective diffusion coefficient, adsorption/desorption rate or equilibrium constants) from the column response peaks. The moment approach in which moments of the outlet peaks are matched to theoretical expressions developed for the system of model (partial) differential equations is widespread because of its simplicity (Schneider and Smith, 1968b).

The today's availability of computers makes matching of column response peaks to model equations the preferred analysis method. Such matching can be performed in the Laplace- (Kamiyanagi and Furusaki, 1983) or Fourier-domain (Zygourakis and Moudgalya, 1987), or, preferably in the time-domain (Wakao et al., 1979; Fahim and Wakao, 1982).

For a T → C system the impulse response (i.e., response to a tracer Dirac pulse) of a column packed with porous spherical particles of radius, R, in which tracer equilibrium adsorption is in the Henry region of the adsorption isotherm (characterized by the adsorption equilibrium constant, K_T) and intraparticle diffusion characterized by an effective diffusion coefficient D_{TC} takes place is given by (Schneider, 1984).

$$h(t) = Q \int_0^\infty \exp\left(\frac{\mathrm{Pe}}{2} - f_1\right) \cos\left(\frac{2\gamma\lambda^2 t}{\delta_o t_{\mathrm{dif}}} - f_2\right) \lambda \mathrm{d}\lambda \tag{14.26}$$

where

$$
\begin{aligned}
f_{1,2} &= \sqrt{\frac{\sqrt{A^2+B^2} \pm A}{2}} \\
A &= \mathrm{Pe}\left(\frac{\mathrm{Pe}}{4} + \frac{3\gamma H_1 t_c}{t_{\mathrm{dif}}}\right) \\
B &= \mathrm{Pe}\frac{t_c}{t_{\mathrm{dif}}}\left(\frac{2\gamma\lambda^2}{\delta_o} + 3\gamma H_2\right) \\
H_1 &= \lambda\frac{\sinh(2\lambda)+\sin(2\lambda)}{\cosh(2\lambda)-\cos(2\gamma)} - 1 \\
H_2 &= \lambda\frac{\sinh(2\lambda)-\sin(2\lambda)}{\cosh(2\lambda)-\cos(2\gamma)}
\end{aligned}
\tag{14.27}
$$

and t_c is the tracer mean residence time in the column of length L ($t_c = L/v$) with carrier gas linear velocity, v. Pe is the Peclet number ($\mathrm{Pe} = vL/E_{TC}$ and E_{TC} is the axial dispersion coefficient), t_{dif} denotes the diffusion time of the tracer in the pore structure of a pellet, $t_{\mathrm{dif}} = R^2\beta/D_{TC}$ (R is the radius of the pellet equivalent sphere), δ_{o} is the tracer adsorption parameter $\delta_{\mathrm{o}} = \gamma(1+K_T)$, and $\gamma = \beta(1-\alpha)/\alpha$. β is the pellet porosity and α is the column void fraction (interstitial void volume/column volume). Thus, γ, is the pore volume per unit interstitial volume. For an inert tracer $K_T = 0$ and $\delta_{\mathrm{o}} = \gamma$. Q is a normalization constant defined so that at the calculated SPSC response maximum the tracer concentration equals unity, $c(t_{\max}) = 1$. Eq. (14.26) assumes no resistance between the bulk stream and the external surface of porous pellets.

Equations (14.26) and (14.27) describe correctly the intracolumn processes but neglect the effects of processes upstream and downstream of the column (extra column effects – ECE) (Šolcová and Schneider, 1996). In the time-domain matching it is possible to include these effects through the application of the convolution theorem. This requires, besides the knowledge of the experimental system response, also the knowledge of the ECE response. The ECE response can be replaced by experimental system responses for two columns with different length. The convolution theorem states that the column response, $c(t)$, is given by the convolution integral

$$c(t) = \int_0^t g(t-u)h(u)\mathrm{d}u \tag{14.28}$$

$h(t)$ is the column impulse response and $g(t)$ describes the shape of the signal entering the column instead of the Dirac impulse. In linear systems it is immaterial if the ECE are distributed in different places of the system or if they are concentrated in one place and in what order they are arranged. Therefore, it is possible to use the experimental responses for the shorter column as $g(t)$. The application of convolution integral

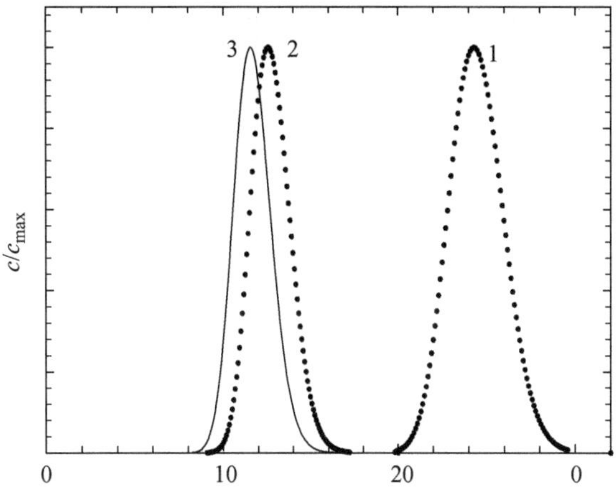

Figure 14.6. Experimental column responses: 1 column 100 cm, 2 column 50 cm, 3 calculated impulse response for column length: 100 − 50 cm = 50 cm. Peaks are normalized to the same height

is demonstrated in Figure 14.6, where 1 is the experimental response of a longer column (100 cm), 2 is the experimental response of a shorter column (100 cm) and 3 is the calculated impulse response for a column with length: 100 − 50 cm = 50 cm. Because the length of longer column is double of the length of shorter column, the difference between peaks 2 and 3 belongs to extra column effects.

There are four unknown parameters in the theoretical impulse response for porous particles, $h(t)$; the pellet diffusion time, t_{dif} (which contains the effective diffusion coefficient of the pair T–C, D_{TC}), the mean residence time of the carrier-gas in the interparticle space, t_c, Peclet number, Pe and the adsorption parameter, δ_o. Because matching with four unknown parameters would give highly correlated parameters, it is better to determine some parameters independently.

Axial dispersion, described by Peclet numbers, can be characterized separately. It was confirmed (Šolcová and Schneider, 2002) that Peclet numbers for identical carrier, and tracer gases and carrier flow-rates are the same for column packed with nonporous and porous particles of identical shape. This permits determination of Peclet numbers from measurements in SPSC with nonporous packing. Then, only two parameters appear in the theoretical impulse response: Pe and t_c; their determination by time-domain matching is very rapid and parameters are not correlated. Use of separately determined Peclet numbers decreases the number of matched parameter, in the analysis of response of SPSC packed with porous particles, from four (t_{dif}, Pe, δ_o, t_c) to three (t_{dif}, δ_o, t_c). Sometimes, Bodenstein numbers, Bo, are used instead of Peclet numbers, Pe. Bodenstein numbers include equivalent diameter of packing particles, d_p, instead of the column length, L_c. Thus $\text{Bo} = vd_p/E_{TC}$ and $\text{Bo} = \text{Pe}(d_p/Lc)$. For spherical pellets the sphere diameter is used as d_p, for pellets with other shapes the equivalent sphere diameter (diameter of a sphere with the same volume to outer surface ratio) is ordinarily applied.

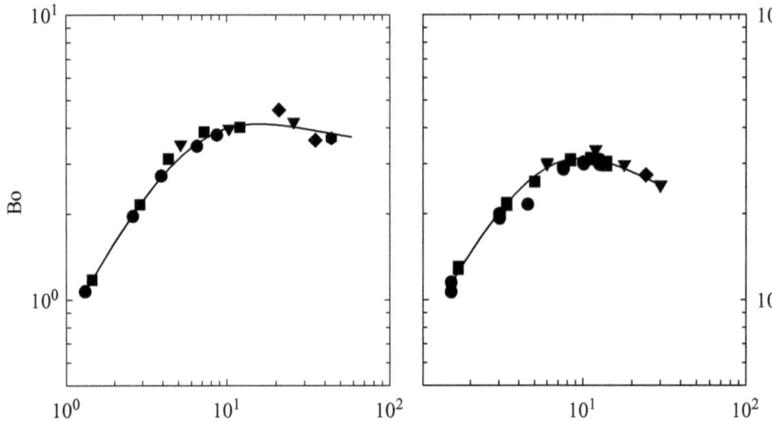

Figure 14.7. Bodenstein numbers for porous catalysts from nonporous packing; points: experimental, line: calculated from Eq. (14.29)

The obtained optimum Bodenstein numbers for two different catalyst pellets are plotted versus the product of Reynolds and Schmidt numbers, ReSc ($\mathrm{ReSc} = v * 2R/\mathcal{D}_{TC}^m$) in Figure 14.7. It is seen that Bodenstein numbers can be obtained with quite good accuracy in a wide range of carrier-gas velocities. The obtained data are correlated by a semi-empirical Equation (14.29) with three adjustable parameters (γ, λ, β).

$$\frac{1}{\mathrm{Bo}} = \frac{\gamma}{\mathrm{ReSc}} + \frac{\lambda \mathrm{ReSc}}{\beta + \mathrm{ReSc}} \tag{14.29}$$

The tracer adsorption parameter, δ_o, can be determined from the difference of first absolute peak moments for two column lengths, $(\mu_1')_{\mathrm{col1}}$, $(\mu_1')_{\mathrm{col2}}$. Thus, only t_{dif} and t_c have to be determined by the time-domain matching; and these parameters are not correlated.

The effective diffusion coefficients, D_{TC}, obtained from the parameter t_{dif}, include contributions from the Knudsen diffusion mechanism and from the bulk diffusion mechanism. Because of the low tracer concentrations the Bosanquet formula (14.30) can be applied.

$$\frac{1}{D_{TC}} = \frac{1}{\langle r \rangle \psi K_T} + \frac{1}{\psi \mathcal{D}_{TC}^m} \tag{14.30}$$

Here $\mathcal{D}_{TC}^m$ is the bulk diffusion coefficient of the pair C–T and $K_T \equiv (2/3)\sqrt{(8R_gT/\pi M_T)}$ is the Knudsen constant for the tracer T, R_g is the gas constant, T temperature, and M_T the tracer molecular weight. Eq. (14.30) can be rearranged into the form (14.31),

$$t_{\mathrm{dif}} K_T = \frac{R^2\beta}{\langle r \rangle \psi} + \frac{R^2\beta}{\psi} \frac{K_T}{\mathcal{D}_{TC}^m} \tag{14.31}$$

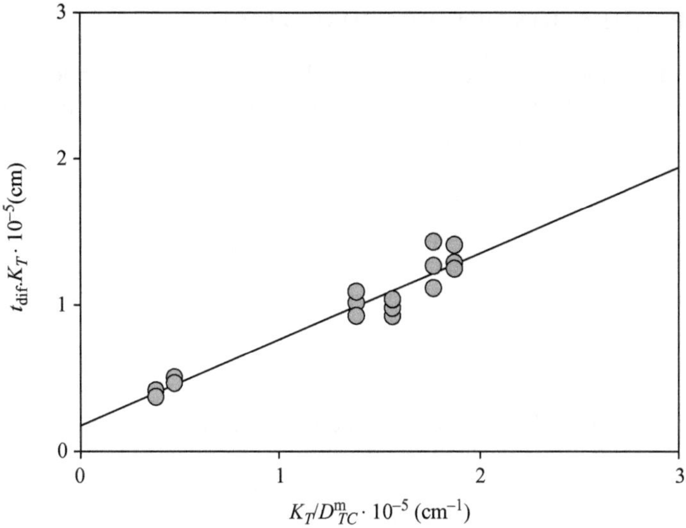

Figure 14.8. Chromatography with SPSC. Correlation of $t_{dif}K_T$ vs. K_T/D^m_{TC} (Eq. (14.31))

Table 14.1. Fraction of the Knudsen diffusion mechanism resistance, Ω_K, from the sum of Knudsen and bulk diffusion resistances for four catalysts with different mean pore radii (a–d)

Pellets	r [nm] mean pore radii	Ω_K [%] for $N_2 \rightarrow He$	Ω_K [%] for $Ar \rightarrow N_2$
a	8	46	43
b	3.5 257	33	30
c	46 630	38	36
d	46 2230	18	16

which permits easy (e.g., graphical) determination of transport parameters ψ and $\langle r \rangle \psi$ from effective diffusion coefficients, D_{TC}. This is shown in Figure 14.8 for one porous catalyst. The obtained diffusion times, t_{dif}, are plotted in coordinates $t_{dif}K_T$ vs. $K_T/\mathcal{D}^m_{TC}$. Transport parameters, $\langle r \rangle \psi$ and ψ, are evaluated from the straight-line slope and intercept.

Furthermore, it is possible to determine the ratio of contributions of Knudsen and bulk diffusion mechanisms. For four catalysts with different mean pore radii the resistance, Ω_K, of the Knudsen diffusion mechanism relative to the sum of Knudsen and bulk diffusion resistances is shown in Table 14.1. It is clearly seen that in agreement with expectation the significance of Knudsen diffusion mechanism decreases with increase of the mean pore radius of catalyst pellets.

14.5 PERMEATION OF INDIVIDUAL GASES

Transport parameters $\langle r \rangle \psi$ and $\langle r^2 \rangle \psi$ can be evaluated from permeation measurements (single gas flow-rate under controlled pressure gradient) performed with pure gases. The corresponding permeation cells can be designed for measurements under steady state as well under dynamic conditions. To gather sufficient experimental information measurements must be repeated for different pressures and different gases. The steady-state permeation requires control of pressures and determination of flow-rates at these pressures. The last task is by no means simple. It is therefore reasonable to use pseudo-stationary or dynamic permeation cells.

14.5.1 Pseudo-Stationary and Dynamic Permeation

Both the pseudo-stationary and dynamic cells (Figure 14.9) consist of upper and lower compartments and a metallic disc with cylindrical holes filled by pellets of porous solid, same as in the Wicke-Kallenbach and Graham diffusion cells. The cells differ mainly in compartment volumes, inlet and outlet gas lines (with valves) and the attached pressure-/differential pressure-transducers.

The number of pellets in the metallic disc and the volumes of cell compartments affect the permeation rate. Thus, their change can be used for obtaining pressure responses compatible with capabilities of the used pressure transducer.

14.5.2 Pseudo-Stationary Cell (Figure 14.9a)

Identical compartment volumes $V = V_L = V_U$ are the basic precondition. After evacuation both compartments are filled with permeation gas up to initial pressure P^o. At run start the gas pressure in the upper cell compartment is increased by a small

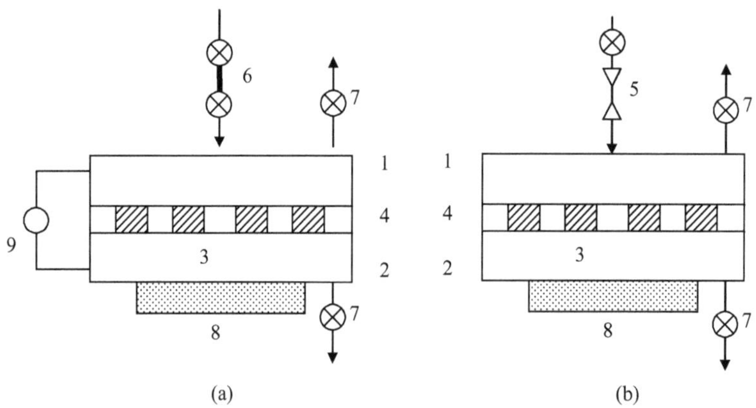

Figure 14.9. Pseudo-stationary (a) and dynamic (b) permeation cells 1, 2 upper and lower cell compartment, 3 cylindrical porous pellets, 4 metallic disc with holes for pellets, 5 capillary gas inlet, 6 100 cm^3 gas volume, 7 vacuum valves, 8 pressure gauge, 9 differential pressure gauge

amount (usually 500–900 Pa), ΔP^o, and the gas inlet is closed. The time-change of the pressure difference between compartments, $\Delta P(t)$, is followed by the differential pressure transducer.

The constitutive law for molar permeation flux density, $N^p(t,\ x)$, of a pure gas in a porous medium under isothermal conditions is given by the single component form of the Darcy equation (14.10) (Dullien, 1979).

$$N^p = -B\frac{\partial p/R_g T}{\partial x} \tag{14.32}$$

where $p(t,\ x)$ is the gas pressure inside the porous medium. For single component permeation the effective permeability coefficient (Eq. (14.11)) simplifies to Weber equation (14.33)

$$B = \langle r\rangle\psi K\frac{\omega + Kn}{1 + Kn} + \langle r^2\rangle\psi\frac{p}{8\eta} \tag{14.33}$$

for both MTPM and DGM models. x the pellets length co-ordinate, T the temperature, R_g the gas constant and t the time. In the majority of cases it appears that $\omega = 1$ is a good approximation. Hence, in the following the simplified form of Eq. (14.33) is used

$$B = \langle r\rangle\psi K + \langle r^2\rangle\psi\frac{p}{8\eta} \tag{14.34}$$

Cell balance. Because the pore volume of porous pellets mounted in the impermeable disk is much smaller than the volume of compartments the gas accumulation in the pores can be neglected, that is, $\mathrm{d}N^p/\mathrm{d}x = 0$.

After sufficiently long time the pressure difference between cell compartment tends to zero and pressures in the upper and lower compartments, P^U, P^L, are obviously

$$P^U(t \to \infty) = P^L(t \to \infty) = \overline{p} = P^o + \Delta P^o/2 \tag{14.35}$$

Using the assumption of negligible gas accumulation in pores ($\mathrm{d}N^p/\mathrm{d}x = 0$) the gas conservation for cell compartments of identical volume can be expressed as $\mathrm{d}P^U/\mathrm{d}t = -\mathrm{d}P^L/\mathrm{d}t$. The material balance then reads

$$\frac{1}{R_g T}\frac{\mathrm{d}\Delta P}{\mathrm{d}t} = -2\frac{SN^p}{V} \tag{14.36}$$

where V is the compartment volume ($V = V_L = V_U$). Using the Darcy equation (14.32) and the simplified Weber law (14.34), integration of Eq. (14.36) (from $x = 0$, where $p = P^U$ to $x = L$, where $p = P^L$) yields

$$N^p R_g TL = \langle r\rangle\psi K\Delta P + \frac{\langle r^2\rangle\psi}{8\mu}\overline{P}\Delta P \tag{14.37}$$

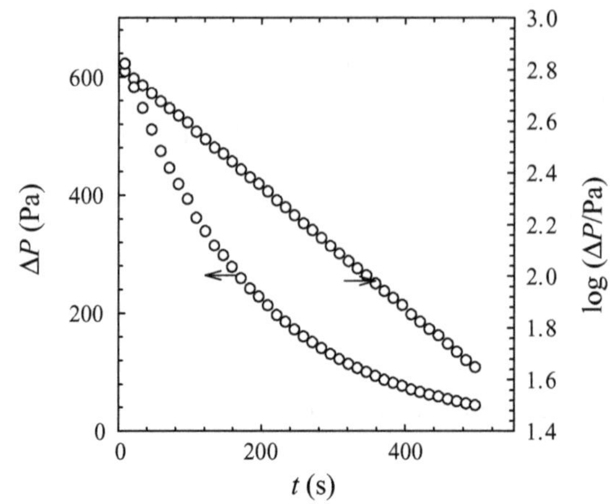

Figure 14.10. Time change of argon pressure difference in the pseudo-stationary permeation cell. $p^o =$ 2000 Pa, $\Delta p^o = 622$ (only a fraction of experimental readings is shown)

with the mean pressure $\overline{P} = (P^U + P^L)/2$. From combination of Eqs. (14.36) and (14.37) it follows

$$\frac{\mathrm{d}\Delta P}{\mathrm{d}t} = -\left(\frac{2S}{LV}\right)\Delta P\overline{B} \tag{14.38}$$

where $\overline{B} = B(\overline{P})$ is the effective permeability coefficient at mean pressure $\overline{P}$. Integration of Eqs. (14.38) with initial conditions: $t \to 0$, $\Delta P \to \Delta P^o$ gives the relation for $\Delta P(t)$:

$$\Delta P = \Delta P^o \exp\left(-\frac{S}{L}\frac{2}{V}\overline{B}t\right) \tag{14.39}$$

Exponential decay of $\Delta P(t)$ is illustrated in Figure 14.10. The obtained effective permeation coefficients, $\overline{B}$, for each gas, change linearly with the mean pressure $\overline{P}$ (Eq. (14.37)). From Eq. (14.34) it also follows that in coordinates $\overline{B}/K$ vs. $\overline{p}/(8K\mu)$ points for different permeation gases should fall on the same straight line with intercept $\langle r\rangle\psi$ and slope $\langle r^2\rangle\psi$.

$$\frac{\overline{B}}{K} = \langle r\rangle\psi + \langle r^2\rangle\psi\frac{\overline{P}}{8K\mu} \tag{14.40}$$

For two porous samples with different pore-size distribution Figures 14.11 and 14.12 illustrate the obtained effective permeabilities, $\overline{B}$, and the ratio $\overline{B}/K$.

14.5.3 Dynamic Cell (Figure 14.9b)

After evacuation both compartments are filled with permeation gas up to initial pressure P^o. At run start the gas pressure in the upper cell compartment is increased to

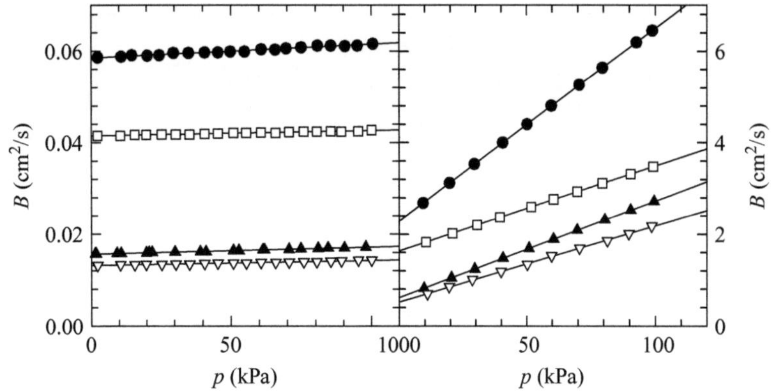

Figure 14.11. Pseudo-stationary permeation cell; correlation of $\overline{B}$ vs. $\overline{p}$ for different gases and porous solids with different pore-size distribution. Points: experimental, line: linear regression

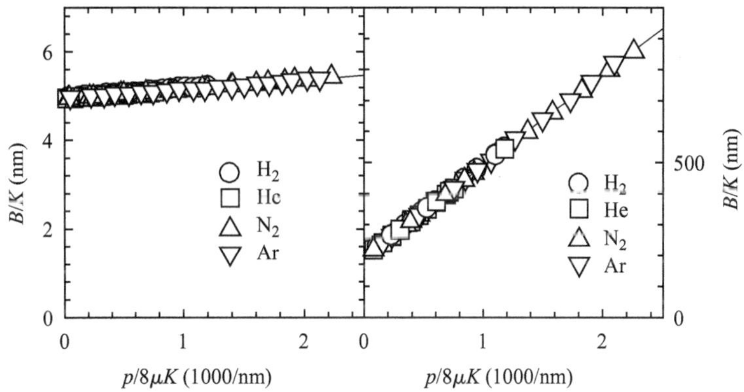

Figure 14.12. Pseudo-stationary permeation cell; correlation of $\overline{B}/K$ vs. $\overline{p}/(8K\mu)$ for porous solids with different pore-size distribution and different gases. Points: experimental, line: linear regression

pressure P^U and kept constant. The pressure transducer in the lower compartment follows the increase of pressure $P(t)$ up to P^o (see Figure 14.13). The procedure is repeated for several pressures P^o and for other permeating gases.

Simplified solution. Similarly as with the pseudo-stationary cell the gas accumulation in the pores can be neglected, that is, $\mathrm{d}N^p/\mathrm{d}x = 0$. Then, the integration of Darcy equation (14.32), together with Eq. (14.34) can provide the expression for the molar permeation flux density, $N^p(t)$,

$$N^p(t)R_gTL = (P^U - P(t))\left[\langle r\rangle\psi K + \frac{\langle r^2\rangle\psi}{16\eta}(P^U + P(t))\right] \tag{14.41}$$

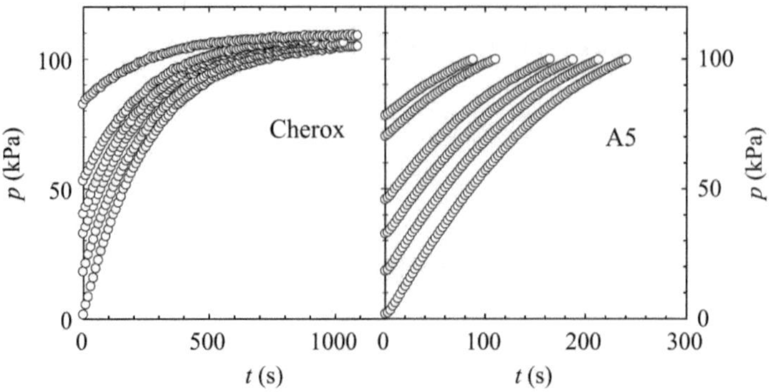

Figure 14.13. Dynamic permeation cell; argon pressure change in the lower cell compartment for different starting pressures and porous solids with different pore-size distribution. Only part of experimental points is shown for clarity

The time change of pressure in the bottom compartment, $P(t)$, obtained by integration of the compartment mass balance (14.42)

$$\frac{\mathrm{d}P}{\mathrm{d}t} = \frac{S}{V_L} N^p R_g T \tag{14.42}$$

together with Eq. (14.41) in the form

$$P(t) = \frac{EP^U - (a + bP^U)(P^U - P^o)}{E + b(P^U - P^o)} \tag{14.43}$$

where

$$E(t) = [a + b(P^U + P^o)] \exp\left[\left(a + 2bP^U\right) \frac{S}{V_L L} t\right] \tag{14.44}$$

and

$$a = \langle r \rangle \psi K \qquad b = \frac{\langle r \rangle \psi K}{16\eta} \tag{14.45}$$

Of course, integration of the system (14.41), (14.42) can be also performed numerically, together with parameter fitting, by any suitable commercial software.

Dynamic cell – full solution. The gas material balance in porous pellets under dynamic conditions reads

$$\varepsilon \frac{\partial p(t, x)/R_g T}{\partial t} = -\frac{\partial N^p(t, x)}{\partial x} \tag{14.46}$$

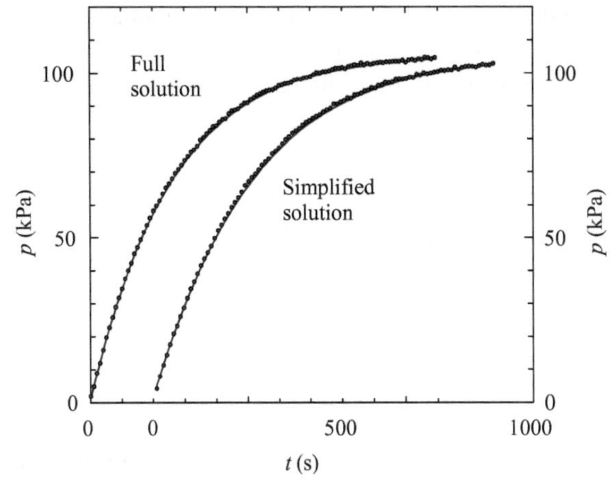

Figure 14.14. Pressure development in the bottom compartment of the dynamic permeation cell; comparison of experiments (points) with full and simplified solution of (lines) of the cell mass balance (Eqs. (14.31), (14.33), (14.45)–(14.48) and Eqs. (14.40), (14.41)). Plots are shifted along the t-axis; only a part of experimental points is shown

with ε porosity of the porous pellet. The boundary condition for Eq. (14.46) at the lower pellets end, $x = L$, is given by the mass balance of the lower cell compartment

$$V_L \frac{\partial P(t)}{\partial t} = SR_g TN^p(t, L) \tag{14.47}$$

where V_L is the volume of the lower compartment with pressure $P(t) \equiv p(t, L), S$ denotes the cross-section of all pellets. The boundary condition for the upper compartment ($x = 0$) is

$$p(t, 0) = P^U \tag{14.48}$$

Initial conditions for the system (14.32), (14.34), (14.46), (14.48) are

$$p(0, x) = P^o \tag{14.49}$$

The initial value boundary problem can be integrated for example, by the method of lines (Sincovee and Madsen, 1975). For parameter fitting the Marquardt-Levenberg procedure can be used with the sum of squared deviations between experimental and calculated $P(t)$ as objective function. The very satisfactory agreement of fit for full and simplified solutions is illustrated in Figure 14.14 (for clarity the plot of simplified solution is shifted along the t-axis).

14.6 COMBINED DIFFUSION AND PERMEATION

The complete set of transport parameters, ψ, $\langle r \rangle \psi$ and $\langle r^2 \rangle \psi$ can be, in principle, evaluated from simple processes in which diffusion and permeation occur at the same time. Of course, simultaneous determination of three mutually correlated parameters leads to deterioration of their confidence. Hence, evaluation of transport parameters from processes of this kind is recommended rather as a check for consistency of parameters from separate diffusion and permeation processes.

14.6.1 Dynamic Wicke-Kallenbach Cell

Combined diffusion and permeation under dynamic conditions can be followed in the adapted Wicke-Kallenbach diffusion cell (Figure 14.15) (Arnošt and Schneider, 1995). The same carrier gas, C, flows steadily through both chambers of the cell with volumes, V^U and V^L. The flow of carrier gas in the upper cell is either substituted by a flow of tracer gas, T, or perturbed by and impulse of the tracer gas, T. Tracer concentration is monitored at the outlet of the lower chamber. This response can be compared with theoretical predictions based on MTPM and DGM. The governing mass balance has the form

$$\varepsilon\frac{\partial \mathbf{c}(x,t)}{\partial t} = \frac{\partial}{\partial x}\left[\mathbf{F}^{-1}(\mathbf{c}(x,t))\frac{\partial \mathbf{c}(x,t)}{\partial x}\right] \tag{14.50}$$

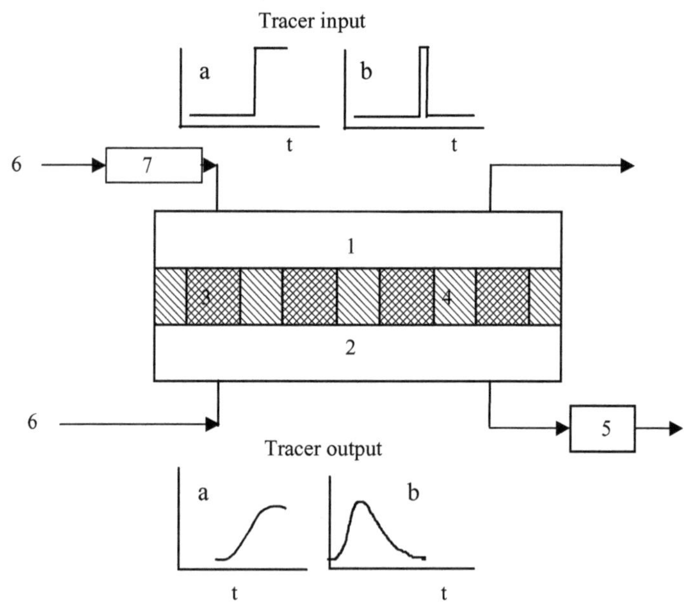

Figure 14.15. Dynamic version of the Wicke-Kallenbach diffusion cell. 1 upper cell compartment, 2 lower cell compartment, 3 cylindrical porous pellets, 4 impermeable pellet holder, 5 tracer detector, 6 carrier gas inlet, 7 carrier gas/tracer-gas switch, a carrier/tracer step change, b tracer impulse

and has to be solved together with the appropriate boundary and initial conditions (see, Arnošt and Schneider, 1995), for example, by the finite difference method (Ehrhardt et al., 1988).

If, instead of Eq. (14.50), the simple Fick's law is used, the resulting set of linear differential mass balances can be solved by Laplace transformation (see e.g., Burghardt and Smith, 1979). For high carrier-gas flow-rates in both chambers, F^U, F^L, and impulse perturbation by tracer gas, T, in the upper cell compartment, the first absolute moment of the tracer response in the lower chamber is given by

$$\bar{\mu}_1' = \frac{L_p^2 \varepsilon}{6 D_{TC}} + \frac{V^U}{F^U} + \frac{V^L}{F^L} \tag{14.51}$$

The effective diffusion coefficient D_{TC} (see Eq. (14.30)) contains the transport parameters ψ and $\langle r \rangle \psi$. Moment expressions for more general situations are summarized in (Arnošt and Schneider, 1994).

14.6.2 Combined Diffusion and Permeation Cell

The diffusion cell shown schematically in Figure 14.16 can be used for determination of dynamic pressure responses to step changes of gas composition (Novák et al., 1988; Novák and Schneider, 1990). The cylindrical porous pellets are mounted in cylindrical holes of the otherwise impermeable metallic disc, which separated the upper flow-through cell chamber from the closed bottom chamber. The closed cell chamber is equipped with a sensitive pressure transducer. By flowing a gas A through the upper compartment, before the start of measurement, *both* cell chambers are filled with A. At the measurement start the in-flow of gas A into the upper compartment is step-wise replaced by gas B (denoted as B → A; if both gases are reversed, then A → B). Output of the pressure transducer is followed until the pressure level prior

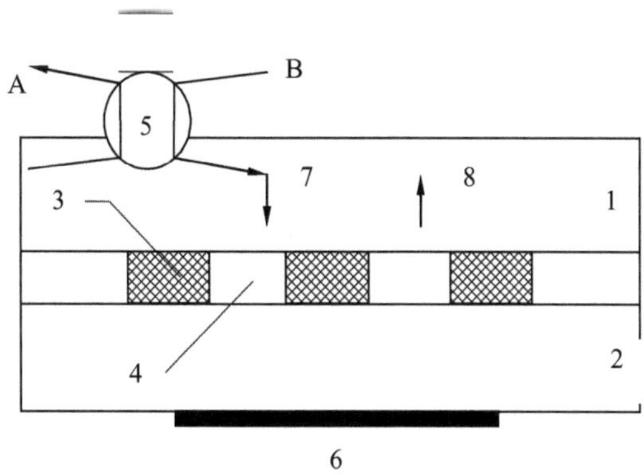

Figure 14.16. Combined diffusion and permeation cell. 1 flow-through upper cell compartment, 2 closed lower cell compartment, 3 cylindrical porous pellets, 4 impermeable pellet holder, 5 four-way valve, 6 pressure transducer, 7 cell inlet of gas A, 8 cell outlet

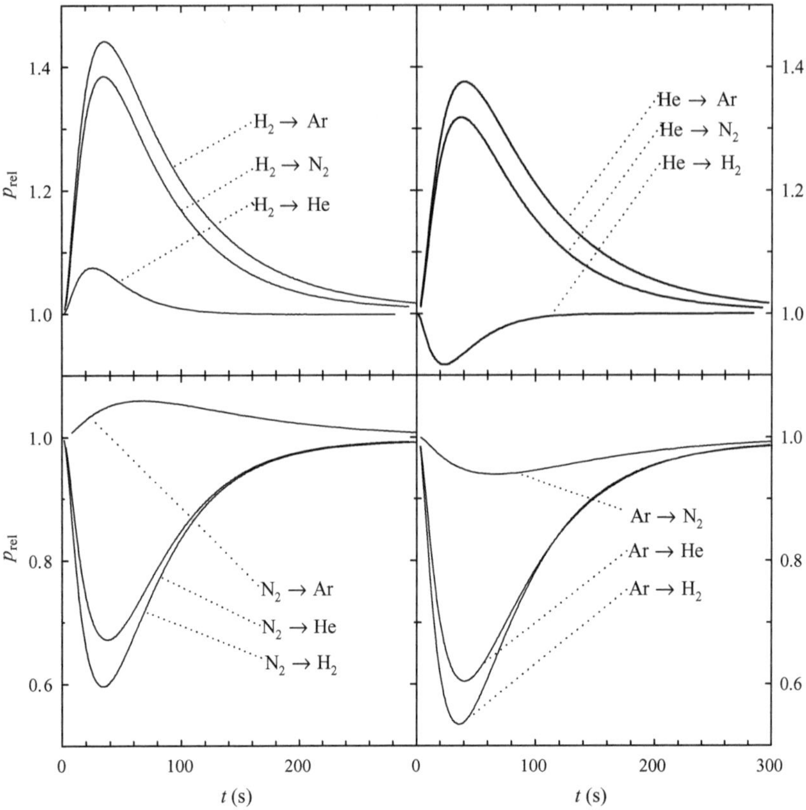

Figure 14.17. Time developments of relative pressure responses for inert binary systems in the combined diffusion and permeation cell for catalyst ICI 52/1 $p_{\rm rel}$ is the relative pressure change in the lower cell compartment ($p_{\rm rel} = p/p_b$, with p_b the atmospheric pressure)

to measurement start is restored. Different pairs B → A can be selected (inert – i.e., non-adsorbable – gases: hydrogen, helium, nitrogen, argon are a good choice since with them the surface transport is absent). Generally, it is possible to use gas mixtures (e.g. $A_1 + A_2$ and/or $B_1 + B_2$) instead of pure A and B.

Figure 14.17 illustrates the obtained pressure responses for pellets of a commercial porous catalyst ICI 52/1 (Imperial Chemical Industries, Ltd., (UK) (Hejtmánek et al., 1998). Textural properties of this catalyst are in Table 14.2. Its pore-size distribution (Figure 14.18) shows that ICI 52/1 is monodisperse.

Modelling of the obtained pressure responses, $p_{\rm rel}(t)$, starts with the governing mass balance (14.50) appended by appropriate boundary and initial conditions (see, Arnošt and Schneider, 1995). The problem can be solved by the finite difference method (see, Ehrhardt et al., 1988). Transport parameters ψ, $\langle r \rangle \psi$ and $\langle r^2 \rangle \psi$ can be obtained by matching the solution to experimental information. Another possibility is to verify if

Table 14.2. Textural properties of a commercial porous catalyst ICI 52/1

Specific surface	$73\ m^2/g$
Skeletal density	$3.833\ cm^3/g$
Apparent density	$1.518\ cm^3/g$
Pore volume	$0.377\ cm^3/g$
Porosity	0.604
Mean pore diameter	16.4 nm

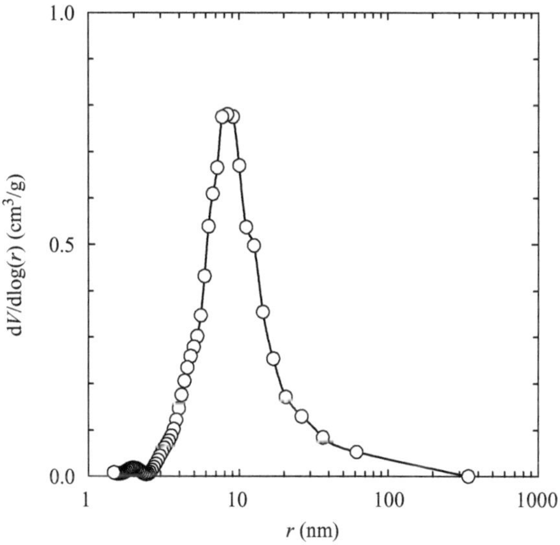

Figure 14.18. Pore-size distribution of catalyst ICI 52/1; combination of mercury porosimetry and physical adsorption of nitrogen

the independently obtained transport paramcter set, for example, from pure diffusion and permeation measurements, reliably predicts the experimental $p_{\mathrm{rel}}(t)$ courses.

NOMENCLATURE

c_T	total molar concentration
c_i	molar concentration of component i
$\mathbf{c}$	vector of molar concentrations of mixture components
D_i^k	effective Knudsen-diffusion coefficient of component i
D_{ij}^m	effective bulk diffusion coefficient of pair i–j
D_{TC}	global effective diffusion coefficient for pair T–C
$\mathcal{D}_{ij}^m$	binary bulk diffusion coefficient of pair i–j
E_{TC}	axial dispersion coeffcient of the Pair T–C

F	volumetric flow-rate
$\mathbf{F}$	$n * n$ matrix with elements $\{f_{ij}\}$
K	Knudsen coefficient $(2/3)(8R_gT/\pi M)^{0.5}$
Kn	Knudsen number $(\lambda/2\langle r\rangle)$
L	column length
L_p	pellet length
M	molecular weight
N^d, N^p, N	diffusion molar flux density, permeation molar flux density, combined molar flux density
$\mathbf{N}$	$n * 1$ vector of combined molar flux densities
P	gas pressure in cell compartments
p	gas pressure in pores
Pe	Peclet number
q_t	tortuosity of transport-pores
$\langle r\rangle, \langle r^2\rangle$	integral mean of transport-pore radii, integral mean of squared transport-pore radii
R	pellet radius (equivalent sphere)
R_g	gas constant
S	total cross-section of pellets in cells
t	time
t_c	convection time (L/v)
t_{dif}	pore-diffusion time $(R^2\varepsilon/D_{TC})$
T	temperature
V^U, V^L	volumes of upper- and lower-compartment, resp.
v	interstitial carrier gas velocity
x	length coordinate in the transport direction
y_i	mole fraction of component i
α	column void fraction
$\varepsilon, \varepsilon_t$	porosity of porous pellet, porosity of transport-pores
η_i, η	viscosity of component i, mixture viscosity
λ	mean free-path length
μ_1'	first absolute moment of the impulse response $c(t) : \mu_1' = \int_0^\infty tc(t)\mathrm{d}t / \int_0^\infty c(t)\mathrm{d}t$
ω	numerical coefficient
ψ	geometric transport parameter (ε_t/q_t)

Subscripts

A, B, i, j for components

Superscripts

U, L upper and lower cell compartment, resp.

REFERENCES

Arnošt D., Schneider P. Effective diffusivities from dynamic diffusion cell: The general moments analysis. Chem. Eng. Sci. 1994; 49:381–393.

Arnošt D., Schneider P. Dynamic transport of multicomponennt mixtures of gases in porous solids. Chem. Eng. J. 1995; 57:91–99.

Burghardt A., Smith J. Dynamic response of a single catalyst pellet M. Chem. Eng. Sci. 1979; 34:267–273.

Dullien F. A. L.. *Porous Media – Fluid Transport and Pore Structure*. New York: Academic Press,. 1979.

Ehrhardt K., Klusáček K., Schneider P. Finite-difference scheme for solving dynamic multicomponent diffusion problems. Comp. Chem. Eng. 1988; 12:1151–1155.

Fahim M. A., Wakao N. Parameter-estimation from tracer response measurements. Chem. Eng. J. 1982; 25:1–8.

Fott P., Petrini G., Schneider P. Transport parameters of monodisperse porous catalysts. Coll. Czech. Chem. Commun 1983; 48:215–227.

Haynes H. W. The experimental evaluation of catalyst effective diffusivity. Jr. Catal. Rev.-Sci. Eng. 1988; 30:563–627.

Hejtmánek V., Čapek P., Šolcová O., Schneider P. Dynamics of pressure buil-up accompanying multicomponent gas transport in porous solids: inert gases. Chem. Eng. J 1998; 70:189–195.

Jackson R., *Transport in Porous Catalysts. Elsevier,* Amsterdam, 1977.

Kamiyanagi K., Furusaki S. Analysis of chromatography by means of transfer-functions. Kag. Kog. Ronbun. 1983; 9:370–375.

Kubín M. Beitrag zur Theorie der Chromatographie. Collect. Czech. Chem. Commun. 1965; 30:1104–1118.

Kučera E. Contribution to the theory of chromatography: linear non-equilibrium elution chromatography. J. Chromatogr. 1965; 19:237–248.

Mason E. A., Maulinauskas A. P., Evans J. W. Flow and diffusion of gases in porous media J. Chem. Phys. 1967; 46:3199–3217.

Mason E. A., Malinauskas A. P. *Gas Transport in Porous Media, The Dusty Gas Model,* Elsevier, Amsterdam, 1983.

Novák M., Ehrhard K., Klusáček K., Schneider P. Dynamics of non-isobaric diffusion in porous catalysts. Chem. Eng. Sci. 1988; 43:185–193.

Novák M., Schneider P. Pressure changes during unsteady-state diffusion of gases in porous catalyst pellets: An experimental study. Appl Catal 1990; 66:209–217.

Rothfeld L. B. Gaseous counterdiffusion in catalyst pellets. A.I.Ch.E.J. 1963; 9:19–24.

Schneider P. Multicomponent isothermal diffusion and forced flow of gases in capillaries. Chem. Eng. Sci. 1978; 33:1311–1319.

Schneider P. Time-domain expression for impulse response (chromatographic) curve for the Kubín-Kučera model of adsorption column. Chem. Eng. Sci. 1984; 39:927–929.

Schneider P., Gelbin D. Direct transport parameters measurements versus their estimation from mercury penetration in porous solids. Chem. Eng. Sci. 1984; 40:1093–1099.

Schneider P., Smith J. M. Chromatographic study of surface diffusion. AICHE J. 1968;14:886–895.

Schneider P., Smith J. M. Adsorption rate constants from chromatography. AICHE J. 1968;14: 762–771.

Scott D. S., Lee W., Papa J. The measurement of transport coefficients in gas-solid heterogeneous reactions. Chem. Eng. Sci. 1974; 29:2155–2167.

Sincovec R. R., Madsen N.K. ACM Trans. Math. Software 1975; 1: 261–267.

Suzuki M., Smith J. M. Transport and kinetic parameters by gas chromatographic techniques. Adv. Chromatogr. 1975; 13:213–263.

Šolcová O., Schneider P. Extra column effects in determination of rate parameters by the chromatographic methods. Collect. Czech. Chem. Commun. 1996; 61:844–855.

Šolcová O., Schneider P. *Studies in surface science and catalysis 144, Characterization of porous solids V.* Amsterdam. Elsevier, 2002.

Šolcová O., Šnajdaufová H., Schneider P. Multicomponent counter-current gas diffusion in porous solids: the Graham's-law diffusion cell. Chem. Eng. Sci. 2001; 56:5231–5237.

Valuš J., Schneider P. A novel cell for gas counter-diffusion measurements in porous pellets. Appl. Catal. 1981; 1:355–366.

Valuš J., Schneider P. Transport characteristics of bidisperse porous alpha-alumina. Appl. Catal. 1985; 16:329–341.

Wakao N., Kaguei S., Smith J. M. Parameter estimation in adsorption chromatography by real-time analysis. J. Chem. Eng. Japan 1979; 12:481–485.

Wicke E., Kallenbach R. Die Oberflächendiffusion von Kohlendioxyd in aktiven Kohlen. Kolloid. Z. 1941; 97:135–151.

Zygourakis K., Moudgalya K. Discrete Fourier transforms and their application to the pulse chromatography method. Chem. Eng. Commun. 1987; 61:107–125.

CHAPTER 15

AIR PERMEABILITY MEASUREMENTS IN POROUS MEDIA

VINCENT C. TIDWELL

Sandia National Laboratories, P.O. Box 5800, MS 0735, Albuquerque, NM 87185, USA

15.1 INTRODUCTION

Fundamental to continuum-based modeling of flow and transport is some knowledge of the permeability of the medium. Permeability quantifies the resistance of the medium to flow and depends only on the characteristics of the porous medium. Permeability is an integrated measure representing the complex relationship between the geometry of the pore system and hydraulics of the flow through that system. Permeability is not measured directly, rather is calculated by inverting an assumed model populated with measured state data (i.e., flux, pressure). As such, permeability has meaning only in the context in which it is measured. That is, permeabilities measured over different scales or using different flow geometries are likely to exhibit very different values.

Below we explore some of the methods available for characterizing the permeability of a porous medium. Our survey focuses on techniques that employ gas as the test fluid. Measurements at both the lab and field scale are reviewed.

15.2 PERMEAMETERS

Simply stated, permeability is determined by applying a pressure gradient across a porous medium while measuring the resulting flux. Both steady and non-steady state methods are routinely employed in these measurements. The non-steady or pressure-decay methods are generally reserved for materials with low permeabilities, where achieving steady-state flow in a reasonable time is impractical.

The constant head permeameter is a common laboratory technique for determining permeability. Its popularity stems from its relative simplicity, both in measurement and permeability calculation. Testing is performed on rock cores or unconsolidated samples loaded in laboratory columns. A constant head difference is applied across

C. Ho and S. Webb (eds.), Gas Transport in Porous Media, 273–278.
© 2006 *Springer*.

the sample producing a steady flow of gas, Q. For a compressible, ideal gas the permeability, k, is determined from the solution of the one-dimensional gas flow equation

$$k = \frac{2\mu Q p_1 L}{A(p_1^2 - p_o^2)} \tag{15.1}$$

where μ is the gas viscosity, L the length of the sample, A the cross-sectional area of the sample, p_1 the pressure at the inlet, and p_o the pressure at the outlet. Note that the gradient is predicated on the squared pressure owing to the compressibility of the gas (e.g., Bear, 1972).

When performing permeability measurements with gas, attention must be paid to gas slippage and inertial effects. Permeabilities measured with gas tend to be larger than equivalent measurements made with liquid. The disparity results from the "slippage" of gas along pore walls that stands in contrast to the no flow boundary layer behavior of liquids. Klinkenberg (1941) provides a thorough explanation of this phenomenon and suggests corrections for its effect. Gas-slippage is most pronounced at low mean pore pressures and for low permeability materials. Alternatively, inertial effects occur when the gas flow is no longer viscous dominated. Such effects can be encountered at relatively low flow rates given the low density to viscosity ratio of gases. A simple check for these effects involves measuring the permeability at several different pressures to verify that the gas flux increases linearly with the gradient.

One disadvantage of the permeameter is that testing requires the destructive sampling of the porous medium. Another disadvantage of the conventional permeameter method is that it is performed on a specific volume of rock or soil, thus aggregating structural detail occurring at smaller scales. These limitations prompted the development of the gas minipermeameter to provide a rapid, inexpensive, and often non-destructive means of sampling permeability (Dykstra and Parsons, 1950; Eijpe and Weber, 1971). Operation involves compressing a tip seal against a flat, fresh rock/outcrop or core surface while injecting gas at a constant pressure. The steady gas flow is directed through the center of the tip seal, discharging through the open faces of the sample beyond the seal (Figure 15.1). Using information on the seal geometry, gas flow rate, gas injection pressure, and ambient (atmospheric) pressure, the permeability is calculated. The solution takes a form similar to that given in Eq. (15.1); however, because of the divergent nature of the flow field L is not known explicitly. Rather, it is replaced by a geometric factor that depends on the ratio of the inner to outer tip seal radii and the shape/size of the test medium. Goggin et al. (1988) provides estimates of the geometric factor as a function of these dimensionless parameters. The active sampling range of the gas minipermeameter is generally from a few milli-Darcys to several hundred Darcys.

Davis and others (1994) modified the gas minipermeameter to facilitate outcrop sampling. The lightweight syringe-based air-minipermeameter (LSAMP) uses a hand-operated syringe to inject gas into the formation whereas the classical minipermeameter depends on a source of compressed gas. The low gas injection pressures used by the LSAMP allow sampling of poorly lithified sediments without damaging

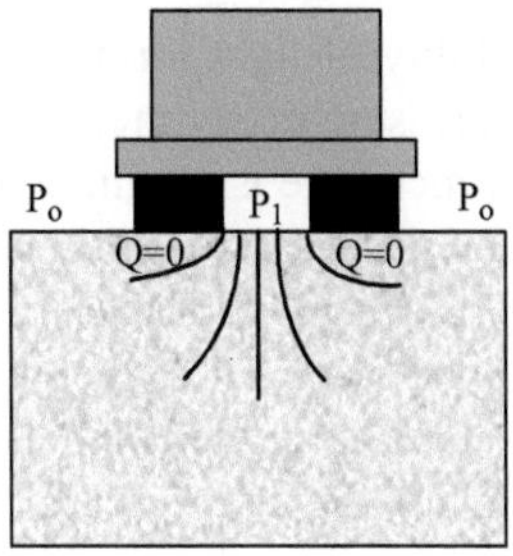

Figure 15.1. Tip seal compressed on porous medium. Shown are the boundary conditions and schematic of flow lines.

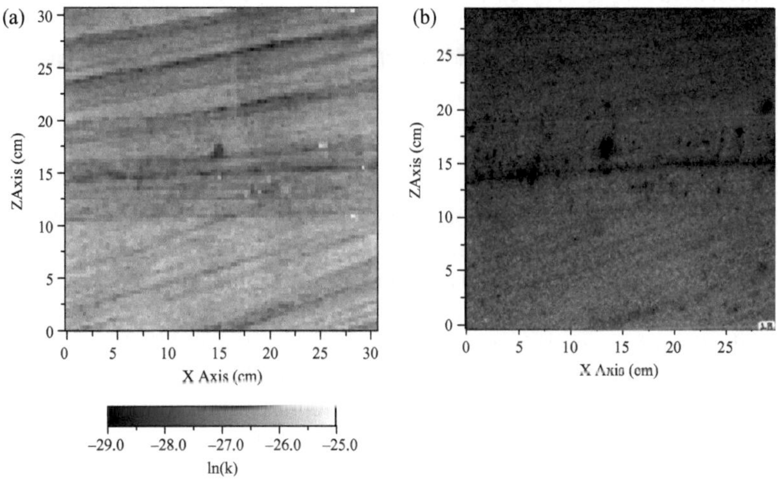

Figure 15.2. Natural log permeability field measured with a gas minipermeameter. Measurements were made on a 2 cm thick slab of cross-bedded Massillon Sandstone. Both the permeability field (a) and the corresponding digital image of the rock slab (b) are shown for comparison

or eroding the formation with compressed gas. The LSAMP has a sampling range of 0.5 to 200 Darcys.

The utility of the gas minipermeameter lies in its ability to map the permeability distribution of heterogeneous geologic formations in detail. Characterization of the distribution of permeability values and the spatial correlation structure is key to developing accurate models of subsurface geologic formations (e.g., Deutsch and Journel, 1997). Gas minipermeameters have found wide spread use in the laboratory (Giordano et al., 1985; Tidwell and Wilson, 2002) and in a variety of field environments, including eolian sandstones (Chandler et al., 1989), fluvial sandstones (Dreyer et al., 1990; Davis et al., 1993), carbonates (Kittridge et al., 1990), and volcanic tuffs (Fuller and Sharp, 1992). Figure 15.2 provides an example of a detailed permeability map collected with a gas minipermeameter.

For measuring the permeability of tight formations, non-steady state (transient) methods are employed. The pressure-decay method was originated by Brace et al. (1968). With this technique a rock core is confined in a permeameter while vessels are attached to either end. The upstream vessel is pressurized and then allowed to equilibrate with the core and the downstream vessel. The pressure decay behavior of the vessels is measured and used to calculate the permeability. Hsieh et al. (1981) extended Brace's analysis to account for the compressive storage effects of the rock sample, while Haskett et al. (1986) used the pressure-decay method to simultaneously determine the porosity and permeability of a porous sample. Jones (1992) modified the gas minipermeameter to operate in a pressure decay mode. The instrument was modified by removing the flow controller and adding reservoirs of different calibrated volumes. Solutions to the pressure decay relations were also enhanced to account for gas-slippage and inertial effects. With these non-steady techniques measurements are possible in less than a minute and can quantify permeabilities at the micro Darcy level.

15.3 BOREHOLE TESTS

A variety of techniques are available for field-scale determination of the permeability of unsaturated and variably saturated media. These techniques are based on both steady state and transient methods, relying on both natural and induced pressure to drive the measurement.

Evaluation of reservoir permeability using gas flow methods dates to the 1930s (Muskat and Botset, 1931; Kirkham, 1946). The traditional single-well pump test modified for the compressibility effects of gas represents a common means of evaluating permeability of unsaturated formations. Matthews and Russell (1967) adapted the Theis solution for one-dimensional radial groundwater flow to analyze airflow data and estimate reservoir permeability. Baehr and Hult (1988) then adapted the Hantush solution for two-dimensional axisymmetric groundwater flow to a partially penetrating well in a confined aquifer to evaluate air flow in the unsaturated zone induced by a vapor extraction well. Later, Baehr and Hult (1991) and Shan et al. (1992) developed steady-state analytical solutions for a partially penetrating well in both a leaky confined reservoir and an unconfined reservoir, while Falta (1995) developed solutions for gas flow through horizontal wells. Analytical and numerical solutions to multiple well tests have also been developed that are analogous to common aquifer tests in the saturated zone. These tests are often conducted in less than a day and require aggressive vacuum and flow rates which may tend to emphasize the effects of preferential flow paths. As such data from these tests may be more characteristic of the preferential flow paths than the overall formation.

In addition to solutions involving flow induced by active vapor extraction techniques, Stallman (1967), Weeks (1978), Massman (1989), Shan (1995), and Rossabi and Falta (2002) have developed analytical solutions for both vertical and lateral permeability based on flow through wells caused by atmospheric pressure fluctuations. These tests generally require long term data logging of pressure and flow. Although still susceptible to measuring the subsurface characteristics of preferential flow paths they are less likely to increase permeability by drying or other processes. Passive tests are inexpensive to conduct using simple pressure transducers and electronic data loggers and the data are comparable to more conventional vadose zone pumping tests (Rossabi and Falta, 2002). The information from these tests are generally specific to the subsurface materials immediately adjacent to the individual test wells and may or may not be accurately extended to the materials between test wells.

From the single well tests it is very difficult to uniquely quantify the heterogeneous structure of the permeability field. Multi-well cross hole tests provide a solution to this limitation. In these tests a series of boreholes are installed (orientation can vary) sufficiently close together to assure communication between the wells. Using multilevel packers, each of the boreholes is then segmented into a series of isolated intervals. A pressure pulse (steady or decay) is then induced in one of the segments while monitoring the response in all other segments. This process is repeated by sequentially moving the source to each of the other individual segments. The three-dimensional heterogeneous permeability field is then estimated through simultaneous inversion of the pressure response data (Huang et al., 1999). This approach has been termed pneumatic tomography (Vesselinov et al., 2001).

REFERENCES

Baehr, A.L., and M.F. Hult, Determination of the air-phase permeability tensor of an unsaturated zone at the Bemidji, Minnesota Research Site, U.S. Geological Survey Program on Toxic Waste, Proceedings of the Fourth Technical Meeting, Phoenix, Arizona, September 25–29, *Water Resour. Invest. Rep.*, 88–4220, 55–62, 1988.

Baehr, A.L., and M.F. Hult, Evaluation of unsaturated zone air permeability through pneumatic tests, *Water Resour. Res.*, 27(10), 2605–2617, 1991.

Bear, J., Dynamics of Fluids in Porous Media, Dover Publications, Inc., New York, 1972.

Brace, W.F., J.B. Walsh, and W.T. Frangos, Permeability of granite under high pressure, *J. Geophysical Res.*, 73, 2225–2236, 1968.

Chandler, M.A., G. Kocurek, D.J. Goggin, and L.W. Lake, Effects of stratigraphic heterogeneity on permeability in eolian sandstone sequence, Page Sandstone, Northern Arizona,*AAPG Bull.*, 73(5), 658–668, 1989.

Davis, J.M., R.C. Lohmann, F.M. Phillips, J.L. Wilson, and D.W. Love, Architecture of the Sierra Ladrones Formation, Central New Mexico: Depositional controls on the permeability correlation structure, *Geol. Soc. Am. Bull.*, 105(8), 998–1007, 1993.

Davis, J.M., J.L. Wilson, and F.M. Phillips, A portable air-minipermeameter for rapid in-situ field measurements, *Ground Water*, 32(2), 258–266, 1994.

Deutsch, C.V. and A.G. Journel, A.G, *GSLIB: Geostatistical Software Library and User's Guide*, Oxford University Press, New York, 1997.

Dreyer, T., A. Scheie, and O. Walderhaug, Minipermeameter-based study of permeability trends in channel sand bodies, *AAPG Bull.*, 74(4), 359–374, 1990.

Dykstra, H., and R.L. Parsons, The prediction of oil recovery by water flood, in *Secondary Recovery of Oil in the United States*, 2nd ed., American Petroleum Institute, New York, pp. 160–174, 1950.

Eijpe, R., and K.J. Weber, Mini-permeameters for consolidated rock and unconsolidated sand, *AAPG Bull.*, 55(2), 307–309, 1971.

Falta, R. W. Analytical solutions for gas flow due to gas injection and extraction from horizontal wells, *Ground Water*, 33(2), 235, 1995.

Fuller, C.M., and J.M. Sharp, Jr., Permeability and fracture patterns in extrusive volcanic rocks: Implication from the Welded Santana Tuff, Trans-Pecos Texas," *Geol. Soc. Am. Bull.*, 104, 1485–1496, 1992.

Giordano, R.M., S.J. Salter, and K.K. Mohanty, The effects of permeability variations on flow in porous media, *SPE 14356*, paper presented at the 60th Annual Technical Conference, Soc. of Pet. Eng., Las Vegas, NV, Sept. 22–25, 1985.

Goggin, D. J., R. L. Thrasher, and L. W. Lake, A theoretical and experimental analysis of minipermeameter response including gas slippage and high velocity flow effects, *In Situ*, 12(1–2), 79–116, 1988.

Haskett, S.E., G.M. Narahara, S.A. Holditch, A method for the simultaneous determination of permeability and porosity in low-permeability cores, SPE 15379 presented at the 1986 SPE Annual Technical Conference and Exhibition, New Orleans, Oct. 5–8, 1986.

Hsieh, P.A., J.V. Tracy, C.E. Neuzil, J.D. Bredehoeft, S.E. Silliman, A transient laboratory method for determining the hydraulic properties of 'tight' rocks-I. Theory, *Int. J. Rock Mech. Min. Sci. and Geomech.*, 18, 245–252, 1981.

Huang, K., Y.W. Tsang, and G.S. Bodvarsson, Simultaneous inversion of air-injection tests in fractured unsaturated tuff at Yucca Mountain, *Water Resour. Res.,* 35(8), 2375–2386, 1999.

Jones, S.C., The profile permeameter – A new, fast, accurate minipermeameter, *SPE 24757*, paper presented at the 67th Annual Technical Conference, Soc. of Pet. Eng., Washington, DC, Oct. 4–7, 1992.

Kirkham, K., Field methods for determination of air permeability of soil in its undisturbed state, *Soil Sci. Soc. Am. Proc.,* 11, 93–99, 1946.

Kittridge, M.G., L.W. Lake, F.J. Lucia, and G.E. Fogg, Outcrop/subsurface comparisons of heterogeneity in the San Andres Formation, *SPE Form. Eval.*, 5(3), 233–240, 1990.

Klinkenberg, L.J., The permeability of porous media to liquids and gases, in *Drilling and Production Practices,* pp. 200–213, American Petroleum Institute, New York, 1941.

Massman, J.W., Applying groundwater flow models in vapor extraction system design, *Journal of Environmental Engineering*, 115(1), 129, 1989.

Matthews, C.S., and D.G. Russell, *Pressure Buildup and Flow Tests in Wells, Soc. Pet. Eng. AIME Monogr.,* vol. 1, 167 pp. Am. Inst. of Min. and Eng., New York, 1967.

Muskat, M., and H.G. Botset, Flow of gas through porous material, *Physics*, 1, 27–47, 1931.

Rossabi, J. and R. W. Falta, Analytical Solution For Subsurface Gas Flow To A Well Induced By Surface Pressure Fluctuations, *Ground Water*, 40(1), 67, 2002.

Shan, C., R. W. Falta, and I. Javandel, Analytical solutions for steady state gas flow to a soil vapor extraction well, *Water Resour. Res.,* 28(4), 1105, 1992.

Shan, C., Analytical solutions for determining vertical air permeability in unsaturated soils, *Water Resour. Res.,* 31(9), 2193, 1995.

Stallman, R.W., Flow in the zone of aeration, in *Advance in Hydroscience*, edited by V.T. Chow, vol. 4, pp. 151–195, Academic, San Diego, CA, 1967.

Tidwell, V.C., and J.L. Wilson, Visual attributes of a rock and their relationship to permeability: A comparison of digital image and minipermeameter data, *Water Resour. Res.*, 38(11),1261–1274, 2002.

Vesselinov, V.V., S.P. Neuman, and W.A. Illman, Three-dimensional numerical inversion of pneumatic cross-hole tests in unsaturated fractured tuff, 1. Methodology and borehole effects, *Water Resour. Res.,* 37(12), 3001–3017, 2001.

Weeks, E.P., Field Determination of Vertical Permeability to Air in the Unsaturated Zone, *U.S. Geol. Surv. Prof. Pap. 1051*, 41, pp. 1978.

CHAPTER 16

ANALYZING BAROMETRIC PUMPING TO CHARACTERIZE SUBSURFACE PERMEABILITY

JOSEPH ROSSABI

Redox Tech, LLC, 1006A Morrisville Parkway, Morrisville, NC 27560, USA, rossabi@redox-tech.com

16.1 INTRODUCTION

The prevalence and persistence of subsurface contamination by volatile organic compounds (VOCs) and the shortage of funds to remediate this contamination below regulatory action levels has led to a determined search for low cost clean up strategies. As an alternative to long term active soil vapor extraction (ASVE), passive soil vapor extraction and injection using atmospheric pressure changes as a driving force is being used at an increasing number of sites. Employing barometric pumping, either for passive injection of nutrients to enhance bioremediation or passive extraction of contaminated soil gas, is an easily verified low cost strategy for remediation. For mass-transfer limited, or remote sites, and also as an interim measure while more aggressive strategies are being selected, barometric pumping may be an appropriate cleanup choice. To verify the viability of these passive techniques, subsurface flow characteristics are vital yet an inexpensive characterization method for acquiring these parameters must be employed to match the low cost of the passive remediation systems.

Figure 16.1 illustrates the scenario for barometric pumping through a vadose zone well. In this figure, the well is screened below a substantial clayey zone. The atmospheric pressure signal will penetrate the subsurface but is delayed and damped by this low permeability material and the other materials between the screen and the ground surface. As a result, there will be a pressure difference between the subsurface at the screen depth (P_z) and the surface (P_{atm}), hence, flow through the well. Primarily the effective permeability of the sandy materials in the screen stratum combined with the magnitude of the pressure difference will control the magnitude of the flow. As the VOCs in the vicinity of the screened zone slowly diffuse out of the fine grain materials into the sandy zone, they are removed during regular periods of flow out of the well. As a result, the magnitude of VOC concentration in the gas phase changes

C. Ho and S. Webb (eds.), Gas Transport in Porous Media, 279–290.
© 2006 *Springer.*

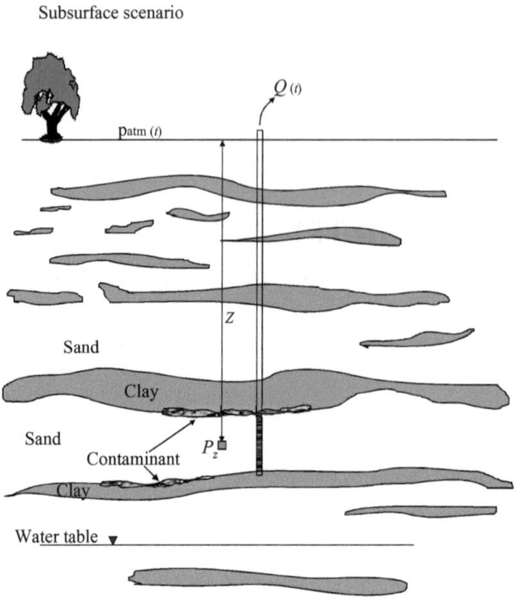

Figure 16.1. Scenario for analytical solution to barometrically induced flow through a well

slowly. If the VOC concentration in the gas is known and the flow out of the well can be predicted, the mass removal of the system can be calculated. Similarly if the nutrient uptake rate (e.g., oxygen utilization rate) is known and the flow into the subsurface can be predicted, the mass of contaminant consumed by bioremediation can be calculated.

Many researchers have devised analytical solutions to the equations describing pressure and flow in porous media with certain boundary and initial conditions. From Theis (1935) to Jacob (1940) to Cooper and Jacob (1946) and to Hantush (1964), analytical solutions to groundwater flow are still generally the first methods used to describe pumping test data. Although numerical methods can more easily solve the equations that describe particular sets of subsurface conditions, heterogeneity and lack of data often limit development of a precise and unique set of descriptive equations on which to apply these methods. In these situations an analytical solution can provide a comparable degree of accuracy and the computational advantages of exact solutions in comparison with iterative methods. When predicting the operation of inexpensive remediation systems, the value of simple analytical solutions is particularly poignant. Analytical solutions to subsurface pressure as a result of atmospheric pressure changes have been developed and successfully applied to field data by several researchers (Weeks, 1978; Rossabi and Riha, 1994; Shan, 1995). These solutions (often adapted from solutions derived for heat conduction) can predict the magnitude and direction of the pressure differential between the surface and subsurface, and have been used for estimating the effective vertical permeability (k_z) of the intervening materials.

The pressure solutions are necessary complements to the analytical flow solution described below.

16.2 PREDICTING SUBSURFACE PRESSURE AND FLOW

When the surface and subsurface are connected by a well, the direction and magnitude of the pressure difference between the surface and subsurface determine whether surface air will flow into the formation through the well or soil gas will flow out of the formation. These factors are important when using barometric pumping forces for removal of volatile contaminants from the subsurface, for injection of oxygen and nutrients to enhance bioremediation, and for understanding and interpreting soil gas measurements from wells. The equation for gas flow in the subsurface (Massman et al., 1989), is given by:

$$\frac{\varphi S_g \mu_g}{P_{avg}} \frac{\partial}{\partial t} P = \nabla \cdot (\tilde{k} \nabla P) \tag{16.1}$$

where φ is the porosity [unitless], S_g is the volumetric gas phase saturation [unitless], μ_g is the viscosity of gas [kg/m sec], P is the pressure [kg/(m sec^2)], and k is the intrinsic permeability tensor including relative permeability effects [m^2]. In this equation the $\tilde{k}P$ term is approximated by $\tilde{k}P_{avg}$, justified by assuming that barometric pressure fluctuations provide a small variation (< 2% in most cases) in gas pressure with respect to the time-average pressure (P_{avg}) of the system. This assumption is equivalent to assuming that the gas density does not vary in space (i.e., it is an incompressible fluid). This equation also neglects the gravity term in Darcy's law.

Although solutions for the subsurface pressure in response to atmospheric pressure are well known, to date few analytical solutions for gas flow through a well in response to surface atmospheric pressure are available in the literature (Rossabi and Falta, 2002; Neeper, 2003). Since the quantitative prediction of flow is one of the most important parameters (along with concentration) in engineered environmental systems, an analytical solution for flow driven by barometric pumping is valuable. The analytical solution can be simply derived with some assumptions obtained from the observation of the behavior of a barometric pumping system.

Beginning with Eq. (16.1), we assume that the vadose zone well is fully penetrating in a relatively thin, horizontally-oriented zone of high gas permeability, and that the zone is radially symmetric with a single value for permeability in the screen stratum (Figure 16.1). Given these conditions, the partial differential equation is posed in a cylindrical coordinate system with symmetry about the azimuth. Since we expect to have mainly radial flow near the well, the local vertical pressure gradients may be neglected in the partial differential equation (PDE) for flow to the well, and this problem becomes one-dimensional. Letting:

$$\lambda_r = \frac{k_r P_{avg}}{\varphi S_g \mu_g} \tag{16.2}$$

the equation for transient radial flow now becomes:

$$\frac{1}{\lambda_r}\frac{\partial}{\partial t}P = \frac{\partial^2 P}{\partial r^2} + \frac{1}{r}\frac{\partial P}{\partial r} \tag{16.3}$$

The pressure difference causing flow in the well is solely a result of the damped and delayed transmission of the surface atmospheric pressure signal (a function of time) to the zone of interest in the subsurface at depth z. The difference between this pressure at depth z, $P_z(t)$ and the surface pressure, $P_{\text{atm}}(t)$, generates a flow when the surface and depth z are connected (e.g., by a vadose zone well).

At this point, it is convenient to define a new pressure variable in which the barometric fluctuations are subtracted out:

$$S(r,t) = P(r,t) - P_z(t) \tag{16.4}$$

where S is the pressure drawdown or build-up relative to ambient pressure at depth. Considering a step change in S at $t > 0$, this problem is equivalent to that of a flowing groundwater well in a nonleaky aquifer. The solution for that problem is given by Jacob and Lohman (1952), and Hantush (1964) [see also Carslaw and Jaeger (1959) for the equivalent heat conduction solution in a region bounded internally by a circular cylinder as with heat flow from buried pipes or cables] as:

$$S = S_{\text{wb}}A(\tau,\varepsilon) \tag{16.5}$$

$A(\tau,r)$ is called the flowing well function for non-leaky aquifers and is defined by:

$$A(\tau,\varepsilon) = 1 - \frac{2}{\pi}\int\limits_{u=0}^{\infty} \mathrm{e}^{-\tau u^2}\frac{J_0(u)Y_0(\varepsilon u) - Y_0(u)J_0(\varepsilon u)}{J_0^2(u) + Y_0^2(u)}\frac{\mathrm{d}u}{u} \tag{16.6}$$

For the gas flow problem,

$$\tau = \frac{k_r P_{\text{avg}} t}{\varphi S_g \mu_g {r_w}^2} \qquad \text{and } \varepsilon = \frac{r}{r_w} \tag{16.7}$$

To solve for the flow rate at the well, Darcy's law is used in the form:

$$Q = -2\pi b \frac{k_r}{\mu_g}\frac{\partial S}{\partial(\varepsilon)} \quad \text{at} \quad \varepsilon = 1 \tag{16.8}$$

where b is the thickness of the screened interval or zone thickness. The derivative, $\partial S/\partial\varepsilon$ is given by Jacob and Lohman (1952), Hantush (1964) and Carslaw and Jaeger (1959) as:

$$\left.\frac{\partial S}{\partial \varepsilon}\right|_{\varepsilon=1} = \frac{-4S_{\text{wb}}}{\pi^2}\int\limits_{u=0}^{\infty} \mathrm{e}^{-\tau u^2}\frac{1}{J_0^2(u) + Y_0^2(u)}\frac{\mathrm{d}u}{u} \tag{16.9}$$

Combining Eqs. (16.8) and (16.9), the equation for flow is defined as:

$$Q = 2\pi b \frac{k_\mathrm{r}}{\mu_\mathrm{g}} S_\mathrm{wb} G(\tau) \tag{16.10}$$

where $G(\tau)$ is known as the flowing well discharge function for nonleaky aquifers (Jacob and Lohman, 1952; Hantush, 1964). This function is defined as:

$$G(\tau) = \frac{4}{\pi^2} \int\limits_{u=0}^{\infty} \mathrm{e}^{-\tau u^2} \frac{1}{J_0^2(u) + Y_0^2(u)} \frac{\mathrm{d}u}{u} \tag{16.11}$$

From Jacob and Lohman (1952), Carslaw and Jaeger (1959), and Hantush (1964), it is known that for large τ, $G(\tau)$ can be approximated by a simpler representation:

$$\text{For} \quad \tau > 500, \;\; G(\tau) \approx \frac{2}{\ln(2.25\tau)}$$

Considering typical vadose zone properties ($k_\mathrm{r} = 10$ E-12 m^2, $\varphi = 0.4$, $S_\mathrm{g} = 0.10$, $r_\mathrm{w} = 0.0254$ m), τ is greater than 500 for times larger than about 0.25 second. Therefore, the approximate solution for calculating flow from a well as a result of a step change in well gas pressure is:

$$Q(t) = 2\pi b \frac{k_\mathrm{r}}{\mu_\mathrm{g}} S_\mathrm{wb} \frac{2}{\ln(2.25\tau)} \tag{16.12}$$

For a succession of step changes in flow and pressure, Eq. (16.12) may be replaced with

$$Q_n(t) = 2\pi b \frac{k_\mathrm{r}}{\mu_\mathrm{g}} S_\mathrm{wb}(k) \frac{2}{\ln\left(2.25\tau\left(1 - \frac{t_k}{t}\right)\right)} \tag{16.13}$$

where n and k are flow and pressure time steps respectively and $S_\mathrm{wb}(k) = P_\mathrm{atm}(t_k) - P_z(t_k)$. This solution can be used directly with surface/subsurface differential pressure data to predict flow from a well screened in a flow zone. The code for a computer model using this solution has been developed and requires surface/subsurface differential pressure data, the thickness of the flow zone, well radius, radial permeability, volumetric moisture content, porosity, and the yearly average pressure as input. Like the subsurface pressure solution, the flow solution is calibrated using field data to calculate the best fit for a single permeability value. In this case, the value is k_r, the radial permeability of the formation. The flow solution (Eq. (16.13)) can also be coupled with an analytical subsurface pressure solution. Using these two analytical solutions, flow from a vadose zone well can be predicted using only surface pressure data.

In many cases, all of the model parameters will not be available so a sensitivity analysis was performed to determine the most important variables. Table 16.1 and Table 16.2 show, respectively, the input values used in the model and the results of this sensitivity analysis. From Table 16.2 it is clear that differential pressure and radial

Table 16.1. Description of input variables and base case values for these parameters

Base case values	Definitions	Units
$k_r = 1.00E\text{-}12$	Radial permeability	m^2
$\mu_g = 1.83E\text{-}5$	Viscosity	kg/m sec
$\varphi = 0.35$	Porosity	unitless
$S_g = 0.25$	Gas phase saturation	unitless
$P_{avg} = 100000$	Average pressure	Pa
$r_w = 0.0254$	Well radius	m
$b = 10$	Zone thickness	m
$S_{wb} = P_{diff} = 100$	Pressure differential	Pa

Table 16.2. Sensitivity analysis of factors affecting flow through well

Radial permeability			Sensitivity analysis Pressure differential			Zone thickness		
$k_r(m^2)$	Q(cfm)	Variance(%)	P_{diff}(Pa)	Q(cfm)	Var. (%)	b(m)	Q(cfm)	Var. (%)
1E-13	0.015	88%	50	0.059	50%	1	0.012	90%
5E-13	0.063	47%	100	0.118	0%	2	0.024	80%
1E-12	0.118	0%	200	0.237	100%	5	0.059	50%
5E-12	0.523	342%	300	0.355	200%	10	0.118	0%
1E-11	0.997	742%	400	0.473	300%	20	0.237	100%
5E-11	4.489	3693%	500	0.592	400%	30	0.355	200%
1E-10	8.610	7174%	600	0.710	500%			
			700	0.829	600%			

Porosity			Gas phase saturation			Average pressure			Well radius		
φ	Q(cfm)	Var.(%)	S_g	Q(cfm)	Var.(%)	P_{avg}(kPa)	Q(cfm)	Var.(%)	r_w(cm)	Q(cfm)	Var.(%)
0.25	0.115	3%	0.1	0.110	7%	98.5	0.119	0.12%	1.25	0.106	10%
0.30	0.117	1%	0.2	0.116	2%	99.0	0.118	0.08%	2.54	0.118	0%
0.35	0.118	0%	0.3	0.120	2%	99.5	0.118	0.04%	3.79	0.127	7%
0.40	0.120	1%	0.4	0.123	4%	100.0	0.118	0%	5.08	0.133	13%
0.45	0.121	2%	0.5	0.125	6%	100.5	0.118	0.04%			
0.50	0.122	3%	0.6	0.127	8%	101.0	0.118	0.08%			
0.55	0.123	4%	0.7	0.129	9%	101.5	0.118	0.12%			
			0.8	0.131	10%	102.0	0.118	0.16%			

permeability are the most important factors in determining the flow. If differential pressure is known (either by field measurements or by using one of the analytical solutions to pressure and surface pressure data), and permeability is known, and as long as estimates of the other parameters are reasonable, the flow through the well can be predicted accurately. It is important to remember that in this derivation, the radial permeability is assumed to have a single value and is independent of changes that might result from the presence of multiple fluid phases. This assumption is generally valid at depths greater than 3 meters where saturation values are not dramatically affected by infiltration events and the assumption of a constant value for S_g is not unreasonable. For shallower depths, however, variable gas saturation values can have a significant impact on flow predictions.

16.3 FIELD RESULTS AND MODEL PREDICTIONS

Figure 16.2 shows a comparison between field measured and model predicted subsurface pressure using an analytical solution for subsurface pressure as a result of atmospheric pressure changes (Rossabi and Falta, 2002). The input parameters are listed in Table 16.3.

The subsurface pressure match is generally good although it does not precisely track the amplitude extents of some of the pressure changes seen in the field data, probably as a result of systematic losses in the field monitoring system.

Figure 16.3 shows a comparison between field measured and model predicted flow using the differential pressure data predicted by the analytical solution to pressure (shown in Figure 16.2) and Eq. (16.13). The input parameters for this model are given in Table 16.4. It is clear from Figure 16.3 that flow out of a vadose zone well can be reasonably predicted using only surface pressure data. Similar to the

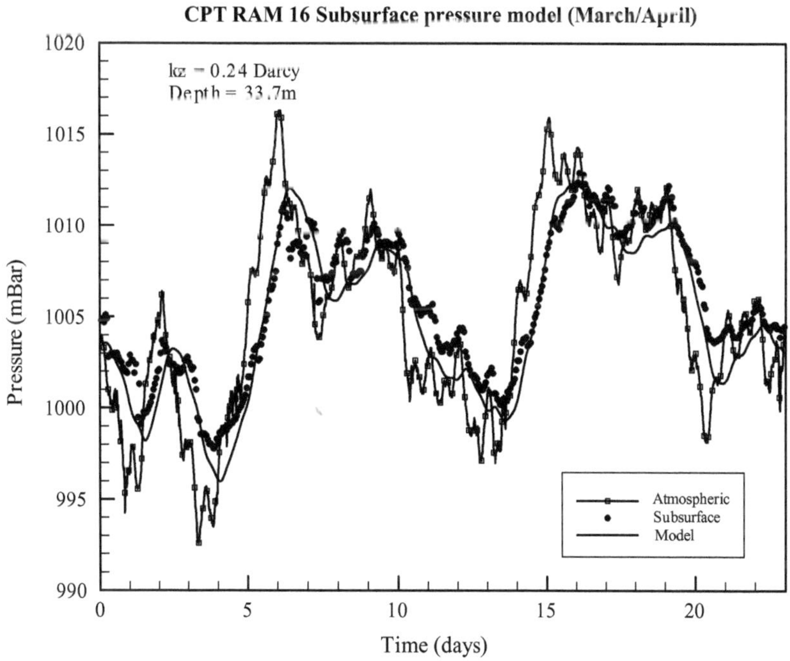

Figure 16.2. Comparison of field and computer modeled pressure data

Table 16.3. List of input parameters for model Barosolv

Porosity = 0.4	Well radius = 1.27 cm
Gas phase saturation (S_g) = 0.25	Depth of well = 32.2 m
	Depth of water table = 39.6 m
Average pressure = 100.4 kPa	Calibrated vertical permeability (k_z) = 2.4E-13m^2
Differential pressure data continuously logged at 10 minute intervals from 3/18/94 to 4/25/94.	

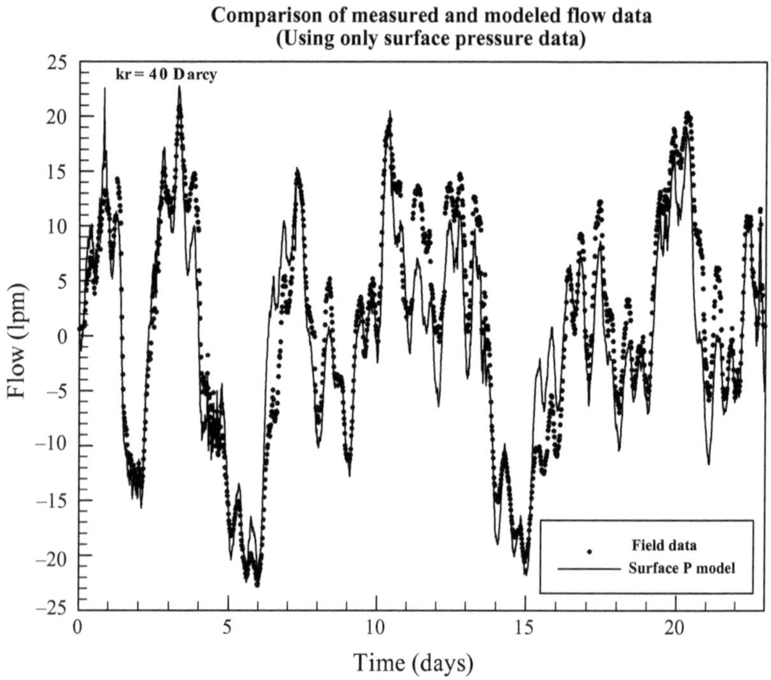

Figure 16.3. Comparison of field measured and computer modeled flow data using measured surface pressure as model input

pressure prediction, the flow prediction is accurate but misses some of the more dynamic flow behavior. Eq. (16.13) was then used with the measured differential pressure data to compare the amount of error introduced by model predicted pressure differential. Figure 16.4 shows a comparison between field measured and model predicted subsurface flow using Eq. (16.13) with measured differential pressure as the input data. The other input parameters were identical to those used for the model in Figure 16.3 and are listed in Table 16.4.

The sum of squared errors (SSE) between the field measured flows and the flows predicted by modeling measured differential pressure and predicted differential pressure (based on surface pressure) were compared. The SSE from the measured

Table 16.4. List of input parameters for model Baroflow

Porosity = 0.4	Well radius = 1.27 cm
Gas phase saturation (S_g) = 0.9	Permeable zone thickness = 30 cm (screen length)
Average pressure = 100.4 kPa	Calibrated radial permeability (k_r) = 40 E-12 m^2
Differential pressure data continuously logged at 10 minute intervals from 3/18/94 to 4/25/94.	

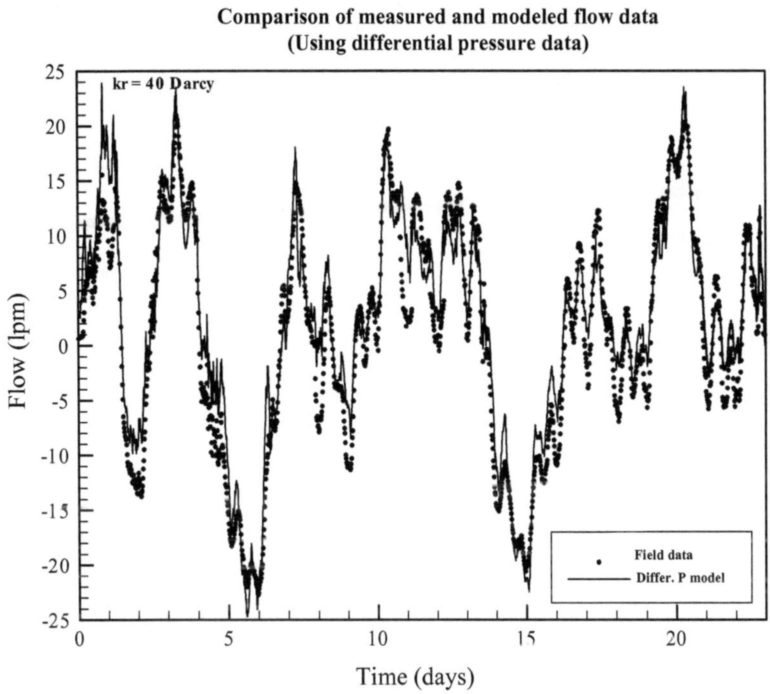

Figure 16.4. Comparison of field measured and computer modeled flow data using measured differential pressure as model input

differential pressure was approximately 30% smaller than the SSE from the surface pressure data set.

Figure 16.5 shows the sensitivity of the model to different values of permeability. As discussed in the sensitivity analysis earlier, the flow prediction is most sensitive to this parameter. Figure 16.5 used the measured differential pressure data as input as well as the same input parameters listed in Table 16.4, varying only the radial permeability. It is clear from this figure that the model is quite sensitive to different permeability values.

These figures show that simple analytical solutions and easily measured data can be used to estimate radial gas permeability values and therefore flow through a well. The passive pumping test may be the most straightforward and least expensive way to

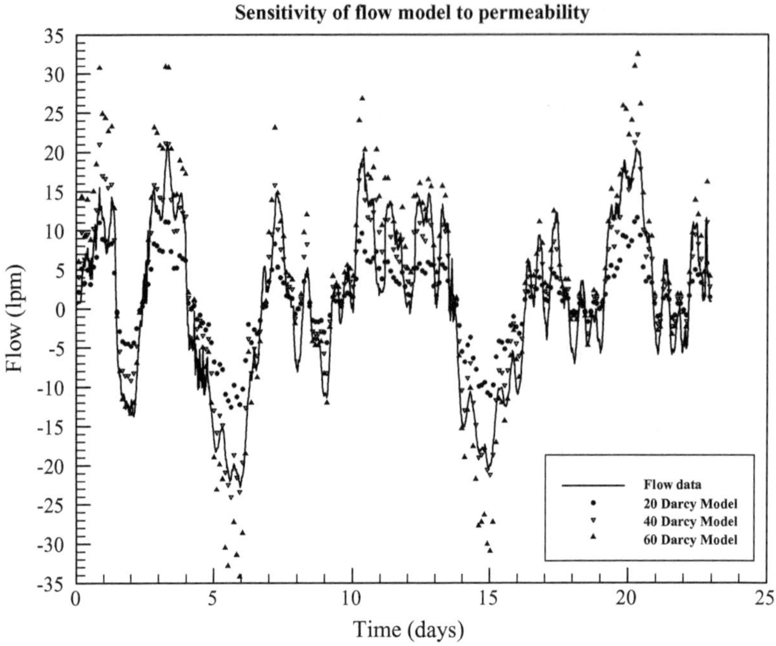

Figure 16.5. Model response to different permeability values

obtain these important data in some situations. For the model shown in Figure 16.3, a value of radial permeability was selected to obtain the best fit to the data, and this value is consistent with other values for permeability in similar zones measured during vadose zone pumping tests conducted in the area (Looney et al., 1991). For practical uses of this model, field flow data should be collected either by a pumping test or by measuring barometrically induced flow for a few weeks to accurately determine the radial permeability. Once the permeability is determined, the system is calibrated for flow predictions. The process of fitting the model to the barometric flow data is an innovative method for determining the radial permeability of the formation around the well screen.

16.4 CONCLUSIONS

Using simple analytical solutions to the governing equations for pressure and flow in the subsurface, subsurface flow through a vadose zone well can be accurately modeled based on measured atmospheric pressure and subsurface properties. With easily collected data, the analytical flow model can determine the radial permeability near a well screen, one of the most important subsurface parameters for developing cleanup strategies. Logging subsurface pressure or flow for relatively short durations (one to two weeks) can accomplish calibration of the subsurface flow model. The calibration process essentially determines values for the radial permeability (k_r) in the flow model case. Once calibrated the model needs only time series differential pressure data to provide accurate predictions of induced flow through a well. If used in conjunction with an analytical solution for the subsurface pressure, only time series atmospheric pressure data are required for flow prediction. The flow model is most sensitive to differential pressure values, intrinsic permeability, and to a lesser extent zone thickness. It is relatively insensitive to values of porosity, average pressure, and well radius. By simple measurements of surface pressure, both the pressure and flow models can be used to predict the performance of a well under the influence of barometric pumping and to estimate the expected range of flow rates and variability to simplify and improve monitoring plans.

REFERENCES

Carslaw, H. S., and J. C. Jaeger, *Conduction of Heat in Solids*, Clarendon Press, Oxford, 1959.

Cooper, H. H., Jr., and C. E. Jacob, A generalized graphical method for evaluating formation constants and summarizing well field history, *Trans. Amer. Geophys. Union*, 27, 526–534, 1946.

Hantush, M. S., Hydraulics of wells, *Adv. Hydrosci.*, 1, 281–432, 1964.

Jacob. C. E., On the flow of water in an elastic artesian aquifer, *Trans. Amer. Geophys. Union*, 2, 574–586, 1940.

Jacob. C. E. and S. W. Lohman, Nonsteady flow to a well of constant drawdown in an extensive aquifer, *Trans. Amer. Geophys. Union*, 33 (4), 559–569, 1952.

Looney, B. B., J. B. Pickett, and J. J. Malot, Pilot Test of a Vacuum Extraction System for Environmental Remediation of Chlorinated Solvents at the Savannah River Site (U), *WSRC RD-91-19*, Westinghouse Savannah River Company, Aiken, SC 29808, 1991.

Massman, J. W., Applying groundwater flow models in vapor extraction system design, *Journal of Environmental Engineering*, 115 (1), 129, 1989.

Neeper, D. A., Harmonic analysis of flow in open boreholes due to barometric pressure cycles, *Journal of Contaminant Hydrology*, 60, 135, 2003.

Rossabi, J. and B.D. Riha, Passive control of VOCs using valved well heads: FY 1994 report (U), *WSRC-TR-94-0524*, Westinghouse Savannah River Company, Aiken, SC 29808, 1994.

Rossabi, J. and R. W. Falta, Analytical solution for subsurface gas flow to a well induced by surface pressure fluctuations, *Ground Water*, 40(1), 67, 2002.

Shan, C., Analytical solutions for determining vertical air permeability in unsaturated soils, *Water Resour. Res.*, 31(9), 2193, 1995.

Theis, C. V., The relation between the lowering of the piezometric surface and the rate and duration of discharge of a well using groundwater storage, *Trans. Amer. Geophys. Union*, 2, 519–524,1935.

Weeks, E. P., Field determination of vertical permeability to air in the unsaturated zone, *US Geological Survey Professional Paper 1051*, 1978.

CHAPTER 17

SUBSURFACE FLOW MEASUREMENTS

JOSEPH ROSSABI
Redox Tech, LLC. 1006A Morrisville Parkway, Morrisville, NC 27560, USA

17.1 INTRODUCTION

Gas movement in the subsurface is predominately driven by diffusion, but there are many important situations during which diffusion is not the principal transport mechanism. Soil vapor extraction or injection systems designed to remove volatile organic contaminants or provide nutrients to subsurface organisms generally use powerful mechanical pumps that induce advective gas flow in the subsurface. Natural pumping caused by pressure differences between the surface and subsurface as a result of atmospheric pressure changes, wind and terrain, or even earth tides can also produce flows on the order of, or significantly higher than diffusion. Biological activity, chemical and thermal changes such as those that occur in landfills can also lead to localized pressure differences that cause subsurface advective flow. During these times especially, knowledge of the gas flow is important for an understanding of the movement of chemicals and contaminants.

The measurement of gas flow in the subsurface like measurement of liquid flow is rarely made in situ or over the entire flow field, rather these measurements are made at endpoints usually accessed via wells although emplaced sensors are occasionally employed. At these endpoints, sensors are placed and are used to sample flow, or tracer materials are introduced and collected. From these point measurements and a conceptual model of the structure and parameters of the subsurface materials, flow through the media is inferred. The range of flow in the subsurface can be very large corresponding to the range of permeability found in subsurface media. The degree of accuracy and precision needed for a particular monitoring situation will also vary. As a result, different methods for measuring flow should be selected that are suitable for the flow range of interest and that satisfy the appropriate data quality objectives. For example, passive injection using barometric pumping operating at a nominal flow rate of 10 to 50 liters per minute will likely require a different measurement technique than a multi-well soil vapor extraction unit operating at 15 m^3 per minute.

C. Ho and S. Webb (eds.), Gas Transport in Porous Media, 291–302.
© 2006 *Springer.*

Because the demand for subsurface flow measurements is relatively low compared with the demand for flow measurements in manufacturing processes (e.g., chemical production and mixing), manufactured applications (e.g., aircraft speed indicators), product distribution (e.g., oil and natural gas distribution) and weather prediction, many of the techniques used for measuring subsurface gas flow were adapted from other applications. Several techniques employ pressure sensors and the application of Bernoulli's equation. These tools include Pitot and Prandtl tubes, venturis, nozzles, and orifice plates and have primarily been advanced by aeronautics and manufacturing needs. Other engineering process monitoring, and weather monitoring needs have advanced additional techniques, such as thermal flow sensors, and ultrasonic flow sensors. The range and suitability of these sensors is often designed for a specific application. Since subsurface flow rate can vary over several orders of magnitude, it is often necessary to have several different types of flow measuring techniques available in a characterization tool kit.

Under field conditions gas flow is most often measured through points accessing some portion of the subsurface, i.e., wells. Wells sample the subsurface flow through the screened portion of the tube contacting the flowing zone. The soil gas flow primarily enters the well screen through angles that are perpendicular or oblique to the axis of the well tube and is then conducted to the surface. The loss due to the angle change of the flow is generally ignored in subsurface gas flow applications as is the loss upon entering a slotted or other screened section. Once gas is flowing through the well tube, flow-measuring methods developed from other disciplines and applications can be employed.

17.2 FLOW METERS

Flow of a fluid that is normally indistinguishable from its background medium is generally measured by its effect on physical objects. People have made observation scales for wind velocity based on deflection of a rising smoke column, the behavior of flags or trees, or other common materials. Weather scientists have commonly used vanes or cups mounted on low friction rotational points. These methods must always account for the effect of the resistance to flow inherent in the measurement system (e.g., the frictional resistance of the bearings of the rotational pivot in cup and vane anemometers) and try to minimize the effects of the measurement (physical resistance and flow obstruction) on the flow system. The resistance represents a loss in the system. If the flow's driving force is large, the effect of the measurement will usually be small but the smaller the flow the more critical is the importance of minimizing obstruction and losses. The ideal flow meter is completely noninvasive, and it uses none of the flow's energy when it measures the flow.

Flow measurement devices generally must balance several conflicting forces. The ideal flowmeter should accurately measure the total flow of the entire flowing system (at the desired time resolution) yet not affect the flow field. Generally gas flow does not occur as a uniform planar front. Over a particular volume, the magnitude of the flow velocity can vary significantly. Although the ideal tool can measure the total flow rate through the conductive medium (i.e., pipe or other conveyance) few accurate, total flow rate methods that do not impede the flow have been devised. It is usually necessary to balance the need for accuracy with the need for unimpeded flow, as well as the need for reasonable cost, and the technology practical for the desired flow range.

Sometimes an approximate flow rate measurement is all that is needed for subsurface systems. A simple total flow rate measurement can be made in many gas flow systems using a lightweight plastic (or similar) bag of known volume and a timer. The flimsy bag is compressed then attached to the flow system and filled until the bag begins to present significant backpressure to the flow system. The time to fill the bag to the measured volume is a direct measure of the flow rate of the system. This method has obvious limitations and sources of error (e.g., accurate volume measurement, backpressure significance, leak-free attachment to the flow system, etc.) however the accuracy and precision of the method may be acceptable for specific applications. This method is analogous to the "calibrated bucket" method often used in the field to verify liquid flow meters or as the primary measurement technique. Obviously this method has many disadvantages with respect to accuracy, time resolution, and repeated monitoring applications; however, it is often accurate enough and simple to apply for first order and scoping measurements. More sophisticated total flow measurements generally rely on physical displacement or flow narrowing/changing devices to make their measurements.

If a total flow measurement is not possible, methods to infer the total flow in a system are used. Techniques that measure the flow at very small sampling points in the flow profile are selected to minimize the effect of the measurement on the system. Many point measurements are made in a particular plane in the flow field and these measurements are used to calculate the total flow through the system (Pesarini et al., 2002). Pitot tubes and thermal mass flow measurements are often used for this method.

The behavior and characteristics of the flowing fluid may also determine the best choice for a flowmeter. Under a uniform force, an ideal frictionless fluid (one with viscosity = 0) will move through a conduit as a uniform planar front. The movement of real fluids deviates from the ideal planar front, however, because they are affected by internal friction (viscosity > 0). At lower velocities, fluids move in layers or laminae as they slide past the conveying structure (which is at rest with respect to the moving fluid), and other fluid layers subsequently affected by the frictional forces. Although gases generally have much lower viscosity than liquids they still are affected by these frictional forces and exhibit a non-uniform velocity profile perpendicular to the vector of motion. Poiseuille's law is an expression that describes the total flow rate of a viscous fluid traveling in laminae through a cylindrical conduit that accounts for the non-uniform velocity of the different fluid laminae.

The Reynolds number of a fluid moving in a physical structure provides a description of the type of flow that is occurring in the system. The two main types of flow are laminar, which occurs at lower flow velocities and is identified as a more homogeneous movement of packets of fluid in an ordered procession, and turbulent flow signified by more chaotic patterns where the individual packets of fluid do not maintain their spatial relation. The type of flow, which is determined by the properties of the fluid as well as the properties and size of the containing structure, in turn determines the resistance to flow in the structure. The type of flow occurring in the system can also affect the choice of measurement techniques as some work better in one flow regime or another. The equation for the dimensionless Reynolds number is provided below:

$$R_e = \frac{\rho v d}{\mu} \tag{17.1}$$

where ρ is the density of the fluid (kg/m^3), v is the average velocity of the fluid (m/sec), μ is the viscosity of the fluid (kg/m sec), and d is the diameter of the structure usually a tube of some sort. For porous media, v is considered the specific discharge and d is a representative dimension of the size of the pores (e.g., mean grain diameter). As it is dimensionless, the Reynolds number can be calculated in other consistent sets of units and provide identical results. The type of flow indicated by the Reynolds number is roughly divided by empirical results which indicate that a Reynolds number of less than 2000 results in laminar flow and greater than 3000 is turbulent flow. Between these values is a transition zone, which may support either flow regime. Ground water flow is generally considered to be laminar, however, soil gas flow especially when induced by external pumping may be turbulent.

17.3 PRESSURE MEASURING DEVICES

These devices are based on the seminal work of Daniel Bernoulli in the early part of the eighteenth century, who first developed the expression relating pressure and velocity of a fluid (Guillen, 1995). The original equation developed by Bernoulli is:

$$P + \rho v^2 = \text{constant} \tag{17.2}$$

where P is the absolute pressure measured in the system (Pa), ρ is the density of the fluid (kg/m^3), and v is the velocity of the fluid (m/sec). When derived to include work defined by changes in potential and kinetic energy, Bernoulli's equation takes the more common form:

$$P + \rho g y + \frac{1}{2}\rho v^2 = \text{constant} \tag{17.3}$$

where g is the gravitational constant (m/s^2), and y is the height above some datum (m). From this equation and the continuity equation first observed by Leonardo DaVinci (Guillen, 1995), many practical flow measurement devices can be devised. One of the simplest is the Venturi meter, which uses the difference in pressure measured in the main body of a cylinder containing a flowing fluid and at the end of a gradual constriction in the cylinder to calculate the velocity of the fluid. An orifice plate is similar in principle to the Venturi meter but uses an abrupt change in the pipe

Table 17.1. Pitot tube measurement range

Pressure	Velocity		Flow rate for different pipe diameters							
			2.5 cm(1")		5 cm (2")		10 cm (4")		15 cm (6")	
Pa	m/s	sfpm	lpm	scfm	lpm	scfm	lpm	scfm	lpm	scfm
0.1	0.4	80.2	12.4	0.44	49.6	1.75	198.3	7.00	446.1	15.75
1	1.3	253.7	39.2	1.38	156.7	5.54	626.9	22.14	1410	49.82
5	2.9	567.3	87.6	3.09	350.5	12.38	1401	49.51	3154	111.4
10	4.1	802.3	123.9	4.38	495.6	17.50	1982	70.01	4460	157.5
100	12.9	2537	392.8	13.84	1567	55.35	6269	221.4	14106	498.2
1000	40.8	8023	1239	43.76	4956	175.04	19825	700.1	44608	1575

diameter to create the pressure difference. Although it produces a significant pressure loss, the orifice plate is an inexpensive, common technique used in many process flow measurements (Urner, 1997; Gan and Riffat, 1997; Krassow et al., 1999; Beck and Mazille, 2002).

Another pressure-based flow meter is the Pitot tube. This instrument measures the difference in pressure at a stagnation point created by inserting a very small tube sealed at its distal end with the proximal end pointing to face flow through the larger flow pipe, and ambient atmospheric pressure outside the pipe. A Prandtl tube is similar to the Pitot tube except it references static pressure in the flowing pipe rather than ambient atmospheric pressure outside the pipe. A critical consideration with these devices is the interference induced in the flow field when the sensor is installed. Obviously, the smaller the device, the less the flow is affected. These devices are accurate, and simple to operate but depend on the capabilities of a differential pressure sensor for their resolution. Table 17.1 provides the practical limitations on flow measurement made with these devices. This table was calculated based on the performance of commercially available pressure transducers, Bernoulli's equation, and commonly used pipe diameters. Differential pressure gauges with sensitivity below 1 Pa are significantly more expensive than other transducers. Finally, a change in density of the fluid through compression in the system may affect the accuracy of the measurement.

The pressure measurement method generally needs to be adjusted for losses incurred for compressibility effects as well as friction and curvature of flow lines (vena contracta). These losses are accounted for by the compressiblity factor and the discharge coefficient that measure the impedance to the flow because of the measurement device (Pnueli and Gutfinger, 1992). The losses generally increase with decreasing Reynolds number. For orifice plates, losses increase with decreasing diameter ratio as well. Often the loss coefficients are empirically determined for a particular instrument. A general understanding under ideal conditions (no compression or discharge losses) is often useful for evaluating the selection of a particular flow measurement device. Derived from Bernoulli's equation, the expression describing flow in a Venturi tube is:

$$Q = \frac{\pi d^2}{4}\sqrt{\frac{2\Delta p}{\rho\left[1-(d/D)^4\right]}} \tag{17.4}$$

where Q is the volumetric flow rate (m^3/sec), Δp is the difference in pressure measured in the pipe and the pipe constriction (Pa), ρ is the density of the fluid (kg/m^3), and d and D are the diameters of the constriction and the main pipe respectively (m).

The expression for fluid velocity as measured by a Pitot tube is:

$$v = \sqrt{\frac{2\Delta p}{\rho}} \tag{17.5}$$

where v is the velocity of the fluid (m/sec), and Δp in this case is the difference between absolute and static pressure measured at the stagnation point (Pa). In general, losses increase from Pitot and Prandtl tubes to Venturi meters to nozzles to orifice plates.

17.4 THERMAL FLOW SENSORS

Because of the prevalence of the use of thermal sensors for gas flow, a convenient way to differentiate flow sensors is in terms of thermal and non-thermal mechanisms. The theoretical basis for thermal flow sensors arose from heat conduction work conducted in the early twentieth century. In particular, the work of King (1914) was seminal to the development of in-line and insertion mass flow meters, two of the most popular process flow meters. King developed the expressions for the conduction of heat from radially symmetric surfaces in a constant flow field that are essentially used today.

Thermal anemometers use the heat transfer properties of the fluid they are measuring to calculate a mass flux rate rather than a volumetric flow rate. The size and number of gas molecules that hit the heated element determine the amount of energy removed from the element. The heated sensor element employed can be a wire, a plate or film, a sphere or another shape that is able to sample the flow yet provide minimal impedance or change to the flow. These devices are deployed in the flow path and are usually referred to as insertion mass flow meters. Their small size also permits easy access to flow fields. Some thermal flow meters use a section of the flow conduit as the measurement device (in-line meters) either as the heated sensor itself or in conjunction with embedded heated elements (Viswanathan et al., 2002).

Thermal sensors measure either the amount of energy (usually as electric power) required to maintain a heating element at a constant temperature in the mass flow path, or they measure the temperature of the element for a particular applied energy (electric power) in the mass flow path (Baker and Gimson, 2001). Although the thermal sensors actually measure the mass of fluid interacting with the heated element, they are often calibrated to the velocity of a particular gas compound flowing by the sensor at a particular point. Multiple measurements of the point velocity across the plane of a flow field can then be converted to the flow rate of the gas.

Thermal anemometers have a dynamic response at lower flow rates where small differences in applied energy can be measured with great accuracy and precision. Measurements of small energy changes are more easily made when lower total energy is required to maintain temperature (i.e., the ability to measure small changes in a low background). At higher flow rates these instruments respond with less resolution because small energy changes are made at high total energy required to maintain temperature. This is the ubiquitous problem of the measurement of a small signal in a high background. At high flow rates where a very large amount of energy is required to fend off the onslaught of heat stealing gas molecules, a different technique using a different physical property of the flow phenomenon may be more appropriate.

Although the strength of thermal anemometers are their ability to measure low flow, at very low flow rates the instruments also have limitations related to the measurement technique that significantly affect their accuracy. At very low flow rates the heated sensor can produce convective flow of the gas near the sensor. The magnitude of this effect will vary with the amount of energy applied to the heated element, the size and thermal conductivity of the element, and the location (proximity and direction) of the background thermal sensor with respect to the heated element. Better response can be obtained if the thermal sensor is out of the convective heat path. For many commercial versions of the thermal anemometer, flow velocities near the sensor of 0.05 m/sec or greater are required to minimize the effects of this self-induced flow although thermal sensors capable of measuring 0.005 m/sec are available (Martin et al., 2002).

Another issue affecting the use of thermal anemometers is phase and compositional differences within the bulk fluid whose flow is being measured. If there is a change in the heat carrying properties of the fluid the sensor will respond differently. For example, if the composition of the gas stream changes from mostly air to mostly carbon dioxide, the flow sensor will produce different values for the same flow rate because of the difference in heat capacity and conduction of the different compounds. In addition, if the fluid is hovering between two different phase states (e.g., liquid and gas) with significant phase transition energy, the sudden occurrence of a different phase on the sensor can produce dramatically different results than if the fluid phase is homogeneous. This last effect is common in subsurface systems as the gas phase is nearly always close to water saturation and condensed water is a common occurrence in both natural and anthropogenic flow structures. If a water droplet falls on a heated sensor wire, the amount of energy required to evaporate the droplet and maintain the temperature of the wire is much higher than even the most dramatic flow changes encountered with single phase soil gas. Often the very high energy in a very short time demanded by the sensor's control mechanism can damage the delicate sensor element, limiting the life of the thermal anemometer. To compensate for this effect, more rugged, thicker sensor elements are made but these have slower response time and less sensitivity because they require more energy to heat the larger sensor mass.

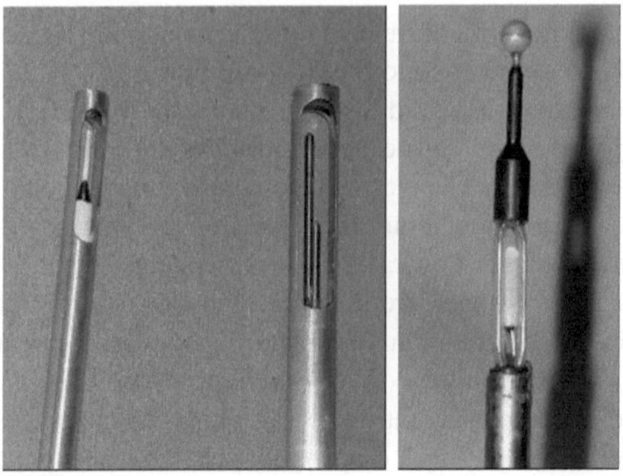

Figure 17.1. Three different types of insertion thermal mass flow meter sensor elements. All elements are less than 1 cm in diameter

Three types of thermal flow sensors are shown in Figure 17.1. There are several manufacturers of thermal anemometers using sensors based on heated wires, heated films, or other elements. Each of these instruments has its appropriate measurement niche.

17.5 OTHER FLOW SENSORS

Relatively inexpensive cup or vane anemometers are excellent tools for wind measurement because they are minimally obtrusive to the flow and use very little energy after they have passed an initial startup threshold. The energy required to overcome the inertial resistance of the bearings is an important loss for these instruments and is critical in low energy flows. The dynamic frictional loss of the bearings is low although these bearings have a limited lifetime and sensitivity to physical effects (impacts, accumulation of debris, warping, etc.). Current limitations on the size of these types of tools also limit their application in many flow regimes, however, the prospect of micromachining very small nearly frictionless bearing systems suggest a promising future for these types of tools in very low flow regimes.

For gas flow through pipe, cup anemometers are not usually used, however, vane anemometers can be adapted based on the size of the pipe and flow rate range of the system. Most vane anemometers have a lower threshold of approximately 0.2 m/sec. The upper range of these sensors is 2 to 5 m/sec. Given a vane diameter of a minimum of approximately 2.5 cm, the flow rate threshold for this tool begins at 7 liters per minute. Vane anemometers are one of the only flow meters that measure the average flow rate through the cross sectional area of the device rather than at a point. They are not susceptible to temperature, pressure or media changes, consume very little power to operate, and can indicate flow direction (Blond and Downing, 2002). Figure 17.2 shows two types of rotary vane anemometers.

Figure 17.2. Two types of rotary vane anemometers

Ultrasonic flow meters use the sound carrying properties of the gas medium to make minimally invasive measurements of the gas flow (Dane, 1997; Brassier et al., 2001; Calogirou et al., 2001). These instruments have only recently been developed and are fast becoming a very popular method for wind speed and direction as well as gas metering applications. Recent developments in electronic instrumentation have enabled these instruments to become reasonably priced and their non-invasiveness is a very important feature. Ultrasonic flow meters incur virtually no flow loss. Ultrasound measurements are made using either a Doppler measurement technique or axial time of flight. The instruments generally require a laminar flow regime to be accurate and are therefore most appropriately applied in lower flow applications.

Diaphragm meters are mechanical devices that use gas pressure as the driving force to alternatively fill and empty volumes within the meter (Nilsson and Delsing, 1998). The gas pressure forces flow into the meter deflecting a diaphragm, which controls a valve that allows flow either into or out of the meter. Each time the chamber is filled and emptied a mechanical register records the event. Generally the meter requires on the order of 2000 Pa gas pressure to operate. The pressure loss in these meters is relatively high as a result of friction and other mechanical causes as well as leakage. Nevertheless, these meters are a cost effective method for measuring gas consumption and could potentially be used for flow monitoring in some situations.

17.6 MICROMACHINED SENSORS

The future holds much promise for the measurement of subsurface gas flow. Actual pore velocities may ultimately be measured in situ using technology that is currently being developed. Nguyen (1997) describes a plethora of micromachined sensors in his review article that are designed for measuring gas flow in the relevant range for vadose zone applications. These sensors are primarily based on integrated circuit technologies and materials using thermal properties and responses of these tools although some cantilever systems using piezoresistive effects are also mentioned. Some problems of these sensors that are particularly important for subsurface applications that still must be solved include long term and drift free operation. The drift problems are not insurmountable, however, and if minimally invasive emplacement and data recovery can be worked out, micro sensors will ultimately be installed for high resolution measurements of subsurface flow behavior.

17.7 CONCLUSIONS

There are several technologies available for measuring subsurface gas flow depending on the desired application. Because of their small size, accuracy and simplicity, Pitot tubes and thermal flow sensors are most often selected for precise flow field measurements. Their inherent physical characteristics can be used to select the appropriate type of sensor for a particular application. For low fluid velocity applications (< 1 m/sec), thermal sensors provide a strong dynamic response while Pitot tubes produce a much weaker response because of the low differential pressure at that flow rate. At high flow rates (> 15 m/sec) and correspondingly high differential pressure, the Pitot tube is superior to a thermal flow sensor which has a lower sensitivity to small changes in flow at those rates (Ismail and Davidian, 1999).

Other flow sensors may also be appropriate for measuring subsurface flow particularly in process-like applications. The user of these instruments must understand her system and measurement objectives well enough to select the right sensor. Additional information can be obtained from several national and international engineering and standards organizations including the International Organization for Standards (ISO) and the American Society for Testing and Materials (ASTM). Figure 17.3 provides a conversion table of flow rate to gas velocity for various size pipes.

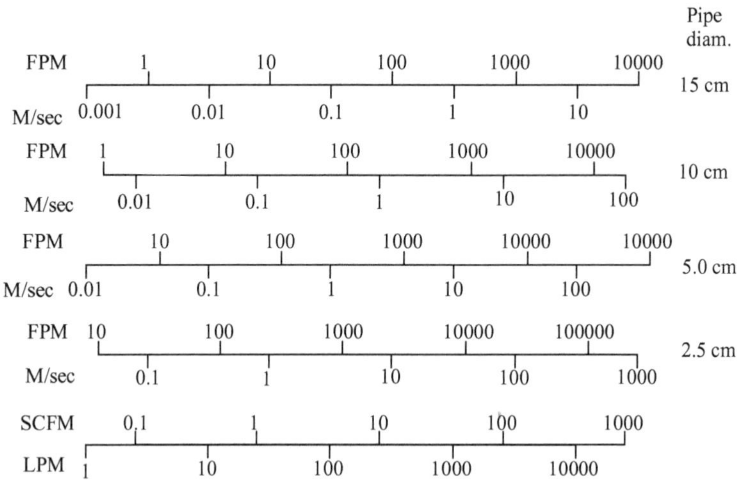

Figure 17.3. Conversion graph for flow rates and pipe velocities

REFERENCES

Airflow Technical Products, Inc., www.airflow.com

Baker, R. C., and Gimson, C. The effects of manufacturing methods on the precision of insertion and in-line thermal mass flowmeters. Flow. Meas. Instrum., 2001; (12):113–121.

Beck, S. B. M., and Mazille, J. A study of a differential flow meter that is insensitive to inlet conditions. Flow. Meas. Instrum., 2002; (12):379–384.

Blond, A. N., and Downing, P. M. Rotating vanes vrs thermal anemometry technology. Technical Topic I003: Airflow Technical Products, 2002.

Brassier, P., Hosten, B., and Vulovic, F. High-frequency transducers and correlation method to enhance ultrasonic gas flow metering. Flow. Meas. Instrum., 2001; (12):201–211.

Calogirou, A., Boekhoven, J., and Henkes, R. A. W. M. Effect of wall roughness changes on ultrasonic gas flowmeters. Flow. Meas. Instrum., 2001; (12):219–229.

Dane, H.J. Ultasonic measurement of unsteady gas flow. Flow. Meas. Instrum., 1997; 8 (3/4):183–190.

Gan, G., and Riffat, S. B. Pressure loss characteristics of orifice and perforated plates. Experimental Thermal and Fluid Science, 1997; (14):160–165.

Guillen, M., *Five equations that changed the world.* New York: MJF Books, 1995.

Ismail, R., and Davidian, S. AVS Technical Document 1001, Cambridge AccuSense, Inc., Shirley, MA., 1999.

King, L.V. On the convection of heat from small cylinders in a stream of fluid. Philos. Trans. R. Soc. A, 1914; 214–373.

Krassow, H., Campabadal, F., and Lora-Tamayo, E. The smart-orifice meter: a mini head meter for volume flow measurement. Flow. Meas. Instrum., 1999; (10):109–115.

Martin, D. P., Grant, J. J., and Ringwood, J. V. Evaluation of a prototype thermal anemometer for use in low airspeed drying measure calculations. Flow. Meas. Instrum., 2002; (12):385–396.

Nguyen, N. T. Micromachined flow sensors - a review. Flow Meas. Instrum., 1997; 8 (1):7–16.

Nilsson, U. R. F., and Delsing, J. In situ detection of inaccurate gas flow meters using a fingerprint technique. Flow Meas. Instrum., 1998; 9:143–152.

Pesarini, A. J., Bigot, R., Nadal Mora, V. J., Di Bernardi, C. A., and Ringegni, P. J. Theoretical analysis related to aspects of the pitot probe methodology in flow rate determination for blowers performance tests. Flow. Meas. Instrum., 2002; (12):373–377.

Pnueli, D., and Gutfinger C., *Fluid Mechanics*. Mass: Cambridge University Press, 1992.

TSI Incorporated, www.tsi.com

Urner, G. Pressure loss of orifice plates according to ISO 5167-1. Flow. Meas. Instrum., 1997; (8) 1:39–41.

Viswanathan, M., Kandaswamy, A., Sreekala, S. K., and Sajna, K. V. Development, modelling and certain investigations on thermal mass flow meters. Flow. Meas. Instrum., 2002; (12):353–360.

CHAPTER 18

MEASUREMENT OF VAPOR CONCENTRATIONS

CLIFFORD K. HO, MICHAEL KELLY, AND MICHAEL T. ITAMURA
Sandia National Laboratories, P.O. Box 5800, Albuquerque, NM 87185, USA

18.1 INTRODUCTION

The measurement of vapor-phase concentrations is important in environmental applications ranging from preliminary site characterization to long-term environmental monitoring and stewardship. Characterizing the extent and potential migration of subsurface volatile organic compounds and other vapor-phase contaminants is needed to address regulatory metrics and to ensure public safety. Current methods of contaminant characterization rely on manual soil and groundwater sampling, which is costly and time consuming. This chapter reviews several technologies that are used to measure vapor-phase concentrations of volatile organic compounds, as well as their potential to be used for continuous in-situ applications.

18.1.1 Background and Scope

Numerous sites containing toxic chemical spills, leaking underground storage tanks, and chemical waste dumps require accurate characterization and long-term monitoring to reduce health risks and ensure public safety. In addition, over two million underground storage tanks containing hazardous (and often volatile) contaminants are being regulated by the United States Environmental Protection Agency (U.S. EPA, 1992), and the tanks require some form of monitoring to detect leaks from the tanks and pipe network. However, current methods are costly and time-intensive, and limitations in sampling and analytical techniques exist. Looney and Falta (2000, Ch. 4) report that the Department of Energy (DOE) Savannah River Site requires manual collection of nearly 40,000 groundwater samples per year, which can cost between $100 to $1,000 per sample for off-site analysis. Wilson et al. (1995, Ch. 36) report that as much as 80% of the costs associated with site characterization and cleanup of a Superfund site can be attributed to laboratory analyses. In addition, the integrity of the analyses can be compromised during sample collection, transport, and storage.

C. Ho and S. Webb (eds.), Gas Transport in Porous Media, 303–318.
© 2006 *Springer.*

Clearly, a need exists for accurate, inexpensive, real-time, in-situ analyses using robust sensors that can be remotely operated.

Although a number of chemical sensors are commercially available for field measurements of chemical species (e.g., portable gas chromatographs, surface-wave acoustic sensors, optical instruments, etc.), few have been adapted for use in geologic environments for long-term monitoring or remediation applications.

The particular focus of this review is limited to the detection and monitoring of volatile organic compounds (VOCs). These include compounds such as aromatic hydrocarbons (e.g., benzene, toluene, xylenes), halogenated hydrocarbons (e.g., trichloroethylene (TCE), carbon tetrachloride), and aliphatic hydrocarbons (e.g., hexane, octane). As a result, sensors and technologies that detect gas-phase constituents in the vadose zone are emphasized because VOCs are most conveniently and economically monitored in the gas phase (Looney and Falta, 2000, Ch. 4).

18.1.2 Categories of Sensors

The sensors reviewed in this chapter have been categorized into four general groups: (1) chromatography and spectrometry; (2) electrochemical sensors; (3) mass sensors; and (4) optical sensors. The categorization of these sensors is based primarily on the principal physics and operating mechanisms of the sensor. For example, chromatography relies on separation of complex mixtures by percolation through a selectively adsorbing medium, with subsequent detection of compounds of interest. Electrochemical sensors, for the purposes of this chapter, include sensors that detect signal changes (e.g., resistance) caused by an electrical current being passed through electrodes that interact with chemicals. Mass sensors rely on disturbances and changes to the mass of the surface of the sensor during interaction with chemicals. Optical sensors detect changes in visible light or other electromagnetic waves during interactions with chemicals. Within each of these categories, some sensors may exhibit characteristics that overlap with other categories. For example, some mass sensors may rely on electrical excitation or optical settings. Nevertheless, these four broad categories of sensors are sufficiently distinct for the purposes of this review.

In the following sections, the four general categories of sensors are reviewed: (1) chromatography and spectrometry; (2) electrochemical sensors; (3) mass sensors; and (4) optical sensors. The following sections are not intended to include exhaustive reviews of all available sensors and technologies. Instead, the following sections provide a brief overview of the relevant features and applications of the different categories of sensors. The intent is to provide guidance for determining the best options for subsurface vapor-phase monitoring applications.

18.2 CHROMATOGRAPHY AND SPECTROMETRY (SEPARATION AND DETECTION)

Chromatography is a method for the separation and analysis of complex mixtures of volatile organic and inorganic compounds. A chromatograph is essentially a highly efficient apparatus for separating a complex mixture into individual components.

When a mixture of components is injected into a chromatograph equipped with an appropriate column, the components travel down the column at different rates and therefore reach the end of the column at different times. A detector is positioned at the end of the column to quantify the concentrations of individual components of the mixture being eluted from the column. Several different types of detectors can be used with chromatographic separation and are discussed later.

Gas chromatography may be classified into two major divisions:

- Gas-liquid chromatography, where the sorbent (material which separates the mixture into individual components) is a nonvolatile liquid called the stationary-liquid phase, coated as a thin layer on an inert, granular solid support, and
- Gas-solid chromatography, where the sorbent is a granular solid of large surface area.

The moving-gas phase, called the carrier gas, is an inert gas such as nitrogen or helium which flows through the chromatographic column packed with the sorbent. The mixture partitions itself between the moving gas phase and the sorbent and individual components of the mixture move through the column at a rate dependent upon its partition coefficient in the liquid phase (gas-liquid chromatography) or upon its adsorption coefficient on the packing (gas-solid chromatography) and the carrier-gas flow rate. Open tubular glass or stainless steel capillary tubes of 0.005–0.02 in. (0.1–0.5 mm) inside diameter and length often as great as 300 ft (90 m), coated on the inside walls with a nonvolatile stationary-liquid phase, are also widely used in the separation of complex mixtures.

The apparatus used in gas chromatography consists of four basic components: a carrier-gas supply and flow controller, a sample inlet system providing a means for introduction of the sample, the chromatographic column and associated column oven, and the detector system.

Although the carrier gas is most commonly nitrogen or helium, other gases such as carbon dioxide, argon, xenon, and hydrogen are occasionally used. Use of a carrier gas of higher molecular weight will improve column efficiency. Therefore nitrogen or perhaps a gas of even higher molecular weight is preferred to helium if a detector other than thermal conductivity is being used.

A rotometer may be used in the carrier-gas system to give an approximate indication of flow rate. A rotometer consists of a graduated tube with slowly increasing inside diameter and a glass or metal ball that is suspended in the gas flow within the tube at a height dependent upon the flow rate. Since the position of the ball is a function of both the flow rate and the column back pressure when positioned at the column inlet, a rotometer can be used only for rough approximations of flow rate. A soap-bubble flowmeter is used for more accurate measurements.

Sample inlets are of two general types depending upon whether the sample is gaseous, liquid, or solid. Liquid samples are generally injected by means of a calibrated hypodermic syringe through a silicon rubber septum into a metal or glass-lined metal injection port, while gaseous samples are introduced by means of a valve and sample loop system. Injection techniques somewhat similar to those used for liquid samples are used for solids. In order to produce sharp chromatographic peaks with minimum peak overlap, solid and liquid samples must be vaporized rapidly upon injection by maintaining the injection port at a temperature greater than the boiling point of the sample.

Solid-phase microextraction (SPME) is a relatively new method that allows trace analytes to be introduced into the chromatographic system without the need for solvents. The method, which was developed in the early 1990s at the University of Waterloo in Ontario, Canada, involves exposing a small segment of fused silica fiber coated with a non-volatile polymeric material. The coated fiber is mounted in a syringe-like device that can expose the fiber to the desired environment and also withdraw it for protection during transfer to the gas chromatograph. The analyte of interest adsorbs on the fiber coating and is thermally desorbed when introduced into the chromatographic injection port. SPME has been commercialized by Supelco, Inc.™ and Varian.™

The column is the heart of the gas chromatograph, and separation of components on packed columns depends more on the choice of liquid phase than on any other factor. Typically the column is a glass or metal tube of 0.125 or 0.25 in. (6 or 13 mm) in diameter and 4–6 ft (1–2 m) in length, packed with an inert diatomaceous earth support coated with a nonvolatile liquid to 3–20% by weight. In open tubular or capillary column technology, the support for the thin film of liquid phase is the wall of the capillary itself. Support-coated open tubular columns are also sometimes used, the sample capacity of the columns being increased by the presence of very loosely packed support or by a roughening of the capillary walls.

The detector produces a response that is proportional to component that is separated by column and is located at the end of the column. Different detectors may be utilized dependant upon the analyte of interest and include a photoionization detector (PID), flame ionization detector (FID), thermal conductivity detector (TCD), electron capture detector (ECD), flame photometric detector (FPD) or far UV absorbance detector (FUV). Some of these detectors, such as the PID, have been commercialized as hand-held units. These portable PID sensors use a small pump to suck vapor through the ionization chamber. Water vapor is completely ignored, but it cannot provide discrimination among different chemicals.

An amplifier, which could be considered part of the detector "package" receives an output from a detector and amplifies it so that the signal can be detected by a recorder or integrator. Subsequently, an integrator takes the signal from the amplifier and produces an output (chromatogram) and peak height or area (used for quantification). The height of the peak measured from the baseline to the peak maximum and the area which is determined by integrating the area underneath the peak are proportional to concentration. Generally integrators will provide both area and height values. At low concentrations with packed columns, peak height may provide a better value.

Pros: The bench-top gas chromatograph can provide superior discrimination capabilities (relative to other devices and sensors) with excellent precision, sensitivity, and reproducibility.

Cons: Not portable. Expensive. Requires training to operate.

18.2.1 Portable Gas Chromatographs

The potential to use an integrated chromatography system as a continuous or in-situ sensor system depends on the miniaturization of its components. Several portable gas chromatography systems that can be carried in a backpack have been commercialized and are suitable for field use. Efforts are also being made to miniaturize the systems even further into hand-held devices or downhole probes. Micro-analytical devices, or micro-chem-lab devices, are being developed at Sandia National Laboratories that are the size of a palm-top computer that incorporate a gas-phase analysis system for detecting volatile organic compounds and chemical warfare agents (e.g., sarin, soman, mustard gas) and a liquid-phase analysis system for detecting explosives.

The micro-chem-lab improves the sensitivity and selectivity to individual chemicals by using a cascaded approach where each channel includes a sample collector/concentrator, a separator, and a chemically selective surface acoustic wave (SAW) array detector. All three components have been developed and demonstrated and can be fit in an area less than that of a U.S. dime. These small components allow miniature systems incorporating multiple analysis channels to be produced. Discrimination of analytes is improved by allowing simultaneous use of more than one channel at a time. Versatility of the micro-chem-lab can be increased by allowing different channels to be tailored to detect different chemical analytes.

The collector/concentrator includes a thin-film porous adsorbent to selectively collect chemical analytes to be detected and a heater for thermal desorption of the collected analyte into a concentrated pulse. The column separator is a long flow channel coated with a stationary phase material. As analytes pass through the column, the analytes are separated based on differences in partitioning behavior into the stationary phase. Detection of the analyte is achieved using an array of SAW sensors acting as sensitive mass detectors. By coating the SAW devices with chemically distinct thin film materials, a unique pattern of responses for different chemical analytes can be used to provide chemically selective detection.

Pros: The bench-top gas chromatograph can provide superior discrimination capabilities (relative to other devices and sensors) with excellent precision, sensitivity, and reproducibility. Portable and micro-chem-lab units provide faster analysis times and potentially increased sensitivity.

Cons: Bench-top gas chromatographs are bulky and not portable. They cannot be used for real-time or in-situ applications. The micro-chem-lab is still under development and requires additional research and testing. Interferences and long-term stability may be issues with the smaller systems.

18.2.2 Ion Mobility Spectrometry

The ion mobility spectrometer (IMS) can be considered a sub-class of chromatographic separators. The principle of every IMS is a time-of-flight measurement. After a gaseous sample has entered the spectrometer it will be ionized by a radioactive source. The resulting positive and/or negative charged species will be accelerated over a short distance in an electrically charged field, and the time-of-flight will be determined. The IMS is different than the mass spectrometer in that it operates under atmospheric conditions and does not need large and expensive vacuum pumps. Because of this, IMSs can be easily miniaturized.

IMS devices are currently used in airports to screen passengers for trace explosives (negatively charged ions). They can also be used by law enforcement personnel to detect narcotics and other drugs (positively charged ions). These devices are suited for the screening of traces in gas and for the detection of toxic industrial compounds and chemical warfare agents down to the ppb-range. Some devices are equipped with automatic polarity switching, and some instruments enable continuous monitoring. The built-in microprocessor evaluates the recorded ion mobility spectra. The results, identified substances and their concentrations, are shown on the display. An integrated alarm function responds according to programmed threshold values.

Pros: IMS systems can be miniaturized because they operate at atmospheric pressure. Can be used to detect toxic industrial compounds and chemical warfare agents in ppb-range. Integrated alarm for threshold detection.

Cons: Cannot currently be used in situ. Contains radioactive ionizing source.

18.2.3 Mass Spectrometry

The principle of the mass spectrometer is similar to the ion mobility spectrometer, except a vacuum is required. Sampled gas mixtures are ionized, and charged molecular fragments are produced. These fragments are sorted in a mass filter according to their mass to charge ratio. The ions are detected as electrical signals with an electron multiplier or a Faraday plate.

Low mass ions are displayed as a vertical line at the left end of a scale while heavy ions are displayed towards the right. The length of a line represents the quantity of that ion in the gas mixture.

Pros: The mass spectrometers have good discrimination capabilities and can detect a wide range of chemicals. Some of the mass spectrometers are portable enough to carry into the field.

Cons: The units are quite expensive and cannot be used in situ. Spectral overlaps can be a problem in detecting mixtures of unknown composition. Contains radioactive ionizing source.

18.3 ELECTROCHEMICAL SENSORS

Electrochemical sensors have been categorized by Wilson et al. (1995, Ch. 36) and Janata (1992) into three groups: (1) potentiometric (measurement of voltage); (2)

amperometric (measurement of current); and (3) conductometric (measurement of conductivity). A summary of gas and vapor phase analytes that can be detected by these groups of sensors is summarized in Table 36.3 of Wilson et al. (1995).

In this section, we first review the conductometric class of sensors. These types of sensors appear to be most relevant for detecting and monitoring VOCs. Then, a brief overview of potentiometric and amperometric sensors is provided.

18.3.1 Conductometric Sensors

Three different types of conductometric sensors are presented in this section. The first is a polymer-absorption sensor that indicates a change in resistance in the conductive polymer electrode when exposed to chemicals. The second is the catalytic bead sensor, which requires elevated temperatures to burn combustible hydrocarbon vapors and change the resistance of an active element. The third sensor is the metal-oxide semiconductor sensor, which responds to changes in the partial pressure of oxygen and requires elevated temperatures to induce combustion of chemical vapors that change the resistance of the semiconductor.

18.3.1.1 Polymer-absorption chemiresistors

The concept of using polymeric absorption to detect the presence of chemicals in the vapor phase has existed for several decades. These polymer-absorption sensors (chemiresistors) consist of a chemically sensitive absorbent that is deposited onto a solid phase that acts as an electrode. When chemical vapors come into contact with the absorbent, the chemicals absorb into the polymers, causing them to swell. The swelling changes the resistance of the electrode, which can be measured and recorded. The amount of swelling corresponds to the concentration of the chemical vapor in contact with the absorbent. The process is reversible, but some hysteresis can occur when exposed to high concentrations. Several companies and organizations have developed chemiresistors, but the specific attributes and types of absorbents, which are generally proprietary, vary among the different applications.

Sandia National Laboratories has developed chemiresistors using polymer films deposited on microelectrodes. Rather than using a single electrode and conductive polymer, the chips used at Sandia can house an array of chemiresistors. Down-hole chemiresistors sensors have been developed and field tested in several applications (see www.sandia.gov/sensor).

Pros: Chemiresistors are small, low power devices that have no moving parts and have good sensitivity to various chemicals. As a result, they are amenable to being placed in situ in monitoring wells. Another advantage for chemiresistors in comparison to the standard electrochemical sensors is that they don't require liquid water to work properly. This will be seen more clearly in the section on electrochemical sensors below, but in brief, standard electrochemistry requires a well controlled liquid environment for the electrodes to work predictably in detecting analytes. That liquid (usually water with controlled ionic strength and pH buffers) must be supplied by the sensing medium (like a well) or a reservoir in the sensor package, which can dry out.

Cons: May not be able to discriminate among unknown mixtures of chemicals. Some polymers react strongly to water vapor. Uncertain durability of polymers in subsurface environments; need to develop robust packaging. May need pre-concentrator to detect very low limits (for regulatory standards). Although reversible, signal may experience hysteresis and a shift in the baseline when exposed to chemicals.

18.3.1.2 Catalytic bead sensors

Catalytic bead sensors are low-power devices (50–300 mW) that have been used for many years in the detection of combustible gases, particularly methane in air (U.S. EPA 1995). They are used widely in portable gas detection instruments. The catalytic bead sensor is comprised of a passive and active element, both made from an embedded coiled platinum wire in a porous ceramic (Figure 18.1). The active element is coated with a catalyst such as platinum, and the passive element is coated with an inert glass to act as a reference element to compensate for environmental conditions. Both elements are heated to a prescribed operating temperature ranging from 300 to 800°C. When a combustible gas such as methane contacts the elements, the vapor combusts on the active element and the active element increases in temperature. As a result, the resistance of the platinum coil changes. The two elements are connected to a wheatstone bridge circuit, so the changes in resistance are measured as a change in voltage.

Portable soil vapor monitors have been commercialized that use the catalytic bead sensor for detecting combustible hydrocarbon gases. These devices may have the ability to eliminate methane from the readings. Methane occurs naturally in the subsurface, so it can provide false readings. Hand-held devices based on catalytic bead sensors can provide discrete readings of real-time gas concentrations. Sampling occurs through a probe, and an internal pump draws the sample.

Pros: Commercial units are portable and can allow the operator to distinguish between methane and other volatile hydrocarbon vapors.

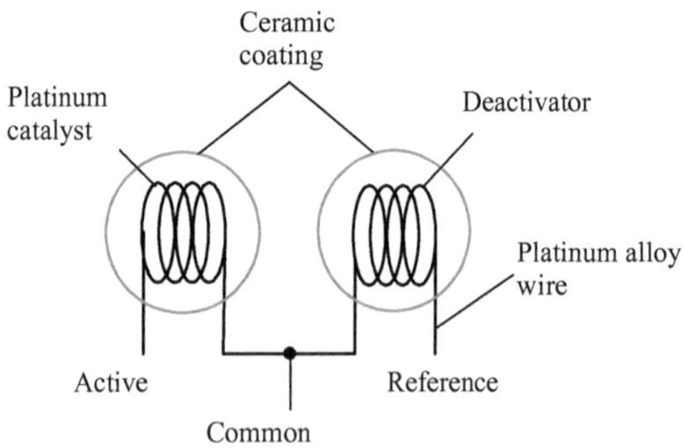

Figure 18.1. Diagram of catalytic bead sensor

Cons: This unit is not amenable for long-term in-situ operation. The catalytic bead sensor requires elevated temperatures for operation. Internal pump is required to sample gas.

18.3.1.3 Metal-oxide semiconductor sensors

The metal-oxide semiconductor (MOS) sensor is comprised of a tin oxide that is sintered on a small ceramic tube. A coiled wire is placed through the center of the ceramic tube to act as the sensor heater. Metal wires provide electrical contact between the tin oxide and the rest of the electronics. The MOS sensor requires between 300 and 600 mW of power to operate the sensor at elevated temperatures between 200 and 450°C. The combination of the sensor operating temperature and the composition of the metal oxide yields different responses to various combustible gases.

When the metal oxide is heated, oxygen is adsorbed on the surface with a negative charge. Donor electrons are transferred to the adsorbed oxygen, leaving a postive charge in the layer. Inside the sensor, electrical current flows through the grain boundary of metal oxide micro crystals. Resistance to this electrical current is caused by negatively charged oxygen at grain boundaries. In the presence of a reducing gas, a surface catalyzed combustion occurs and the surface density of negatively charged oxygen decreases, thereby decreasing the resistance of the sensor. The relationship between the amount of change in resistance to the concentration of a combustible gas can be expressed by a power-law equation.

Pros: The MOS sensors have high sensitivity to combustible gases (e.g., hydrogen, carbon monoxide, methane, ethane, propane, alchohols, etc.). They are compact and durable. The cost is also relatively inexpensive.

Cons: U.S. EPA (1995) performed tests on some Figaro MOS sensors and found that they had more drift during exposure to xylene than the polymer-absorption sensors. The MOS sensor has a fair amount of sensitivity to water humidity, which may be problematic in subsurface environments. Sensitivity to aromatic and halogenated hydrocarbons is questionable.

18.3.2 Potentiometric and Amperometric Sensors

Potentiometric and amperometric sensors employ an electrochemical cell consisting of a casing that contains a collection of chemical reactants (electrolytes or gels) in contact with the surroundings through two terminals (an anode and a cathode) of identical composition. For gas sensors, the top of the casing has a membrane which can be permeated by the gas sample. Oxidization takes place at the anode and reduction occurs at the cathode. A current is created as the positive ions flow to the cathode and the negative ions flow to the anode. Gases such as oxygen, nitrogen oxides, and chlorine, which are electrochemically reducible, are sensed at the cathode while electrochemically oxidizable gases such as carbon monoxide, nitrogen dioxide, and hydrogen sulfide are sensed at the anode. Potentiometric measurements are performed under conditions of near-zero current. Amperometric sensors are usually operated by imposing an external cell voltage sufficiently high to maintain a zero oxygen concentration at the cathodic surface; therefore, the sensor current response is diffusion controlled (Tan and Tan, 1996). According to Tan and Tan (1996), sensitivity of amperometric sensors is better than potentiometric sensors. In addition, U.S. DOE (1994, p. A-1) reports that amperometric methods are used in high-performance liquid chromatography because of its enhanced sensitivity. Additional details of potentiometric and amperometric sensors can be found in Janata (1990).

A common application for potentiometric and amperometric sensors is for water analysis. The most common is the pH sensor system. The basic principal of these devices is that they require two separated, carefully controlled liquid reservoirs with two different chemically unstable electrodes (called reference electrodes), for example a silver wire with a coating of silver chloride. The pH is measured by the voltage difference between the two reference electrodes, so the unknown sample must be in electrochemical connection with both solutions through a glass membrane. However, these thin porous membranes can break, the solutions can leach out or dry out, or the chemistry of the reference electrode itself can change giving a slightly different voltage. Small changes in the chemistry can result in large changes in output voltage. Consequently these systems require constant attention and calibration against known pH solutions.

Many so-called ion selective electrodes for particular ions are sold using basically the same system described above but with special membranes taking the place of the pH-sensitive glass that give potential differences for different ions. The same maintenance and calibration problems exist, as well as interference problems from other ions. Some gases that can be detected using potentiometric methods include carbon dioxide, oxygen, carbon monoxide, hydrogen, chlorine, arsenic oxides, and oxidizable pollutants.

Commercial potentiometric cells for VOCs are less common; most are used for toxic gases and oxygen. There are several research papers describing how to measure different VOCs in liquid electrochemical cells (Sawyer et al., 1995). The same corrosion and drift problems exist for these experiments as described above. In addition, a membrane or porous plug must be used to provide the diffusion of the vapor phase VOC molecules from the gas phase into the electrolyte and electrode surface. To speed up the process, the working electrode is placed virtually on top of the gas-permeable membrane. A recent research example of trace detection of explosive molecules is given in Berger (2000) with some discussion of the difficulties (interference of electroactive O_2 is important) and virtues (under controlled conditions, very low vapor-phase concentrations (ppb) of TNT can be detected).

Pros: These devices can be specific for a particular gas or vapor and are typically very accurate. They do not get poisoned and can monitor at ppm levels.

Cons: Primary sensitivity is for toxic gases and oxygen, not VOCs. Not amenable for in-situ applications. Membranes are sensitive and may degrade with time. Devices are not very durable and have short shelf lives. Subject to interfering gases such as hydrogen.

18.4 MASS SENSORS

18.4.1 Surface Acoustic Wave Sensors/ Portable Acoustic Wave Sensors

Surface Acoustic Wave Sensors (SAWS) are small miniature sensors used to detect VOCs. A SAW device consists of an input transducer, a chemical adsorbant film, and an output transducer on a peizoelectric substrate (see Figure 18.2). The piezoelectric substrate is typically quartz. The input transducer launches an acoustic wave which travels through the chemical film and is detected by the output transducer. The device runs at a very high frequency 100MHz. The velocity and attenuation of the signal are sensitive to the viscoelasticity as well as the mass of the thin film which can allow for the identification of the contaminant. Heating elements under the chemical film can also be used to desorb chemicals from the device. A signal pattern recognition system that uses a clustering technique is needed to identify various chemicals.

SAWS have been able to distinguish organophosphates, chlorinated hydrocarbons, ketones, alcohols, aromatic hydrocarbons, saturated hydrocarbons, and water.

Pros: Small, low power, no moving parts other than the high-frequency excitation, good sensitivity to various chemicals. Can detect chemicals in very low concentrations.

Cons: May not be able to discriminate among unknown mixtures of chemicals. Some polymers react strongly to water vapor; uncertain durability in subsurface environments. Requires high-frequency excitation.

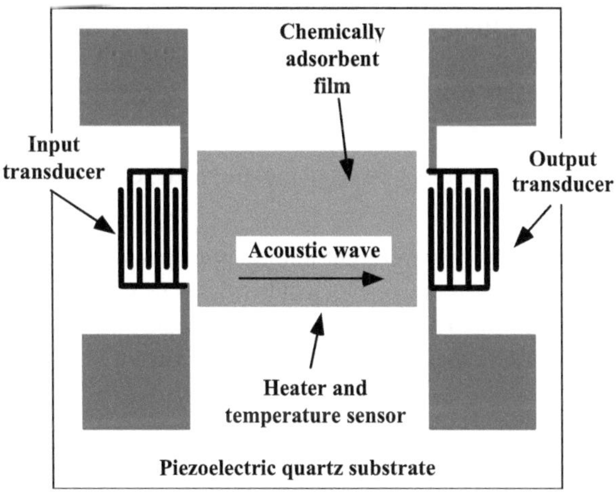

Figure 18.2. Schematic of SAW device

18.4.2 Microcantilever Sensors

Microelectromechanical systems can be composed of multiple micron-thick cantilevers (visualize miniature diving boards) that respond by bending due to changes in mass. Appropriate coatings are applied to the cantilevers to adsorb chemicals of interest. The adsorption of the chemical causes the microcantilever to bend, which can be recorded either electrically or optically. This particular technology has been used for developing infrared sensors to "see" images in darkness, but commercial devices using microcantilever sensors to detect volatile organic chemicals are less common. Active research in this area is being performed by Oak Ridge National Laboratory.

18.5 OPTICAL SENSORS

18.5.1 Fiber Optic Sensors

Fiber optic sensors are a class of sensors that use optical fibers to detect chemical contaminants. Light is generated by a light source and is sent through an optical fiber. The light then returns through the optical fiber and is captured by a photo detector. Some optical fiber sensors use a single optical fiber while others use separate optical fibers for the light source and for the detector. There are three general classes of fiber optic sensors (see Figure 18.3). The first type is completely passive. A spectroscopic method can be used to detect individual types of contaminants. This method involves sending a light source directly through the optical fiber and analyzing the light that is reflected or emitted by the contaminant. The refractive index of the material at the tip of the optical fiber can be used to determine what phases (vapor, water, or NAPL) are present. A second class of fiber optic sensors consist of a fiber optic sensor with a chemically interacting thin film attached to the tip. This film is formulated to bind with certain types of chemicals. Contaminant concentration can be found by measuring the color of the thin film, the change in refractive index, or by measuring the flourescing of the film. The third type of fiber optic sensors involves injecting a reagent near the sensor. This reagent reacts either chemically or biologically with the contaminant. The reaction products are detected to give an estimate of the contaminant concentration.

Pros: Low power; several types have no moving parts; can detect various chemicals at very low concentrations.

Cons: Limited ability to transmit light through the optical fiber over long distances. Some organic pollutants are not easily differentiated using UV-visible spectroscopy. Concentration range sensitivity may be limited. Sensors that use chemically sensitive coatings may degrade with time.

18.5.2 Colorimetry

Pocket colorimeter test kits can be used to measure trace levels of contaminants. They work by analyzing the color of contaminated water that has been mixed with

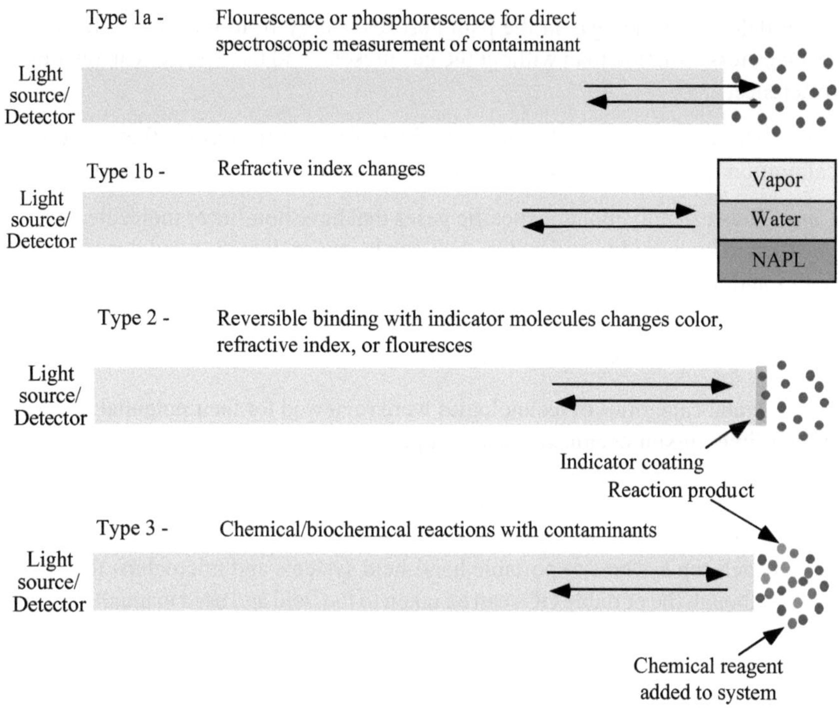

Figure 18.3. Three types of fiber-optic sensors

a particular chemical reagent. Hach sells pre-measured, unit-dose reagent that react with water samples. To test water samples, the pocket colorimeter compares a reacted sample with a sample blank and yields results in concentration units.

Pros: Portable, simple to use. Visual evidence of gas detection event. Not prone to interferences.

Cons: Limited chemical sensitivity to individual VOCs; needs actual water samples (cannot be used in situ); most kits do not meet U.S. EPA method requirements and may not be used for compliance monitoring. Requires visual inspection and is not amenable to long-term in-situ applications.

18.5.3 Infrared Sensors

Infrared sensors can be used to detect gases, which, in general, have unique infrared absorption signatures in the 2–14 μm range. The uniqueness of the gas absorption spectra enables identification and quantification of chemicals in liquid and gas mixtures with little interference from other gases. These devices are typically comprised of a source of infrared radiation, a detector capable of seeing the infrared radiation, and a path between the detector and source that is exposed to the gas being detected. When gas in the path absorbs energy from the source, the detector receives less radiation than without the gas present, and the detector can quantify the difference.

Pros: These devices can be made to identify specific gases; they require less calibration than other sensors; good durability with minimal maintenance.

Cons: They can only monitor specific gases that have non-linear molecules; they can be affected by humidity and water; they can be expensive; dust and dirt can coat the optics and impair response, which is a concern in in-situ environments.

18.6 SUMMARY AND DISCUSSION

Four general categories of technologies were reviewed for their potential application in real-time, in-situ chemical sensing applications. The first category reviewed was chromatography/spectrometry, which included ion-mobility spectrometry and mass spectrometry. The gas chromatographs reviewed provide excellent discrimination among various chemicals of interest. The size of gas chromatographs range from large bench-top systems to portable hand-held systems and microchips the size of a coin. Although the portable GCs can be taken to the field and used manually to sample monitoring wells, the majority of these devices are not yet amenable to real-time, in-situ downhole applications. The micro-chem-lab might be a potential candidate, but it requires micro-pumps to circulate gas through the system, and these moving parts may not be able to withstand long periods in geologic environments. The ion-mobility and mass spectrometers also have excellent discrimination capabilities, but like the gas chromatographs, they are not currently amenable for in-situ applications.

The second category reviewed was electrochemical sensors, which included conductometric, amperometric, and potentiometric sensors. The amperometric and potentiometric devices traditionally are used to monitor oxygen, carbon monoxide, chlorine, and other constituents for air quality purposes. Water quality parameters such as pH can also be measured with these devices. However, the amperometric and potentiometric devices are not widely used for detection of VOCs. The conductometric sensors reviewed include polymer-absorption chemiresistors, catalytic bead sensors, and metal oxide semiconductors. These devices are sensitive to VOC exposure, resulting in large changes to resistance in the device. However, current commercial devices intended for use in situ (primarily polymer-absorption sensors) cannot discriminate different constituents in a mixture. Some hand-held polymer absorption devices *can* discriminate different species because of the use of arrays of chemiresistors, but they are not amenable to in-situ applications. Catalytic bead sensors and metal-oxide semiconductors require elevated temperatures for operation, and they may not be amenable for prolonged periods in situ.

The third type of technology reviewed was the mass sensor. These devices typically absorb the chemical of interest onto a surface, and the device detects the change in mass. The detection can be accomplished through changes in acoustic waves propagated along the surface (SAW devices) or by actual bending or a change in shape of the device as mass is accumulated (micro-cantilever devices). These devices are very sensitive to the presence of various volatile and semi-volatile chemicals. However, the sensitivity of the device depends on high-frequency excitation or extremely small changes in shape; these aspects have not been tested in prolonged geologic environments.

Finally, the fourth category of devices reviewed was optical sensors. These include fiber optical sensors, colorimetry, and infrared sensors. These sensors rely on changes in electromagnetic radiation (e.g., visible, infrared) to detect and identify the presence of chemicals. The sensitivity of these sensors to VOCs can be good, and a TCE fiber optical sensor integrated with a cone penetrometer already exists. Its use in long-term applications still requires testing. Colorimetry is a simple and quick method to detect changes in color in solutions mixed with the sample, but it requires manual intervention. The infrared sensor appears to be useful for detecting combustible hydrocarbons (e.g., methane, propane), but the devices reviewed were not amenable for real-time, in-situ applications involving other lower volatility VOCs.

The most viable sensors for in-situ chemical sensing appear to be electrochemical sensors, specifically conductometric sensors, based on their simplicity and robustness. Reports from the U.S. EPA (1992, 1995) have indicated that polymer-absorption and metal-oxide-semiconductor sensors are viable candidates for use at underground storage tanks. In addition, fiber-optic sensors and mass sensors (SAW devices in particular) also appear to be viable candidates for in-situ applications. The general issue among all of these sensors is that few, if any, have been tested and demonstrated in long-term geologic environments.

In U.S. EPA (1992, 1995), they report that interference from water and methane are concerns for sensor technology because they can trigger false positives in geologic environments; therefore, tests on potential candidate sensors should be performed to determine if methane and water vapor significantly impact the signal from the sensors. In addition, the ability to discriminate among different chemical species needs further investigation among the polymer-absorption and SAW sensors, especially in geologic environments and applications. The ability to retrieve quantified information from the in-situ sensors such as contaminant characteristics and location may prove useful for the end user to make informed decisions regarding treatment and remediation.

ACKNOWLEDGMENTS

Sandia is a multiprogram laboratory operated by Sandia Corporation, a Lockheed Martin Company for the United States Department of Energy's National Nuclear Security Administration under contract DE-AC04–94AL85000.

REFERENCES

Berger, T., H. Ziegler, and M. Krausa, 2000, "Development of Electrochemical Sensors for the Trace Detection of Explosives and for the Detection of Chemical Warfare Agents," SPIE Proceedings, Vol. 4038, p. 452, AeroSense 2000, April 24–28, 2000, Orlando, Florida.

Hughes, R. C., S. A. Casalnuovo, K. O. Wessendorf, D. J. Savignon, S. Hietala, S. V. Patel, and E. J. Heller, 2000, Integrated Chemiresistor Array for Small Sensor Platforms, SPIE Proceedings paper 4038–62, p. 519, AeroSense 2000, April 24–28, 2000, Orlando, Florida.

Janata, J., 1990, Potentiometric Microsensors, Chem. Rev., 90, 691–703.

Janata, J., 1992, Chemical Sensors, Analytical Chemistry, 64:196R.

Looney, B. B. and R. W. Falta (eds.), 2000, Vadose Zone Science and Technology Solutions, Battelle Press, Columbus, OH, 1540 pp.

Sawyer, D. T., A. Sobkowiak, and J. L. Roberts. Jr., 1995, Electrochemistry for Chemists, 2nd Edition, Wiley-Interscience, New York.

Tan, Y. and T. C. Tan, 1996, Modelling and Sensing Characteristics of an Amperometric Hydrocarbon Sensor, Chemical Engineering Science, Vol. 51, No. 16, p. 4001–11.

U.S. DOE, 1994, Literature Search, Review, and Compilation of Data for Gas Chromatography Sensors and Electrochemical Sensors: Task 5 Report, prepared by the Hazardous Waste Remedial Actions Program, Oak Ridge, TN.

U.S. Environmental Protection Agency (EPA), 1992, Measurement and Analysis of Adsistor and Figaro Gas Sensors Used for Underground Storage Tank Leak Detection, Report #EPA/600/R-92/219.

U.S. Environmental Protection Agency (EPA), 1995, Measurement and Analysis of Vapor Sensors Used at Underground Storage Tank Sites, Report #EPA/600/R-95/078.

Wilson, LG., L. G. Everett, and S. J. Cullen (eds.), 1995, Handbook of Vadose Zone Characterization & Monitoring, CRC Press, Boca Raton, FL.

CHAPTER 19

IN SITU MEASUREMENT OF INDUCED CONTAMINANT FLUX

ANDRE TARTRE
EcoRemediation Inc., 70 South Winooski, Suite 201, Burlington, VT 05401, USA

19.1 INTRODUCTION

Estimates of *in situ* migration of contaminants are typically inferred from discrete samples of soil or groundwater concentrations. Based on the spatial distribution of concentrations, migration of contaminants is then estimated based on transport models and site-specific hydrogeologic parameters, which are obtained from laboratory analyses or field tests (pump or tracer tests). This chapter introduces an alternative method to directly measure the *in situ* mass flux of contaminant vapors directly.

The proposed sampling technique starts by purging the subsurface with a non-contaminated gas for several minutes. The same fluid is collected away from the injection point, and concentrations are monitored. Stabilized concentrations reached during the purge are used to estimate the flux of compounds (analytes) immediately available for migration. We term this flux the "*advective flux*." When concentrations stabilize (i.e., reach a chemical equilibrium), the purge is stopped. Rebound concentrations are monitored over a short period of time to evaluate the migration of compounds induced by the artificial chemical gradient. If there is no *advective* flux, rebounds are used to estimate migration due to molecular diffusion. We define this flux as a " *diffusive flux*." This sampling technique may be used as an extension of conventional soil-gas investigations. These *in situ* flux measurements may, in turn, be used to estimate contaminant migration or biodegradation rates.

19.2 THEORY

Development of the "induced flux measurement technique" occurred during two-years of research conducted in Canada for investigating releases from underground storage tanks (USTs) (Tartre, 1997). This technique is built on the same theoretical

C. Ho and S. Webb (eds.), Gas Transport in Porous Media, 319–330.
© 2006 *Springer.*

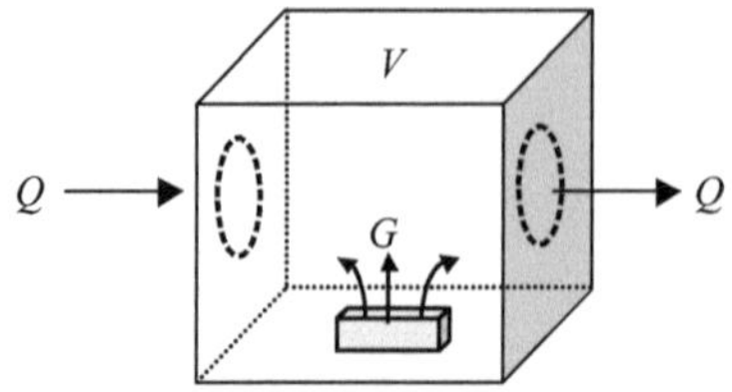

Where

G = Generation rate

Q = Flow rate, or extraction rate

V = Volume

Figure 19.1. Basic concept for the induced In Situ flux method : concentrations fluctuate depending on generation rates and extraction rates

foundation as well-established procedures for assessing the ventilation efficiency within buildings. The procedure involves locally ventilating the UST backfill with fresh atmospheric air over a range of flow rates, while concurrently measuring increases or decreases in gas or vapor concentrations due to flux from the surrounding environment. Fluctuations in concentrations under these imposed dynamic conditions permit an estimate of gas or vapor flux within the backfill.

When a closed (or reasonably confined) space is ventilated, gas concentrations fluctuate depending on both the generation rate and the applied ventilation flow rate (extraction rate) (National Safety Council, 1995). This relationship between ventilation flow rates and gas generation is illustrated in Figure 19.1. If the generation rate is greater than the extraction rate, gas concentrations increase. Conversely, if the flow rate is greater than the generation rate, gas concentrations decrease.

19.3 DESCRIPTION OF THE TECHNIQUE IN THE VADOSE ZONE

The proposed technique purges the vadose zone in the vicinity of a sampling probe with a non-contaminated gas during soil-gas investigation. The equipment used to perform this task is presented in Figure 19.2. Purging the soil gas with nitrogen or other inert compounds affects the gas-liquid-soil equilibrium, causing compounds (analytes) that are present in sorbed or dissolved phases to partition into the gas phase. During the ventilation period, chemical equilibrium among soil phases is disrupted such that the rate at which vapor contaminants and biogenic gases are transferred to the soil gas may be estimated. We define this mass transfer as a generation rate. When a void is created *in situ* by the probe, this mass transfer is divided by the exposed surface of the void to obtain the flux of gas and vapors immediately available for migration. We term this flux "*advective flux*."

After conditions begin to stabilize, the ventilation is significantly reduced or stopped altogether. Subsequent increases in analyte concentrations, as well as the length of the lag time between purge cessation and concentration rebounds, are used

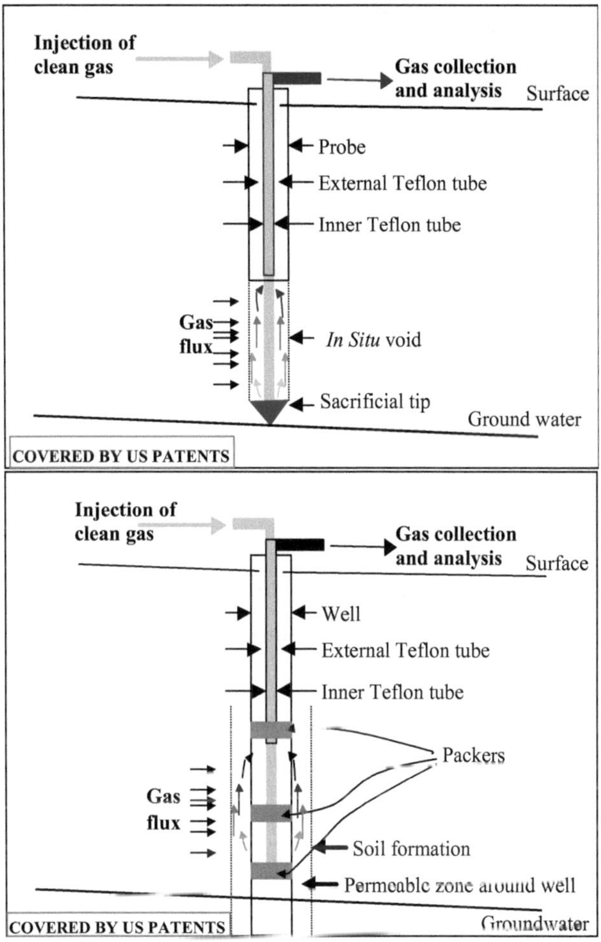

Figure 19.2. Schematic of the *In Situ* flux method and associated equipment. Top: probe configuration. Bottom: existing well configuration

to identify analytes that diffuse due to an artificial chemical gradient. We term this flux "*diffusive flux*." This process is shown schematically as Figure 19.3.

Using the same sampling event, the method has been further developed to measure *in situ* flux of biogenic gases (e.g., those produced during the degradation of organic contaminants). When *advective* flux of daughter compounds within "active" subsurface locations is superior to their *in situ flux* at "inactive" locations, production rate is estimated. Basically, mapping of *advective and diffusive* flux of fluids are used to evaluate whether the migration process is primarily controlled by advection, diffusion, or production. An experimental setting is shown on Figure 19.2 to apply the proposed sampling method in existing wells.

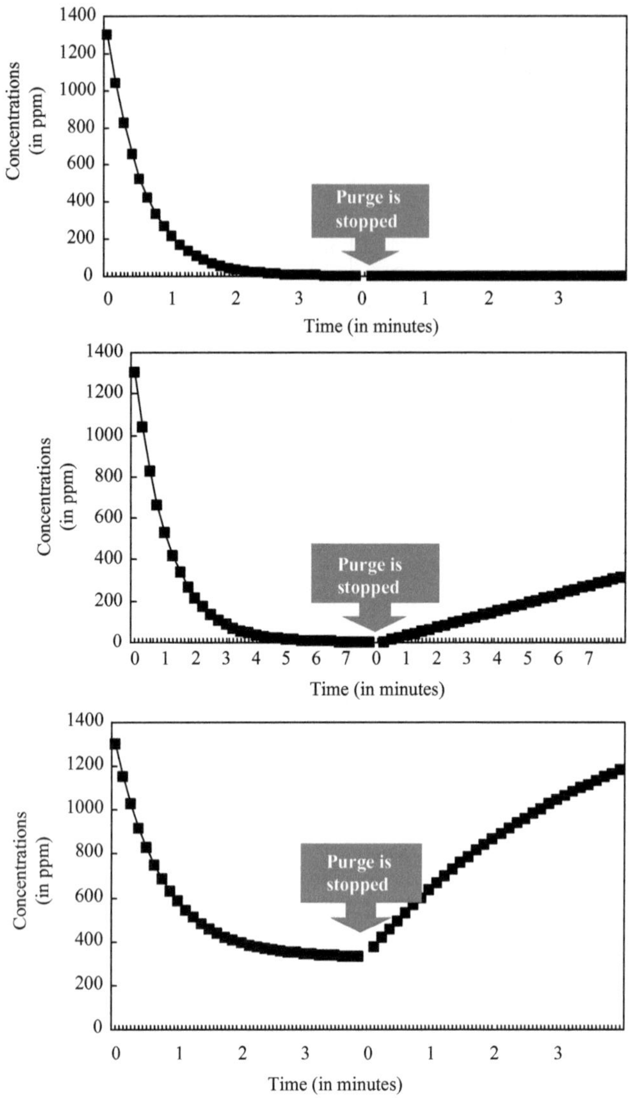

Figure 19.3. Typical transient concentration profiles during and after the purge for three different flux conditions of a volatile contaminant. Top: no flux (e.g., at plume edge, where no sorbed, solubilized or free contaminants are near sampling location. Middle: diffusion of vapors in the plume. Bottom: advection (e.g., near the source of the plume)

In essence, this new technique incorporates a dynamic approach into traditional "static concentration" approach, expanding the interpretation of results and their application. Whereas traditional methods rely on contaminant concentrations and geological parameters to estimate chemical fate, this technique focuses on the direct and more relevant measure of mass transfer and mass production rates for a limited volume of soil, as small as few cubic feet.

19.4 EVALUATION OF ADVECTIVE FLUX

The proposed sampling technology is the only one so far that evaluates the potential for migration directly, *in situ*, using the contaminant of interest. This technology offers an interesting alternative method to standard sampling practices.

During the purging period, *advective* flux is quantified when stable concentrations are reached due to the presence of free products or due to immediate desorption of contaminants from the soil matrix.

In the vadose zone, *advective* flux of volatile contaminants correlated well with concentrations of specific compounds in the surrounding soil (Tartre, 1997) (e.g., : $r = 0,88$ for BTEX). The proposed *In Situ* flux method may then be used in the vadose zone for most of volatile compounds (petroleum and chlorinated) to identify on site sources of contamination and plume extent. This method is particularly cost effective with gas and vapor sampling comparing with standard "soil/groundwater" sampling methods.

With a minimal cost, the proposed method can be used with direct reading instruments as a preliminary investigative tool to delimit plume and to identify sources. For instance, the difference between gas and vapor rebounds after the purging period is obtained with specific reading instruments. This sampling method is a powerful tool to delimit a plume of volatile compounds at low cost.

Currently, the method may be applied in two (2) ways. Clean gas may be injected, along with a tracer, directly into the subsurface under positive pressure so that *advective* flux may be evaluated through porous material. Alternatively, clean gas may be injected into a controlled volume at an inflow rate that equals the outflow rate for more impervious material, which is used to evaluate *advective* migration via channeling.

The steady state concentration reached during the purging period (concentration (A) in Figure 19.1) is multiplied with the effective flow rate of the clean gas in order to evaluate the rate at which contaminants are partitioning (from non-aqueous or sorbed phases). This rate divided by the surface of the soil void provides *in situ* flux of compounds immediately available for migration, identify as a potential *advective* flux.

19.5 EVALUATION OF DIFFUSIVE FLUX

The proposed sampling technique can also be used to directly evaluate *diffusive* flux. When the effective flow rate of the purging fluid approaches the generation rate of analytes, the purge is halted. Rebound concentrations are monitored within the first few minutes. This monitoring permits a direct estimate of analytes migrating in soil gas under an artificial chemical gradient. This monitoring provides a direct evaluation of molecular diffusion for a specific subsurface location.

The diffusive flux is then estimated by considering the slope obtained at a specific induced chemical gradient. As an example, the induced *diffusive* flux may be evaluated

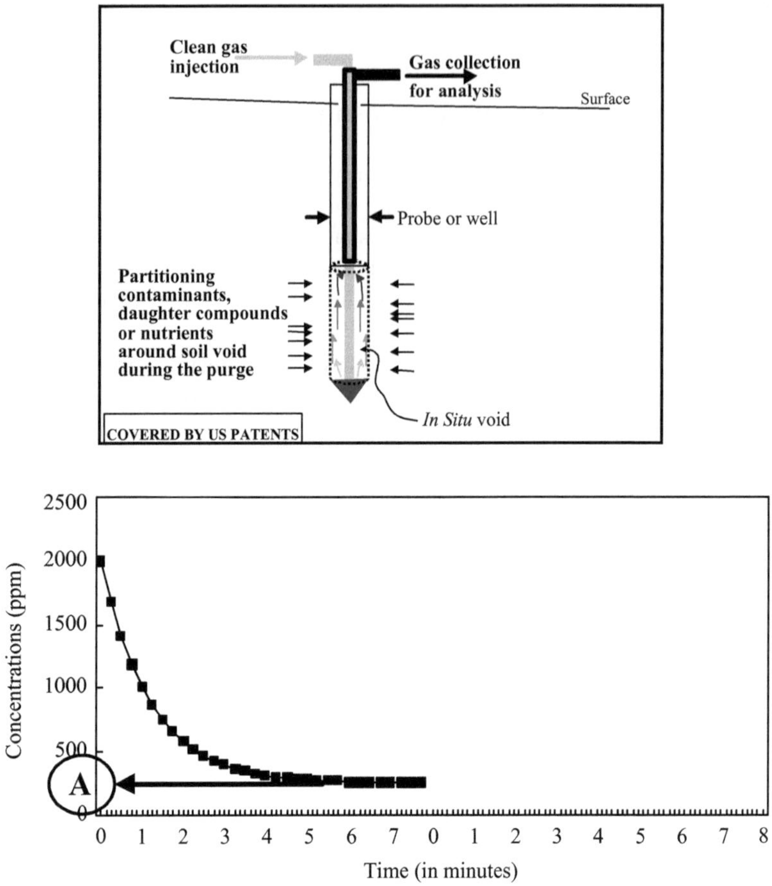

Figure 19.4. Steady state concentrations reached during the purging period as an estimate of *advective* flux of compounds surrounding the soil void

with the last example by the slope on Figure 19.5. The chemical gradient is the difference between the initial concentration and the concentration "B", where the slope is calculated (Figure 19.6). Diffusive flux may be evaluated for different chemical gradients from the same set of data.

19.6 EVALUATION OF BIODEGRADATION PROCESSES

The success of the natural attenuation application depends on both hydrogeological and geochemical parameters. Documenting decreased contaminant concentrations over time is not sufficient to demonstrate that natural attenuation processes are occurring (Commission on Geosciences, Environment and Resources, 2000). A reduction

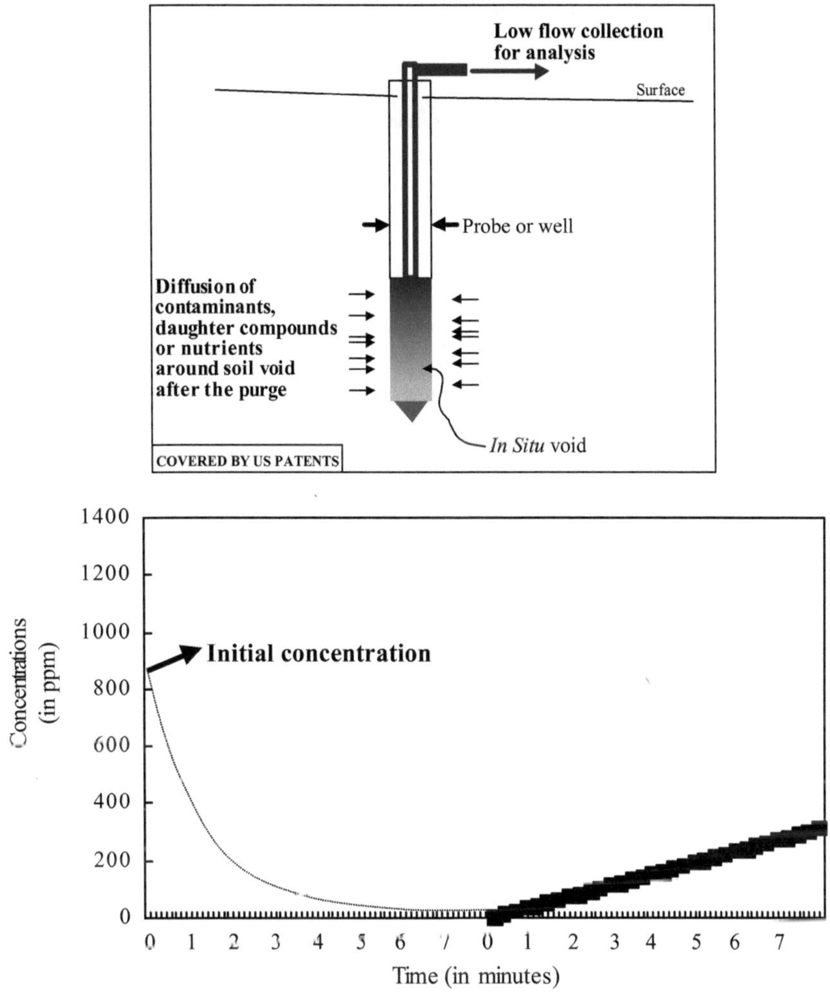

Figure 19.5. Concentrations reached after the purging period as an estimate of migrating compounds due to diffusion surrounding soil void

in contaminant concentration at any given location may result from spatial heterogeneity in soil properties or hydrogeological conditions. Natural attenuation processes may fluctuate as a function of season or of changing geochemical conditions.

MNA is demonstrated in most cases by the occurrence of specific biodegradation processes. Conventional monitoring techniques assess biodegradation processes via the comparison of product and reactant concentrations in up gradient locations with those in the plume. Relative concentrations of these analytes in up gradient versus plume locations are used to identify which natural attenuation processes are actually occurring (Wiedemeier et al., 1995). A common problem with using such a

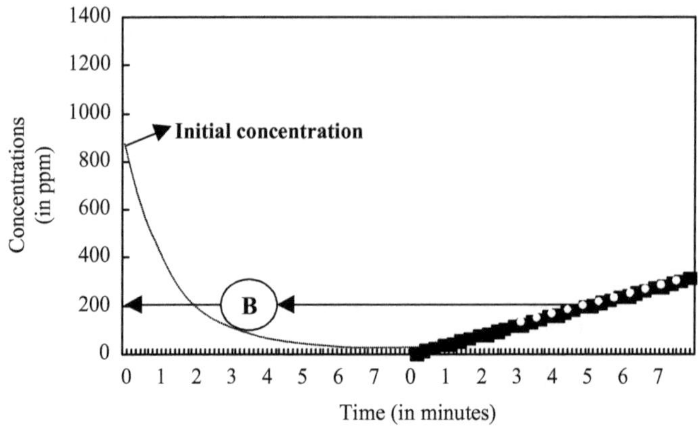

Figure 19.6. Diffusive flux evaluated from the slope of the rebound at a specific concentration gradient (initial concentration-"B")

mass balance approach is that advection may not represent the dominant transport mechanism.

The effectiveness of biodegradation processes should be evaluated with respect to reactant consumption or daughter compound generation. The best measure of these processes is a method that is able to generate data on a small-scale basis so that a comparison of analyte concentrations in up gradient and plume areas is influenced minimally by spatial and temporal variations in hydrology or geochemistry. The measurement of *in situ* flux rates for reactants (e.g., contaminants or nutrients) and products (e.g., daughter compounds or biogenic gases) represents such a method.

The traditional method for calculating field biodegradation rates requires integration of conservative tracer results, an estimate of retarded solute migration in the saturated zone, and hydrogeological calculations (Wiedemeier et al., 1996). An alternative method proposes calculation of biodegradation rates using direct monitoring data (Buscheck and Alcantar, 1995). With either method, the retarded solute velocity is generally overestimated to account for the uncertainty in model parameters. To our knowledge, there is no commercial method that directly measures either destruction rates of contaminants or consumption rates of reactants (Crumbling, 1999; Haggerty et al., 1998; Lahvis and Baehr, 1996; Reinhard and Hopkins).

The induced *in situ* sampling technique brings offers of a different approach to assess the production of daughter compounds. For example, rates of biogenic gas production from petroleum products (e.g., carbon dioxide and methane) are evaluated to predict biodegradation rates across a site. These production rates are estimated from *advective* flux from active subsurface locations compared to diffusive flux at inactive locations (e.g., near sources or beyond plume edge). *In situ* flux measurements are relatively inexpensive especially for daughter products that partition into the gas phase from either the groundwater or the vadose zone.

19.7 OPTIMIZATION OF EXTRACTION PROCESSES

The induced *in situ* flux sampling technique may also be used to optimize extraction processes during remediation efforts. Often, extraction processes are limited after a short period by diffusion of the contaminant from less permeable zone to more permeable zone. Knowledge on exact locations of high *advective* flux is helpful to target more precisely subsurface location where extraction will be driven mainly by advection. The proposed equipment for existing well may be used to define clearly subsurface zone with highest *advective* flux.

19.8 CASE STUDIES

19.8.1 Case Study "1"

An initial investigation to characterize subsurface conditions was conducted beside a gas station where a gasoline leak was suspected. Some smells and vapors were previously detected in subsurface utilities under the street just beside this gas station. The preliminary investigation consisted to dig beside the gas station nearby the street to localize potential contamination. At one location, fresh gasoline was observed at about 8 feet below grade surface (BGS). A soil gas survey was then proposed to the city to determine the extent of the contamination and possible migration pathways under the street.

The "*In Situ* flux method" was conducted on few sampling points during this soil gas survey. The result at point "1" where fresh gasoline was observed is presented in Figure 19.7. The presence of fresh gasoline beside the sampling probe induced an immediate rebound of vapor concentrations after the purge with nitrogen. Considering other results from adjacent sampling locations, it was clear during investigation that the source of the contamination under the street was originating from location "1".

The results at point "16" is presented also in Figure 19.7. This sampling location was someway apart from the plume. Obviously, rebound of vapors is less important, which indicates clearly that dissolved or adsorbed gasoline is at some distance from this sampling location.

A few months after this sampling effort, an extensive investigation with standard soil/groundwater samples shows exactly the same extent of contamination with the same migration pathways than the soil gas survey and the "*In Situ* flux method".

19.8.2 Case Study "2"

The "New Hampshire Department of Environmental Services" wanted to determine if the contamination at a site was spread uniformly, or if some hot spot could be determined to reduce cost of the remediation effort. A soil gas survey was conducted and few "*in situ* flux measurement" were performed.

The site was mainly contaminated with a petroleum product, nearby a former "Varsol" distilling process used since World War Two. During the investigation,

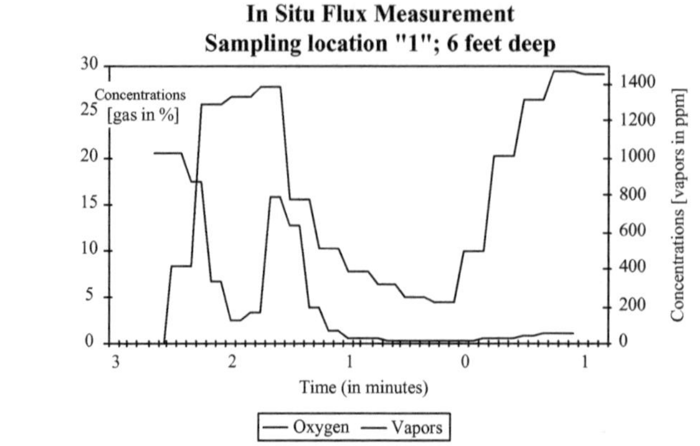

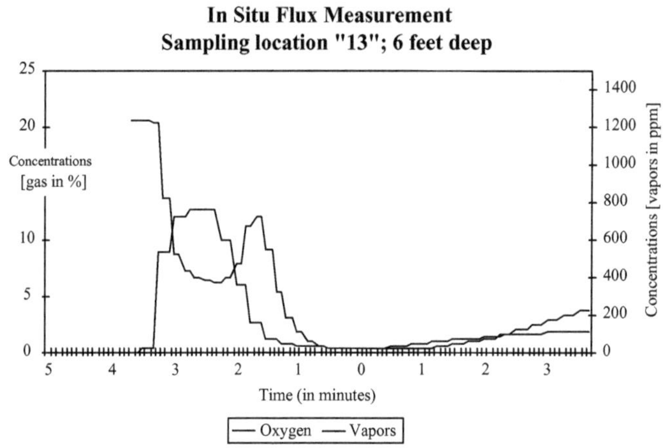

Figure 19.7. Induced *in situ Flux Measurements* for gasoline vapors at two sampling locations. Top: Sampling location "1" (3 to 2 minutes: reading of the analytical equipment; 2 minutes: start of the purge with nitrogen; 0 minutes: purging is stopped; rebound of vapor concentrations after the purging period shows that gasoline is present near the probe). Bottom: Sampling location "13" (4 to 3 minutes: reading of the analytical equipment; 2.5 minutes: start of the purge with nitrogen; 0 minute: purging is stopped; rebound of vapor concentrations after the purging period shows that gasoline is at some distance from the probe) (Bump of oxygen at the beginning of the purge is due to air contained in tubes, which is injected in the subsurface. This bump serves as a quality control indicator).

analysis of gas samples was performed with direct reading instruments for carbon dioxide, methane, oxygen and organic vapors.

Two (2) results during this investigation are presented Figure 19.8. At sampling point "16", we get advective flux of methane during the purging period. Furthermore, rebound of methane was important enough to get back original concentration on a relative short period of time. This result may be explained only if we consider an higher generation rate of methane than its migration rate at this sampling location.

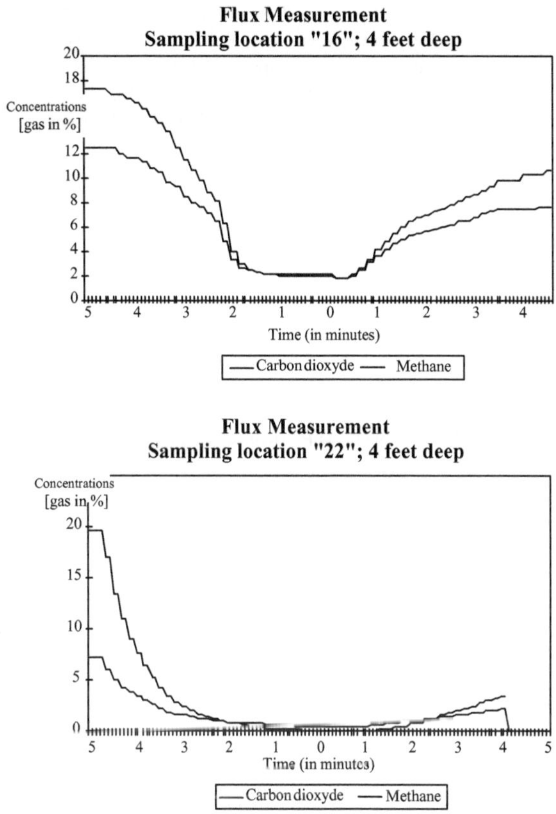

Figure 19.8. Induced *in situ Flux Measurements* for daughter compounds. Top: sampling in an active methanogenesis zone. Bottom: sampling outside the active methanogenesis zone.

Conversely, we find no advective flux at sampling point "22", just at some distance away from sampling point "16". We get only a weak diffusive flux after the purging period.

A few months after this investigation, the site was dug out. During remediation activities, it was clear that methanogenesis was occurring at the site. Characterization of the subsurface during digging confirms that sampling point "16" was in the center of the methanogenesis active zone, while sampling point "22" was at the edge of this zone.

REFERENCES

Buscheck, T.E. and C.M. Alcantar, "Regression Techniques and Analytical Solutions to Demonstrate Intrinsic Bioremediation", in Intrinsic Bioremediation, ed. R.E. Hinchee, J.T. Wilson and D.C. Downey, 109–116, Columbus, Ohio: Battelle Press, 1995.

Commission on Geosciences, Environment and Ressources, "Natural Attenuation for Groundwater Remediation", National Academy Press, Washington D.C., 2000, 287 pages.

Crumbling, D.M., "Improving the Cost-effectiveness of Hazardous Waste Site Characterization and Monitoring", USEPA Technology Innovation Office, Washington DC, 1999.

Haggerty, R., Schroth, M.H., Istok, J.D., "Simplified Method of Push-pull Test Data Analysis for Determining In situ Reaction Rate Coefficients", Ground Water, Vol. 36, March/April 1998.

Lahvis, A.M., Baehr, A.L., "Estimation of Rates of Aerobic Hydrocarbon Biodegradation by Simulation of Gas Transport in the Unsaturated Zone", Water Resources Research, Vol. 32, No.7, pages 2231–2249, July 1996.

National Safety Council, "Fundamentals of Industrial Hygiene", 4th ed.,1995, p. 579–580.

Reinhard, M., Hopkins, G.D., "In situ Demonstration of Anaerobic BTEX Biodegradation Through Controlled Release Experiments", Applied Bioremediation of Petroleum Hydrocarbons, pp. 263–270.

Tartre, A., "Development of an Assessment Method to Monitor Soils Around Underground Petroleum Storage tanks: In situ GEN-VOC Technology - Project Summary", The National Contaminated Sites Remediation Program, Development and Demonstration of Site Remediation Technology (DESRT), Environment Canada, 1997.

Wiedemeier, T., Wilson, J.T., Kampbell, D.H., Miller, R.N., Hansen, J.E., "Technical Protocol for Implementing Intrinsic Remediation with Long – Term Monitoring for Natural Attenuation of Fuel Contamination Dissolved in Groundwater", Air Force Center for Environmental Excellence, USAF, 1995.

Wiedemeier, T., Swanson, M.A., et al., "Approximation of Biodegradation Rate Constants for Monoaromatic Hydrocarbons (BTEX) in Ground Water", GWMR, pp. 186–194, Summer 1996.

PART 3

APPLICATIONS

CHAPTER 20

RADON TRANSPORT

BILL W. ARNOLD
Sandia National Laboratories, P.O. Box 5800, Albuquerque, NM 87185, USA

20.1 INTRODUCTION

Radon is a naturally occurring, chemically inert, radioactive gas that exists in soil and rock of all geological environments. Several radioisotopes of radon occur naturally, but the most environmentally significant is ^{222}Rn (radon), which has a half-life of 3.8 days. This value of the half life and the gaseous nature of ^{222}Rn allows relatively rapid migration over considerable distances in both soils and the atmosphere before decay.

The radioisotope ^{222}Rn occurs in the decay chain of ^{238}U (uranium), with ^{226}Ra (radium) as its immediate parent. The relatively long half life of ^{226}Ra (1600 years) means that radon will be produced at a fairly constant rate by material containing ^{226}Ra, over time periods relevant to environmental concerns. ^{222}Rn decays by alpha particle emission to ^{218}Po (polonium), and to other solid progeny, including ^{214}Po and ^{214}Pb (lead).

The alpha emitting progeny of radon are the primary agents with direct impact on human health. Although alpha particles do not penetrate deeply into tissue, radon progeny that have been inhaled into the lungs are in close contact with lung tissue and substantial radiation dose is deposited by the decay of these progeny. Exposure to high concentrations of radon gas is known to cause lung cancer in underground uranium miners (Chaffey and Bowie, 1994). In addition, chronic exposure to more moderate concentrations of radon in dwellings may be a concern as a significant cause of lung cancer in the general population (Lubin and Boice, 1997).

Exposure to radon gas is ubiquitous, both indoors and outdoors. Radon present in dwellings and other structures is largely the result of natural contamination from soils over which the building is constructed. Indoor concentrations of radon gas are generally greater than outdoor concentrations because of limited air exchange and lack of dilution by mixing with uncontaminated air from higher in the atmosphere. Concentrations of radon gas within homes vary significantly, depending on radon

C. Ho and S. Webb (eds.), Gas Transport in Porous Media, 333–338.
© 2006 *Springer.*

emanation rates from underlying soils, transport pathways into the home, and the amount of air exchange with outdoor air.

Uranium mill tailings are a potentially significant source of radon gas contamination, if they are not sufficiently isolated from the environment. Uranium is extracted from the ore in the milling process, but the decay products of uranium, including radium, remain in high concentrations within the mill tailings. Consequently, the mill tailings are a long-term source of ^{222}Rn from their ^{226}Ra content.

20.2 RADON TRANSPORT IN SOILS

Radon transport in soils begins with the release of radon into the gas phase present in the soil. Because radium is present in the solid phase of mineral grains, radon must be released following decay of its parent. This process may be enhanced by crystallographic dislocations resulting from alpha recoil following decay of the radium. The efficiency at which radon is released by soils is often expressed as the radon emanation coefficient, which is simply the fraction of radon released to the gas phase. The radon emanation coefficient for uranium mill tailings typically varies from 0.1 to 0.4 (Rogers et al., 1984).

Migration of radon in geologic media can occur by both diffusion and advection. Although advective gas migration may be significant in some geologic settings, strong gradients in gas pressure are generally not present under ambient conditions in the shallow subsurface. However, advective barometric pumping of radon gas into dwellings may be a factor in the case of this exposure mechanism. Diffusion is the dominant process for radon migration in most subsurface environments, particularly in landfill systems specifically designed for the disposal of uranium mill tailings.

The equation for steady-state, one-dimensional, vertical diffusive transport of radon in soils is:

$$D\frac{\mathrm{d}^2C}{\mathrm{d}z^2} - \lambda C + Q = 0$$

where D is the effective diffusion coefficient, C is the concentration of ^{222}Rn (activity/volume), λ is the decay constant of ^{222}Rn (2.1×10^{-6} s^{-1}), and Q is the source term for ^{222}Rn. The source term for radon in the soil can be estimated by the equation:

$$Q = \frac{R\rho_b\lambda E}{\phi}$$

where R is the specific activity of ^{226}Ra in the soil, ρ_b is the dry bulk density of the soil, E is the radon emanation coefficient, and ϕ is the porosity.

Radon gas is moderately soluble in water that is present in variably saturated soils of the vadose zone. Because the diffusion coefficient in the liquid phase is much lower than in the gas phase, diffusion of radon is dominated by migration in the gas phase. However, partitioning of radon into the relatively immobile liquid phase may inhibit diffusive migration somewhat.

The effective diffusion coefficient for radon transport in soils is a function of soil type, compaction, and moisture content. Finer grained and more compacted soils have generally lower values of the diffusion coefficient. The effective diffusion coefficient of radon in the gas phase decreases significantly with increasing moisture content of the soil, being on average about two orders of magnitude lower as the liquid saturation approaches 90% (Rogers et al., 1984). The variability in measured values of the effective diffusion coefficient also increases dramatically with increasing moisture content of the soil.

The moisture content of soils effects diffusive radon transport in two offsetting ways. As noted above, the effective diffusion coefficient is significantly lower at higher moisture contents. The volume of the gas phase also decreases with increasing moisture content, leading to an increase in the concentration of radon in the gas phase of the soil. This increased radon concentration and higher concentration gradients relative to the essentially zero concentration in the atmosphere above the soil, tends to enhance the diffusive radon migration. As stated above, these two effects tend to counteract one another, but the relatively higher dependency of effective diffusion on moisture content dominates. Consequently, net radon diffusion decreases with increasing moisture content of the soil.

20.3 RADON MIGRATION FROM URANIUM MILL TAILINGS

The disposal of uranium mill tailings is regulated to limit the environmental exposure of human populations to radon gas. The federal standard (EPA 40 CFR 192) states that the average ^{222}Rn flux to the atmosphere at the landfill cover shall be less than 20 $\text{pCi/m}^2 \cdot \text{s}$. Landfill cover systems for uranium mill tailings are typically designed with a layer of compacted clay as a barrier to diffusive radon migration. Design requirements for the thickness of the clay layer are generally calculated based on the estimated values of material properties (e.g., effective diffusion coefficient), the radium content of the mill tailings, and other design features in the landfill cover.

Uncertainties in the disposal system for uranium mill tailings may have an impact on the assessments of landfill cover performance and on the risks associated with a particular cover design. Uncertainties exist in the parameter values for various components of the disposal system. In addition, considerable uncertainty exists in future climatic conditions and potential degradation of the disposal system.

A risk-based probabilistic performance assessment of the uranium mill tailings disposal site at Monticello, Utah was conducted with regard to radon migration (Ho et al., 2002). The landfill cover at this site consists of a 24 inch clay layer above the tailings, overlain by a 60 mil high-density polyethylene liner, a 12 inch sand drainage layer, and 66 inches of soil. This analysis considered uncertainty in the radon emanation rate of the tailings, the effective diffusion coefficients in multiple layers of the system, and the moisture content in the layers above the tailings. Scenarios for present and future conditions were constructed, in which uncertainty in infiltration and moisture content was considerably higher under future conditions that could be influenced by climatic changes. Also, degradation of the polyethylene geomembrane layer was considered for future conditions.

The RAECOM software program (Rogers et al., 1984) was used to simulate steady-state, one-dimensional diffusion of ^{222}Rn through a five layer cover system at the Monticello site. The uncertainty analysis was conducted by coupling the RAECOM code with the FRAMES software program, which supports Monte Carlo analyses with automated parameter sampling using the Latin Hypercube method (Wyss and Jorgensen, 1998). Uncertainty distributions were assigned to the parameters for each of the layers in the landfill system based on variations in site measurements. A negative correlation was assigned to the sampling of effective diffusion coefficient and the moisture content, based on the relationship reported in Rogers et al. (1984).

For the Monte Carlo analyses of uncertainty in ^{222}Rn flux through the surface of the landfill cover system, 100 realizations of the diffusion model were run for present conditions and for future conditions. The results of these simulations are shown as a cumulative distribution of surface ^{222}Rn flux in Figure 20.1. For present climatic conditions and essentially undegraded geomembrane performance, the uncertainty in simulated ^{222}Rn flux at the land surface spans more than two orders of magnitude with a maximum value of about $1\,pCi/m^2 \cdot s$. For future conditions (greater than 100 years), the simulated ^{222}Rn flux shows a similar span in uncertainty and has a maximum value of $29\,pCi/m^2 \cdot s$. Overall, these results indicate a high degree of confidence (greater than 97%) that the landfill cover design at Monticello meets the performance objective of $20\,pCi/m^2 \cdot s$ of ^{222}Rn flux for both present and future conditions.

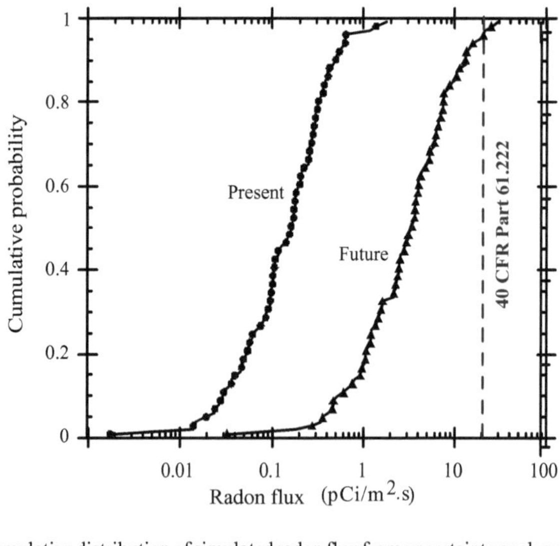

Figure 20.1. Cumulative distribution of simulated radon flux from uncertainty analyses of the Monticello uranium mill tailings cover, for present and future conditions

The difference in the results for present conditions and future conditions is due primarily to potentially degraded performance of the geomembrane as a diffusion barrier in the future. Uncertainty exists with regard to the impact of numerous degradation processes for high-density polyethylene geomembranes and estimates suggest a maximum lifetime of several hundred years (Koerner et al., 1991; Hsuan and Koerner, 1998). It should be noted that in the model the geomembrane layer functions as a significant barrier to gaseous diffusion only at relatively low values of defect density.

REFERENCES

Chaffey, C.M. and C. Bowie, 1994, Radon and health: an update, Journal of Public Health Medicine, vol. 16, no. 4, 465–470.

Environmental Protection Agency (EPA), 40 CFR 192, Health and Environmental Protection Standards for Uranium and Thorium Mill Tailings.

Ho, C.K., B.W. Arnold, J.R. Cochran, and R.Y. Taira, 2002, Development of a Risk-Based Probabilistic Performance-Assessment Method for Long-Term Cover Systems – 2nd Edition, SAND2002–3131, Sandia National Laboratories, Albuquerque, NM.

Hsuan, Y.G. and R.M. Koerner, 1998, Antioxidant depletion lifetime in high density polyethylene geomembranes, Journal of Geotechnical and Geoenvironmental Engineering, vol. 124, no. 6, pp. 532–541.

Koerner, R.M., Y. Halse-Hsuan, and A.E. Lord, 1991, Long-term durability of geomembranes, Civil Engineering, April, 1991, pp. 56–58.

Lubin, J.H. and J.D. Boice, 1997, Lung cancer risk from residential radon: Meta-analysis of eight epidemiological studies, Journal of the National Cancer Institute, vol. 89, no. 1, pp. 49–57.

Rogers, V.C., K.K. Nielson, and D.R. Kalkwarf, 1984, Radon Attenuation Handbook for Uranium Mill Tailings Cover Design, NUREG/CR-3533, U.S. Nuclear Regulatory Commission, Washington, D.C.

Wyss, G.D. and K.H. Jorgensen, 1998, A user's guide to LHS: Sandia's Latin Hypercube Sampling Software, SAND98–0210, Sandia National Laboratories, Albuquerque, NM.

CHAPTER 21

GAS TRANSPORT ISSUES IN LANDMINE DETECTION

JAMES M. PHELAN
Sandia National Laboratories, Albuquerque, NM, 87185, USA

Principles of gas and vapor transport have been applied to evaluate the processes affecting the chemical signature from buried landmines. Due to the shallow burial depth of landmines, environmental factors play a significant role in the chemical signature available for trace chemical detection by sensors or animals. Figure 21.1 depicts the complex interdependencies affecting landmine chemical transport in the soil.

Chemical vapors emanate from a buried landmine by permeation through the case materials or leakage through seals and seams, and from surface contamination on the case. Transport through the surrounding soil occurs in the liquid and gas phases by diffusion and advection. Liquid phase advection is driven by precipitation and evaporation of water from the soil, and gas advection can occur due to barometric pressure changes. Partitioning among the phases is important for explosive compounds, which tends to concentrate the explosives on the solid and in the liquid phase. The explosive compounds are also subject to biodegradation.

Trace chemical detection of buried landmines is a challenging task due to the extremely low concentrations available for sensing. Understanding the nature of the trace chemical detection process has matured through evaluation of the individual landmine-soil-atmosphere processes, and through evaluation of the interdependencies among these individual processes using sophisticated computer simulation tools (Phelan and Webb, 2002). Computational simulation models articulate the fundamental processes mathematically to provide a representation of the chemical signature emissions from the buried landmine, movements through the soil, and vapor release at the ground surface. Demonstration calculations have shown the dramatic impact of surface weather conditions on the vapor emanations from soil that are available for detection by technology or canines. Comparisons of simulation model predictions to laboratory data (Phelan et al., 2000, 2001) provide confidence that the model represents well the reality of certain situations. Simulations were extended

C. Ho and S. Webb (eds.), Gas Transport in Porous Media, 339–352.
© 2006 *Springer.*

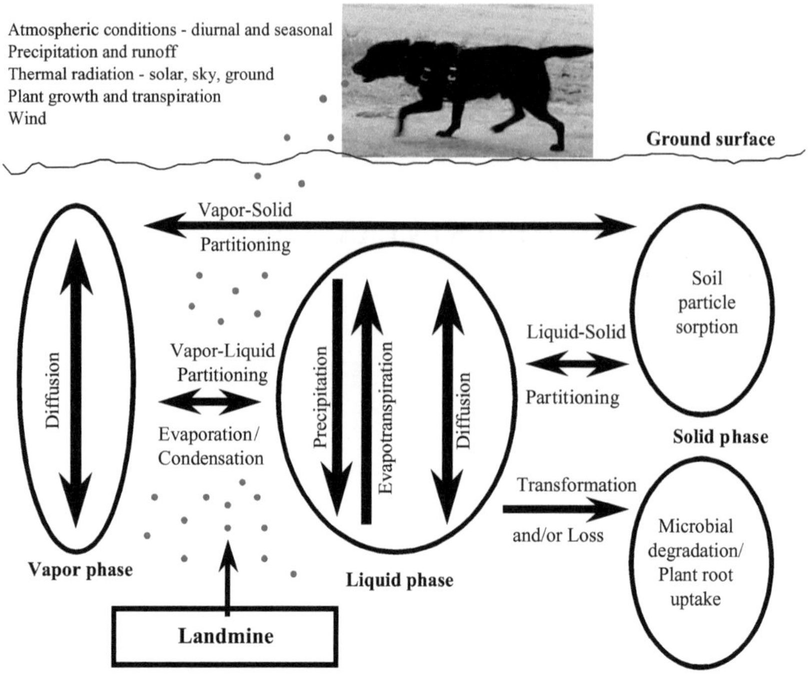

Figure 21.1. Complex interdependencies affecting landmine chemical signatures in soil

to location-specific weather impacts on laboratory determined landmine chemical emissions to provide a comprehensive time-series analysis of the changes in explosive vapor emanations at the ground surface (Webb and Phelan, 2003). Data-model comparisons between simulation results and landmine soil residue chemical data showed good correlation and affirmed that this approach can provide great insight into the art of landmine detection (Phelan and Webb, 2003).

A number of soil physics based simulation models have been used to evaluate landmine chemical vapor transport in soils. The models evolved from simple screening methods to more complex numerical approaches as insight was gained in the phenomenon involved.

The Behavior Assessment Model (BAM) (Jury et al., 1983, 1984a, b, c), which was later modified to become the Buried Chemical Model (BCM) (Jury et al., 1990) uses a one-dimensional screening model to compare the pollution potential of various agricultural chemicals. The BAM and BCM models were very useful in categorizing the relative mobility, volatility and persistence of the landmine signature chemicals TNT, DNT and DNB in relation to other well-characterized agricultural chemicals. These models only require a simple set of input parameters: soil-water partitioning coefficient (K_d), soil-air partitioning coefficient (Henry's constant, K_H), diffusion coefficient in air (D_G), and the bio-chemical half-life ($T_{1/2}$).

The Buried Chemical Model was also used to evaluate the effect of differing soil properties, water flux conditions and sequences on the behavior of TNT, DNT and DNB (Phelan and Webb, 1997; Phelan and Webb, 1998a; Phelan and Webb, 1998b; Webb et al., 1998). The surface vapor flux was evaluated because this parameter was considered to be the principal pathway for detection of buried landmines by dogs.

While the Buried Chemical Model was valuable for an initial assessment, the assumptions of constant and uniform liquid content and temperature, as well as a single boundary layer thickness, were obviously great simplifications. In order to address these and other issues, a multidimensional mechanistic code was modified for application to this problem. This code, which is based on the TOUGH code from Lawrence Berkeley Laboratory (Pruess, 1987, 1991), considers air, water vapor, and explosive chemical mass transport and heat flow in a porous media, and has been named T2TNT (Webb et al., 1999).

Modifications to TOUGH2 to produce T2TNT included the following:

(1) chemical components for landmine signatures (TNT, DNT, and DNB),
(2) gas diffusion – gas diffusion can be an important transport mode for explosive vapors in the subsurface,
(3) liquid diffusion – liquid diffusion can be a dominant transport mode for explosive vapors in the subsurface, especially for moderate and high moisture content conditions,
(4) liquid-solid sorption,
(5) vapor-solid sorption – vapor-solid sorption is significant for explosive vapors at low soil moisture contents,
(6) biodegradation,
(7) surface boundary conditions – due to the shallow burial depth of many landmines, the fluid conditions surrounding the landmine are strongly influenced by the surface conditions. The parameters necessary to adequately model the surface boundary conditions include: solar and long-wave radiation, the surface boundary layer that is a function of wind speed and soil-air temperature differences, precipitation and evaporation at the surface, plants and their root systems, and the diurnal and seasonal variation of these parameters.

To partially validate chemical transport in surface soil, an experimental soil column was fitted with an atmospheric air plenum configured to collected vapor emissions by solid phase micro extraction followed by chemical quantification with gas chromatography (Phelan et al., 2000). A liquid DNT source was injected into the soil and upward water and solute gradients were established with low humidity air flowing across the soil surface. Tests were performed to evaluate the effects of wetting and drying phenomena on the vapor flux of DNT at low liquid saturation, which include vapor-solid sorption phenomena (Phelan et al., 2001).

The data and model comparison for the surface flux of DNT is shown in Figure 21.2 that reflect the variation in soil column conditions. The initial relative humidity of the air was ~50%. At Day 35, the relative humidity was changed to 0%, which increased the evaporation rate and the DNT vapor flux. At Day 44, a drying event was imposed that dramatically lowered the soil saturation and the DNT vapor flux. A wetting event

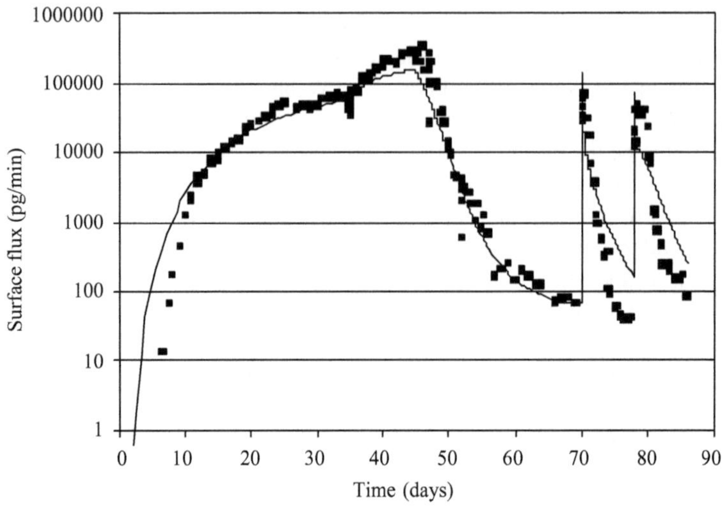

Figure 21.2. T2TNT Data model comparisons

at Day 69 significantly increased the DNT vapor flux and the soil saturation and another wetting-drying cycle was imposed after Day 69.

The simulation results show excellent agreement with the data, especially considering the three order-of-magnitude variations during wetting-drying cycles. The maximum differences are about a factor of 3, which is excellent considering the 5000-fold change in DNT surface flux.

The results from this test show how important soil-vapor partitioning can be to the vapor released by surface soils as indicated by the dramatic rise in the surface flux after wetting. In addition, the soil-water partition coefficient must be modeled with a Freundlich isotherm rather than a linear one, and the soil-water partition coefficient must be weighted for soil saturation. These test results give confidence in the predictive capability of the T2TNT code.

Demonstration calculations were then performed to estimate the influence of surface weather conditions. These calculations assumed a constant chemical source flux and a constant biodegradation rate for each chemical and used actual weather data. The results indicate the variability in the chemical concentrations on the soil surface over the long-term (1 year) and the short-term (daily). Details on the input data requirements and simulation results can be found in Webb and Phelan (2000). The weather data from a standard weather station consisted of the following: atmospheric pressure, air temperature, relative humidity, solar radiation, precipitation, wind speed and wind direction at four elevations. Long-wave radiation from the atmosphere was estimated from measured weather parameters.

Figure 21.3 show the diurnal variation in T2TNT simulation results showing key factors of precipitation and resulting soil saturation, surface radiation balance and

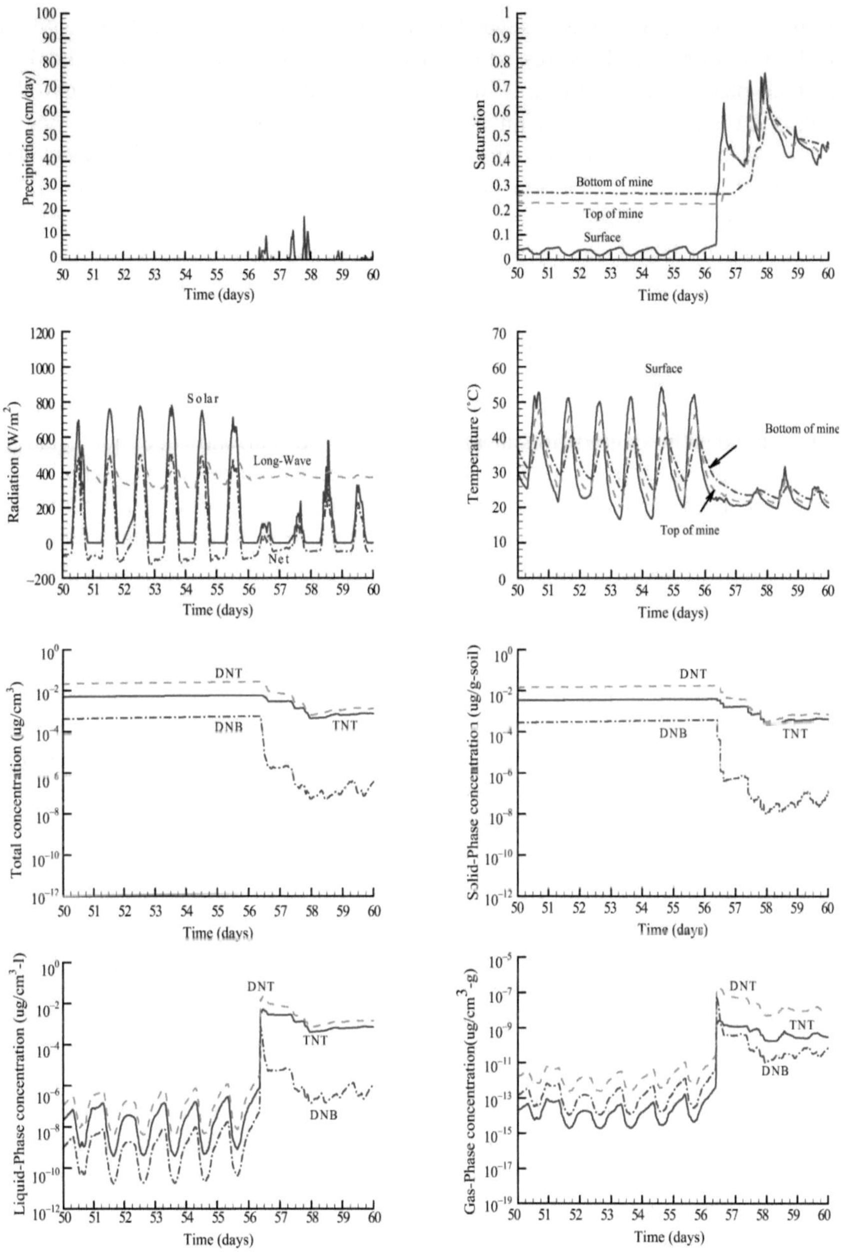

Figure 21.3. Diurnal variation of various surface parameters for the period 50–60 days

resulting soil temperatures at several depths, as well as the chemical concentrations of TNT, DNT and DNB expressed as total concentrations and as separate solid, liquid and gas phase concentrations. Of note is the dramatic increase in surface gas-phase concentrations of all three chemicals following a rainfall event.

The demonstration calculations showed that the complex interdependent interactions that occur in the soil could be simulated to produce estimates of the vapor concentrations and surface soil residues for comparison to mine dog performance capabilities. This tool was then used to evaluate various scenarios – certain combinations of particular mine types (leakage), soil properties, and weather patterns.

The impact of weather conditions on the movement of landmine signature chemicals through the subsurface and into the atmosphere was evaluated using the T2TNT code with specific improvements since initial development (Webb and Phelan, 2003). These included temperature dependent mine flux and temperature/soil moisture content dependent biodegradation rates. The T2TNT code currently does not include the effects of plants, so a bare soil version was used. Weather data from three diverse sites were used: Kabul, Afghanistan, an arid to semi-arid climate, Ft. Leonard Wood, Missouri, USA, a moderately wet climate with significant seasonal temperature variations, and Napacala, Mozambique, a wet climate with minimal seasonal temperature variations.

The results for the Kabul and Ft. Leonard Wood TNT gas-phase and solid-phase surface concentrations for the final year of the simulations are shown in Figure 21.4 and Figure 21.5. For Kabul, the year is from January to December. For Ft. Leonard Wood, the year is from mid-July to mid-June because of the weather data. Both the surface gas-phase concentrations shown in Figure 21.4 and the surface concentrations in Figure 21.5 vary widely due to seasonal weather patterns, most importantly, by precipitation. For Kabul, the large increases in gas-phase concentration occur after a rainfall event. The gas-phase concentration is predicted to increase up to 6–7 orders of magnitude. This increase is due to reduced vapor-solid sorption. When the soil saturation increases, the amount of vapor that can be sorbed to the solid decreases, resulting in a large increase in the gas-phase concentration. The Ft. Leonard Wood variation is similar to Kabul. However, because of the increased annual precipitation (62 cm vs. 18 cm), the dry-period values are generally lower for Ft. Leonard Wood because the chemical signature has been transported deeper by infiltrating rainfall and by accelerated biodegradation.

The surface solid-phase concentrations shown in Figure 21.5 for Kabul also decrease after rainfall, again primarily due to "washing down" of chemical signature. The values decrease during rain, and increase during dry periods. The range of solid-phase concentration is much smaller than the range in gas-phase values. For Ft. Leonard Wood, the variation is similar. With more frequent rain, there are much more frequent decreases in solid-phase concentration, and the values are generally lower.

Differences in soil vapor emanations for scenarios in regional areas (Afghanistan, Cambodia, Mozambique, Angola) demonstrates the critical nature of maintaining optimal mine dog performance and the futility of certain situations where vapor levels are well below typical mine dog vapor sensing capabilities.

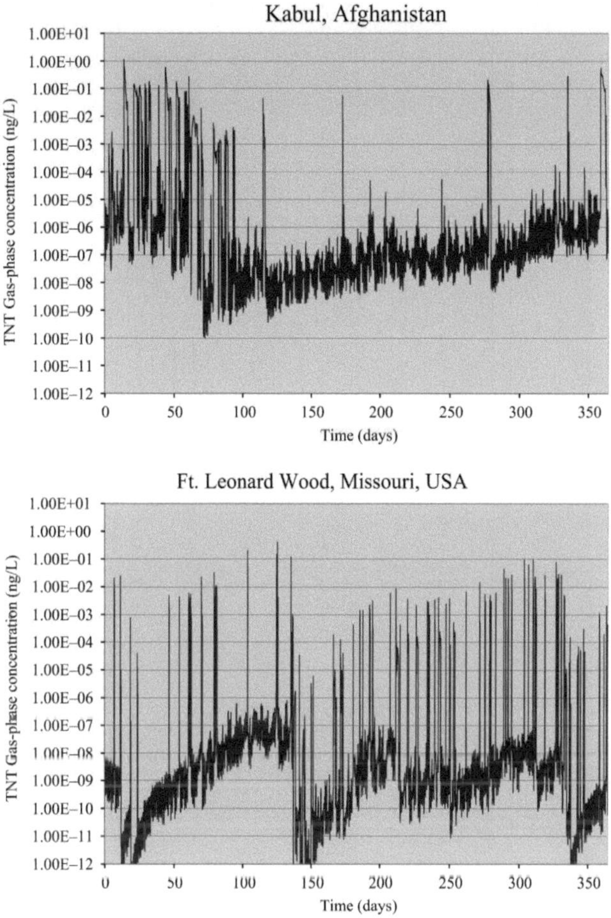

Figure 21.4. TNT surface gas-phase concentrations for two locations

Interpreting model predictions depends on how well the model represents actual field conditions. The process of model validation is difficult because neither the data available nor the mathematics representing the physical processes are complete enough to accurately represent reality. Understanding the nature of the actual situation and the constraints of the computational representation will provide a tremendous amount of knowledge in which to interpret new scenarios. With the confidence of the partial laboratory validation tests and recent landmine flux experimental data (Phelan et al., 2003) the T2TNT model was compared to historic field measured surface and sub-surface landmine signature chemical soil residues (Phelan and Webb, 2003).

The DARPA Dog's Nose Program developed a landmine chemical detection test facility at Fort Leonard Wood, Missouri, USA in 1998. Replicates of five types of

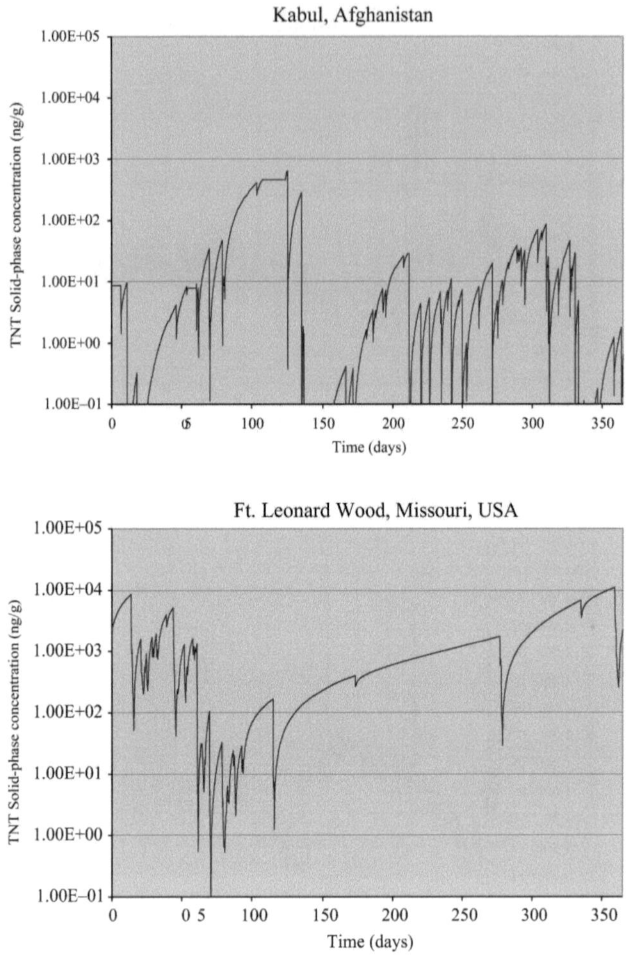

Figure 21.5. TNT surface solid-phase concentrations for two locations

landmines were buried and soil samples were collected during six campaigns over 21 months for analysis of the explosive signature residues (Jenkins et al., 2000; Hewitt et al, 2001). The landmines selected for this data-model comparison include only the PMA1A and PMA2 types because mine chemical emission data were only available for these two.

Using actual weather data, soil properties and mine flux rates, T2TNT estimates of the soil residues were compared to actual measured values from soil samples collected in the field. One must recognize that field sampling at discrete depth intervals is not exact. The soil surface is uneven, sampler penetration may vary plus/minus several millimeters or more, and corehole sloughing may transfer soil particles from

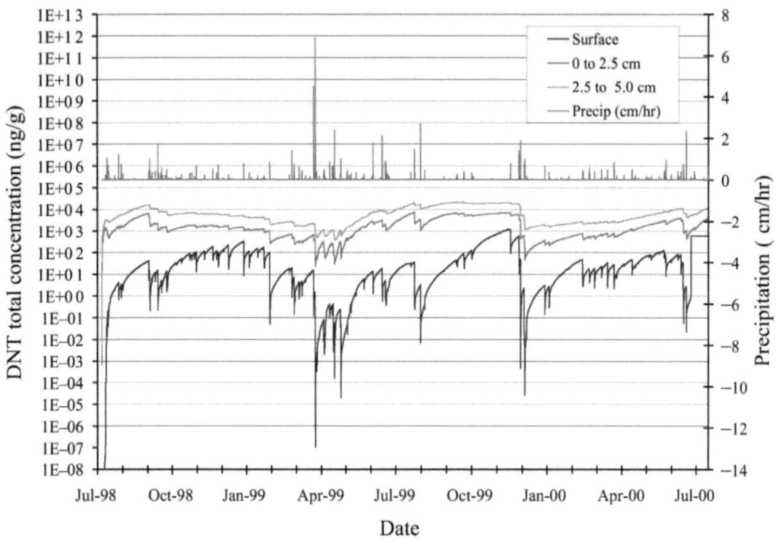

Figure 21.6. T2TNT Simulation Model Results – PMA1A Landmine, DNT Total Concentration

upper horizons to lower ones. Concentration profiles depicted by T2TNT have shown significant gradients especially near the surface (Webb and Phelan, 2002); therefore, comparison of model estimates to field soil samples was performed by qualitative visual comparison.

Figure 21.6 shows the total predicted concentration (sum of solid, aqueous and vapor phases) of DNT for the three field sample depth intervals and the precipitation for the entire two-year simulation period for the PMA1A landmine. The surface soil residues show the greatest impact of precipitation, with variable decreases in concentration directly proportional to the magnitude of the rain.

Periods of significant precipitation (Feb–Jun 1999 and Dec 1999) caused very large (10^4 to 10^6) decreases in surface soil residues. Depending on the amount of rainfall, recovery took several months. The greatest surface soil residues were about 1000 ng/g in November 1999 after a prolonged dry period. More typical values were from 1 to 100 ng/g, which approximate the summary statistical values (mean 7.7 ± 52.7 ng/g; median 32 ng/g; maximum 227 ng/g) of field samples for the entire period in Jenkins et al. (2000).

Figure 21.7 shows detailed plots of the T2TNT results for the field-sample intervals and precipitation for a 55-day period that includes the field sample campaign. Each field sampling campaign typically occurred over a 3-day period. The boxes on each plot represent the low to high values from the field samples.

These data-model comparisons represent very good to excellent correlation when considering the limitation of a 1-D simulation and constraints in formulation of the T2TNT model. The over and underestimations ranged from a factor of 10 to 100.

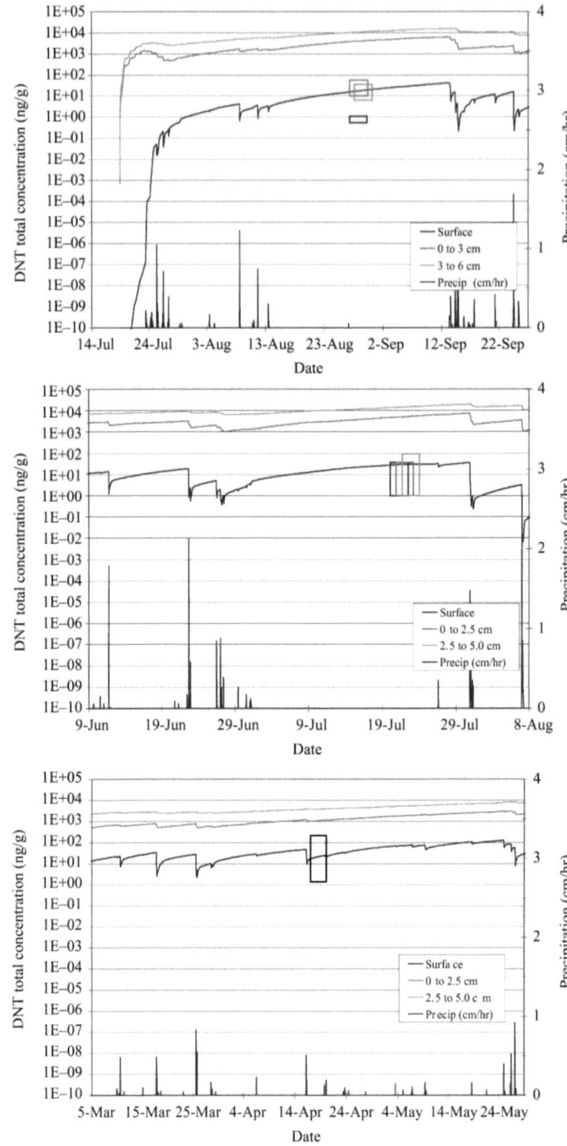

Figure 21.7. T2TNT simulation results (lines) and field-data samples (boxes) for PMA1A landmine, DNT, in 1998 (top), 1999 (middle), and 2000 (bottom)

Overestimates of model values would be expected because field samples are obtained adjacent to the landmine. The 1-D nature of the simulation also would contribute to overestimates because mass transfer takes place only vertically, with no lateral transport. The T2TNT simulation model does not include the effects of plants, which will significantly influence the near surface soil water and heat balance, or freezing conditions due to a limitation in the equation of state. In order to avoid freezing conditions, the air weather temperature was constrained to a minimum of 12°C.

This effort provides evidence that the T2TNT model can be used to estimate field conditions for trace chemical detection of buried landmines by comparison to total soil concentration data. Point sampling and chemical analysis of field soil residue data were adequately represented by the T2TNT model within the constraints of several factors: analytical detection limits, 1-D representation of a 3-D field data set, homogeneous model representation of heterogeneous soil properties, minimum temperature constraints of the model, and a bare soil representation of a low grass field.

With careful consideration of these factors, the T2TNT model can be used to evaluate combinations of weather, soil, and landmine flux to evaluate optimum conditions for use of trace chemical detection technology. This is supported by recent analysis by the US Government, which determined that methods for detecting the chemical components of explosives found in landmines and techniques for modeling the environmental effects on various mine sensors needs priority research (MacDonald et al., 2003). Cooperative efforts between model improvements, scenario analyses and feedback from field demining campaigns will further improve the utility of trace chemical detection technology for buried landmines.

REFERENCES

Hewitt, A.D., Jenkins, T.F., and Ranney, T.A. 2001, Field Gas Chromatography/Thermionic Detector System for On-Site Determination of Explosives in Soils. US Army Corps of Engineers, Engineer Research and Development Center, Cold Regions Research and Engineering Laboratory. ERDC/CRREL TR-01–9. May 2001.

Jenkins, T.F., Walsh, M.E., Miyares, P.H., Kopczynski, J., Ranney, T., George, V., Pennington, J., and Berry, T., 2000, Analysis of Explosives-Related Chemical Signatures in Soil Samples Collected Near Buried Landmines. U.S. Army Corps of Engineers, Engineer Research and Development Center, Report ERDC TR-00–5. August 2000.

Jury, W.A., Spencer, W.F., and Farmer, W.J., 1983, Behavior Assessment Model for Trace Organics in Soil: I. Model Description. *J. Environ. Qual.*, 12, 558–564.

Jury, W.A., Farmer, W.J., and Spencer, W.F., 1984a, Behavior assessment model for trace organics in soil: II. Chemical classification and parameter sensitivity. *J. Environ. Qual.*, 13, 567–572.

Jury, W.A., Spencer, W.F., and Farmer, W.J., 1984b, Behavior assessment model for trace organics in soil: III. Application of screening model. *J. Environ. Qual.*, 13, 573–579.

Jury, W.A., Spencer, W.F., and Farmer, W.J., 1984c, Behavior assessment model for trace organics in soil: IV. Review of experimental evidence. *J. Environ. Qual.*, 13, 580–586.

Jury, W.A., Russo, D., Streile, G., and Abd, H., 1990, Evaluation of volatilization by organic chemicals residing below the surface. *Water Resour. Res.*, 26, 13–20, January 1990.

MacDonald, J., Lockwood, J.R., McFee, J., Altshuler, T., Broach, T., Lawrence Carin, Russell Harnon, Carey Rappaport, W. Scott and R. Weaver, 2003. Alternatives for Landmine Detection. RAND Science and Technology Policy Institute for the Office of Science and Technology Policy. © RAND. February 2003.

Phelan, J.M., and Webb, S.W., 1997, Environmental Fate and Transport of Chemical Signatures from Buried Landmines-Screening Model Formulation and Initial Simulations. Sandia National Laboratories. SAND97–1426, June 1997.

Phelan, J.M., and Webb, S.W., 1998a, "Chemical Detection of Buried Landmines in". *Proceeding of the 3rd International Symposium on Technology and the Mine Problem*, April 6–9, 1998. Mine Warfare Association.

Phelan, J.M., and Webb, S.W., 1998b, "Simulation of the Environmental Fate and Transport of Chemical Signatures from Buried Landmines," in *Proceedings of the SPIE 12th Annual Conference on Detection and Remediation Technologies for Mines and Minelike Targets III, Orlando, FL, April 13–17, 1998.*

Phelan, J.M., and Webb, S.W., 2002. Chemical Sensing for Buried Landmines – Fundamental Processes Influencing Trace Chemical Detection. Sandia National Laboratories Report SAND2002–0909. May 2002.

Phelan, J.M., and Webb, S.W., 2003, "Data-Model Comparison of Field Landmine Soil Chemical Signatures at Ft. Leonard Wood", in *Proceedings of the SPIE Vol 5089, AeroSense Photonics for Defense and Security, Detection and Remediation Technologies for Mines and Minelike Targets VIII*, April 21–25, 2003, R.S. Harmon, J.H. Holloway and J.T. Broach, eds.

Phelan, J.M., Gozdor, M., Webb, S.W., and Cal. M., 2000. "Laboratory Data and Model Comparisons of the Transport of Chemical Signatures From Buried Landmines/UXO" in *Proceedings of the SPIE 14th Annual International Symposium on Aerospace/Defense Sensing, Simulation and Controls, Detection and Remediation Technologies for Mines and Minelike Targets V*, April 24–28, 2000, A.C. Dubey, J.F. Harvey, J.T. Broach, R.E. Dugan, eds.

Phelan, J.M., Barnet, J.L., Romero, J.V., Kerr D.R., and Griffin, F.A., 2003. Explosive Chemical Emissions from Landmines. in *Proceedings of the SPIE Vol 5089, AeroSense Photonics for Defense and Security, Detection and Remediation Technologies for Mines and Minelike Targets VIII,* April 21–25, 2003, R.S. Harmon, J.H. Holloway and J.T. Broach, eds.

Phelan, J.M., Webb, S.W., Gozdor, M., Cal M., and Barnett, J.L., 2001. Effect of Wetting and Drying on DNT Vapor Flux – Laboratory Data and T2TNT Model Comparisons. in *Proceedings of the SPIE 15th Annual International Symposium on Aerospace/Defense Sensing, Simulation and Controls, Detection and Remediation Technologies for Mines and Minelike Targets VI*, April 16–20, 2000, Orlando, FL. A.C. Dubey, J.F. Harvey, J.T. Broach, V. George, eds.

Pruess, K. 1987. TOUGH User's Guide, NUREG/CR-4645, SAND86–7104, LBL-20700, US Nuclear Regulatory Commission.

Pruess, K. 1991. TOUGH2 – A General-Purpose Numerical Simulator for Multiphase Fluid and Heat Flow. LBL-29400, Lawrence Berkeley Laboratory.

Webb, S.W., Finsterle, S.A., Pruess, K., and Phelan, J.M., 1998. Prediction of the TNT Signature from Buried Landmines. Proceedings of the TOUGH '98 Workshop. Berkeley, CA.

Webb, S.W., and Phelan, J.M., 2000. Effect of Diurnal and Seasonal Weather Variations on the Chemical Signatures from Buried Landmines/UXO. in *Proceedings of the SPIE 14th Annual International Symposium on Aerospace/Defense Sensing, Simulation and Controls, Detection and Remediation Technologies for Mines and Minelike Targets V,* April 24–28, 2000, A.C. Dubey, J.F. Harvey, J.T. Broach, R.E. Dugan, eds.

Webb, S.W., and Phelan, J.M., 2002. Effect of Weather on the Predicted PMN Landmine Chemical Signature for Kabul, Afghanistan. Sandia National Laboratories Report SAND2002–3779, Albuquerque, NM. November 2002.

Webb, S.W., and Phelan, J.M., 2003. Effect of Weather on Landmine Chemical Signatures for Different Climates. in *Proceedings of the SPIE Vol 5089, AeroSense Photonics for Defense and Security, Detection and Remediation Technologies for Mines and Minelike Targets VIII,* April 21–25, 2003, R.S. Harmon, J.H. Holloway and J.T. Broach, eds.

Webb, S.W., Pruess, K., Phelan J.M., and Finsterle, S., 1999. Development of a Mechanistic Model for the Movement of Chemical Signatures from Buried Landmines/UXO. in *Proceedings of the SPIE 14th Annual International Symposium on Aerospace/Defense Sensing, Simulation and Controls, Detection and Remediation Technologies for Mines and Minelike Targets IV,* April 5–9, 1999, A.C. Dubey, J.F. Harvey, J.T. Broach, R.E. Dugan, eds., Vol. 3710, pp. 270–282.

CHAPTER 22

ENVIRONMENTAL REMEDIATION OF VOLATILE ORGANIC COMPOUNDS

RON FALTA
Clemson University, Clemson, SC, USA

22.1 INTRODUCTION

Over the past 25 years, groundwater and subsurface contamination by volatile organic compounds (VOCs) has emerged as a widespread problem in industrialized countries. This class of compounds is broad, and it includes a variety of common organic chemicals such as aromatic hydrocarbons (e.g., benzene, toluene, and xylenes) aliphatic hydrocarbons, (e.g., pentane, hexane, and cyclohexane) and halogenated hydrocarbons (e.g., carbon tetrachloride, tetrachloroethylene, and trichloroethylene). These chemicals are widely used as chemical feedstocks, as cleaning solvents and degreasers, and, in the case of hydrocarbons, as fuels.

These chemicals, besides having a high vapor pressure, tend to have low aqueous solubilities, on the order of 50 to 2000 mg/l. The low solubility of most VOCs tends to result in high Henry's constants, and strong partitioning into the gas phase from the aqueous phase.

Unfortunately, many of these common chemicals are also toxic, and have very low maximum contaminant levels (MCLs) in drinking water. For example, benzene, which is one of the most common industrial chemicals in the world, has an aqueous solubility of about 1760 mg/l, yet the United States MCL is only 5 ug/l.

Due to their low aqueous solubility, VOCs typically exist as nonaqueous phase liquids (NAPLs), and they are often released to the environment in this form. The hydrocarbon fuel mixtures such as gasoline, kerosene, diesel, jet fuel, and fuel oil form LNAPLs (lighter than water NAPLs), while the chlorinated cleaning solvents form DNAPLs (denser than water NAPLs). Due to capillary trapping, complete removal of NAPLs from the subsurface by mobilization is nearly impossible, and conventional fluid pumping operations typically leave behind NAPL saturations of 5 to 25%.

C. Ho and S. Webb (eds.), Gas Transport in Porous Media, 353–370.
© 2006 *Springer.*

Removal of the trapped NAPL generally requires mass transfer of the NAPL components into either the gas phase or the aqueous phase, where they can be pumped out, or destroyed by aqueous or gas phase chemical or biological reactions. Arguably, the most successful NAPL/VOC remediation schemes have used the gas phase to transfer contaminants out of the subsurface, or to deliver beneficial chemicals or nutrients to the subsurface. Generally speaking, remediation methods that rely on gas phase extraction are most effective for chemicals that have high vapor pressures and Henry's constants, while methods that rely on gas phase delivery of chemicals or nutrients can attack a wider range of contaminants.

Obviously, remediation schemes based on gas phase transport are applicable in the vadose zone, above the water table, however there are also effective remediation methods that involve injecting air or steam below the water table. The following sections describe some of the more popular gas-based remediation methods for VOCs.

22.2 SOIL VAPOR EXTRACTION

22.2.1 Introduction to SVE Applications, Induced Gas Flow

Soil vapor extraction is probably the simplest, most common, and most successful of all of the in-situ techniques for removing VOC's from the vadose zone. SVE systems use one or more wells screened above the water table. A blower connected to the well normally provides the vacuum (typically from 0.05 to 0.2 atmospheres) that induces flow to the well. Typical gas flow rates are in the range of 50 to 300 cfm (0.024 m^3/s to 0.14 m^3/s) depending on the vacuum in the well and the gas phase permeability. Figure 22.1 from Shan et al. (1992) shows the calculated gas streamlines for a single SVE well in a homogeneous, isotropic system that is open to the atmosphere. Each of the streamtubes in this figure originate at the ground surface, and they each contribute 5% of the total gas flow to the well. As clean soil gas flows through the contaminated zone, NAPLs evaporate, and dissolved or adsorbed VOCs partition into the moving gas phase. Once the gas is removed from the well, it is treated, often by granular activated carbon adsorption. The gas travel time from a location inside the SVE well capture zone to the well determines how effectively that location is flushed by gas. Considering that several thousand pore volumes of gas are often needed for

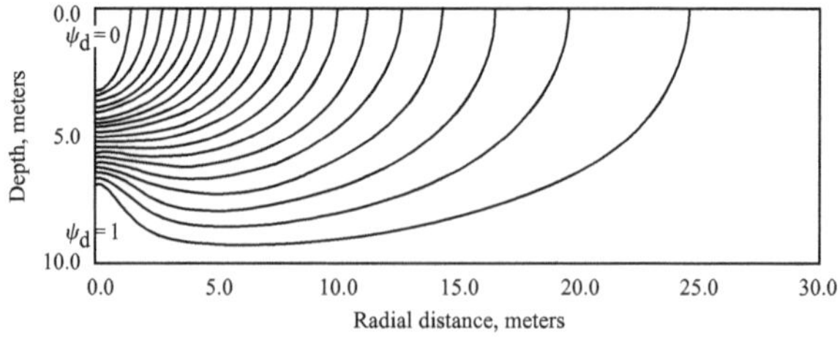

Figure 22.1. Calculated gas streamlines around an SVE well for homogeneous, isotropic conditions, with an open ground surface (from Shan et al. (1992))

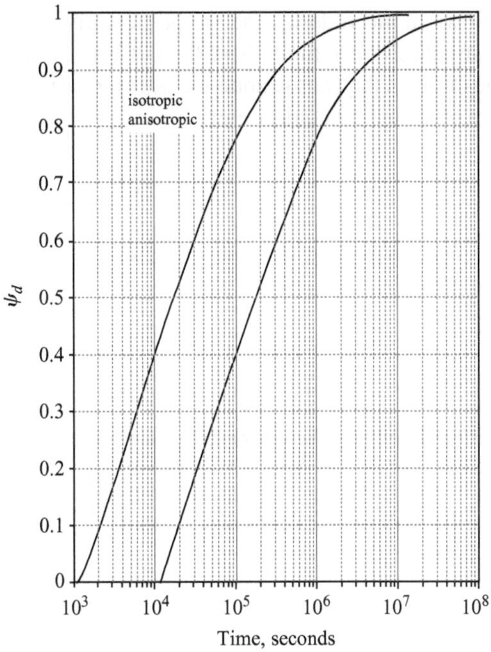

Figure 22.2. Calculated gas travel time from the ground surface to the SVE well for the streamlines shown in Figure 22.1 (from Shan et al. (1992))

complete clean up at real sites (Murdoch, 2000), it is clear that the true "radius of influence" for an SVE well should be determined by gas travel time, rather than by the pressure field induced by the well. Figure 22.2 shows the calculated gas travel times from the ground surface to the well for each of the streamlines shown in Figure 22.1 assuming a gas flow rate of 0.081 m^3/s(170 cfm), a horizontal gas phase permeability of 1×10^{-11} m^2, and a vertical permeability of either 1×10^{-11} m^2 or 1×10^{-12} m^2 (the anisotropic case). From this figure it can be seen that while the streamlines that start near the well have very short travel times, and are flushed by many pore volumes of gas, those farther from the well have much larger travel times, and are not flushed as effectively. Contour plots of the computed gas travel time distribution or gas velocity give a better idea of the effective cleanup zone for an SVE well (see, e.g., Falta et al., 1993; DiGiulio and Varadhan, 2001).

22.2.2 Heterogeneous Nature of Gas Flow, Effects of Pore Water

The preceding discussion of the SVE gas flow patterns was based on a simplified conceptual model of a homogeneous system. Real subsurface systems always exhibit some degree of heterogeneity, and it is common for SVE induced gas flows to be strongly affected by these heterogeneities. Compared to groundwater flow below the water table, gas flows in the vadose zone are particularly sensitive to variations in the porous media structure due to multiphase flow effects. As discussed in Chapter 2, Darcy's law for gas flow in the vadose zone includes a relative permeability term k_{rg} to account for pore blockage by liquid water:

$$v_g = -\frac{kk_{rg}}{\mu_g}(\nabla p_g + \rho_g g \nabla z) \tag{22.1}$$

and the relative permeability term is typically a nonlinear function of the water saturation, such as (Fatt and Klickoff, 1959)

$$k_{rg} = (1 - S^*)^3 \tag{22.2}$$

where $S^* = (S_l - S_{lr})/(1 - S_{lr})$ and S_{lr} is the residual water saturation. Thus, the water content of the soil can have a large effect on the effective gas velocity, and variations in the water content can have as much effect on the gas flow as variations in the intrinsic permeability. Using the idea of Leverett scaling of capillary pressure (Leverett, 1941), it is expected that capillary pressure should vary with the inverse square root of the intrinsic permeability:

$$p_{c\,lg} \propto \frac{1}{\sqrt{k}} \tag{22.3}$$

so layers or regions with low permeability also tend to have high capillary pressure at a given water saturation. Therefore, under gravity-capillary equilibrium conditions above the water table, low permeability zones tend to have high water saturations, leading to very low gas phase relative permeabilities. This compound effect on the effective gas phase permeability dominates the gas flow in heterogeneous systems.

Consider, for example, the simplified layered vadose zone shown in Figure 22.3. This relatively homogeneous formation consists of three different sands, with corresponding capillary pressure curves, and an intrinsic permeability variation of a factor of 30. By neglecting infiltration and assuming gravity-capillary equilibrium the individual media capillary pressure curves can be used to construct the profile of water saturation above the water table. This is shown as the dark line in the right hand part of the figure. Due to it's proximity to the water table, the bottom medium sand layer has a very high water saturation, about 0.98 or so. Using Eq. (22.2) this would lead to a gas phase relative permeability of only about 0.00001, so this unit would be virtually impermeable to gas flow unless the water table dropped. The coarse sand above the medium sand has a lower water content, roughly 0.37, giving a relative permeability of about 0.4. Therefore, this coarse sand layer would have an effective gas phase permeability of 4. The fine sand layer just above has a high water saturation of about 0.75, and it's gas phase relative permeability would only be about 0.025, for an effective permeability of only 0.0075. Comparing these last two layers, the contrast in intrinsic permeability is a factor of 30, while the contrast in effective gas permeability is a factor of more than 500. The coarse sand layer just above the fine sand has an effective gas permeability that is more than 1000 times higher than the fine sand.

22.2.3 Diffusive Mass Transfer Limitations

Contaminant concentrations leaving SVE wells are typically well below saturation values predicted by the chemical's vapor pressure or Raoult's Law. While this is partly due to dilution effects, a number of experimental studies (Gierke et al., 1992; Wilkins et al., 1995; Fischer et al., 1996) have evaluated the local scale interphase mass transfer between the NAPL or aqueous phase, and the flowing gas phase. At the small scale of these studies (cm to tens of cm) it does not appear that gas phase transport is the limiting factor in the mass transfer. For example, Baehr et al. (1989) observed the evaporation of gasoline in an experimental column to be an equilibrium process. Similarly, Wilkins et al. (1995) observed relatively small deviations from expected equilibrium values in their column NAPL evaporation tests at realistic gas velocities, although at higher gas velocities, the effluent concentrations were only about 60 to 80% of the equilibrium value.

For cases where the contaminant is dissolved in pore water, the local mass transfer into the gas phase can be slower, depending on the type of soil, and the length of time over which the experiment is operated (Gierke et al., 1992; Fischer et al., 1996). At large times, it appears that the local interphase mass transfer is limited by aqueous diffusion, but at early times, an equilibrium model usually provides a good fit with the data.

As discussed previously in Section 2.6, gas flow patterns that occur in heterogeneous media give rise to larger scale gas diffusion limited mass transfer in SVE systems. Referring again to Figure 22.3, for a given horizontal gas pressure gradient, the gas flow in the coarse sand units would be hundreds of times faster than in the fine

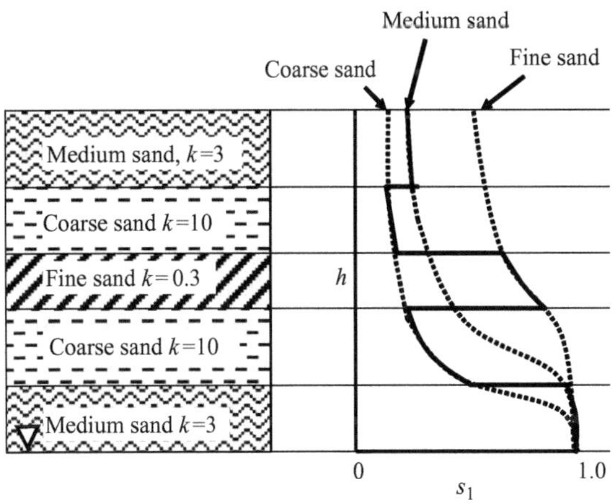

Figure 22.3. Gravity-capillary equilibrium in a layered vadose zone (after Looney and Falta, (2000))

sand unit. This strong advective sweeping would tend to quickly remove any volatile contaminants from the coarse sand zones (as would natural processes like barometric pumping). However contaminants trapped in the fine sand would not be subject to much gas flushing, and the rate of contaminant removal would be limited by the rate of gas diffusion out of the fine sand, into the coarse sand. This conceptual model has been validated by the experimental and analytical work by Ho and Udell (1991,1992) (see Section 2.6).

22.3 AIR SPARGING

22.3.1 Introduction to Air Sparging

Air sparging refers to the injection of a noncondensible gas (usually air) below the water table for the purpose of evaporating NAPLs and volatilizing dissolved VOCs. Air sparging wells are almost always combined with SVE wells, and they typically inject at a rate of about one-fifth to one-tenth of the SVE flow rate to insure capture of the injected gas. Air sparging is inexpensive and easy to implement at a site but the efficiency in the field seems highly variable, and sometimes disappointing. The apparent inefficiency of some sparging operations probably results from the nature of the gas flow below the water table, which can lead to poor contact with contaminants and poor mass transfer into the gas phase.

22.3.2 Nature of Noncondensible Gas Flow Below Water Table, Bubbles Versus Channels, Capillary Barriers, Unstable Flows

The flow of noncondensible gases below the water table is completely different from gas flow in the vadose zone. Although Darcy's law still applies to the gas flow, below the water table the gas movement is dominated by gravitational and capillary effects, and it is almost always unstable.

Some early studies of mass transfer during air sparging assumed that the injected gas moved upwards through the water by gravity as discrete bubbles. However, experimental studies have shown that this is only true in coarse gravels (Ji et al., 1993). In typical systems, the pore geometry is such that a nonwetting gas phase bubble would be trapped by capillary forces. These capillary forces are only exceeded when the vertical length of the gas bubble becomes very large, such as when the gas forms a channel. Figure 22.4 shows a schematic diagram of a trapped gas bubble below the water table. In order for this gas to flow upwards as a bubble, the gravitational buoyancy force must exceed the capillary force needed to squeeze through the pore throat. Following Hunt et al. (1988), a force balance may be used to analyze the conditions under which a gas bubble could migrate. Referring to the figure, the bubble has a vertical length of l_v, and the radius of curvature at the top of the bubble, r_1 is approximately equal to the pore throat radius. The radius of curvature at the bottom of the bubble, r_2 is approximately equal to the pore body radius. At static

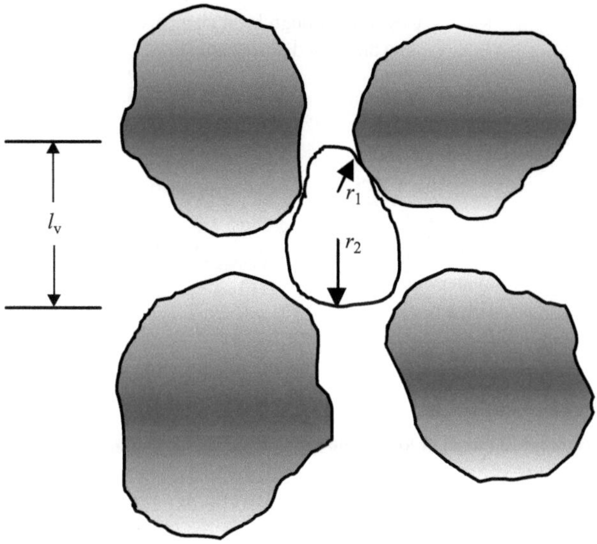

Figure 22.4. Geometry of a trapped gas bubble below the water table

equilibrium, the buoyancy forces are balanced by the capillary forces:

$$(\rho_l - \rho_g)gl_v = 2\sigma_{lg}\left(\frac{1}{r_1} - \frac{1}{r_2}\right) \tag{22.4}$$

where σ_{lg} is the surface tension of the wetting fluid (72 dynes/cm for pure water). Eq. (22.4) may be used to predict the conditions where gas bubble flow is possible in porous media. Suppose that the average pore body radius is twice the average pore throat radius ($r_2 = 2r_1$), and that the average vertical bubble length is equal to the pore body diameter ($l_v = 2r_2 = 4r_1$). Rearranging (4.4), and solving for r_1 we find that a pore throat diameter of 2.7 mm would be necessary for gas bubble mobility. Therefore, except in unusually coarse media, gas flow below the water table occurs in gas channels. Figure 22.5 shows a conceptual diagram of the local scale gas channeling that occurs during air sparging. Depending on the average spacing of these channels, the mass transfer of dissolved VOCs into the gas phase can be limited by aqueous phase diffusion (Clayton, 1998; Elder and Benson, 1999; Semer and Reddy, 1998).

At a macroscopic scale, gas flows below the water table are strongly affected by heterogeneous capillary barrier effects caused by variations in the air entry pressure of different materials. This phenomena is in some ways analogous to the multiphase flow capillary effects described in the SVE section. The air entry pressure of a porous material is usually related to the overall strength of the capillary pressure in that

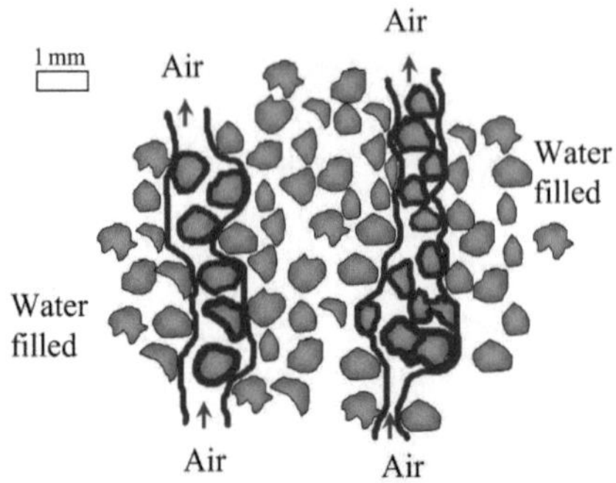

Figure 22.5. Conceptual model of local scale gas flow during air sparging (from Falta, (2000a))

material. Therefore, Leverett scaling could also be expected to apply to the gas entry pressure:

$$p_{\text{entry}} \propto \frac{1}{\sqrt{k}} \tag{22.5}$$

A gas phase will only penetrate a water filled material when the gas pressure exceeds the sum of the water pressure and the entry pressure. The high entry pressure of lower permeability materials often diverts the gas flow. These types of capillary barrier effects during air sparging can be clearly seen in the experiments of Ji et al. (1993) and in the numerical simulations by McCray and Falta (1996, 1997).

Injection of noncondensible gases below the water table tends to cause unstable displacements that lead to undampened fingering at various scales. The low viscosity of gases compared to water (about a factor of 50) leads to viscous instability when the gas is displacing water, while the low density of gas compared to water leads to buoyancy instability when the gas is moving upward through the water. The fact that both stability criteria are not met means that the gas flow during air sparging is theoretically unconditionally unstable (Slobod and Howlett, 1964). However it should be noted that several numerical modeling attempts have been able to reproduce the experimentally observed gas flow patterns (Larson and Falta, 1996; Hein et al., 1997; McCray and Falta, 1997).

22.3.3 Expected Average Gas Saturations during Sparging

It has been observed in both laboratory and field studies that sparging gas saturations tend to be higher in fine grained, low permeability formations, and lower in coarse grained, high permeability formations (Clayton, 1998). This observation can be explained by an analysis of one-dimensional vertical flow of gas at steady state (Falta, 2000). Numerical simulations by McCray and Falta (1996, 1997) have shown that once the sparging system stabilizes, the water inside the sparge zone is essentially static, and at hydrostatic pressure. Therefore, the vertical water pressure gradient is $\partial p_l/\partial z = \rho_l g$. Because the gas pressure is related to the water pressure by the capillary pressure, the vertical gas pressure gradient is

$$\frac{\partial p_g}{\partial z} = \frac{\partial p_l}{\partial z} + \frac{\partial p_{c\,\mathrm{lg}}}{\partial z} \tag{22.6}$$

If the vertical capillary pressure gradient is small compared to hydrostatic pressure gradient, then the vertical gas pressure gradient can be approximated as the hydrostatic water pressure gradient. Substituting into the multiphase Darcy's law and neglecting the gas density gives the approximate vertical sparge gas velocity

$$v_g \approx \frac{kk_{rg}}{\mu_g} \rho_l g \tag{22.7}$$

rearranging, the gas phase relative permeability is found to be linearly proportional to the gas Darcy velocity, and inversely proportional to the intrinsic permeability:

$$k_{rg}(S_g) = \frac{v_g \mu_g}{k \rho_l g} \tag{22.8}$$

Thus, for a given darcy velocity, the relative permeability, and hence gas saturation must be large if the intrinsic permeability is small, and the relative permeability and gas saturation must be small if the intrinsic permeability is large.

22.3.4 Mass Transfer Issues at Different Scales

A number of laboratory air sparging experiments using dissolved VOCs have shown strong nonequilibrium effects (Braida and Ong, 1998; Hein et al., 1998; Semer and Reddy, 1998; Adams and Reddy, 1999). Attempts to simulate these laboratory experiments assuming local chemical equilibrium between the gas and aqueous phases tend to greatly overpredict the rate of removal of the dissolved VOC. Figure 22.6 shows the effluent gas concentration of trichloroethylene (TCE) measured during a column air sparging experiment by Hein et al. (1998). The experimental data are represented by the open circles, and they show an initial concentration peak, followed by extensive tailing. An attempt to simulate this experiment using a conventional multiphase flow approach assuming local chemical equilibrium produces the solid line in the figure. The local equilibrium model predicts rapid and complete removal of the TCE from the column with no tailing.

The tailing of VOC concentrations during column sparging tests is believed to be due to aqueous phase diffusion limitations [Clayton, 1998; Semer and Reddy, 1998; Elder and Benson, 1999; Falta, 2000a]. A common approach for modeling this type of kinetic interphase mass transfer is through a first order mass transfer reaction in each gridblock (see, e.g., Sleep and Sykes, 1989; Gierke et al., 1992; Braida and Ong, 1998; Elder et al., 1999). Considering a two-phase gas/aqueous (no NAPL) system, the rate of interphase mass transfer can be calculated by:

$$Q_{\mathrm{imt}} = k_{\mathrm{imt}} a (C_l H - C_g) \tag{22.9}$$

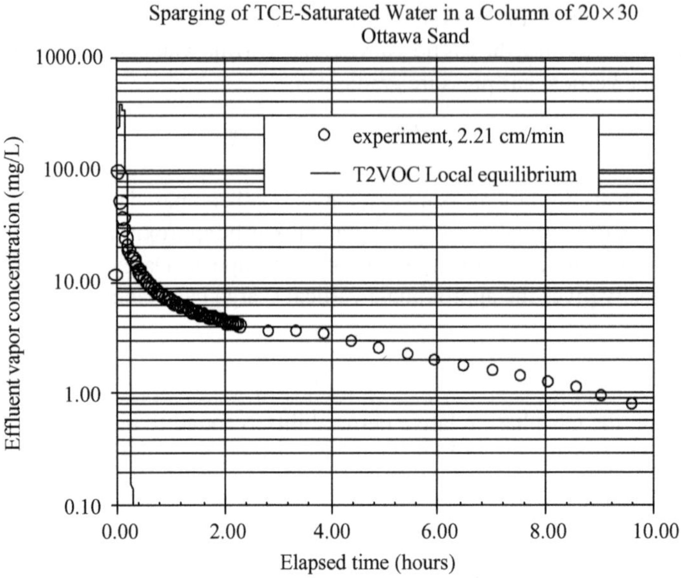

Figure 22.6. Simulation of an air sparging experiment assuming local chemical equilibrium (from Falta, (2000a))

where Q_{imt} is the rate of chemical mass transfer from the water to the gas phase per unit volume of porous media (kg/m^{3}s), C_l is the aqueous phase chemical concentration (kg/m^3), C_g is the gas phase chemical concentration (kg/m^3), H is the dimensionless Henry's constant for the chemical in water (assumed to be constant), and $k_{imt}a$ is the mass transfer coefficient-interfacial area product (s^{-1}).

Figure 22.7 shows a comparison of the experimental sparging data with a simulation that considers the local mass transfer from the pore water into the flowing gas phase by diffusion (Falta, 2000a). Although this numerical result was calculated using a dual domain approach, it is essentially equivalent to assuming local equilibrium within the gas channels (which still contain some pore water around the soil grains) with a first order mass transfer between the channels and the stagnant pore water.

Interestingly, in simulations of a hypothetical homogeneous field scale air sparging system, Falta (2000a) found that the mass transfer coefficients calibrated from the small laboratory experiments produced a result that was almost the same as simulations that assumed local equilibrium. This implies that the mass transfer rate limitations that are important in small column experiments may not be all that important in field scale systems that operate on larger length and time scales. Instead, large scale fingering due to heterogeneous capillary barrier effects may dominate the field scale mass transfer of real air sparging systems.

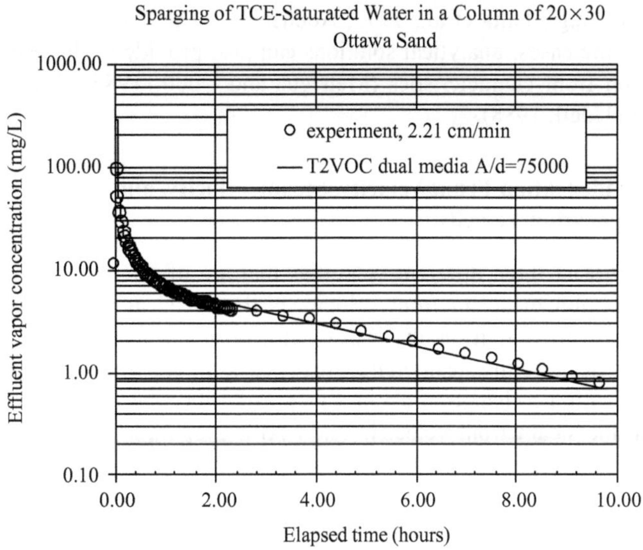

Figure 22.7. Simulation of an air sparging experiment using a dual-domain kinetic interphase mass transfer approach (from Falta, (2000a))

22.4 STEAM INJECTION

22.4.1 Introduction to Steam Injection

Subsurface steam injection can be used to heat both the vadose zone and areas below the water table to water boiling temperatures. Because chemical vapor pressures are a strong function of temperature, heating can greatly increase the rate of evaporation of a NAPL. A typical steam injection operation involves one or several steam injection wells, combined with multiphase extraction wells. When steam is injected in the vadose zone, it forms an expanding steam zone where the temperature is about 100°C. A condensation front exists at the edge of the steam zone, where the temperature quickly drops from the steam temperature to the ambient temperature (Menegus and Udell, 1985; Hunt et al., 1988; Stewart and Udell, 1988). As the steam contacts NAPL, it evaporates or boils, and the vapors move in the gas phase by advection to the condensation front, where they can condense back into NAPL. Because the NAPL has a tendency to accumulate at the condensation front, it often becomes mobile as a separate phase there (Stewart and Udell, 1988).

Below the water table, the steam process is similar, except that much of the resident pore water is hydraulically displaced by the advancing steam front. The movement of the water/NAPL condensation front is controlled by a complex interplay of multiphase flow and thermodynamic effects, but it is nonetheless fairly predictable by numerical methods (see, e.g., Falta et al., 1992; Adenekan, 1992; Falta, 2000b; Ochs et al., 2002). In some cases, analytical solutions can also provide useful estimates of the condensation front characteristics (Menegus and Udell, 1985; Hunt et al., 1988; Stewart and Udell, 1988).

22.4.2 Condensation Front Velocity, Steam Velocity, and NAPL Evaporation Velocity

There are at least three different velocities that are of interest in steam flooding. The condensation or steam front velocity is the speed at which the steam zone grows. The steam velocity is the Darcy velocity of the steam behind the condensation front. Because the steam condenses back to liquid water at the condensation front with a volume change of about a factor of 1600, the steam velocity is much larger than the steam front velocity. The NAPL evaporation velocity is the speed at which an evaporation front moves through immobile NAPL trapped at residual saturation.

Considering a one-dimensional system without any NAPL, Menegus and Udell, (1985) developed an analytical solution for the condensation front velocity:

$$v_{sf} = \left(\frac{\dot{m}_{in}}{\rho_l}\right)\left[\frac{\frac{Xh^{w}_{vap}}{C_{pl}(T_s - T_0)} + 1}{(1-\phi)\frac{\rho_R C_R}{\rho_l C_{pl}} + \phi \bar{S}_l}\right] \tag{22.10}$$

where $\dot{m}_{in}$ is the injected steam mass flux (kg/m^{2}s), X is the steam quality (the mass fraction of water vapor in $\dot{m}_{in}$), C_{pl} is the heat capacity of water, h^{w}_{vap} is the heat of vaporization of water, T_s and T_0 are the steam zone and ambient temperatures, respectively, and $\bar{S}_l$ is the water saturation behind the condensation front. Condensation front velocities in field applications are in the range of a few meters per day, depending on the well geometry, formation thickness, steam injection rates, and the steam quality. From Eq. (22.10), it is apparent that the one-dimensional steam front velocity is controlled by the rate of heating of the porous media and residual water.

The steam darcy velocity behind the condensation front may be estimated from a simple mass balance as:

$$v_s = \frac{\dot{m}_{in} X}{\rho_g} \tag{22.11}$$

where ρ_g is the steam density, about 0.6 kg/m^3 at 100°C. The steam darcy velocity behind the steam front is about two orders of magnitude larger than the condensation front velocity. This very high gas velocity gives the steam a tremendous capability for evaporation of NAPLs in the steam zone. It is useful to note that noncondensible gases behave differently from steam because the leading edge of the noncondensible gas advances at a rate similar to the gas pore velocity behind the leading edge. For this reason, the addition of air or other gases to steam can substantially change the characteristics of the condensation front, causing it to spread out over a larger distance. This spreading of the condensation front by noncondensible gas injection can be used to minimize the accumulation and mobilization of NAPL at the condensation front (Farber, 1997).

If the NAPL boiling point is lower than the steam temperature, then any NAPL not mobilized by the steam front will boil as the condensation front passes through it. If however, the NAPL has a higher boiling point, some of the trapped NAPL can remain behind the condensation front in the steam zone. The rate at which this trapped NAPL evaporates is of interest because in some cases the trapped high boiling point NAPL can still evaporate fast enough to keep up with the advancing steam condensation front (Falta et al., 1992; Yuan and Udell, 1993).

The speed of the NAPL evaporation front in the steam zone can be estimated by assuming a constant, immobile NAPL saturation, local chemical equilibrium between the gas and NAPL phases, and the ideal gas law. Using the chemical vapor pressure at the steam temperature, p^c_{vap}, the saturated vapor concentration is calculated, and multiplied by the steam darcy velocity to get the NAPL evaporation rate. This is divided by the mass of NAPL per unit volume to get the NAPL evaporation front velocity (Falta et al., 1992):

$$v_{ef} = \frac{v_s (p^c_{\mathrm{vap}} M^c_{wt} / R T_s)}{\phi \rho_n S_{nr}} \tag{22.12}$$

where M^c_{wt} is the molecular weight of the chemical, and S_{nr} is the residual NAPL saturation. It appears that NAPLs with boiling points below about 175°C will not tend to exist in the steam zone based on the analyses by Falta et al. (1992) and Yuan and Udell (1993), and on the experiments of Stewart and Udell (1988), Basel (1991) and Yuan and Udell (1993).

22.4.3 Stripping of Dissolved Chemicals during Boiling

One of the interesting phenomena that can occur during steam injection operations is boiling of pore water in hot zones during depressurization (Udell, 1994). This boiling can be stimulated in field operations by stopping the steam injection once the treatment zone is hot, and starting active vapor extraction operations to create a vacuum. As the pore water boils into vapor, the associated volume change of a factor of about 1,600 leads to a very strong vapor stripping mechanism that can remove dissolved contaminants from the liquid water. Following Udell (1994), this steam stripping mechanism can be analyzed using a mass balance on the dissolved contaminant. It is assumed that the porous media is already at the steam temperature, and that depressurization is causing boiling of the pore water. The mass of chemical in the pore water can be written as

$$M_c = \frac{M_l C_l}{\rho_l} \tag{22.13}$$

where M_l is the mass of liquid water. The volumetric rate at which vapor is created and leaves the system is equal to the rate at which the liquid mass is depleted by boiling, divided by the gas (steam) density. The rate at which vapor phase chemical leaves the system is then equal to the product of the gas concentration, and the volumetric rate of vapor creation. Setting up a mass balance on the chemical,

$$\frac{1}{\rho_l}\frac{\mathrm{d}(M_l C_l)}{\mathrm{d}t} = \frac{C_g}{\rho_g}\frac{\mathrm{d}M_l}{\mathrm{d}t} \tag{22.14}$$

Assuming local equilibrium with Henry's Law gives

$$M_l \frac{\mathrm{d}C_l}{\mathrm{d}t} = \left(\frac{H\rho_l}{\rho_g} - 1\right) C_l \frac{\mathrm{d}M_l}{\mathrm{d}t} \tag{22.15}$$

This equation may be solved with the initial condition of a known concentration and liquid water mass to get (Udell, 1994)

$$\frac{C_l}{C_l^o} = \left(\frac{M_l}{M_l^o}\right)^{\left(\frac{H\rho_l}{\rho_g} - 1\right)} \tag{22.16}$$

Eq. (22.16) gives the dissolved concentration reduction that results from boiling of a fraction of the initial water mass. Because the ratio of the liquid to the gas density appears in the exponent, very large reductions in the dissolved concentration occur with only small amounts of boiling. For example, Udell (1994) shows that using the properties of TCE, a reduction in the water mass of 2% leads to a five order of magnitude reduction in the dissolved concentration, while a 5% loss of water leads to a twelve order of magnitude reduction in the dissolved concentration.

22.5 BIOREMEDIATION

22.5.1 Introduction to Aerobic Degradation of Hydrocarbons

In-situ destruction of contaminants through biological reactions is an important component of the overall chemical fate and transport process. Bacteria called organotrophs use organic compounds for their energy source in an oxidation reaction. The organic chemical loses electrons as it is oxidized, and is thus referred to an electron donor. The biological oxidation–reduction reaction causes the reduction of an electron acceptor. If the electron acceptor is oxygen (O_2), then the process is called aerobic degradation. If the electron acceptor is anything other than oxygen, such as nitrate (NO_3^-), iron (Fe^{3+}), or sulfate (SO_4^-), the reaction is anaerobic (Hughes, 1999). It is generally believed that almost all of the biological activity takes place in the aqueous phase, and not in the gas or NAPL phases. For this reason, biodegradation can be limited by the rate of interphase mass transfer of contaminants from the NAPL or gas phase into the aqueous phase.

Many of the hydrocarbon compounds are readily biodegradable in water under aerobic conditions. The most highly regulated components of hydrocarbon fuels, benzene, toluene, ethylbenzene, and xylenes, (BTEX) are degraded quickly under aerobic conditions (Alvarez and Vogel, 1991). While the details of the aerobic reactions, and their kinetics are fairly complex, the oxygen demand can be estimated from straightforward stoichiometric relationships. For example, an estimate of the amount of oxygen needed to aerobically degrade hydrocarbons is calculated using a representative compound such as hexane (Looney and Falta, 2000):

$$C_6H_{14} + 9.5O_2 = 6CO_2 + 7H_2O \tag{22.17}$$

Therefore, for every mole of hexane (84 g), 9.5 moles of O_2 (304 g) are consumed, producing 6 moles of CO_2 (264 g). In other words, destruction of 1 mg/l of hexane in water requires a dissolved oxygen concentration of 3.6 mg/l, and it produces a carbon dioxide concentration of 3.1 mg/l. Measurements of the rates oxygen depletion and carbon dioxide production are commonly used to infer the overall hydrocarbon biodegradation rate in the field.

The dissolved concentration of oxygen in water equilibrated with air is small, ranging from 10.9 mg/l at 10°C to 8.1 mg/l at 25°C (Thibodeaux, 1979). Considering that some of the aromatic hydrocarbons have solubilities in the hundreds of mg/l, it is apparent that aerobic degradation of hydrocarbons is often limited by the transport of oxygen into the aqueous phase in the contaminated zone. In fact, a common approach for modeling aerobic degradation assumes that the reactions are instantaneous, and limited by amount of dissolved oxygen (see, e.g., Borden and Bedient, 1986; Bedient et al., 1999).

22.5.2 Bioventing and Biosparging

Bioventing and biosparging approaches are modifications to SVE and air sparging designed to maximize the amount of aerobic biodegradation by providing oxygen to the system. With bioventing, a blower can be used to provide a vacuum as in SVE, or air can be injected. Typical flow rates for bioventing systems are lower than for SVE operations in order to maximize the in-situ chemical destruction. Similarly, low flow rate air injection in the vadose zone can used without extraction in cases where complete biological destruction is expected in the subsurface. This design has the significant advantage of eliminating the need for gas treatment at the surface (Looney, 2000). Bioventing is widely used in practice, and the reader is referred to Leeson and Hinchee (1996) and EPA (1995a,b) for design concepts and practices.

Biosparging usually involves the injection of air or oxygen in sparge wells, typically at a lower rate than is used in conventional air sparging applications. As in the case of bioventing, at some sites it is possible to use the sparge wells alone, without SVE, provided that complete degradation takes place before the gases reach the ground surface. The advantage of using pure oxygen rather than air in the sparge well is that a higher oxygen concentration in water can be achieved (approximately a factor of 5 times higher because air is only about 20% oxygen). Case studies and design information for biosparging can be found in Hinchee et al. (1995).

REFERENCES

Adams, J.A. and Reddy, K.R., Laboratory study of air sparging of TCE-contaminated saturated soils and ground water, *Ground Water Monitoring and Remediation*, 19, 182–190, 1999.

Adenekan, A.E., Numerical modeling of multiphase transport of multicomponent organic contaminants and heat in the subsurface. Ph.D. dissertation, University of California, Berkeley, 1992.

Alvarez, P.J.J. and Voger, T., Substrate interactions of benzene, toluene, and para-xylene during degradation by pure cultures and mixed culture aquifer slurries, *Appl. Environ. Microbiol.*, 57: 2981–2985, 1991.

Baehr, A.L., Hoag, G.E., and Marley, M.C., Removing volatile contaminants from the unsaturated zone by inducing advective air-phase transport, *J. Contam. Hydrol.* 4, 1–26, 1989.

Basel, M.D. 1991. Two-dimensional propagation of steam through partially saturated porous media. Ph.D. dissertation, University of California, Berkeley.

Bedient, P.B., Rifai, H.S., and Newell, C.J., *Ground Water Contamination Transport and Remediation*, Prentice Hall PTR, 604, 1999.

Borden, R.C. and Bedient, P.B., *Transport of dissolved hydrocarbons influenced by oxygen-limited biodegradation: 1. theoretical development, Water Resources Research*, 13, 1973–1982, 1986.

Braida, W.J. and Ong, S.K. Air sparging: air-water mass transfer coefficients, *Water Resour. Res.*, 34, p. 3245–3253, 1998.

Clayton, W.S., A field and laboratory investigation of air fingering during air sparging, *Ground Water Monitoring and Remediation*, 18, 134–145, 1998.

DiGiulio, D.C. and Varadhan, R., Limitations of ROI testing for venting design: description of an alternative approach based on attainment of a critical pore-gas velocity in contaminated areas, *Ground Water Monitoring and Remediation*, 21, 97–114, 2001.

Elder, C.R. and Benson, C.H., Air channel formation, size, spacing, and tortuosity during air sparging, *Ground Water Monitoring and Remediation*, 19, 171–181, 1999.

Elder, C.R., Benson, C.H. and Eykholt, G., Factors affecting mass removal during in-situ air sparging, *J. Geotech. Geoenviron. Engin.*, 125, 11, 1999.

EPA, Bioventing principals and practice, volume I: bioventing principals, *EPA/540/R-95/534a, USEPA*, 1995a.

EPA, Principles and Practices of Bioventing, volume II: bioventing design, *EPA/540/R-95/534a, USEPA*, 1995b.

Falta, R.W., Pruess, K., Javandel, I., and Witherspoon, P.A., Numerical modeling of steam injection for the removal of nonaqueous phase liquids from the subsurface, 2. code validation and application. *Water Resour. Res.*, 28, 451–465, 1992.

Falta, R.W., Pruess, K., and Chesnut, D.A., Modeling advective contaminant transport during soil vapor extraction, *Ground Water*, 31, November–December, 1993.

Falta, R.W., Numerical modeling of kinetic interphase mass transfer during air sparging using a dual-media approach, *Water Resour. Res.*, 36, 3391–3400, 2000a.

Falta, R.W., Steam flooding for environmental remediation, in *Groundwater Contamination by Organic Pollutants, Analysis and Remediation*, Kaluarachchi, J.J., ed., *ASCE*, Reston, VA, 238 pp., 2000b.

Farber, A., Wärmetransport in der ungesättigenten Bodenzone: Entwicklung einer thermischen in-situ Sanierungstechnologie, *Mitteilungen Institut für Wasserbau Universität Stuttgart* (96), 1997.

Fatt, I. and Klikoff, W.A., Effect of fractional wettability on multiphase flow through porous media, *AIME Transactions*, 216, 246, 1959.

Fischer, U., Schulin, R., Keller, M., and Stauffer, F., Experimental and numerical investigation of soil vapor extraction, *Water Resour. Res.*, 28, 3413–3427, 1996.

Gierke, J.S., Hutzler, N.J., and McKenzie, D.B., Vapor transport in unsaturated soil columns: Implications for vapor extraction, *Water Resour. Res.*, 28, 323–335, 1992.

Hein, G.L., Hutzler, N.J., Gierke, J.S., and Falta, R.W., Three-dimensional experimental testing of a two-phase flow-modeling approach to air sparging, *Ground Water Monitoring and Remediation*, 17, 3, 1997.

Hein, G.L., Hutzler, N.J., Gierke, J.S., Stright, L.E., and Wolfe, A.N.C., Quarterly status report for contract number DE-AR021–96MC33082, June 22, 1998–September 22, 1998, submitted to the U.S. Department of Energy, Federal Technology Center, Pittsburgh, PA, 1998.

Hinchee, R.E., Miller, R.N., and Johnson, P.C., *In Situ Aeration: Air Sparging, Bioventing, and Related Remediation Processes*, Battelle Press, 1995.

Ho, C.K. and Udell, K.S., A mass transfer model for the removal of a volatile organic compound from heterogeneous porous media during vacuum extraction, *ASME Heat Transfer in Geophysical Media*, 172, 55–62, 1991.

Ho, C.K. and Udell, K.S., An experimental investigation of air venting of volatile liquid hydrocarbon mixtures from homogeneous and heterogeneous porous media, *J. Contam. Hydrol.*, 11, 291–316, 1992.

Hughes, J.B., Contaminant fate processes, in *Groundwater Contamination Transport and Remediation* by P.B., Bedient, H.S., Rifai, and C.J., Newell (eds.), Prentice Hall PTR, 604 pp., 1999.

Hunt, J.R., Sitar, N., and Udell, K.S., Nonaqueous phase liquid transport and cleanup 1. Analysis of mechanisms, *Water Resour. Res.*, 24, 1247–1258, 1988.

Ji, W., Dahmani, A., Ahlfeld, D.P., Lin, J.D., and Hill, E., Laboratory study of air sparging: Air flow visualization, *Ground Water Monitoring and Remediation*, 13, 115–127, 1993.

Larson, A., and Falta, R.W., Numerical Simulation of Field Air Sparging Operations, in *Non-Aqueous Phase Liquids (NAPLS) in Subsurface Environment*, Proceedings of the specialty conference held in conjunction with the ASCE National Convention, Washington, D.C., November 12–14, 551–562, 1996.

Leeson, A. and Hinchee, R.E., *Soil Bioventing: Principals and Practice*, Lewis Publishers, 1996.

Leverett, M.C., Capillary behavior in porous solids, *Trans. Soc. Pet. Eng. AIME*, 142, 152–169, 1941.

Looney, B.B. and Falta, R.W., eds., *Vadose Zone Science and Technology Solutions*, Battelle Press, 1542 pp., 2000.

McCray, J.E. and Falta, R.W. Determining the air sparging radius of influence, *Journal of Contam. Hydrol.*, 24, p. 25–52, 1996.

McCray, J.E. and Falta, R.W., Numerical simulations of air sparging for subsurface NAPL remediation, *Ground Water*, 35, 1, 1997.

Menegus, D.K. and Udell, K.S., 1985., A study of steam injection into water saturated capillary porous media. *Heat Transfer in Porous Media and Particulate Flows, ASME HTD*, Vol. 46, p. 151–157.

Murdoch, L., Remediation of organic chemicals in the vadose zone, in *Vadose Zone Science and Technology Solutions*, Looney, B.B., and Falta, R.W., eds., Battelle Press, 1542 pp., 2000.

Ochs, S.O., Hodges, R.A., Falta, R.W., Kmetz, T.M., Kupar, J.J. and Brown, N.N., Predicted heating patterns during steam flooding of coastal plain sediments at the Savannah River Site, in press, *Environmental and Engineering Geoscience*, 2002.

Shan, C., Falta, R.W., and Javandel, I. Analytical solutions for steady state gas flow to a soil vapor extraction well, *Water Resour. Res.*, 28, 4, April, 1992.

Semer, R. and Reddy, K.R., Mechanisms controlling toluene removal from saturated soils during in-situ air sparging, *J. Hazard. Mater.*, 57, 209–230, 1998.

Sleep, B.E., and Sykes, J.F., Modeling the transport of volatile organics in variably saturated media, *Water Resour. Res.*, 25, 81–92 1989.

Slobod, R.L., and Howlett, W.E., The effects of gravity segregation in laboratory studies of miscible displacement in vertical unconsolidated porous media, *Trans, AIME*, 231, 1–8, 1964.

Stewart, L.D. and Udell, K.S., Mechanisms of residual oil displacement by steam injection. *SPE Reservoir Engineering*, 3, 1233–1242, 1988.

Thibodeaux, L.J., *Chemodynamics: Environmental Movement of Chemicals in Air, Water, and Soil*, John Wiley and Sons, 501p., 1979.

Udell, K.S, Heat and mass transfer in cleanup of toxic waste, *Advances in Heat Transfer Research*, Tien, C.L., ed., *Environmental and Engineering Geophysical Society*, Englewood, CO, p. 195–211, 1994.

Wilkins, M.D., Abriola, L.M., and Pennell, K.D., An experimental investigation of rate-limited nonaqueous phase liquid volatilization in unsaturated porous-media: steady state mass transfer, *Water Resour. Res.*, 31, 2159–2172, 1995.

Yuan, Z.G. and Udell, K.S., Steam distillation of a single component hydrocarbon liquid in porous media. *Int. J. Heat Mass Transfer*, 36, 887–897, 1993.

CHAPTER 23

YUCCA MOUNTAIN HEATER TESTS

YVONNE Y. W. TSANG
Lawrence Berkeley National Laboratory, Berkeley, CA 94720, USA

23.1 INTRODUCTION

Yucca Mountain, approximately 135 km northwest of Las Vegas, Nevada, has been designated as the site of the U.S. national geological repository for the permanent disposal of spent fuel and high-level radioactive waste. The geological formation in which the proposed repository is situated is densely fractured welded tuff in the unsaturated zone a few hundred meters above the water table. The welded tuff has very small pores with strong capillarity effect, holding a significant amount of water in the matrix, and the matrix permeability of the welded tuff is low, on the order of microdarcies. However, this welded tuff is intensely fractured, with the permeability of these well-connected fractures many orders of magnitude larger than that of rock matrix. At thermodynamic equilibrium with the matrix at steady state, the fractures are essentially drained of water. Consequently, liquid and gas tend to flow in different parts of the rock formation: liquid predominantly in the low-permeability matrix and gas predominantly in the high-permeability fractures. At ambient conditions, the gas phase consists predominantly of air. However, in a nuclear waste repository, the heat generated from the decay of radioactive waste will cause water in the matrix pores to boil and greatly increase the vapor constituent in the gas phase. The pressure increase from vaporization drives the gas-phase transport by advection mainly through the fractures with its higher permeability and lower water saturation. Gas-phase diffusion driven by concentration gradient also occurs, because of the increase in water vapor (from boiling) and carbon dioxide (exsolved from the pore water with increasing temperature).

Heat from the anticipated decay of the radioactive waste would set in motion various coupled thermal, hydrological, mechanical and chemical processes. These thermally-driven coupled processes would affect the characteristics of the waste package environment, characteristics that control the corrosion of waste canisters and the mobilization rate of radionuclides. These coupled processes would also affect the rate

C. Ho and S. Webb (eds.), Gas Transport in Porous Media, 371–384.
© 2006 *Springer.*

of possible migration of radionuclides in the fractured rock away from the proposed repository into the environment. To assess the impact of such heat-induced coupled processes on the safe performance of the repository, the United States Department of Energy (DOE) has been conducting *in situ* heater tests in an underground facility built into the unsaturated tuff at Yucca Mountain – the Exploratory Studies Facility (ESF). The primary objective of the ESF heater tests is to develop a better understanding of the coupled thermal, hydrological, mechanical, and chemical processes likely to exist in the Yucca Mountain rock mass surrounding the proposed geological repository.

Gas-phase transport, the subject of this monograph, occurs in the heater tests as part of the coupled thermal-hydrological processes. We shall therefore, in the remainder of this chapter, focus our discussion on the observation of these thermal-hydrological processes, following a brief introduction of the general setting and configuration of the ESF heater tests, the Single Heater Test (SHT) and the Drift Scale Test (DST).

The ESF consists of an 8 km long tunnel with seven side alcoves for conducting underground experiments. The almost 8 m diameter tunnel descends gently through the different stratigraphic units of Yucca Mountain: from the alluvium at the surface, down through the Tiva Canyon welded and the Paintbrush nonwelded tuffs, and finally reaching the proposed repository horizon, the Topopah Spring welded tuff. As shown in Figure 23.1, both the SHT and the DST have been conducted in one of the side alcoves (Alcove 5) at the proposed repository horizon in the ESF. For a detailed discussion of the SHT and DST, the readers are referred to two Yucca Mountain project reports: Single Heater Test Final Report (CRWMS M&O 1999) and Drift Scale Test Progress Report No. 1 (CRWMS M&O 1998).

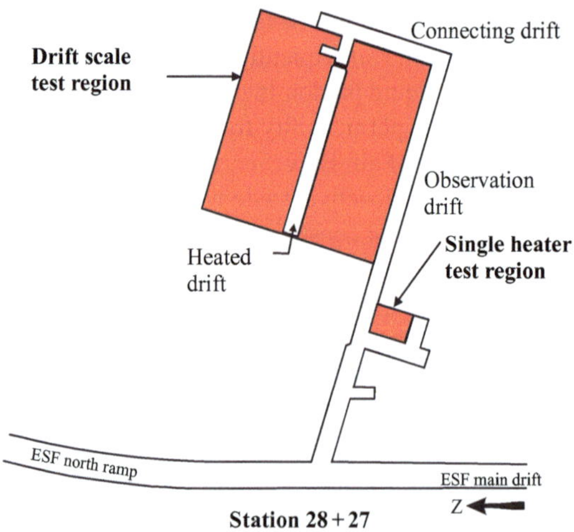

Figure 23.1. Schematic plan view of ESF Thermal Test Facility including the Single Heater test (SHT) and the Drift Scale Test (DST)

The first heater test, the SHT, consists of a 5 m long rod heater, at a nominal heating rate of 4 kW, emplaced horizontally among 30 instrumental boreholes spanning a rock block of approximately 13 m by 10 m by 13 m. Figure 23.2 shows a schematic of the SHT borehole layout. The boreholes are numbered and color coded according to

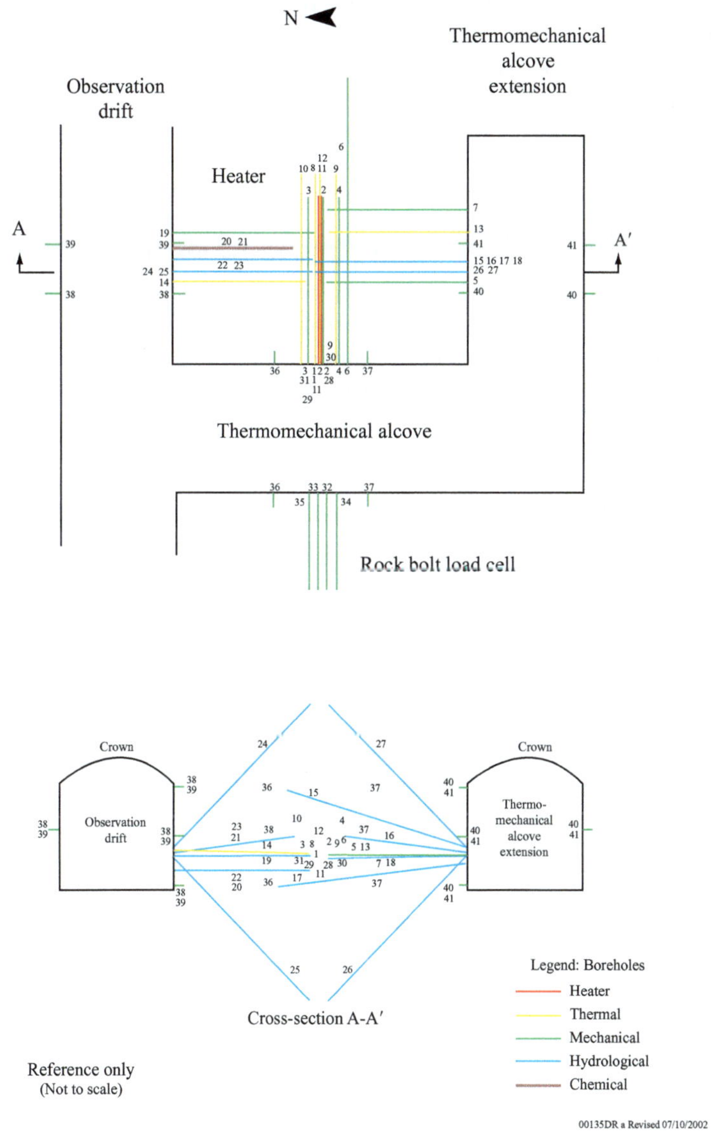

Figure 23.2. Schematic SHT layout of the instrumental boreholes, color coded according to their functions

their functions: thermal, hydrological, mechanical, and chemical. The heater in the SHT was activated on August 26, 1996, for a heating phase of nine months (until the end of May 1997). Following a cooling phase of seven months, post-cooling field characterization activities began in January 1998. All testing activities and analyses have been completed for the SHT.

The second heater test, the DST, began its heating phase in December 1997 at a thermal rate of 187 kW for a heating period of four years. The heat in the DST was turned off on January 14, 2002, for a planned cooling phase of another four years. The DST is much larger than the earlier SHT. It has nine canister heaters placed in a 5 m diameter Heated Drift, running approximately east-west over a length of 47.5 m (Figure 23.1 and Figure 23.3). Flanking the north and south sides of the Heated Drift are drilled fifty nominally 11.5 m long horizontal boreholes, twenty five drilled on each side of the Drift. Each of these boreholes houses a rod heater, composed of an inner and outer segment joined together by an accordian-like flexible joint. Each nominally 10 m long heater is termed a wing heater. The diameter of the Heated Drift is comparable to that of the waste emplacement drifts for the current repository design (BSC, 2001). The DST heating rate and the inclusion of the wing heaters were designed to heat a large volume of rock mass in a reasonably short testing period of a few years, so as to simulate the thermal-hydrological conditions reached only after 50 to 100 years of waste emplacement in the repository. Close to 100 instrumented boreholes in the DST test block have been installed, with thousands of sensors monitoring thermal, mechanical, hydrological, and chemical conditions/parameters and data recorded on an hourly basis. These boreholes are color-coded according to their functions in Figure 23.3. Most boreholes labeled as "hydrological" and "chemical"

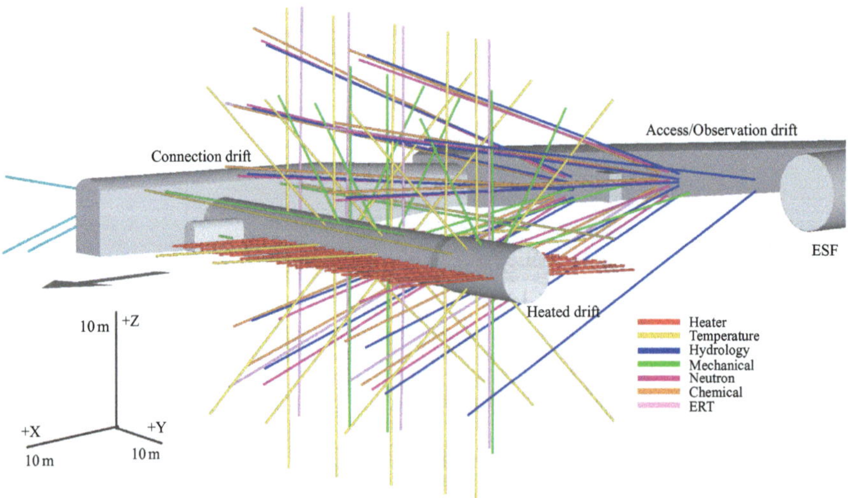

Figure 23.3. Perspective view showing drifts and boreholes of the DST, boreholes are color coded according to their functions

do not originate from the Heated Drift; but rather from the Observation Drift that runs parallel to the Heated Drift at a distance of 30 m from the Heated Drift axis. They are clusters of 40 m long boreholes forming vertical fans that bracket the Heated Drift and the wing heaters, primarily used for periodic active testing (geophysical logging and air-injection tests) to monitor time evolution and spatial distribution of drying and condensation zones in the test block. They are also used to collect water and gas samples for chemical and isotopic analysis.

23.2 GAS-PHASE TRANSPORT IN HEATER TESTS

Drying and wetting in different locations within the test block is the primary manifestation of the thermal-hydrological processes involving gas-phase transport (mainly vapor transport). As the formation temperature around the heater(s) approaches boiling (nominal boiling temperature is about 96°C at the prevailing ambient pressure and elevation of the heater tests site), pore water in the rock matrix boils and vaporizes. Most of the vapor that is generated moves into the fractures, where it is driven with the gas pressure gradient away from the heat source. The design of the DST with a permeable bulkhead implies that a considerable fraction of the generated vapor may be transported via the Heated Drift and lost through the bulkhead. The remainder of the generated vapor is transported within the test block, and when the vapor encounters cooler rock, it condenses. However, the increase in liquid saturation in the fractures from the condensate is likely to be small, since, because of the disparity of capillary suction forces in the fractures and matrix, majority of condensate will be imbibed into the matrix. In addition, because capillary suction is relatively weak in the fractures, we anticipate the condensate remaining in the fracture to drain downward by gravity. This drained condensate above the heater(s) would flow toward the heaters, boil again and then be transported away from the heaters to condense in the cooler rocks. The thermal-hydrological processes of continuous vaporization, gas transport, and condensation leads to a decrease of rock-mass liquid saturation close to the heat source and an increase in rock-mass liquid saturation from condensate immediately outside of the drying zone. Over time, the drying zone expands during heating, with vapor transport being the key process that gives rise to the formation of drying and wetting zones in the fractured rock. The zones where most pore water has vaporized would correlate closely with temperature measurements above boiling. Prior to this, as long as mobile water is still present while vaporization is taking place, two-phase conditions prevail and the temperature will remain at the nominal boiling point.

23.3 MEASUREMENTS TO VALIDATE VAPOR TRANSPORT PROCESSES

Observations of the gas-transport-driven thermal-hydrological processes in the heater tests are made in the following ways:

1. Temperature is continuously monitored at various locations with respect to the heat source. Although vapor pressure continuously increases with rising temperature,

thermal-hydrological processes do not become prominent until the temperature approaches boiling. Once a particular location within the test block reaches boiling, it is expected to remain at boiling for some time until all the mobile water surrounding the location of this particular temperature sensor is vaporized. Then the temperature will rise above boiling. It is safe to assume that the rock mass within the boiling isotherm (closer to the heat source) is "dry." That is, the liquid saturation in the rock mass is greatly reduced from the ambient values, though it may not be identically zero because of the strong capillary forces in the rock matrix. Hence, the passive monitoring temperature as a function of time at different locations within the rock mass provides indirect observation of the gas-transport processes.

2. Geophysical methods sensitive to rock moisture content are applied in the heater tests. In both the SHT and DST, neutron logging, electrical resistance tomography, and cross-hole radar tomography have been carried out in different "hydrological" boreholes, at appropriate times throughout the tests, to track the spatial distribution and time evolution of the drying and condensation zones. Since most water resides in the rock matrix in the welded tuff, changes in moisture content (detected by geophysical methods) reflect changes in the rock matrix.
3. Air-injection tests are employed to probe slight changes in the very small fraction of water residing in the fractures. Air-injection tests are carried out periodically in borehole sections isolated by inflatable packers (in a subset of boreholes labeled "hydrological" in Figure 23.2 and Figure 23.3). Any increase in liquid saturation from condensate results in increased resistance to air flow; hence, "wetting" of fractures is indicated by a decrease air permeability. In addition to the thermal-hydrological processes, thermal-hydrological-mechanical effects can also cause changes in air permeability decrease near the heat source resulting from the closure of fracture apertures, and increase in regions away from the heat source resulting from the opening of fracture apertures. In this chapter we focus only on the effect of wetting on the air permeability changes in order to illustrate vapor transport processes. We present below the air-permeability data collected over time from the SHT and DST.

23.3.1 Air-Permeability Measurements in the SHT

In the SHT, air-injection tests were carried out in boreholes 16 and 18 (see Figure 23.2) prior to heating, and also during heating and cooling phases, to track the changes in fracture liquid saturation. Note that boreholes 16 and 18 make up one of many pairs of boreholes labeled "hydrological" lying in a vertical plane orthogonal to the heater, intersecting the 5 m long heater at its midpoint. Two pairs of boreholes making the largest angle from horizontal (boreholes 24-25 and 26-27) are used for conducting electrical resistance tomographic (ERT) measurements, while two other pairs of boreholes closer to the heater horizon are used for neutron logging and crosshole radar measurements. A pair of hydrological boreholes (holes 16 and 18) intended for air-injection tests make the smallest angle with respect to the heater horizon. Boreholes

16 and 18 were each installed with four high-temperature inflatable packers to separate the borehole into isolated sections, with an air-injection line and pressure sensor for each zone to facilitate air-permeability measurements. A typical test consisted of injecting air in one chosen zone at a constant flow rate while monitoring pressure responses in this and all other isolated zones. The steady-state pressure response in the injection borehole itself can be used to estimate the local permeability, specific to the injection zone, averaged over the length of the packed-off zone. Assuming that air behaves as an ideal gas and that air flow through fractures is governed by Darcy's law, and that a finite line source can represent a borehole injection section, an analytical solution for ellipsoidal flow of incompressible fluid from a finite line source in an infinite medium is used to estimate permeability (Hvorslev, 1951; LeCain, 1995; Guzman et al., 1996):

$$k = \frac{P_{sc} Q_{sc} \mu \ln \frac{L}{r_w} T_f}{\pi L \left(P_2^2 - P_1^2\right) T_{sc}} \tag{23.1}$$

where

k	permeability (m^2)
L	length of air injection zones (m)
P_1, P_2	initial pressure and final steady state pressure for air injection tests (Pa)
P_{sc}	pressure at standard conditions, 1.013×10^5 Pa
Q_{sc}	flowrate at standard conditions in air injection tests (m^3/s)
r_w	radius of boreholes (m)
T_f	temperature of formation in air injection tests (°K)
T_{sc}	temperature at standard conditions (273.15°K)
μ	dynamic viscosity of air (Pa.s), 1.81×10^{-5} at 20°C

Figure 23.4 shows the air-permeability measurements for zone 3 in boreholes 16 and 18. Zone 3 is the isolated borehole section between the fourth packer farthest from the borehole collar and the bottom of the borehole. Zone 3 of borehole 16 spans a radial distance of 1.1 m to 3.2 m from the heater, and zone 3 of borehole 18 spans a radial distance of 1.5 m to 3 m from the heater. The horizontal axis of Figure 23.4 is the time line, with the heater activation and termination dates marked; the vertical axis is the measured air permeability. Graphs in Figure 23.4 show a decrease in air permeability (from the preheating values) by a factor of four and two, respectively, for zone 3 of boreholes 16 and 18, during the entire heating phase of the SHT. The air permeability returns to its preheating value[1] after the termination of heating.

Air-injection tests in the other zones in boreholes 16 and 18 closer to the borehole collar (or farther away from the heater) show no air-permeability decrease from their preheating values throughout the heating phase. These observations are consistent with the expected thermal-hydrological processes discussed in Section 23.2. This follows from the fact that temperature measurements in the SHT test block indicate that the boiling isotherm is at about 0.8 m radial distance from the heater after three

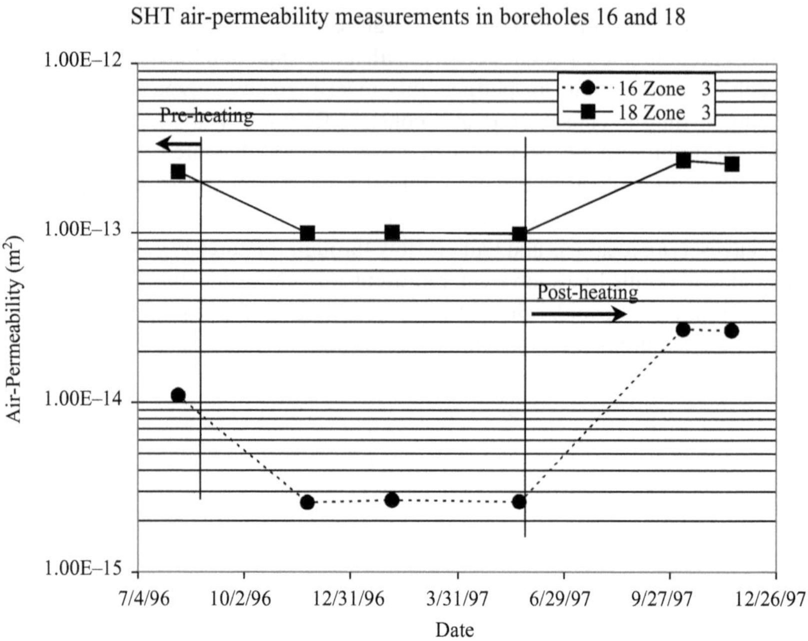

Figure 23.4. Air permeability measurements in zone 3 of boreholes 16 and 18 for before, during and after heating phase of the SHT

months of heating, and moves out to about 1.2 m at the end of nine months of heating. Zone 3 of boreholes 16 and 18 are located within the condensate zone outside of the boiling isotherm, and thus they experienced a drop in air permeability during the heating phase. Other zones in boreholes 16 and 18 are too far removed from the boiling isotherm to experience any increase in liquid saturation that would result in a decrease of air permeability. During the cooling phase, air-permeability values return to their preheating values, since the cooling phase differs from the heating phase in that vaporization ceases while imbibition of condensate from the fractures into the rock matrix and gravity drainage continues. Without the continuous supply of vapor (during the cooling phase) to replenish the accumulation of condensate in the fractures that imbibes into the matrix or drains by gravity, little water enters the fractures, and the air- permeability values return to preheating values during the cooling phase.

23.3.2 Air-Permeability Measurements in the DST

Air-injection tests were carried out during the preheating and heating phases of the DST. Figure 23.5 shows the preheating baseline air-permeability measurements in the twelve (hydrological) boreholes that form three vertical fans. These three clusters are orthogonal to the Heated Drift, intersecting it at y = 10 m (cluster of five

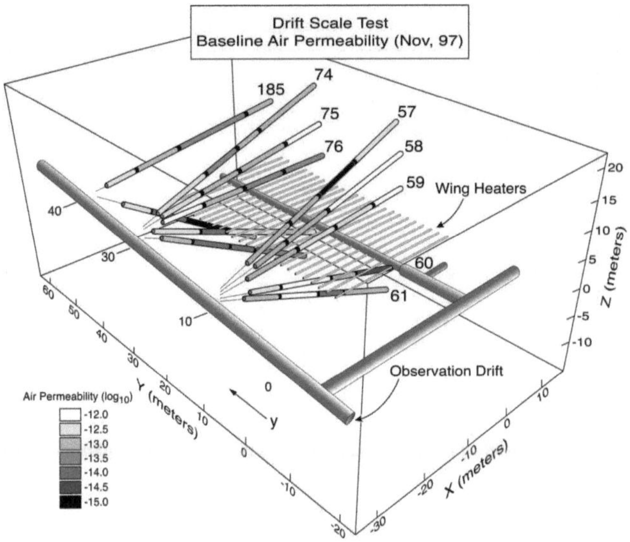

Figure 23.5. Preheating baseline permeability measurements in zones isolated by packers in twelve 40 m long boreholes in the DST, the permeability values range over three orders of magnitude

boreholes labeled 57–61), y = 30 m (cluster of five boreholes labeled 74–78), and y = 40 m (pair of two boreholes labeled 185–186). Each 40 m long borehole was installed with high-temperature inflatable packers to isolate the borehole into four[2] 8 to 10 m sections. Each isolated zone (zone 1 being closest to the collar and zone 4 closest to the bottom of each borehole) was equipped with an air-injection line and pressure sensor. Air-permeability measurements were carried out in each of the 46 isolated zones in the 12 boreholes. The estimated preheating permeability for different zones of the hydrology holes ranged over three orders of magnitude, from 1.6×10^{15} m^2 to 9.7×10^{13} m^2.

During the heating phase of the DST, air-injection tests were repeated at approximately three-months intervals in the 12 boreholes. The air permeability in each zone was computed according to Equation 23.1, and normalized[3] to its respective preheating baseline value. These normalized values as a function of time with respect to heat activation are plotted in Figure 23.6 through Figure 23.8 for the three clusters of boreholes. A very clear trend of decrease in air permeability is seen in many of the 46 borehole sections. The time and magnitude of the decrease bears a clear correlation to the respective zone's proximity to the heat source. That is, as heating continues, drying around the Heated Drift and the wing heaters expands, and the condensation front moves outward. The condensate causes a locally increased liquid saturation in the fractures and a corresponding decrease of permeability. It is largest in magnitude and occurs earliest in time in those zones closest to the wing heaters (e.g., zones 60-2, 60-3, 186-2, and 186-3). Those zones that are farthest from the heaters show no change or a slight increase in permeability from their preheating baseline values.

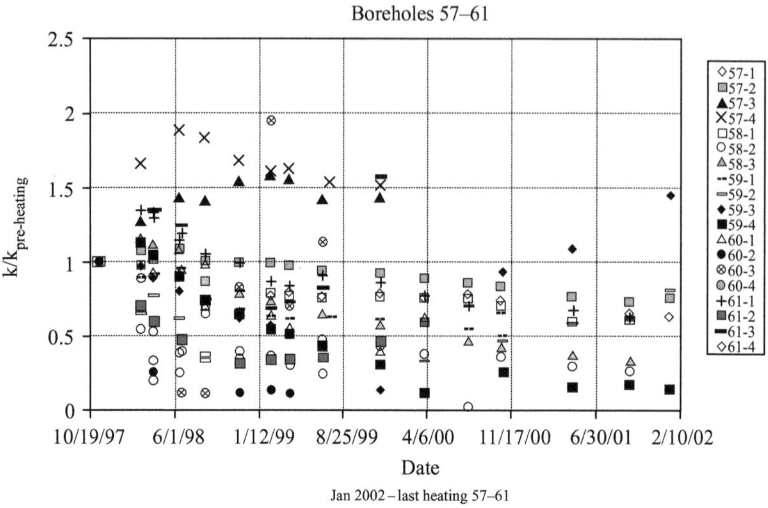

Figure 23.6. Changes in permeability displayed as ratio to the preheating permeability estimate for DST boreholes 57–61

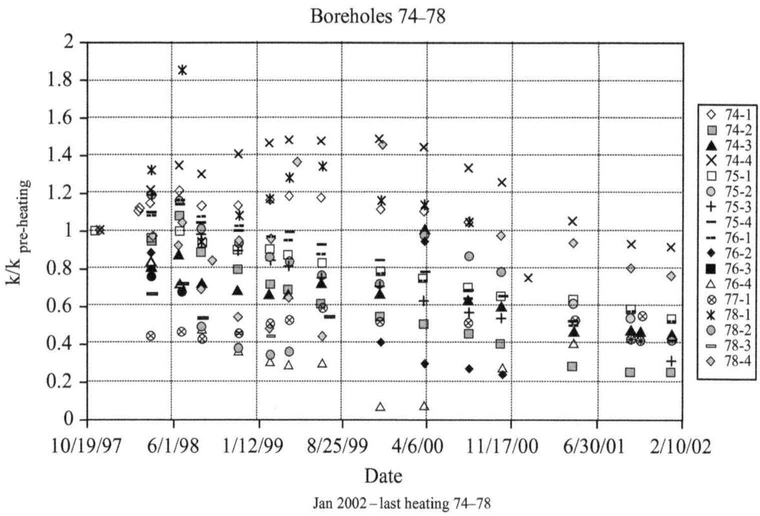

Figure 23.7. Changes in permeability displayed as ratio to the preheating permeability estimate for DST boreholes 74–78

No change in air permeability indicates that these zones are too far from the heated zone to experience an increase in liquid saturation from condensation. The slight increase in permeability in those zones far from the heat source may be attributed to the opening of fractures from thermal-mechanical coupled processes. Note also

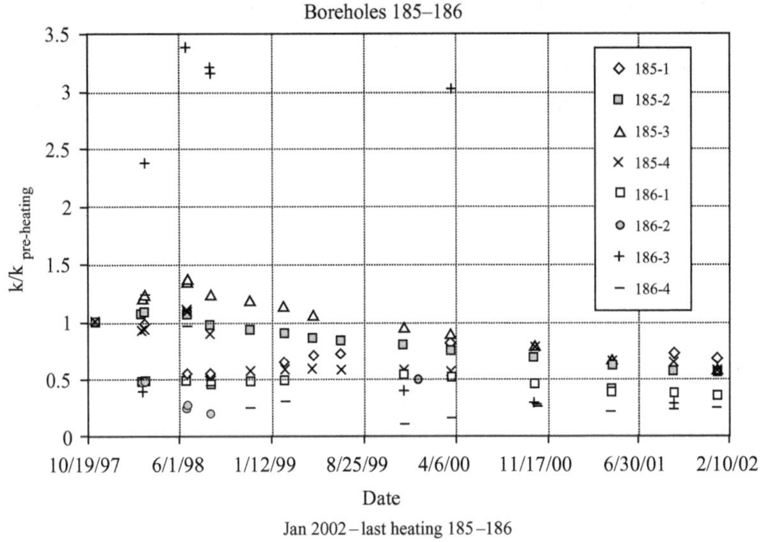

Figure 23.8. Changes in permeability displayed as ratio to the preheating permeability estimate for DST boreholes 185–186

that with time, the drying zone expands, and those zones that at earlier times show a decrease of air permeability (because of condensate accumulation) become drying zones at later times, with permeability returning towards the preheating values. This is illustrated by the data for zones 60-3 and 59-3 in Figure 23.6. For zone 60-3, after a sharp drop in permeability at three months after heating, the permeability at nine months of heating returns to preheating baseline value. For 59-3, the permeability value decreases steadily until November 2000, after which the permeability increases, indicating drying.

As mentioned earlier, thermally induced stress release in regions farther from the heat source can cause opening of fracture apertures and give rise to increase in air permeability. This is illustrated by those values of $k/k_{\text{preheating}}$ larger than one in early times shown in Figure 23.6 through Figure 23.8. These correspond to boreholes section 57-3 and 57-4 in Figure 23.6, 74-1 and 74-4 in Figure 23.7, and 185-3 and 186-3 in Figure 23.8.

23.4 SUMMARY

We have carried out modeling of the thermal-hydrological processes of the SHT and DST, using the integral-finite-difference numerical simulator TOUGH2, developed by Pruess (1987, 1991), for simulating multidimensional transport of water, vapor, air, and heat in heterogeneous porous and fractured media. For both the SHT and DST, site-specific 3-D numerical models have been constructed, with the fractured welded tuff modeled as a dual continuum. The conceptual model described in Section 23.2 is confirmed by these simulations, and the measured trends in air-permeability changes with time discussed in Sections 23.3.1 and 23.3.2 are consistent with simulated moisture redistribution of moisture performed by the 3D thermal-hydrological numerical model of the SHT (Tsang and Birkholzer, 1999) and the DST (Birkholzer and Tsang, 2000).

ACKNOWLEDGMENTS

This work was supported by the Director, Office of Civilian Radioactive Waste Management, U.S. Department of Energy, through Memorandum Purchase Order EA9013MC5X between Bechtel SAIC Company, LLC and the Ernest Orlando Lawrence Berkeley National Laboratory (Berkeley Lab). The support is provided to Berkeley Lab through the U.S. Department of Energy Contract No. DE-AC03–76SF00098. The author thanks Chin-Fu Tsang and Dan Hawkes of Berkeley Lab, Steve Sobolik of Sandia National Laboratories, and Bob Levich of Department of Energy for critical review of the manuscript and suggestions for improvement.

NOTES

[1] In fact, the postheating values are slightly higher than the preheating values. Such postheating air-permeability increase is found generally in many other boreholes, which is hypothesized as indicative of the opening of fractures from thermal-hydrological-mechanically coupled processes (Tsang et al., 1999).

[2] For two of the boreholes, boreholes 58 and 77, only three packers could be installed because of numerous fractures and lithophysal cavities in the borehole walls.

[3] Missing data in Figures 23.6, 23.7, and 23.8 can be from two sources. One is when steady state pressure response is not attained, and the other is when packers separating certain zone became permanently damaged so that a comparison with baseline is no longer possible.

REFERENCES

Birkholzer, J.T. and Tsang, Y.W., 2000. Modeling the thermal-hydrologic processes in a large-scale underground heater test in partially saturated fractured tuff, Water Resources Research, 36 (6), 1431–1447.

BSC 2001. Subsurface facility system description document SDD-SFS-SE-000001 REV 01 ICN 02. Las Vegas, Nevada. Bechtel SAIC Company. ACC: MOL.20011217.0015

CRWMS M&O, 1998. Drift Scale Test progress report no. 1. BAB000000–01717–5700–00004 REV 01. Las Vegas, Nevada. ACC: MOL.19990209.0240

CRWMS M&O, 1999. Single Heater Test final report BAB000000–01717–5700–00005 REV 00 ICN 1. Las Vegas, Nevada. ACC: MOL.20000103.0634

Guzman, A.G., A.M. Geddis, M.J. Henrich, C.F. Lohrstorfer, and S.P. Neuman, 1996. Summary of air permeability data from single-hole injection tests in unsaturated fractured tuffs at the Apache Leap Research Site; Results of steady-state interpretation, U.S. Nuclear Regulatory Commission Report NUREG/CR-6360, NRC, Washington DC.

Hvorslev, M.J., 1951. Time Lag and Soil Permeability in Ground-Water Observations. AEWES Bulletin 36. Vicksburg, Mississippi: U.S. Army Corps of Engineers, Waterways Experiment Station.

LeCain, G.D., 1995. Pneumatic testing in 45-degree-inclined boreholes in ash-flow tuff near Superior, Arizona, USGS Water Resources Investigations Report 95–4073, USGS, Denver, Colorado.

Pruess K., 1987. TOUGH user's guide, Technical Report LBL-20700, Lawrence Berkeley National Laboratory, Berkeley, California.

Pruess K., 1991. TOUGH2 -A general purpose numerical simulator for multiphase fluid and heat flow, Technical Report LBL-29400, UC-251, Lawrence Berkeley National Laboratory, Berkeley, California.

Tsang Y. W., Apps J., Birkholzer J. T., Peterson J., Sonnenthal E., Spycher N. and Williams K., 1999. Yucca Mountain Drift Scale Test progress report, Technical Report LBNL-42538, Lawrence Berkeley National Laboratory, Berkeley, California.

Tsang, Y.W. and Birkholzer, J.T., 1999. Predictions and Observations of the Thermal-hydrological conditions in the Single Heater Test. J. of Contaminant Hydrology, *38(1–3)*, 385–425.

CHAPTER 24

IMPACT OF GAS GENERATION ON THE PERFORMANCE OF THE WASTE ISOLATION PILOT PLANT

PALMER VAUGHN
Sandia National Laboratories, Albuquerque, New Mexico, 87185, USA

24.1 INTRODUCTION

The Waste Isolation Pilot Plant (WIPP) is the U.S. Department of Energy's (DOE) mined geologic repository located 42 km (26 miles) east of Carlsbad, in southeastern New Mexico, USA.

In May 1998, the United States Environmental Protection Agency (EPA) certified the United States Department of Energy's (DOE) Waste Isolation Pilot Plant (WIPP) as being in compliance with applicable long-term regulations governing the permanent disposal of spent nuclear fuel, high-level and transuranic radioactive waste. Sandia National Laboratories is DOE's Scientific Advisor for the WIPP and developed a Performance Assessment (PA) to quantify potential long-term risks associated with the disposal of these waste. This PA supported DOE's application, the WIPP Compliance Certification Application, United States Department of Energy, 1996a.

The core of Sandia's capability is a probabilistic approach based on "Monte-Carlo" techniques for assessment of health risks posed by contaminant mixtures and their potential migration. The result from a probabilistic risk assessment is a range and probability distribution of risk. This goes well beyond the traditional deterministic or single value risk result in that uncertainty associated with the result is also presented. Sensitivity analysis of the results of the probabilistic approach enables prioritization of site characterization information and experimental programs so that maximum benefit can be achieved from available resources while meeting site performance goals.

This chapter walks through the performance assessment methodology used to support the WIPP CCA but focuses on the gas generation related aspects. The eight steps

C. Ho and S. Webb (eds.), Gas Transport in Porous Media, 385–406.
© 2006 *Springer.*

that make up the probabilistic PA are:
(1) Define the goal: level of acceptable performance,
(2) Characterize the system: waste, facility, and site,
(3) Identify the relevant features, events, and processes (FEPs)
(4) Construct scenarios,
(5) Build models (conceptual, mathematical, and numerical),
(6) Use model parameters to characterize the uncertainty,
(7) Perform the uncertainty analysis using Monte Carlo simulations, and
(8) Perform a sensitivity analysis to guide the program.

24.2 STEP 1: DEFINE THE GOAL AND LEVEL OF ACCEPTABLE PERFORMANCE

The overall process of assessing whether or not a waste disposal system meets a set of performance criteria is known as a *performance assessment.*A performance assessment (PA) provides important input to performance criteria on the safety of a plan of action using a detailed procedure and scientific knowledge. For radioactive wastes, a computationally demanding set of risk-based performance criteria is specified in the EPA's *Environmental Radiation Protection Standards for Management and Disposal of Spent Nuclear Fuel, High-Level and Transuranic Radioactive Wastes,*40 CFR 191 (EPA, 1985, as amended, 1993). These are quantitative criteria that specify probabilistic limits that must be met for the first 10,000 years of operation of a waste repository. It is not sufficient to develop an accurate scientific understanding of the *current* status of a disposal system. Calculations illustrating possible behavior well into the future are required.

In addition to the transuranic radionuclides, the wastes contain other hazardous constituents, such as heavy metals and volatile organic compounds (VOCs). Therefore, a full suite of regulatory criteria had to be satisfied before the WIPP could be certified as an acceptable repository for federal wastes. The primary goal of the WIPP PA is to quantitatively evaluate the performance of the WIPP repository and risks to the public within the context of uncertainty and as specified in the regulation.

24.3 STEP 2: CHARACTERIZE THE SYSTEM: WASTE, FACILITY, AND SITE

Constructed in bedded salt deposits 655 m (2150 feet) below the ground, the site is designed for the permanent burial of transuranic radioactive waste generated by defense related activities. WIPP destined waste consists primarily of 208L (55 gal) drums containing radioactive contaminated tools, rags and other materials. The repository is designed to dispose of approximately 175,000 cubic meters of transuranic waste (materials containing more than 100 nanocuries [3700 becquerels] per gram of transuranic isotopes with half lives greater than 20 years) derived from defense-related activities of the United States government. Total activity of the waste (as of 1995) is estimated to be 7.4 × 10^6 curies (2.7 × 10^{17} becquerels), with the largest component of this activity coming from isotopes of plutonium.

The WIPP repository (Figure 24.1) is constructed on one level in the 600 m-thick bedded salt of the Salado formation, which dips slightly to the south. The salt is interspersed with thin horizontal interbeds of anhydrite. The thicker and more laterally persistent interbeds are referred to as marker beds. The closest marker beds to the repository are MB 139 one to two below and MB 138 some 10m above The thinner anhydrite interbeds are referred to as layers. Anhydrite layers a and b are within a few meters of the repository. Ultimately, eight panels of seven rooms each will be mined. The total facility covers an area approximately 775 × 1585 m, and the tunnels are initially 4.0 m in height having a total excavated volume of 583, 400 $m^{3.}$ Before the repository is closed permanently, each panel will be backfilled with MgO and

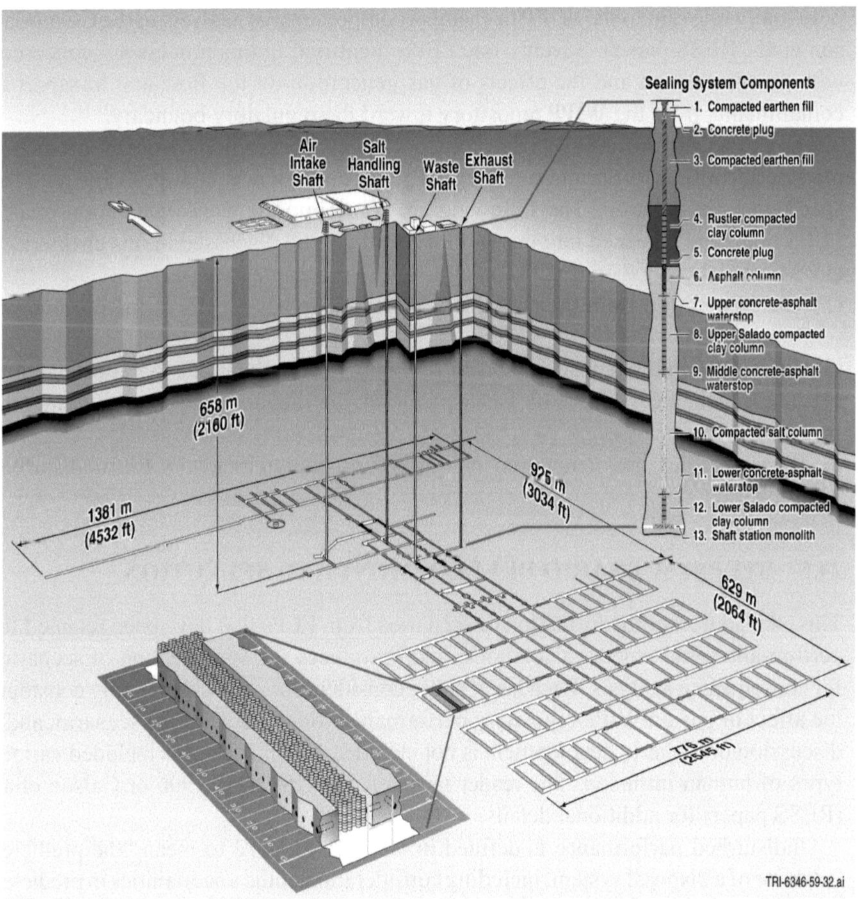

Figure 24.1. WIPP repository in thick-bedded salt

sealed. Waste is placed in the drifts between panels; the access ways will be sealed off from the shafts, and the four shafts (waste handling, salt handling, air intake and air exhaust shafts) will be sealed.

The waste contains, in part, steel, cellulosics, plastics, and rubbers, which produce gases (mainly hydrogen from corrosion and carbon dioxide or methane from biodegradation) when exposed to brine. This gas generation may result in increased pressures over time. The presence of gas and the potential for increased pressure have profound implications on the performance of the WIPP. These impacts were evaluated as part of the CCA, (US DOE, 1996).

24.4 STEP 3: IDENTIFY THE RELEVANT FEATURES, EVENTS, AND PROCESSES (FEPS)

The DOE used a systematic approach to the identification and screening of FEPS potentially relevant to the WIPP to ensure a comprehensive analysis in the CCA (Galson et al., RESS paper). Seventy-one FEPS identified in this process are concerned with gas generation and the effects of gas generation on the flow and transport of contaminants from the WIPP repository toward the regulatory boundary.

The FEPS are screened in a formal process and may be excluded from the predictive models according to the three criteria of low consequence, low probability, or as specified by regulation. The following are some important gas generation related FEPS that were screened into the WIPP PA and will be discussed in this chapter:

(1) Microbial gas generation from degradation of organic material.
(2) Gas generation from the anoxic corrosion of ferrous metals in the presence of brine.
(3) Fluid flow in the repository and surrounding host rock do due to gas production.
(4) The mechanical effect of gas generation on the host rock surrounding the repository.
(5) The effect of gas generation on release of contaminants in the undisturbed scenario.

24.5 STEP 4: SCENARIO DEVELOPMENT AND SELECTION

This step addresses the formation of scenarios from FEPs that have been retained for performance assessment calculations, and introduces the specification of scenarios for consequence analysis. Because of space consideration, this chapter only considers the affect of gas on WIPP repository performance for an undisturbed scenario, and a discussion of scenario development is not included. Other scenarios included various types of human intrusion. The reader may refer to US DOE, 1996 or Galson et al. (RESS paper) for additional details.

Undisturbed performance is defined in 40 CFR §191.12 to mean "the predicted behavior of a disposal system, including consideration of the uncertainties in predicted behavior, if the disposal system is not disrupted by human intrusion or the occurrence of unlikely natural events." Consideration of only undisturbed performance is required for compliance assessments with respect to the Individual and Groundwater Protection Requirements (40 CFR §191.15 and 40 CFR §191.24). Undisturbed performance is also considered, together with disturbed performance, for performance assessments with respect to the Containment Requirements (40 CFR §191.13).

24.6 BUILD MODELS

The models used in the WIPP performance assessment, as in other complex analyses, exist at three different levels:

(1) Conceptual models are a set of qualitative assumptions used to describe a system or subsystem for a given purpose.
(2) Mathematical models are developed to represent the processes at the site. The conceptual models provide the context within which these mathematical models must operate and define the processes they must characterize.
(3) Numerical models and computer codes are developed to provide approximations of mathematical model solutions because most mathematical models do not have closed-form solutions. These models are often encoded into computer codes. For the undisturbed simulation, the numerical models BRAGFLO, United States Department of Energy, 1996b and NUTS, United States Department of Energy, 1996c are used for prediction of radionuclide migration.

24.6.1 Conceptual Model

Conceptually, there are several pathways for radionuclide transport within the undisturbed disposal system that may result in releases to the accessible environment.

Figure 24.2 is a schematic of the undisturbed scenario. Contaminated brine may migrate away from the waste-disposal panels if pressure within the panels is elevated by the generation of gas from corrosion or microbial degradation. Identified are two potential pathways to the accessible environment. The first is upward through the sealed shaft system directly to the surface or into the overlying water-bearing unit, the Culebra, and laterally to the accessible environment. The second is through the disturbed rock zone (DRZ) surrounding the waste disposal regions, into the more permeable anhydrite interbeds, and laterally to the land withdrawal boundary, where regulatory compliance is evaluated.

The coupled processes of deformation of the rock surrounding the excavation, fluid flow, and waste degradation dominate the behavior of the disposal system. Each of these processes can be described independently, but the extent to which each process occurs will be affected by the others. Figure 24.3 depict a progression of what may occur conceptually (and what we observed from our simulations) in the repository during undisturbed conditions. In Figure 24.3(a), the waste drums are shown in the excavation shortly after emplacement. The stress field produces a DRZ around the excavation. The surrounding host rock is salt (halite) of low porosity, and the voids are saturated with brine. The anhydrite interbeds are similarly of low porosity, but more transmissive, and are also saturated with brine.

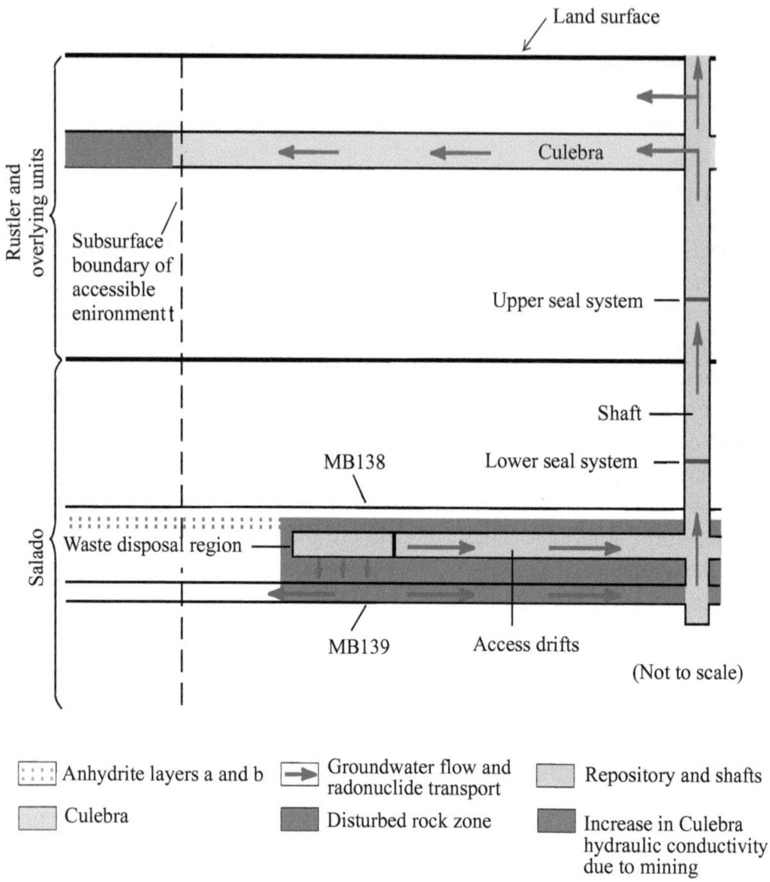

Figure 24.2. Undisturbed scenario showing possible release pathways

Characterization of the Salado Formation indicates that fluid flow does not occur on time scales of interest in the absence of an artificially imposed hydraulic gradient. Lack of fluid flow is a result of the extremely low permeability of the evaporite rocks that make up the Salado. Excavation of the repository has disturbed the natural hydraulic gradient and rocks properties and has resulted in fluid flow. After the repository is sealed, brine flows toward the excavation from the far field because of the low-pressure associated with the excavation, Figure 24.3(b). This brine primarily flows through the anhydrite interbeds, through the DRZ, and into the excavation. The halite has a strong tendency to creep under the lithostatic load and over time encapsulates the waste. In the waste disposal region, waste consolidation continues until back-stresses imposed by the compressed waste resist further closure or until fluid pressures become sufficiently high. Within the first few hundred years, the

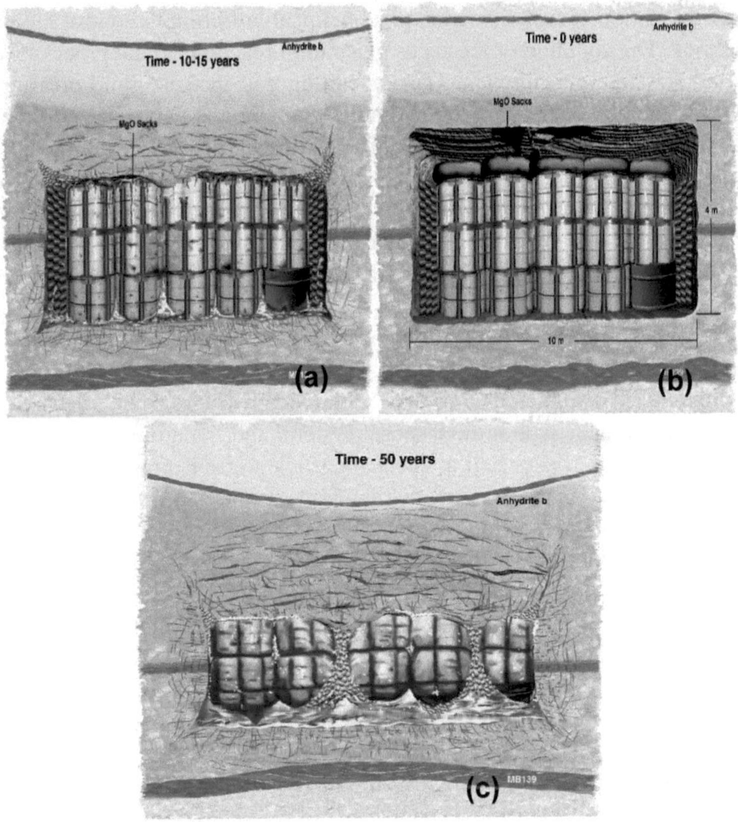

Figure 24.3. Time sequence in the WIPP repository showing conceptualization of processes occurring in the mined repository

excavation collapses to 1 to 2 m in height from its original 4-m height and the porosity is reduced. The amount of waste consolidation that occurs and the time it takes to consolidate are governed by properties of the waste (waste strength, modulus, etc.), properties of the surrounding rock, the dimensions and location of the room, and the quantities and types of fluids (gas and brine) present in the room. The waste drums are crushed, exposing their contents. Brine flows through the excavation in contact with the metal drums and their contents. The radionuclides and other soluble contaminants dissolve in the brine according to their solubility properties. Additionally, the brine contacts the metal and biodegradable constituents of the waste resulting in the generation of gases and consumption of brine.

Initially, the gas phase will consist primarily of air trapped at the time of closure, although other gases will form as a result of waste degradation. Pressure will rise due to creep closure, gas generation, and brine inflow, creating the potential for flow outward from the excavated region. Emplaced MgO backfill will remove the CO_2 generated by biodegradation as well as maintain a chemical buffering to control radionuclide solubilities. The coupling of these gas generation reactions to the processes of fluid flow and salt creep is complex. Gas generation will increase fluid pressure in the repository, thereby decreasing the hydraulic gradient and deviatoric stress between the far field and the excavated region and inhibiting the processes of brine inflow and salt creep. Anoxic corrosion will also consume brine as it breaks down water to oxidize iron and release hydrogen gas. Thus, corrosion has the potential to be a self-limiting process, in that as it consumes all water in contact with iron, it will cease. Microbial reactions are also considered to be dependent on the presence of water to occur, although their net effect is uncertain. It is assumed that microbial reactions will result in neither the consumption nor creation of water. The total volume of gas that may be generated by corrosion and microbial degradation may be sufficient to result in repository pressures that approach lithostatic. Sustained pressures above lithostatic are not physically reasonable within the disposal system, and fracturing of the more brittle anhydrite interbeds is expected to occur if sufficient gas is present. The conceptual model implemented in the performance assessment causes permeability and porosity of the anhydrite interbeds to increase rapidly as pore pressure approaches and exceeds lithostatic. This conceptual model for pressure-dependent fracturing approximates the hydraulic effect of pressure-induced fracturing and allows gas and brine to move more freely within the anhydrite interbeds at higher pressures. In Figure 24.3(c), continued gas generation occurs resulting in pressure that may approach lithostatic, 14.7 MPa. Gas is forced out of the repository, either driving brine outward or retarding its rate of inflow. As the pressure approaches lithostatic, the host rock responds. As the pore pressure increases, an increasing amount of overburden stress is supported by pore fluid pressure and less overburden stress is supported by the strength of the waste matrix. Waste consolidation will cease when the sum of the stresses felt by the waste matrix and fluid pressure reaches lithostatic pressure. There is some reversal of the creep consolidation, and fractures begin to develop. This relieves pressure and provides for pathways for fluid flow and potential release to the accessible environment. In the long term, the pressure in the repository is capped naturally by the lithostatic load.

24.6.2 Mathematical Model

24.6.2.1 Gas generation

A summary of the gas generation treatment is discussed in this section. Additional details may be found in Wang and Brush (1996). Gas will be produced in the repository because of a variety of chemical reactions, primarily those occurring between brine, metals, microbes, cellulosics and similar materials, plastics, and rubber materials, and via liberation of dissolved gases to the gas phase. Radiolysis has been demonstrated by laboratory experiment and model calculations to be insignificant. During the undisturbed scenario, gas generation will affect repository pressure, which is important in other submodels of the disposal system, such as those calculating creep closure, interbed fracturing, and two-phase flow.

Since the repository will become anoxic shortly (within 50 years relative to the 10,000 year regulatory period) after waste emplacement and sealing, only anoxic steel corrosion is considered. Two reaction paths are possible depending on the presence of gaseous products from the biodegradation process.

Steel Corrosion in the Absence of CO_2 and H_2S

In this case, 2 steel corrosion reactions are possible (Telander and Westerman, 1993, 1995):

$$Fe + 2H_2O \rightarrow Fe(OH)_2 + H_2 \tag{24.1}$$

$$3Fe + 4H_2O \rightarrow 4H_2 + Fe_3O_4 \tag{24.2}$$

Measured corrosion rate on specimens exposed to inundated conditions under a range of WIPP relevant conditions provide the basis for the rates used (Telander and Westerman, 1993, 1995). The corrosion rate observed on specimens exposed to humid conditions is negligible, based on essentially non-existent presence of corrosion product and lack of apparent H_2 generation (Telander and Westerman, 1995). Reaction (24.2) does not occur in the WIPP environment.

Steel Corrosion in the Presence of CO_2 and H_2S

In the event of significant microbial gas generation, steel corrosion can proceed via the following reactions in addition to Reaction (24.1) (Telander and Westerman, 1993, 1995):

$$Fe + CO_2 + H_2O \rightarrow FeCO_3 + H_2 \tag{24.3}$$

$$Fe + H_2S \rightarrow FeS + H_2 \tag{24.4}$$

One possible effect of CO_2 and H_2S on steel corrosion is that they may cause passivation of the steel. However, passivation is unlikely to occur under the repository conditions. This is because the microbial production rate of CO_2 and H_2S is too slow and it will take an exceedingly long time period (relative to the experimental time scale) for these gases in the repository to reach their concentration levels required for passivation under the experimental conditions.

In the absence of passivation, the microbial generation of CO_2 and H_2S will increase steel corrosion rates in the repository either by lowering the repository pH or by initiating additional reaction pathways (Reactions (24.3) and (24.4)) (Telander and Westerman, 1995). This is accounted for by modifying the sampling range of steel corrosion rate. Under humid conditions, experimental results show a negligible effect of CO_2 and H_2S on steel corrosion (Telander and Westerman, 1995).

Microorganism will consume cellulosics mainly via the following reaction pathways in the repository (Brush, 1995):

$$C_6H_{10}O_5 + 4.8\,H^+ + 4.8\,NO_3^- \rightarrow 7.4\,H_2O + 6\,CO_2 + 2.4\,N_2 \tag{24.5}$$

$$C_6H_{10}O_5 + 6\,H^+ + 3\,SO_4^{-2} \rightarrow 5\,H_2O + 6\,CO_2 + 3\,H_2S \tag{24.6}$$

$$C_6H_{10}O_5 + H_2O \rightarrow 3\,CH_4 + 3\,CO_2 \tag{24.7}$$

Reaction with MgO added to the repository consumes the CO_2 produced. Thus, the net quantity of gas developed by microbial degradation is correlated with constituents of the waste disposal region and included in the stoichiometry. It is assumed that the microbial degradation process neither produces nor consumes water, but its rate is dependent on the amount of liquid present in a computational cell. The rate of cellulosics biodegradation was measured by incubating representative cellulose materials (filter paper, paper towels, and tissue) in WIPP brine with microbes enriched from various WIPP environments (Francis and Gillow, 1994; Francis et al., 1995a and b). The incubation experiments were conducted under various conditions: aerobic or anaerobic, inundated or humid, with or without bentonite, amended or unamended with nutrients or NO_3^-.

Microbial degradation may also consume plastic and rubber materials in the repository. The rates of plastics and rubber biodegradation under expected WIPP conditions were measured by Francis et al. (1995a and b). The experimental data show that plastics and rubbers are much less biodegradable than cellulosics, although the data themselves are not sufficient for us to constrain the long-term biodegradation rate for plastics and rubbers. The DOE assumes that in half of those simulations in which microbial degradation of cellulosics occurs, microbial degradation also acts on plastic and rubber materials in the waste disposal region.

24.6.2.2 Multi-phase flow

For miscible flow, the fluids will diffuse within each other and there is not a defined interface between the fluids. Specifically, there is no capillary pressure effect. Immiscible fluids are not capable of mixing and have interfaces across which pressure discontinuities exist. This interfacial tension effect produces a capillary pressure between phases. Flow in and around the repository is conceptualized as the two-phase immiscible flow of brine and gas through porous media. This is based on heuristic extensions of Darcy's law (Darcy, 1856).

Fluid flow modeling in the undisturbed scenario is concerned with (1) flow in the waste, (2) flow to, from, and through the Salado and shafts, and (3) fluid flow between the repository and intrusion boreholes. These are important in assessing gas generation rates, repository pressure, and the mobility of radionuclides in the disposal system. Conceptually, a DRZ surrounds the repository and the shafts.

The Salado is the principal natural barrier to fluid flow between the waste disposal panels and the accessible environment. Two rock types, impure halite and anhydrite, are used to represent the intact Salado. For several meters above and below the repository, a DRZ has increased permeability compared to intact rock and offers little resistance to flow between anhydrite interbeds and the repository. In all rock units, porosity can vary from initial values due to compressibility.

Gas may not be able to flow through the intact Salado under realistic conditions for the repository. Gas flow in liquid-saturated rock depends on the gas pressure required to overcome capillary resistance to initial gas penetration and development of interconnected gas pathways that allow gas flow (threshold pressure). While the permeability of the halite units is known to be low (approximately 10^{-23} m^2 or lower), its threshold pressure has never been measured. An empirical relationship between threshold pressure and permeability in non-WIPP rocks (Davies, 1991) suggests that threshold pressure will be sufficiently high that gas will not be able to flow through the halite-rich strata of the Salado under any conditions foreseeable for the WIPP.

Three distinct anhydrite interbeds are represented in the CCA, MB138, anhydrite layers a and b, and marker bed 139 (MB139), Figure 24.2. Porosity and permeability can vary temporally and spatially during a simulation depending on the extent of interbed fracturing. The three interbeds are included because they could communicate hydrologically with the repository through the DRZ. It is expected that, because of density differences between gas and brine and their stratification within the repository, brine outflow will be dominantly in MB139, and gas outflow will occur in anhydrite a and b or MB138. However, the model does not preclude other flow patterns.

The following system of two mass balance equations and two constraint equations is a description of the two-phase flow in the WIPP PA (Peaceman, 1977):

Gas Mass Balance [kg $H_2/(s \cdot m^3)$]:

$$\nabla \cdot \left[\frac{\alpha \rho_g [k] k_{rg}}{\mu_g} \left(\nabla P_g + \rho_g g \nabla h \right) \right] + \alpha q_g + \alpha q_{rg} = \alpha \frac{\partial \left(\phi \rho_g S_g \right)}{\partial t} \tag{24.8}$$

Brine Mass Balance [kg brine/$(s \cdot m^3)$]:

$$\nabla \cdot \left[\frac{\alpha \rho_b [k] k_{rb}}{\mu_b} \left(\nabla P_b + \rho_b g \nabla h \right) \right] + \alpha q_b + \alpha q_{rb} = \alpha \frac{\partial \left(\phi \rho_b S_b \right)}{\partial t} \tag{24.9}$$

Saturation Constraint [m^3 phase i/m^3 pore volume]. Note: Phase saturation is the fraction of the pore space occupied by the fluids in a given phase:

$$S_g + S_b = 1 \tag{24.10}$$

Capillary Pressure (P_c) Constraint [Pa]:

$$P_c = P_g - P_b \tag{24.11}$$

Additional terms and parameters are listed under Notation.

24.6.2.3 Creep closure

Creep closure is accounted for in BRAGFLO numerical model by changing the porosity of the waste disposal area according to a look-up table of porosity (porosity surface) generated using the SANTOS code, a quasistatic, large deformation finite-element structural analysis code. The porosity surface approach is used because current codes are not capable of fully coupling creep closure, waste consolidation, brine availability, and gas production and migration with computational efficiency. The porosity surface method incorporates the results of closure calculations obtained from the SANTOS code. The porosity surface is constructed from a set of nonlinear finite element analyses in which the gas generation potential is varied to generate porosity time history curves. Disposal room porosities and gas pressures are calculated for each of the assumed histories as a function of time. SANTOS modeling results in a three-dimensional porosity surface representing changes in gas pressure and porosity over the 10,000-year simulation period.

The completed porosity surface is compiled in tabular form and used to solve the gas and brine mass balance equations presented in the preceding section. Porosity is interpolated from the porosity surface corresponding to the calculated gas pressure and fluid saturations. Because of the tightly coupled and non-linear nature of the processes and equations, the equations are solved iteratively in a fully implicit fashion. The closure data provided by SANTOS can be viewed as a surface, with any gas generation history computed by BRAGFLO constrained to fall on this surface. The adequacy of the method is documented in Freeze et al. (1995), who concludes that the approximation is valid so long as the rate of room pressurization in final calculations is bounded by the room pressurization history that was used to develop the porosity surface.

Halite deformation is predicted using a multi-mechanism deformation steady-state creep model with work hardening and recovery transient response. The focus of the mechanistic part of the model is definition of steady-state creep strain, with transient creep strain described through a multiplier on the steady-state rate, thus accommodating both transient changes in stress loading and unloading.

Computation of repository creep closure is a particularly challenging structural engineering problem, because the rock surrounding the repository continually deforms with time. Not only is the deformation of the salt inelastic, but it also involves larger deformations than are customarily addressed with conventional structural deformation codes. In addition, the formation surrounding the repository is not homogeneous in composition, containing various parting planes and interbeds with different properties than the salt. Deformation of the waste is also nonlinear, with large strains, and its response is complicated by the presence of gas. These complex characteristics of the materials comprising the repository and its surroundings require the use of highly specialized constitutive models that have been built into the SANTOS code over a number of years. Details of the mathematical model used to predict the halite creep around the WIPP repository is beyond the scope of this chapter. The interested reader is referred to Stone (1995b) for these details.

24.6.2.4 Interbed dynamic fracture model

Interbeds contain natural fractures that may be partially healed. If high pressure, resulting from gas generation, is developed in an interbed, its preexisting fractures may dilate or new fractures may form, altering its porosity and permeability. Pressure-dependent changes in permeability are supported by experiments conducted in the WIPP underground and in the laboratory (Beauheim et al., 1993). Accordingly, the DOE has implemented in BRAGFLO a porous-media model of interbed dilation and fracturing that causes the porosity and permeability of a computational cell in an interbed to increase as its pore pressure rises above a threshold value. To the extent that it occurs, dilation or fracturing of interbeds is expected to increase the transmissivity of interbed intervals.

In the model a fracture initiation pressure is specified above the initial pressure and above which local fracturing takes place, so that increases changes in permeability and porosity occur above this pressure. Below the fracture initiation pressure, an interbed has the permeability and compressibility assigned by LHS and representative of intact rock. Below the fracture initiation pressure, the initial sampled porosity is modified slightly with pressure caused by compressibility. Above the fracture initiation pressure, the local compressibility of the interbed is assumed to increase linearly with pressure. This increases the rate at which porosity increases with increasing pore pressure. Additionally, permeability increases by a power function of the ratio of altered porosity to initial porosity. Changes of 10 orders of magnitude in permeability with time are possible.

24.6.3 Numerical Model

The numerical model, BRAGFLO, BRine and Gas FLOw, (DOE, 1996 and SNL, 1996) simulates the gas generation process, the flow process, and the simultaneous interaction of the gas generation process on the two-phase flow of brine and gas, on salt creep, and host rock fracturing. The complex interactions among these reactions with brine and gas inflow and outflow in each grid cell are all accounted for and sorted out by means of the implicit solution of the mass balances, Equations (24.8) and (24.9).

24.6.3.1 Multi-phase flow

At the core of the BRAGFLO model is the finite difference numerical representation and solution of the immiscible multi-phase flow equations for brine and gas.

The above system of four equations, (Equations 1 to 4), describes the time and space behavior of the four variables S_g, S_b, P_g, P_b. The finite difference method requires the introduction of a grid in one, two, or three dimensions and then a discretization of the gas and brine mass balance equations with respect to this grid and with respect to time. The resulting system of nonlinear algebraic equations is solved over a time step at each of the discretized numerical grid blocks by the Newton-Raphson method. For each grid block there are two material balance equations, Equations (24.8) and (24.9). The saturation constraint is used to eliminate brine saturation and the capillary pressure equation is used to eliminate the gas pressure. The remaining unknowns for each grid block are gas saturation, S_g, and brine pressure, P_b.

In regions that contain waste, the terms q_{rg} and q_{rb} represent the rates of gas production and the consumption of brine collectively from corrosion and biodegradation. In the repository, the porosity, ϕ, varies with time and pressure and is interpolated from the results of the mechanical deformation code, SANTOS (Stone, 1995a,b). In regions, which permit pressure induced fracturing, such as the anhydrite interbeds, the permeability tensor and the porosity vary with pressure.

The relative permeabilities, k_{rg} and k_{ro} of Equations (24.8) and (24.9) and the capillary pressure, P_c, are defined empirically by two-phase "characteristic curves." The ability of a fluid to flow is affected by the presence of another fluid. The relative permeability is the ratio of the effective permeability of a given fluid phase to the permeability at 100% saturation.

Relative permeability and capillary pressure are computed from several different empirical relations (Brooks and Corey, 1964; van Genuchten, 1978; WIPP PA, 1992). The choice of the characteristic curves is controlled by input directives.

24.6.3.2 Gas generation

Performance assessment uses the average stoichiometry model to estimate gas generation occurring in the waste disposal region. This model was developed for WIPP performance assessment based on gas generation experiments performed for the WIPP. The average stoichiometry model accounts for the formation of gas by anoxic corrosion of steels and microbial degradation of cellulosics, including plastics and rubbers. Gas is assumed to be generated at a rate dependent on the availability of brine in the computational cell. Because of uncertainties in viability of the microbial colonies in the WIPP environment, microbial degradation occurs in half of the Monte Carlo realizations.

Anoxic corrosion is represented by a generic equation BRAGFLO that includes the possibility of corrosion via the reactions (24.1) and (24.2).

$$\mathrm{Fe} + \frac{4+2x}{3}\mathrm{H_2O} \rightarrow \frac{4-x}{3}\mathrm{H_2} + x\mathrm{Fe(OH)_2} + \frac{1-x}{3}\mathrm{Fe_3O_4} \tag{24.12}$$

With $x = 1.0$, this equation accounts for corrosion only of the steel content in the repository by the reaction expected to dominate, Reaction (24.1). Because the total quantity of aluminum and aluminum alloys is a small compared to the quantity of iron base metals, corrosion of aluminum is omitted for simplicity. The reaction proceeds until one of the reactants brine or iron is depleted or the simulated time is exceeded.

Like anoxic corrosion, microbial gas generation is abstracted into a single general overall reaction:

$$\frac{1}{6}\mathrm{C_6H_{10}O_5} + \text{unknown} \rightarrow y\ \text{gas} + \text{unknown} \tag{24.13}$$

The stoichiometric factor y depends on the extent of the progress of each individual reaction pathway (Reactions (24.1), (24.2), and (24.12)). It can be estimated based on the inventory estimates of the transuranic waste to be emplaced in the Waste Isolation Pilot Plant and it depends on whether CO_2 and H_2S is consumed by reactions with steel and steel-corrosion products. As indicated in Equation 14, y is an imprecisely known parameter whose value is calculated once other imprecisely known parameter values are selected.

$$y = y_{\min} + \beta(y_{\max} - y_{\min}) \tag{24.14}$$

with $0 \leq \beta \leq 1.0$. y_{min} and y_{max} are complex functions of the rates of corrosion and biodegradation as well as the mass of iron, cellulosics, nitrates, and sulfates initially present in the repository at the time of closure (see pages 6, 7, and 8 of Brush, 1995 for more detail). The calculational scheme implemented automatically correlates y with waste inventory estimates as well as with reaction rates.

24.6.3.3 Pressure-induced fracture treatment

BRAGFLO allows for coupling to the mechanical system (rock matrix) through the introduction of rock compressibility, which is defined as the relative rate of change of porosity with respect to pressure. The fracture treatment in BRAGFLO (Key, 1994) allows for pressure-induced alterations to the porosity by introducing a pressure-dependent porosity. This is accomplished be considering the rock compressibility to vary with pressure above a rock dependent threshold pressure. The porosity/pressure dependence is obtained by integrating the mathematical definition of rock compressibility. The permeability/pressure dependence is obtained by applying a power law relationship between permeability and porosity. This relationship is suggested by extending the parallel plate analogy for flow in fractured rock. The parameter values for this sub-model and their uncertainty are obtained by fitting the relationships to fracture experiments in the vicinity of the WIPP repository.

In a system where gas is displacing water, the fracture treatment can influence the gas migration distance. At elevated pressure, the increased porosity will produce more storage with corresponding shorter gas migration distance, while the increase in permeability will enhance the fluid mobility with resulting longer gas migration. The net effect will depend on the relative magnitudes of the storage and permeability effects.

24.6.3.4 Creep closure

The excavation of the WIPP will result in the plastic deformation of the salt material (creep) and resultant closure (creep closure) of excavated areas. The principal effect of this closure for performance assessment is its effect on repository pressure due to the reduction in void volume accompanying creep closure. Accordingly, creep closure is represented in BRAGFLO by changing the porosity of the waste disposal area in a manner consistent with expectations developed from detailed modeling studies. Detailed modeling studies used the code SANCHO (Butcher et al., 1995), to develop time- and pressure-dependent predictions of waste disposal area void volume. Pressure variation was consistent with variation in the gas generation rates possible in the waste disposal areas. The results were used to develop a look-up table of porosity as a function of time and waste disposal area pressure that is used in BRAGFLO to vary waste disposal area porosity.

24.6.3.5 Klinkenberg effect

For gases in a tight porous material at low pressures the boundary condition of zero gas velocity at the solid surface is not satisfied (Klinkenberg, 1941; Corey, 1990). This phenomenon has been called gas slippage and results in a larger gas velocity.

This slipping of the gas flow contributes to an apparent pressure-dependent increase in permeability, which was originally observed by Klinkenberg. A correction to the formation permeability to the gas phase is related to pressure in BRAGFLO.

24.7 USE MODEL PARAMETERS TO CHARACTERIZE UNCERTAINTY

Once scenarios, conceptual models, and the computational modeling system have been defined and developed, parameters are selected to be treated as either imprecisely known (with uncertainty) or precisely known. If the parameter is treated with uncertainty, then its values are defined by a range and distributions for their possible values. Table 24.1 presents gas generation parameters values including those for both the inundated and humid rates of cellulose and steel degradation. They can be converted to units consistent for use in Equations (24.8) and (24.9) using molecular weights, stoichiometry, and waste surface area data.

In BRAGFLO, an average gas-generation rate is determined from by linear interpolation that combines humid and inundated rates based on the effective liquid saturation, which is one of the dependent variables from the BRAGFLO solution to Equations (24.8) and (24.9). The effective liquid saturation in a computational cell in BRAGFLO for the purpose of gas generation is the computed liquid saturation in

Table 24.1. Gas-generation parameters for the long-term WIPP performance assessment[a]

Parameter	Value
Probability of occurrence of significant microbial gas generation	50%
Probability of occurrence of plastics and rubber biodegradation in the event of significant gas generation	50%
Rate of inundated cellulosics biodegradation	0.01–0.3 mole C/(kg-year)
Rate of humid cellulosics biodegradation	0.0–0.04 mole C/(kg-year)
Rate of inundated steel corrosion for the cases without microbial gas generation	0.0–0.5 μm/year[b]
Rate of humid steel corrosion for the cases without microbial gas generation	0.0 μm/year
Rate of inundated steel corrosion for the cases with microbial gas generation	0.0–6.5 μm/year
Rate of humid steel corrosion for the cases with microbial gas generation	0.0 μm/year
Stoichiometric factor x in Reaction (24.12)	1.0
Stoichiometric factor y in Reaction (24.13)	Calculated from Eqn. (24.14)
Factor β in Equation (24.14)	0–1.0
NO_3^- initially present in the waste	2.6×10^7 moles
SO_4^{-2} initial present in the waste	6.6×10^6 moles

[a] From Wang and Brush (1996).

[b] Multiplying 0.141 mole/(μm − m^2) will convert the unit of steel corrosion rate from μm/*year to mole*/(m^2-year), Telander and Westerman (1993).

that cell, plus an adjustment to account for uncertainty in the capillary rise (wicking) characteristics of the waste.

These gas generation parameters, along with many others that characterize the repository and formation hydrology and radionuclide transport, are treated with uncertainty in the WIPP PA.

24.8 PERFORM THE UNCERTAINTY ANALYSIS USING MONTE CARLO SIMULATIONS

24.8.1 Modeling Process

Uncertainty analyses evaluate uncertainty in performance estimates that results from uncertainty about imprecisely known input parameters. The probabilistic approach uses a linked system of deterministic consequence models (coupled source-term, multiphase flow, gas generation, and contaminant transport models) to represent the repository and associated geologic system, and a Monte Carlo technique that relies on multiple simulations using sampled values for selected input parameters to propagate uncertainty through to the calculation of system performance measures. A full analysis includes selecting imprecisely known parameters to be sampled, constructing distribution functions for each of these parameters (incorporating available data and subjective information to capture the uncertainty), generating a sample set by selecting a parameter value from each distribution function, and calculating consequences for the sample set. This procedure is repeated many times to produce distributions of model outcomes (consequences) that represent the spectrum of potential system consequences.

By performing many simulations using differing parameter values drawn from the distribution of possible values, we translate the uncertainty in input to a range of consequences (predictions of radionuclide migration), any of which is possible. Many of the results are displayed as time histories of entire simulation sets (3 replicates of 100 simulations in the case of the WIPP evaluation). Three replicates, each using a different random seed to generate differing random samplings of the input parameter distributions, were used to demonstrate that adequate coverage of parameter uncertainty is capture by using a sample size of 100. These time histories show the range of possible consequences given the uncertainties present in the various measured input parameters.

24.8.2 Results

Among the 100 realizations from the first replicate, the volume of gas generated varies over more than an order of magnitude, from 1.5×10^6 m^3 to 2.8×10^7 m^3 of hydrogen, at standard conditions Figure 24.4(a). This variation is associated with uncertainty over the 10,000-year regulatory period in repository chemistry, in the rate of corrosion and biodegradation, in what constituents will biodegrade, and in uncertainty in the rate at which brine contacts the waste. The corrodible metal is never completely

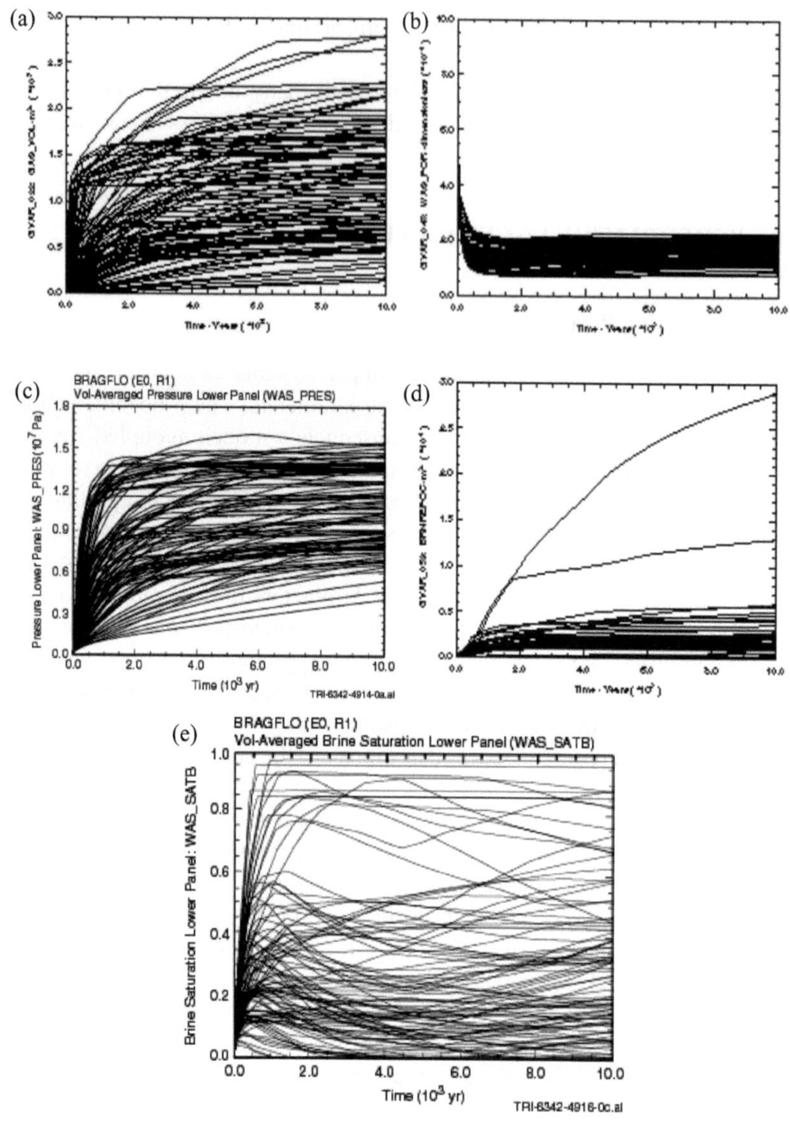

Figure 24.4. (a) Gas generation (m^3) vs. time (yr); (b) repository porosity (%) vs. time (yr); (c) repository pressure (Pa) vs. time (yr); (d) cumulative flow of brine out of the repository vs. time (yr); and (e) repository brine saturation vs. time (yr) for the undisturbed repository.

consumed in 10,000 years. In realizations where corrosion ceases, it is a lack of brine in the waste that causes corrosion to stop. In contrast, biodegradation (if it occurs), completely consumes the inventory of cellulosics.

Halite creep causes the pore volume of the repository to decrease over time, Figure 24.4(b). The porosity of the waste drops from its initial porosity of 84.8% during the first few hundred years as the repository creeps shut. The porosity reaches a minimum between 7 and 22% of the initial excavated volume, depending on parameter uncertainty.

Pressures in the repository increase from their initial value of 1 atmosphere, Figure 24.4(c). The pressure increases are driven by a number of competing processes (1) brine inflow and pressure equilibration with the surrounding host rock, (2) gas generation, (3) creep consolidation of the surrounding halite, and (4) pressure relief through developing fractures.

Cumulative brine flows out of the repository are shown in Figure 24.4(d). This brine cannot migrate significant distances in halite because of the low permeability of halite. To get to the land withdrawal boundary, brine from the repository must first flow through the DRZ into one of the permeable anhydrite interbeds or up the sealed shafts. Not all of the brine leaving the repository enters the anhydrite interbeds. Most of the brine leaving the repository during the undisturbed scenario remains in the disturbed rock zone surrounding the repository. Additional analysis indicates a maximum of 3700 m^3 enters the interbeds over 10,000 years. The cumulative flows across the LWB in the anhydrite interbeds are small. Brine volumes crossing the LWB during the 10,000-year regulatory period range up to 216 m^3.

The amount of brine contained within the pore space of the repository helps determine the rate of gas generation from corrosion and microbial degradation. In Figure 24.4(e) repository brine saturation histories as determined from BRAGFLO are present for a set of 100 realizations that span the uncertainty associated with WIPP conditions over 10,000 years.

The flow results from BRAGFLO indicate that in some cases brine flows across the subsurface land withdrawal boundary into the accessible environment. The results from the transport calculations, which use the BRAGFLO flow results, indicate cumulative releases of all radionuclides in this brine are on the order of 3.4×10^{-15} to 3.4×10^{-8} curies. None of these releases is considered meaningful, United States Department of Energy (1996a).

24.9 PERFORM A SENSITIVITY ANALYSIS TO GUIDE THE PROGRAM

Sensitivity analysis is performed on the results of the consequence analysis to evaluate which processes, parameters, and combinations of parameters are most significant in influencing the predicted performance of the system. The primary goals of this step are to identify and rank, in order of importance, data and assumptions that influence predicted results. This information is used to prioritize site characterization activities and make model improvements and simplifications.

For undisturbed repository performance there are no releases of any consequence so that identification of important drivers to contaminant release is impossible. Sensitivity analysis can be done on secondary or sub-system metrics, such as repository pressure. Results documented in the CCA, United States Department of Energy (1996a) and in Helton et al. (1998) indicate the important role of gas generation.

24.10 CONCLUSIONS

Gas generation is an important process that has the potential to influence the flow of brine and transport of radionuclide contaminants from the WIPP repository. Regardless of gas generation, no meaningful release of contaminants from WIPP is expected over 10,000 years under undisturbed conditions. Gas generation influences release of contaminants during potential human intrusion scenarios.

NOTATION

g	acceleration of gravity [$9.79\ m/s^2$ at WIPP site]
h	elevation [m]
$[k]$	permeability [m^2]
k_{rb}, k_{rg}	relative permeability to brine (b) or gas(g) [dimensionless]
P_c	capillary pressure [Pa]
P_b, P_g	brine (b) or gas (g) pressure [Pa]
q_b, q_g	rate of injection (+) or production (–) of brine (b) or gas (g) [kg $i/(s \cdot m^3)$]
q_{rb}, q_{rg}	rate of brine (b) or gas (g) production (or consumption, if negative) due to chemical reaction [kg brine/($s \cdot m^3$)]
S_b, S_g	brine (b) or gas (g) saturation [m^3 brine or gas/m^3 void volume]
t	time [s]
x	stoichiometric coefficient for $Fe(OH)_2$ in anoxic corrosion reaction [Moles $Fe(OH)_2$/Mole Fe]
y	gas stoichiometric coefficient in microbial degradation reaction [Moles gas/Mole of C]
α	geometric factor dependent on dimensionality of the problem: $\alpha = \Delta y \Delta z$ for 1-D problems, $\alpha = \Delta z$ for 2-D problems, and $\alpha = 1$ for 3-D problems
ϕ	porosity [m^3 pore volume/m^3 grid block volume]
ρ_b, ρ_g	density of brine (b) or gas (g) [kg/m^3]
μ_i	viscosity of brine (b) or gas (g) [$Pa \cdot s$]

REFERENCES

Bean, J.E., Berglund, J.W., Davis, F.J., Economy, K., Garner, J.D., Helton, J.C., Johnson J.D., MacKinnon, R.J., Miller, J., O'Brien, D.G., Ramsey, J.L., Schreiber, J.D., Shinta, A., Smith, L.N., Stockman, C., Stoelzel, D.M., and Vaughn, P. (1998). "Uncertainty and Sensitivity Analysis Results Obtained in the 1996 Performance Assessment for the Waste Isolation Pilot Plant." SAND98–0365. Sandia National Laboratories, Albuquerque, NM.

Beauheim, R.L., Roberts, R.M., Dale, T.F., Fort, M.D., and Stensrud, W. A., (1993). "Hydraulic Testing of Salado Formation Evaporites at the Waste Isolation Pilot Plant Site: Second Interpretive Report". SAND92–0533 (WPO#23378). Sandia National Laboratories, Albuquerque, NM.

Brooks, R.H., and Corey, A.T., (1964). "Hydraulic Properties of Porous Media", Hydrology Paper No. 3. Fort Collins, CO: Colorado State University.

Brush, L.H., (1995). "Systems Priority Method – Iteration 2 Baseline Position Paper: Gas Generation in the Waste Isolation Pilot Plant". Sandia National Laboratories, Albuquerque, NM.

Butcher, B.M., Webb, S.W., and Berglund., J.W., (1995). "System Prioritization Method – Iteration 2. Baseline Position Paper: Disposal Room and Cuttings Models". Volume 2, Appendix F. Sandia National Laboratories, Albuquerque, NM.

Corey, A.T., (1990). *Mechanics of Immiscible Fluids in Porous Media*. pp. 98, Water Resources Publications, Littleton, CO.

Darcy, H., (1856). *Les fontaines publiques de la ville de Dijon*. Dalmont, Paris.

Davies, P.B., (1991). "Evaluation of the Role of Threshold Pressure in Controlling Flow of Waste-Generated Gas into Bedded Salt at the Waste Isolation Pilot Plant". pp. 17–19, SAND90–3246 (WPO# 26169). Sandia National Laboratories, Albuquerque, NM.

EPA (U.S. Environmental Protection Agency), (1985). "40 CFR Part 191: Environmental Standards for the Management and Disposal of Spent Nuclear Fuel, High-Level, and Transuranic Radioactive Wastes; Final Rule". *Federal Register*, Vol. 50, No. 182, pp. 38066 – 38089, September 19,1985. Office of Radiation and Air, Washington, DC.

Francis, A.J and Gillow J.B., (1994). "Effects of Microbial Processes on Gas Generation under Expected WIPP Repository Conditions: Progress Report through 1992". SAND93–7036. Sandia National Laboratories, Albuquerque, NM.

Francis, A.J., Gillow J.B., and Giles M.R., (1995a and b). "Effects of Microbial Processes on Gas Generation under Expected WIPP Repository Conditions: Data Report for the Period 1993 – 1995". (draft)

Freeze, G.A., Larson, K.W., and Davies, P.B., (1995). "Coupled Multiphase Flow and Closure Analysis of Repository Response to Waste-Generated Gas at the Waste Isolation Pilot Plant (WIPP)". SAND93–1986 (WPO# 29557). Sandia National Laboratories, Albuquerque, NM.

Key, S., Fewell, M., Vaughn, P., Lord, M., and Labreche, D., (1994). "Background and Basis for the Fluid Flow Model Incorporated in BRAGFLO for Representing Pore Pressure-Induced Alterations in Permeability and Porosity in the Anhydrite Interbeds Above and Below the WIPP Repository Horizon". SAND94–0381. Sandia National Laboratories, Albuquerque, NM.

Klinkenberg, L.J., (1941). "The Permeability of Porous Media to Liquids and Gases". *Drilling and Production Practices*. pp. 200–213. American Institute of Petroleum Engineers (API).

Peaceman, D.W., (1977), *Fundamentals of Numerical Reservoir Simulation*. Elsevier Scientific Publishing Co., Amsterdam.

Stone, C.M., (1995a). "*Transfer of Porosity and Gas Curves from SANTOS for the Final Porosity Surface*", Memorandum to B.M. Butcher, December 19, 1995. Sandia National Laboratories, Albuquerque, NM.

Stone, C.M., (1995b). "SANTOS - A Two-Dimensional Finite Element Program for the Quasistatic, Large Deformation, Inelastic Response of Solids". WPO# 35674. Sandia National Laboratories, Albuquerque, NM.

Telander, M.R., and Westerman, (1993). "Hydrogen Generation by Metal Corrosion in Simulated Waste Isolation Pilot Plant Environments: Progress Report for the Period November 1989 through December 1992". SAND92–7347. Sandia National Laboratories, Albuquerque, NM.

Telander, M.R. and Westerman, (1995). "Hydrogen Generation by Metal Corrosion in Simulated Waste Isolation Pilot Plant Environments: Final Report". (draft)

United States Department of Energy, (1996a). "Title 40 CFR Part 191 Compliance Certification Application for the Waste Isolation Pilot Plant, Chapter 6: Containment Requirements". DOE/CAO-1996–2184. U.S. Department of Energy, Carlsbad Area Office, Carlsbad, NM.

United States Department of Energy, (1996b). "Title 40 CFR Part 191 Compliance Certification Application for the Waste Isolation Pilot Plant, Appendix NUTS: Containment Requirements". DOE/CAO-1996-2184. U.S. Department of Energy, Carlsbad Area Office, Carlsbad, NM.

United States Department of Energy, (1996c). "Title 40 CFR Part 191 Compliance Certification Application for the Waste Isolation Pilot Plant, Appendix BRAGFLO: Containment Requirements". DOE/CAO-1996-2184. U.S. Department of Energy, Carlsbad Area Office, Carlsbad, NM.

van Genuchten, R., (1978). "Calculating the Unsaturated Hydraulic Conductivity with a New Closed-Form Analytical Model". Research Report 78-WR-08. Princeton, NJ: Princeton University, Department of Civil Engineering. Princeton, NJ.

Wang, Y. and Brush, L.H., (1996). "Estimates of Gas Generation Parameters for the Long-Term WIPP Performance Assessment". Memorandum to Martin S. Tierney, January 26, 1996. Sandia National Laboratories, Albuquerque, NM.

WIPP PA (Performance Assessment) Department, (1992). "Preliminary Performance Assessment for the Waste Isolation Pilot Plant, December 1992, Volume 3: Model Parameters". SAND92–0700/3 (WPO# 23529). Sandia National Laboratories, Albuquerque, NM.

WIPP PA (Performance Assessment) Department, (1996). "WIPP PA User's Manual for BRAGFLO, Version 4.00, Document Version 1.01". WPO# 30703. Sandia National Laboratories, Albuquerque, NM.

CHAPTER 25

OIL AND GAS INDUSTRY APPLICATIONS OF GAS FLOW IN POROUS MEDIA

DAVID J. BORNS
Sandia National Laboratories, P.O. Box 5800, Albuquerque, NM 87185, USA

25.1 INTRODUCTION

Gas flow in porous media and multiphase flow in porous medium are essential processes of oil and gas production. As discussed in this chapter, multiphase flow and the porous media properties that may control flow are dynamic throughout production. This chapter is a basic overview of oil and gas reservoir behavior and an introduction to several applications in traditional oil and gas production and storage. The interested reader is referred several overviews of reservoir engineering for more detailed information (Ahmed, 2000; Chierici, 1994; Satter, 2000).

Natural gas and oil reservoirs exhibit complex multiphase behavior of the accumulated hydrocarbon. The hydrocarbon phases present may be solid, liquid or gas. The hydrocarbons are also associated within other fluids such as water, CO_2, N_2, and H_2S. Enhanced production of hydrocarbon reservoir may also introduce other liquids such as flue gas, surfactants, steam, and CO_2.

A basic overview of hydrocarbon reservoirs is that they contain the following components (see Figure 25.1):

- *Source* rocks (the zone where petrogenesis occurred)
- *Migration pathway* (permeable paths: permeable rocks, faults, fractures, variations in relative permeability) to a *trap*
- *Trap* (geologic, relative permeability or hydrodynamics structure favorable for the accumulation of migrating hydrocarbons, sealed by a cap)
- *Cap* (geologic, relative permeability or hydrodynamics structure the limits the migration of hydrocarbons).

The architecture for reservoir fluid flow develops in multiple rocks types such as:

- Porous media: Sandstones, diatomites, limestones
- Fractured media: carbonates, volcanic rock.

C. Ho and S. Webb (eds.), Gas Transport in Porous Media, 407–418.
© 2006 *Springer.*

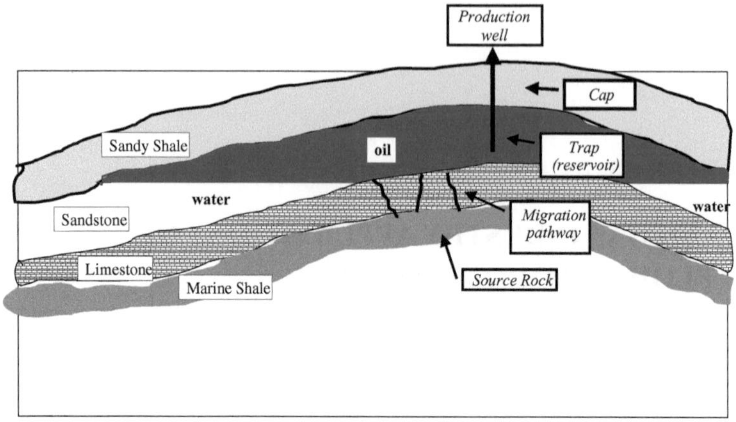

Figure 25.1. Idealized geologic cross-section of an oil reservoir form in an anticlinal structural trap at the start of production, with the components of the reservoir identified as in the text

With this basic overview of oil and gas reservoirs, this chapter will now discuss some of the issues of gas flow in natural gas and oil production and present examples of applications of an improved understanding of gas flow in porous will affect the efficiency of oil and gas production; such as, enhanced production and gas storage.

25.2 MULTIPHASE BEHAVIOR OF OIL AND GAS MIXTURES

During production oil and gas reservoirs become highly dynamic in response to production-induced pressure and temperature gradients. Such dynamic behavior includes evolving changes in saturation, relative permeability, viscosity, pressure and temperature gradient. Parameters that are often assumed to be static, such as porosity and compressibility, wettability and capillarity, can change during production due to mechanical and chemical changes in the formation.

Figure 25.2 shows the multiphase behavior of typical oil and gas reservoir. This diagram contains a *critical point* at which all of the oil and gas properties are the same and a *two-phase* region within a two-phase envelope. Within the two-phase region, as temperature increases the gas saturation increases as marked by the dashed contours with percentages of gas phase. Outside the two-phase envelop, a reservoir is oil saturated at high pressures and low temperature and is gas saturated at high temperatures and low pressures. This two phase behavior is important in pressure (P)-temperature (T) space since during reservoir production different P–T paths can develop (see Figure 25.3). Following these P–T paths the reservoirs can develop different oil-gas saturations and condensation of secondary phases. These processes will affect fluid flow through changes in relative permeability and entry pressure.

Based on the idealized reservoir in Figure 25.1, a production scenario can be hypothesized (Figure 25.3, Figure 25.4, and Figure 25.5). At point A, which would represent the P and T conditions of the structural trapped reservoir in Figure 25.1. At

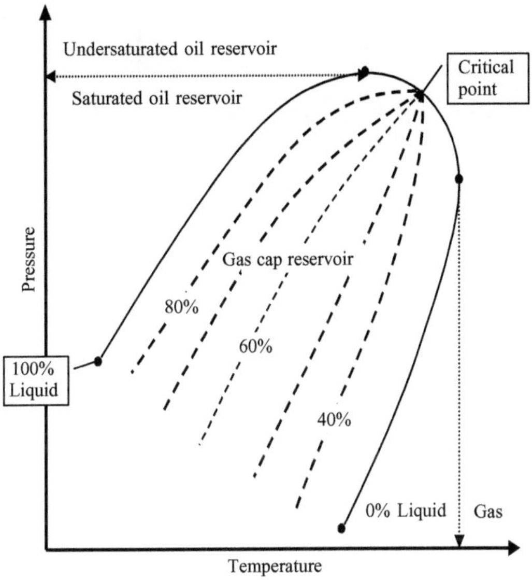

Figure 25.2. Pressure-temperature diagram for multicomponent oil and gas system

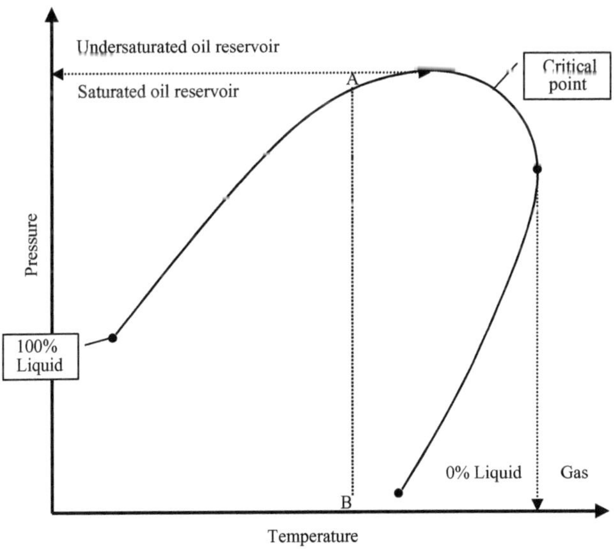

Figure 25.3. *P–T* diagram for multiphase behavior with isothermal depressurization following production in idealized reservoir from Figure 25.1, *P–T* path will follow line A-B

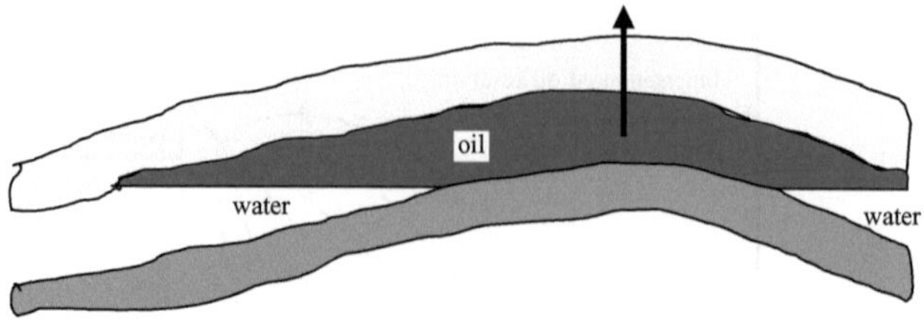

Figure 25.4. Simplified geologic cross-section of a aniclinical oil reservoir at start of production, point A in Figure 25.3

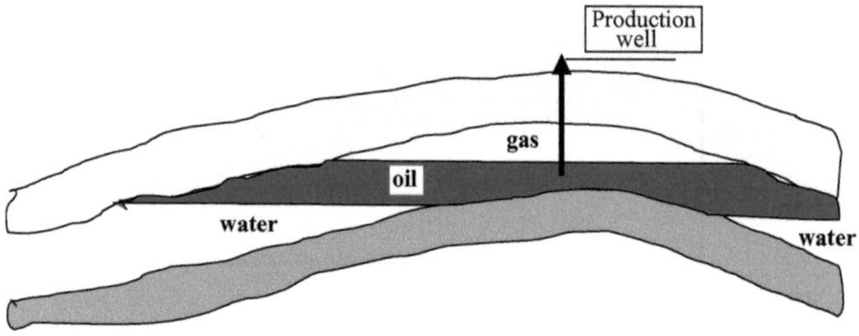

Figure 25.5. Simplified geologic cross-section of a aniclinical oil reservoir after production, point B in Figure 25.3

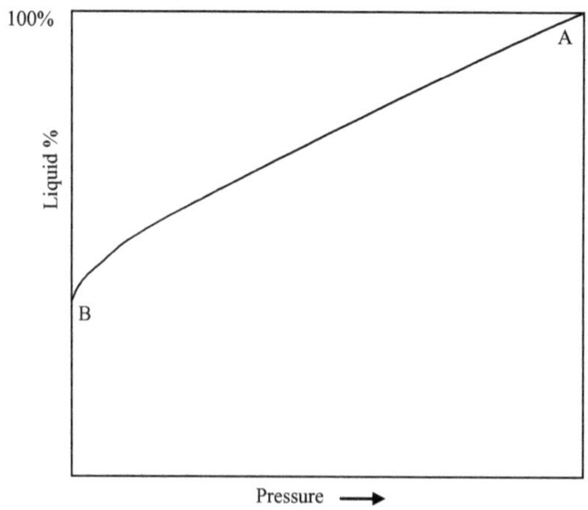

Figure 25.6. Pressure versus % Liquid plot for isothermal production path A-B in Figure 25-3

this point, the reservoir is at or near oil-saturation. As oil is produced (Figure 25.5), the pore pressure in the reservoir around the producing well decreases following the path A-B in Figure 25.3. Natural gas evolves along this *P–T* path, thus, changing the relative saturation of oil and gas in the reservoir, as shown in the *P* versus % saturation plot in Figure 25.7.

25.3 INDUSTRY APPLICATIONS OF GAS FLOW WITHIN A MULTIPHASE RESERVOIR

- Effects of oil, gas and water saturation changes, e.g., relative permeability changes, dynamic traps were gas or liquid (i.e., condensate) where saturation changes inhibit reservoir flow.
- Effective stress changes during production resulting in changes in effective porosity of the reservoir and failure of wells (this stresses the need of coupling gas flow with mechanical changes).

25.4 APPLICATION EXAMPLE: ENHANCED OIL RECOVERY

Primary production of an oil reservoir may only extract 10 to 20% of the oil in the reservoir. Primary production encompasses drilling and completion, initial gas or water driven flow, and artificial lift (e.g., pumping). To obtain more oil from the reservoir (20 to 40%), secondary and tertiary recovery methods are required. Oil from enhanced recovery provides a major fraction of the nation's proven reserves. For example, in a typical West Texas well (Moritis, 2001):

- Primary production can result in 11,000,000 bbl of oil recovered
- Secondary recovery (water injection) can result in 10,000,000 bbl of oil recovered
- Tertiary recovery (water alternating with CO_2 gas –WAG) can result in 10,100,000 bbl of oil recovered.

The production of more oil out of new and existing fields has largely resulted from several technologies initiated in the 1950s. Hydraulic fracturing was one, in which pumps at the surface inject a fluid into the oil-producing reservoir rocks. The pressure exerted by the fluid, is great enough to fracture the rocks around the well, and proppants injected with the fluid keep the cracks propped open following cessation of the pumping. The open fractures add effective porosity to the reservoir, and oil can flow more easily into the well. Although hydraulic fracturing will not be used as an example in this chapter, the process has interesting gas flow process, especially considering the non-Darcy flow may be a significant process in propped fractures. Water flooding is another methods, in which water is injected into the reservoir rocks to maintain reservoir pressure as the oil is withdrawn and to sweep the oil out of the reservoir rocks and toward the well (see Figures 25.7, 25.8, and 25.9). In the most commonly used technique, the five-spot pattern (see Figure 25.6), water is pumped into the reservoir rocks at four wells arrayed around a central producing well.

Figures 25.7, 25.8, and 25.9 show an idealized sequence of enhanced secondary and tertiary recovery. The production infrastructure in based on the fore-mentioned five-spot pattern with a central producing well surrounded by injection wells. The injection of water, natural gas or flue gas at the reservoir horizon (Figure 25.9) characterizes secondary recovery. Flow of the injected fluid can be non-homogeneous; thus, the understanding of this heterogeneous flow is important in optimizing production (e.g., sweep efficiency). An idealized Water-Alternating-with-Gas (WAG) tertiary recovery process is portrayed in Figure 25.8. In this process successive injections of gas such

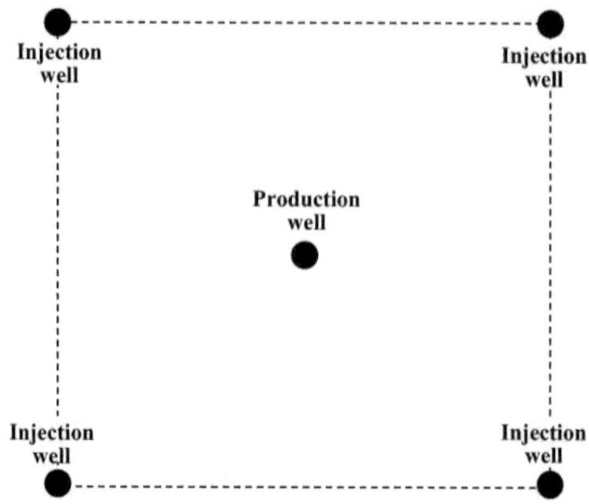

Figure 25.7. Plan view: five spot pattern of outer injector wells and central production Well

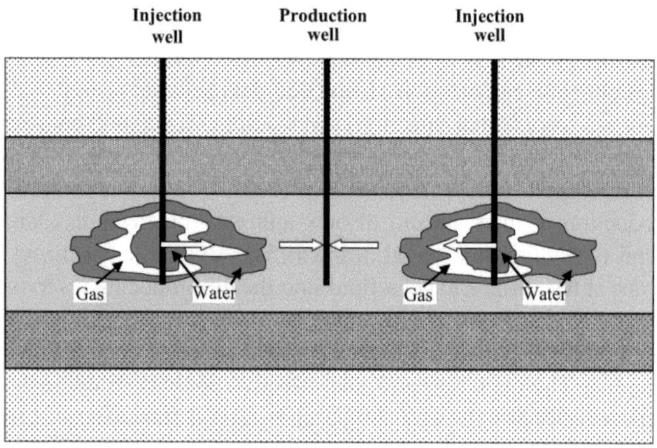

Figure 25.8. Cross-section view of a water alternating with gas (WAG) process

as CO_2 are in injected followed by an injection of water to control the injected gas and the move the resultant lower viscosity oil.

25.5 INDUSTRY APPLICATIONS OF GAS FLOW TO ENHANCED RECOVERY

- Optimization of Production
  - *Sweep efficiency:* From the micro- to reservoir scale, sweep efficiency describes the portion of the multiphase pores to portion of reservoir scale compartments or structures that the injected water, gas, surfactants or other recovery fluids affect. At the micro-scale, capillary barriers, surface cohesion, and relative permeabilties

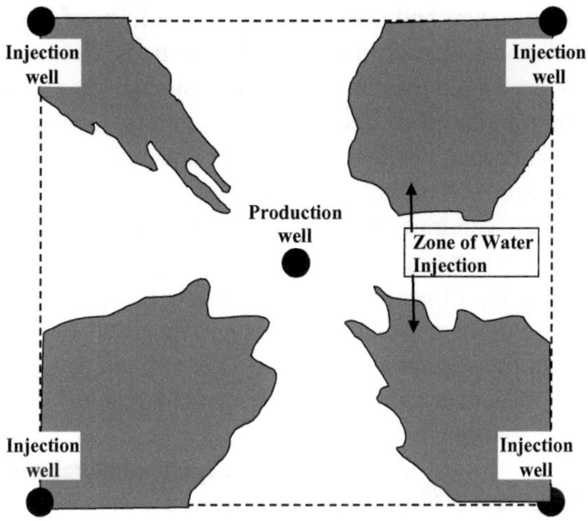

Figure 25.9. Plan view of zone of water injection at target horizon in reservoir

may affect the sweep efficiency. At the reservoir scale, heterogeneities such as fingering, anisotropic flow architectures (such as fractures, sedimentary structures, relative permeability structures, etc.) may also affect the sweep efficiency.

- *Control of water and gas breakthrough into the production well:* The goal of enhanced recovery is to maximize the hydrocarbon production from the extraction well. The understanding of gas flow allows the production engineer control the throughput of the water or gas injected. This control in turn affects the sweep efficiency and economics of the enhanced recovery process.
- *Minimization of borehole and formation damage*: The injection and withdrawal of fluids in a production zone and the zone adjacent to the well bores needs to be optimized. Again, the understanding of gas flow in the formation and the well bore zone provides input to the understanding of pressure and flow effects on formation damage through mechanism such as irreversible compaction in response to changing deviatoric stresses and mobilization of formation particles (sand production).

25.6 APPLICATION EXAMPLE: UNDERGROUND GAS STORAGE

Demand for natural gas is rising in the United States (Energy Information Agency and Gas Technology Institute). Current forecasts from sources like the National Petroleum Council suggest consumption could grow from an annual 22 trillion cubic feet (Tcf) at the end of the twentieth century to more than 30 Tcf annually by 2015. The increased demand for natural gas is expected to quickly outstrip existing pipeline capacity. Even though pipeline expansion is anticipated to proceed, a significant increase in storage capacity will be needed to meet the peak deliverability needs, to maximize pipeline efficiency, and to provide security of energy availability.

The safest, most secure and economically viable form of natural gas storage is storage in natural or man-altered underground formations (Beckman and Determeyer, 1997). The three principal types of underground storage sites in the U.S. today are: (1) depleted reservoirs in oil or gas fields, (2) aquifers, (3) salt cavern formations. Two of the most important characteristics of an underground storage facility are its capability to hold large volumes of natural gas for future use and the rate at which the gas inventory can be withdrawn (its delivery rate). There is presently about 3 Tcf of underground natural gas storage working gas capacity in the U.S., with an associated maximum deliverability of nearly 70 Billion cubic feet per day (Bcfpd). There are currently about 415 storage facilities operating. Based on capacity, 85 percent are depleted reservoirs, 12 percent aquifers and 3 percent salt caverns. Salt represents a larger percentage of deliverability (11 percent) with depleted reservoirs accounting for 79 percent and aquifers for 10 percent.

Most existing gas storage in the U.S. is in depleted natural gas or oil fields (348 sites 3,368 Bcf – billion cubic feet– capacity) that are close to consumption centers. Conversion of a field from production to storage takes advantage of existing infrastructure (wells, gathering systems, and pipeline connections). Depleted oil and gas reservoirs are the most commonly used underground storage sites because of their wide availability.

Aquifers are used in some areas to store natural gas (40 sites, 392 Bcf capacity) if an impermeable layer overlies the formation. While aquifers are similar to depleted fields, their use for storage requires more base (cushion) gas and greater monitoring of withdrawal and injection performance. Delivery rates may be enhanced by the presence of an active water drive.

Salt caverns (27 sites, 139 Bcf capacity) provide very high withdrawal and injection rates compared with their working gas capacity and base gas requirements are relatively low. The majority of salt cavern storage facilities have been developed in salt dome formations in the Gulf Coast states. Vertical salt caverns leached from bedded slat formations in Northeastern, Midwestern, and Western states are also being developed to take advantage of the high volume and flexible operations possible with a cavern facility. Cavern construction is more costly than depleted field conversions, but the ability to perform several withdrawal/injection cycles per year reduces the per-unit cost of gas injected. The largest growth in daily withdrawal capability has been from salt cavern storage. These facilities can cycle their inventories more rapidly than the other types of storage.

Before subsurface injection of natural gas, gas storage operators look for reservoirs that will accept, hold and release gas most efficiently. Therefore, they target porous and permeable zones that are well sealed and not overly complex or compartmentalized (DOE/NETL). However, these zones are natural reservoir, and the wells do not always behave as planned. Heterogeneities and other variations in the reservoir will impact storage field performance. For example, gas may migrate from the field through uncharacterized conduits to areas from which it cannot be extracted.

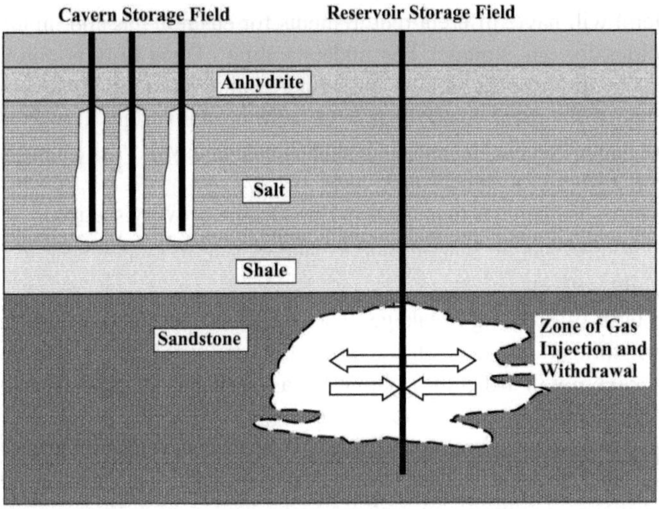

Figure 25.10. Cross-section of approaches to underground storage.

Gas reservoir storage is one of purest applications of gas flow in porous media within the oil and gas industry (Tek, 1987). To improve the understanding and operation of gas storage reservoirs, gas storage operators are increasingly utilizing the advanced reservoir modeling, visualization, and simulation techniques that are commonplace in gas exploration and production operations (Albrecht, 1998).

25.6.1 Industry Applications of Gas Flow to Gas Storage

- Injectivity: The understanding of gas flow is porous media in needed to predict the amounts of cushion gas and working gas that a specific field can store. For a reservoir storage field both the processes of flow and dispersion are important. Some of the gas injected may be irreversible lost. System design needs to minimize such loss.
- Deliverability
  - Amount recovered (base load): Base load is the capability to store a large enough volume of gas to provide the bulk of the demand above the gas that long haul pipelines can deliver. This base load storage consists of a long withdrawal period and along injection period. The gas injected into a reservoir has two components: *cushion gas* and *working gas* (Beckman and Determeyer, 1997). Cushion gas is the volume of gas needed to fill a reservoir such that the reservoir pressure provides sufficient flow of gas as needed. Working gas is the volume of gas that is injected and produced during a storage cycle. Based on our understanding of gas flow, different types of reservoir fields (i.e., depleted oil and gas reservoir versus

aquifers) will have different requirements for cushion gas and, in turn, different capacities for gas storage. The understanding of gas flow is porous media in needed to optimize the storage and withdrawal cycles to maintain deliverability.
  - Rate recovered (peak load): Peak load facilities deliver gas at high rates to meet urgent needs for a short period of time (hours or days). Peak storage can sustain deliverability for a limited time.
  - Formation Damage: as discussed in Enhanced Recovery section.
  - Well bore damage: as discussed in Enhanced Recovery section.
- Permeability effects
  - Long term effects, e.g., leakage: These effects apply equally to reservoir and cavern storage. The flow of gas farther into the reservoir or into the wall rock of the cavern will affect the volume of gas available for delivery and the rate of delivery if the storage pressure decays.
  - Maximum operating pressures: The effects on operating pressure are linked to leakage and relative permeabilties. The hydrostatic pressure of the formation bound the operating pressure of a storage cavern or a storage field. If higher pressures are desired then the host formation or reservoir seal has to have a lower permeability to support the pressures desired . Maximum safe operating pressures for a reservoir depend on four geomechanical factors that may be influenced by gas flow (Terratek):
    - The mechanical properties of the reservoir and overburden;
    - The in situ stresses and fracture pressure in the reservoir;
    - Stresses induced in the reservoir by gas cycling; and,
    - Stresses induced in the caprock material by gas cycling.

25.7 SUMMARY

This chapter is a brief overview of the applications of gas flow in porous medium. The two examples given, enhanced recovery and underground gas storage, barely begin to describe the need for the understanding of gas flow as applied to the industry. New applications emerge steadily, especially with the emphasis of domestic natural gas production. Coal bed methane is new gas frontier that is coming on line and may provide 50 percent of the new natural gas production until 2020. In the future, the production of gas from methane hydrates and the sequestration of CO_2 in reservoir and aquifers may emerge as new applications.

REFERENCES

Ahmed, T. H., 2000, Reservoir Engineering Handbook, Gulf Publishing Company, Houston, 863p.

Albrecht, J, 1998, Underground gas storage: Improving the process, Enhancing the Resource, Gas Research Institute Digest 21, Summer (No. 2), 12–15.

Beckman, K. L., and P. L. Determeyer, 1997, Natural gas storage: Historical development and expected evolution, Gas TIPs, Spring (No. 2), 13–22.

Chierici, G. L., 1994, Principles of Petroleum Reservoir Engineering, Springer-Verlag, Berlin, 419p.

Moritis, G. 2001, New Companies reshape landscape for CO_2 EOR in US, Oil and Gas Journal, May 14, 68–73.
National Petroleum Council, 1999, Meeting the Challenges of the Nation's Growing Natural Gas Demand, National Petroleum Council, Washington, D.C.
Satter, A., 2000, Computer-Assisted Reservoir Management, PennWell, 278p.
Tek, M. R., 1987, Contributions in Petroleum Geology and Engineering Volume #: Underground Storage of Natural Gas, Gulf Publishing Company, Houston, 399p.

CHAPTER 26

GEOLOGIC CARBON SEQUESTRATION: CO_2 TRANSPORT IN DEPLETED GAS RESERVOIRS

CURTIS M. OLDENBURG
Lawrence Berkeley National Laboratory, Berkeley, CA 94720, USA

26.1 INTRODUCTION

The correlation between increased atmospheric carbon dioxide (CO_2) concentrations and increased global surface temperatures during the last century suggests that anthropogenic CO_2 emissions are linked to global warming and climate change through the role of CO_2 as a greenhouse gas. Anthropogenic CO_2 emissions, primarily from the burning of fossil fuels for energy production and transportation, are on the order of 2.7×10^{10} tonnes/year (Reichle et al., 1999). To put this number in perspective, the total annual volcanic CO_2 emissions are estimated at 5.6×10^8 tonnes/year (Gerlach, 1991). In order to mitigate CO_2 emissions, the idea of capturing CO_2 from flue gases (e.g., from coal- and gas-fired power plants) and injecting the CO_2 into deep subsurface formations is receiving considerable attention. Called geologic carbon sequestration, this process aims to prevent CO_2 from entering the atmosphere by storing it permanently in three main target formations, specifically, depleted oil and gas reservoirs, brine formations, and deep unminable coal seams.

Of the potential target formations for geologic carbon sequestration, depleted natural gas (CH_4) reservoirs appear to be obvious targets for initial utilization for several reasons. First, depleted gas reservoirs have a proven record of long-term gas containment and capacity as evidenced by their exploitation for CH_4 production. Second, there is an infrastructure of wells and pipelines, knowledge of the local subsurface conditions, and history of industrial activity at gas fields that favor implementation of CO_2 injection. Third, natural gas reservoirs are typically depleted and uneconomic even though 10 to 40% of the original gas in place is still in the reservoir. Thus the possibility exists to inject CO_2 into depleted gas reservoirs to increase reservoir pressure and produce additional CH_4 for economic benefit. Termed Carbon Sequestration with Enhanced Gas Recovery (CSEGR), this potential dual use of depleted gas reservoirs

C. Ho and S. Webb (eds.), Gas Transport in Porous Media, 419–426.
© 2006 *Springer.*

appears promising (van der Burgt et al., 1992; Blok et al., 1997; Oldenburg et al., 2001; Mamora and Seo, 2002; Oldenburg et al., 2004).

The economic feasibility of CSEGR depends critically on the extent to which the CO_2 and CH_4 will mix in the reservoir. Complex diffusive and advective transport phenomena will control gas mixing as injected CO_2 migrates through the formation as controlled by formation heterogeneity and by the properties of the gases. If extensive mixing occurs, the value of the natural gas will be diminished. While CH_4-CH_4 mixing has been documented in gas reservoirs by carbon isotopic analyses (Buzek, 1992), inert gas cushions (e.g., N_2) appear not to mix excessively with CH_4 in natural gas storage projects (Carriére et al., 1985; Laille et al., 1988). In this chapter, we discuss transport and mixing issues relevant to the technical feasibility of CSEGR. Because CSEGR has not yet been attempted in any actual gas reservoir, our approach relies on theory and numerical simulation.

26.2 PHYSICAL PROPERTIES OF CO_2 AND CH_4

Gas mixing in a natural gas reservoir under CO_2 injection will depend strongly on the properties of the gases. In Figure 26.1, we show the phase diagram of CO_2 with a subsurface $P - T$ path assuming hydrostatic pressure and a geothermal gradient of 25°C/km. Natural gas reservoirs will typically be at temperatures above the critical temperature for CO_2 (T_{crit} = 31.8°C) . Methane is supercritical above –82.4°C, therefore both gases will be supercritical in typical gas reservoirs. The pressure in depleted gas reservoirs may fall below the critical pressure of CO_2 (P_{crit} = 73.8 bar), but typical pressures in deep gas reservoirs will be above 73.8 bar. While CH_4 behaves much like an ideal gas as pressure changes in its supercritical state, CO_2 deviates strongly from ideal-gas behavior and undergoes large changes in density and viscosity as the pressure increases through its critical region.

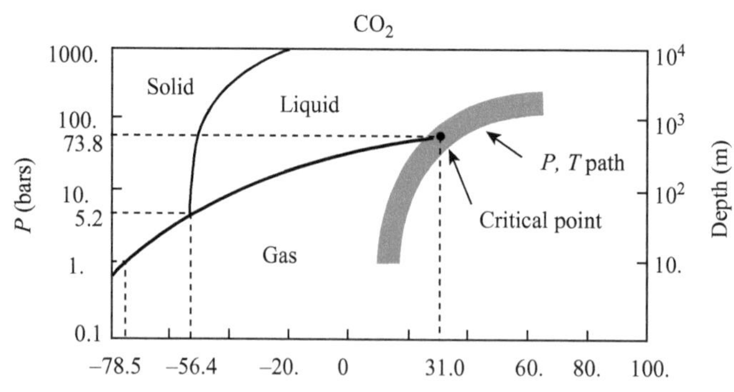

Figure 26.1. Phase diagram showing CO_2 will normally be supercritical in natural gas reservoirs (Reprinted with permission from Oldenburg, 2003. Copyright 2003 American Chemical Society)

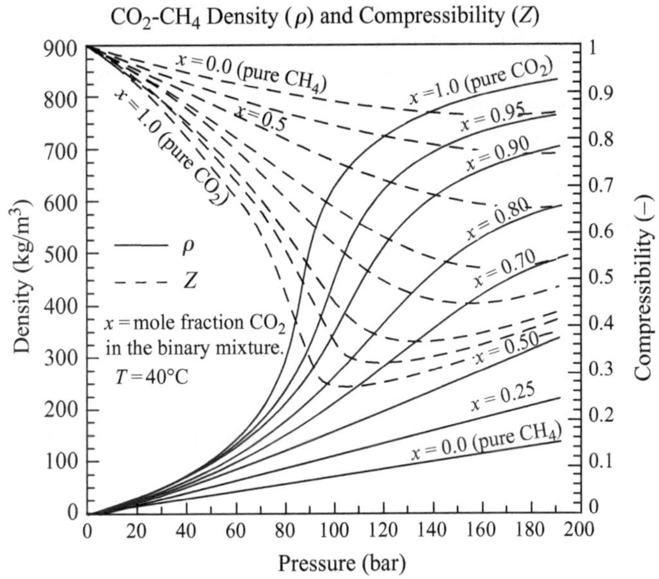

Figure 26.2. CO_2-CH_4 density and compressibility as function of pressure at 40°C (Source: NIST, 1992). (Reprinted with permission from Oldenburg, 2003. Copyright 2003 American Chemical Society)

Shown in Figure 26.2 are CO_2-CH_4 mixture density (ρ) and compressibility factor (Z) as a function of pressure at 40°C. As shown in the figure, the density of supercritical CO_2 approaches that of water at high pressures, while CH_4 induces more gas-like behavior in CO_2-CH_4 mixtures. Note in Figure 26.2, that the addition of only a small amount of CH_4 to supercritical CO_2 (e.g., at 100 bars) produces a large decrease in gas density which corresponds to an increase in pressure at constant volume. For example, consider the hypothetical mixing of one cubic meter of CO_2 with one cubic meter of CH_4 at 100 bar (10^7 Pa), 40°C. From the densities shown in Figure 26.2 ($\rho_{CH_4} = 70$ kg m^{-3}, $\rho_{CO_2} = 625$ kg m^{-3}) and the respective molecular weights, we can calculate the number of moles (n) of CH_4 (4375 moles) and CO_2 (1.420×10^4 moles) to find the long-term fully mixed mole fraction of CO_2 in the mixture ($x_{CO_2} = 0.765$). Estimating the Z factor from Figure 26.2 for a CO_2-CH_4 gas mixture with $x_{CO_2} = 0.765$ at 100 bar, 40°C, we find $Z = 0.61$. Using this Z factor as the initial guess in the relation

$$P = \frac{ZnRT}{V} \tag{26.1}$$

where the fixed volume (V) is 2 m^3, we calculate a pressure of 147 bar (1.47×10^7 Pa). Iterating repeatedly using the last-calculated pressure to find an updated Z from Figure 26.2, we arrive at an estimate of the final pressure equal to 123 bar (1.23×10^7 Pa) and $Z = 0.51$ for this CO_2-CH_4 mixture. This calculation demonstrates that the pressure in a CH_4 reservoir into which CO_2 is injected will tend to increase due to mixing, without any further injection whatsoever. Such pressure increases will effectively decrease the potential CO_2 storage capacity of the reservoir.

In addition to being denser than CH_4, CO_2 is also more viscous than CH_4 at all relevant pressures and temperatures. Therefore displacement of CH_4 by CO_2 will be a favorable mobility ratio displacement diminishing the tendency of the two gases to interfinger. The contrasting density and viscosity of CO_2 and CH_4 suggest that strategies can be devised to minimize CO_2-CH_4 mixing during CSEGR, for example, injecting CO_2 at large distances from production wells and at low levels in the reservoir to maximize recovery of existing CH_4 with minimal CO_2 mixing and breakthrough.

26.3 SIMULATION OF THE CSEGR PROCESS

Numerical simulations of CSEGR in a quarter five-spot domain have been carried out using the multicomponent and multiphase numerical simulation code TOUGH2 (Pruess et al., 1999) with the research module EOS7C for modeling flow and transport of real gas mixtures of CO_2, CH_4, and H_2O. Because of the relatively high pressures involved, we have used the Advective Diffusive Model for gas transport rather than Dusty Gas Model (e.g., Webb, 1998; Oldenburg et al., 2004). The properties of the system are presented in Table 26.1, with a schematic of the system shown in Figure 26.3. The domain was discretized into 23,120 gridblocks. All boundaries are closed, and the initial condition is constant pressure of 50 bars. Injection of CO_2 is specified in the front lower corner of the domain, while CH_4 production occurs simultaneously at the rear upper corner. The objective of the simulation is to demonstrate that CO_2 can be injected at low levels in a reservoir while CH_4 is produced from high levels in the reservoir with limited mixing while significant repressurization and associated enhanced gas recovery can be carried out. Further

Table 26.1. Properties of three-dimensional quarter five-spot domain

Property	Value	Units
Quarter five spot size	200 × 200 (~ 10 acres)	m^2
Thickness	50	m
Porosity	0.20	–
Permeability (isotropic)	1.0×10^{-12}	m^2
Residual liquid saturation	0.20	
Relative permeability		
liquid	Immobile	–
gas	Equal to gas saturation ($k_{rg} = S_g$)	–
Molecular diffusivity in gas, liquid	1.0×10^{-5}, 1.0×10^{-10}	m^2s^{-1}, m^2s^{-1}
Temperature	75	°C
Initial pressure	50	bars
CO_2 injection rate	0.10	$kg\ s^{-1}$
Gas production rate	0.009	$kg\ s^{-1}$
Final pressure (after 10 years)	71	bars

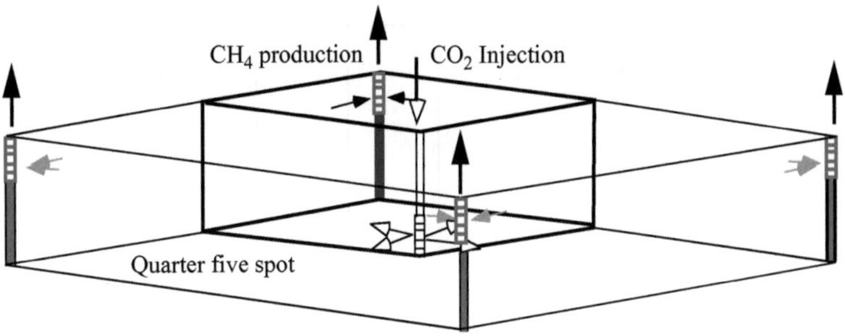

Figure 26.3. Schematic of 3-D five-spot domain

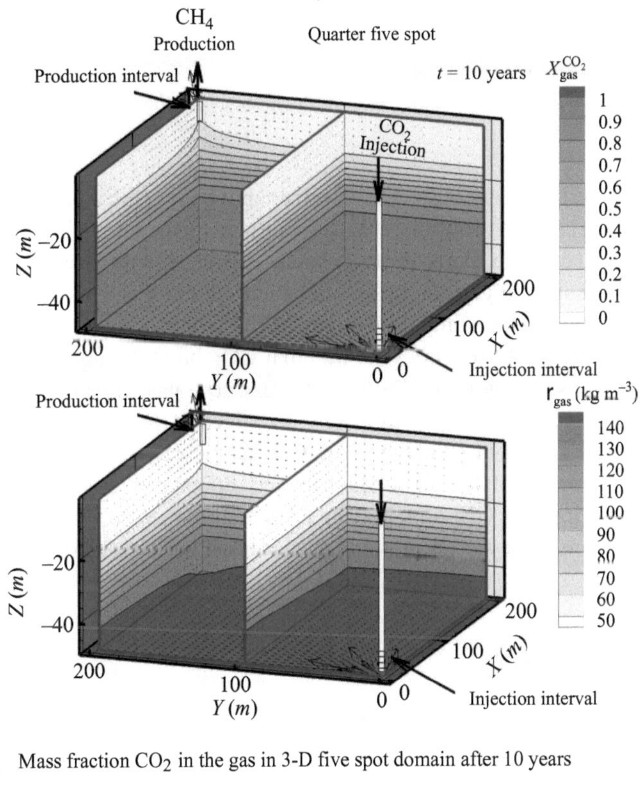

Figure 26.4. Mass fraction CO_2 in the gas in 3-D five spot domain after 10 years

details of this simulation along with discussion of the economics of CSEGR can be found in Oldenburg et al. (2004).

We present in Figure 26.4 the mass fraction of CO_2 in the gas and gas-phase density after 10 years of CSEGR. As shown, the injected CO_2 remains largely at the bottom while CH_4 fills the top of the reservoir. The injected CO_2 effectively sweeps from the bottom upwards delaying CO_2 mixing and breakthrough. The tendency for density

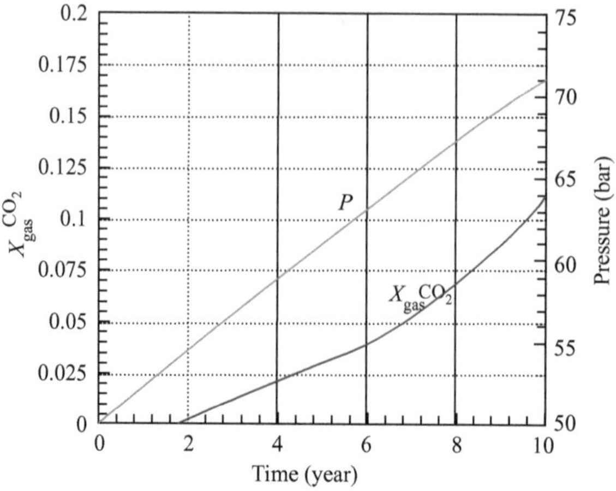

Figure 26.5. Mass fraction CO_2 in the gas and pressure vs. time at the production well

stratification will be even larger in a reservoir above the critical pressure where CO_2 density is much larger. Furthermore, a reservoir with lower permeability in the vertical direction might also be expected to enhance this stratification effect. Figure 26.5 shows the pressure and gas composition at the production well as a function of time for the simulation. This simulation suggests that enhanced gas recovery could be carried out for at least 10 years with minimal CO_2 breakthrough. If the economics would allow for separation of CO_2 from the produced gas, enhanced recovery could be carried on even longer. Regardless, CO_2 injection could continue for decades as the reservoir is shut in and filled with CO_2 as the reservoir is repressurized. We note that care must be taken to account for the ultimate long-term mixing that will occur between residual CH_4 that may be present and injected CO_2, in particular the associated effects on reservoir pressure as discussed in Section 26.2.

26.4 CONCLUSIONS

The large density and viscosity of supercritical CO_2 relative to CH_4 can be exploited to limit mixing in gas reservoirs and favor CSEGR. However, supercritical CO_2 properties are strongly dependent on composition, and self-pressurization can occur due solely to mixing when supercritical CO_2 is contaminated by CH_4. Three-dimensional simulations of CSEGR suggest that enhanced gas recovery can be carried out for approximately 10 years prior to CO_2 breakthrough. While simple models for diffusive and advective transport are adequate for high-permeability and high-pressure reservoirs, more sophisticated models such as the dusty gas model may be needed for modeling diffusive migration through low-permeability cap rocks and at low pressures near the ground surface (e.g., Webb, 1998; Oldenburg et al., 2004).

ACKNOWLEDGMENTS

This work was supported by the Assistant Secretary for Fossil Energy, Office of Coal and Power Systems through the National Energy Technology Laboratory, and by Lawrence Berkeley National Laboratory under Department of Energy Contract No. DE-AC03–76SF00098.

REFERENCES

Blok, K., Williams, R.H., Katofsky, R.E., and Hendricks, C.A., Hydrogen production from natural gas, sequestration of recovered CO_2 in depleted gas wells and enhanced gas recovery, *Energy*, 22, 161–168, 1997.

Buzek, F., Carbon isotope study of gas migration in underground gas storage reservoirs, Czechoslavakia, *Appl. Geochem.*, 7, 471–480, 1992.

Carriére, J.F., Fasanino, G., and Tek, M.R., Mixing in underground storage reservoirs, *Soc. Petrol. Eng. SPE 14202*, 1985.

Gerlach, T.M., Present-day CO_2 emissions from volcanoes, *EOS Trans. AGU*, 72, 249, 254–255, June 4, 1991.

Laille, J-P., Molinard, J.E., and Wents, A., Inert gas injection as part of the cushion of the underground storage of Saint-Clair-Sur-Epte, France, *Soc. Petrol. Eng. SPE 17740*, 343–352, 1988.

Mamora, D.D., and Seo, J.G., Enhanced gas recovery by carbon dioxide sequestration in depleted gas reservoirs, *Soc. Petrol. Engineers SPE-77347*, SPE Annual Technical Conference and Exhibition, San Antonio TX, 29 Sept. –2 Oct. 2002.

NIST (National Institute of Science and Technology), "NIST Database 14 Mixture.Property Database, version 9.08," U.S. Dept. of Commerce, Oct. 1992.

Oldenburg, C.M., Carbon dioxide as cushion gas for natural gas storage, *Energy Fuels*, 17, 240–246, 2003.

Oldenburg, C.M., Stevens, S.H., and Benson, S.M., Economic Feasibility of carbon sequestration with enhanced gas recovery (CSEGR), *Energy*, 29, 1413–1422, 2004.

Oldenburg, C.M., Pruess, K., and Benson, S.M., Process modeling of CO_2 injection into natural gas reservoirs for carbon sequestration and enhanced gas recovery, *Energy Fuels*, *15*, 293–298, 2001.

Oldenburg, C.M., Webb, S.W., Pruess, K., and Moridis, G.J., Mixing of stably stratified gases in subsurface reservoirs: A comparison of diffusion models, *Trans. Porous Med.*, 54, 323–334, 2004.

Pruess, K., Oldenburg, C.M., and Moridis, G.J., TOUGH2 User's Guide Version 2.0, Lawrence Berkeley National Laboratory Report *LBNL-43134*, 197 pp., Nov. 1999.

Reichle, D., Houghton, J., Kane, B., Ekmann, J., Benson, S., Clarke, J., Dahlman, R., Hendrey, G., Herzog, H., Hunter-Cevera, J., Jacobs, G., Judkins, R., Ogden, J., Palmisano, A., Socolow, R., Stringer, J., Surles, T., Wolsky, A., Woodward, N., and York, M., Carbon sequestration research and development 2000, U.S. Department of Energy, Washington, D.C. *DOE/SC/FE-1*, 1999.

van der Burgt, M.J., Cantle, J., and Boutkan, V.K., Carbon dioxide disposal from coal-based IGCC's in depleted gas fields, *Energy Convers. Mgmt.*, 33, 603–610, 1992.

Webb, S.W., Gas-phase diffusion in porous media – Evaluation of an advective-dispersive formulation and the dusty-gas model for binary mixtures, *J. Porous Media*, 1, 187–199, 1998.

CHAPTER 27

INDUSTRIAL GAS TRANSPORT PROCESSES INVOLVING HEAT TRANSFER

O.A. "GUS" PLUMB
University of Wyoming, Laramie, Wyoming, USA

27.1 INTRODUCTION

Gas transport in porous media accompanied by heat transfer occurs in a number of commonly utilized industrial processes. These include drying processes, packed bed heat and mass exchangers and thermal storage devices, and porous heat transfer enhancement technology. To remain consistent with the focus of the book this chapter focuses on those processes that do not involve liquid or two-phase flows. In what follows, the most commonly used models for predicting such processes are discussed along with theoretical and empirical models for the necessary transport parameters The focus is on two processes – drying and packed beds.

27.2 DRYING AT LOW LIQUID CONTENT

A large number of materials are dried for a variety of reasons including preservation, weight reduction to reduce shipping costs, and performance. Examples include food products, paper, wood products, pharmaceuticals, concrete, and ceramic materials. In addition the natural drying process is relevant to the earth's surface (water balance and irrigation schedules) and the natural life cycle of many plants.

Much of the focus in studying drying is typically on liquid phase transport since the bulk of the weight loss is liquid and preservation/performance constraints are often satisfied when the liquid is removed. Thus, for many drying problems of practical importance gas phase transport is not an important issue.

Gas phase transport is important in the freeze-drying process and in drying processes where removal of all or nearly all of the moisture is important. Good examples of the latter are the drying of pharmaceuticals and ceramic powders processed using liquid-phase precipitation.

C. Ho and S. Webb (eds.), Gas Transport in Porous Media, 427–438.
© 2006 *Springer.*

Gas phase transport of interest in drying can be classified as diffusive or convective. Diffusion processes are those driven by gradients in concentration (or partial pressure). In the case of drying, the gradient in water vapor concentration in air provides the driving force. For the case of convection, the transport is driven by gradients in total pressure. A complete model, in the absence of liquid phase transport, would include the gas phase continuity equation, vapor phase continuity equation, air phase continuity equation and the energy equation. These, together with Darcy's law, Fourier's law, Fick's law, and the necessary volumetric constraints, result in a complete model upon assuming ideal gas behavior and the Clausius-Clapeyron equation to establish vapor-liquid equilibrium conditions.

Some of the fundamental issues that must be resolved in order to predict gas phase transport by diffusion and convection during the drying process will be discussed in the next sections.

27.2.1 The Diffusion Coefficient for a Dry Porous Medium

The first issue that faces the engineer or scientist who wishes to predict diffusion in a porous medium is how to adjust the diffusion coefficient that is presumed to be known for the constituents of interest–water vapor in air for example. As a first estimate one is inclined to adjust the diffusion coefficient for the continuum by multiplying by the porosity of the material of interest. This is perfectly logical. Consider a simple geometry, such as a cube, with side of unit length and a hole of diameter, d, as a representative elementary volume of a porous medium. In this case the porosity is $\pi d^2/4$, precisely the ratio of area available for gas phase diffusion divided by the total area. In what follows we will explore the validity of this simple estimate for the diffusion coefficient and attempt to determine the conditions for which it might be valid.

The most obvious constraint on the validity is the ratio of the pore diameter, d, to the mean free path for the molecules that compose the system. For extremely small pore sizes or alternatively low pressures, we would assume that if the pore size is not much larger than the mean free path then collisions with the wall will significantly effect the diffusion process. This is termed Knudsen diffusion and at atmospheric pressure it is not an important effect for pore sizes greater than approximately 1 micron (Kaviany, 1991). Hence, Knudsen effects will not be considered in the ensuing discussion.

27.2.2 Theoretical-Volume Averaging to Determine the Diffusion Coefficient

The application of volume averaging, as first developed by Whitaker (1977) for the diffusion process, leads to a mathematical formalism through which to explore the validity of the simple first estimate presented above. Applying volume averaging to the diffusion equation as presented by Whitaker in Chapter 6 leads to

$$\phi_g \frac{\partial \langle c_g \rangle^g}{\partial t} = \nabla_g \left[D_g \left(\phi_g \nabla \langle c_g \rangle^g + \langle c_g \rangle^g \nabla \phi_g + \frac{1}{V} \int_{A_{gs}} \vec{n}_{gs} c_g \mathrm{d}A \right) \right] \tag{27.1}$$

where $\langle c_g \rangle^g$ is the phase-averaged concentration. This result contains two terms that imply deviation from the simple first order estimate of the effective diffusion coefficient as the product of the porosity and the molecular diffusion coefficient, D_g. The first (second term in the brackets on the RHS of the equation) is clearly the result of spatial variations in porosity. Hence, if the porosity is not uniform, which is the case for most real systems, then the accuracy of the simple first order estimate should be questioned. The second term that implies a deviation from the first order estimate (last term in brackets on the RHS of Eqn. (27.1)) is commonly referred to as the tortuosity effect. This accounts for the fact that in a real porous medium the path length for diffusion is greater than the shortest path between two points as in the case of a continuous fluid.

For the simple case of a straight cylindrical hole in a unit cell, it is clear that the porosity is constant and that the tortuosity effect is zero. However, if we add the next level of complexity by making the hole other than straight then it is clear that the tortuosity takes on a value other than zero.

The next question that must be addressed is how do we solve the volume averaged equation or how do we determine an appropriate "effective" diffusion coefficient that will allow us to collapse the volume averaged result to the "normal" diffusion equation?

There are two possible approaches to address this question. First, we note that the term resulting from the spatial variation in porosity lends itself to direct computation providing that the spatial variation in porosity is well defined for the system of interest. The tortuosity term is more difficult since it involves the point concentration as opposed to the volume averaged concentration. Thus, a closure scheme must be developed for this term.

The approach, originally developed by Crapiste et al. (1986), decomposed the point concentration into the sum of the volume average concentration and a deviation from the volume average

$$c_g = \langle c_g \rangle^g + \tilde{c}_g \tag{27.2}$$

A governing equation for the deviation, $\tilde{c}_g$, is then derived and closure obtained by developing order of magnitude arguments for the neglect of higher order terms. Upon the solution for $\tilde{c}_g$, the effective diffusion coefficient is determined and the volume averaged diffusion equation can be solved. The model for predicting $\tilde{c}_g$ as developed by Ryan et al. (1981) can be expressed

$$\frac{\partial \tilde{c}_g}{\partial t} = D_g \nabla^2 \tilde{c}_g \tag{27.3}$$

$$\begin{aligned} -\vec{n}_{gs} \cdot D_g \nabla \tilde{c}_g &= \vec{n}_{gs} D_g \nabla \langle c_g \rangle^g \\ \tilde{c}_g (r + l_i) &= \tilde{c}_g (r) \\ \tilde{c}_g &= h(r)\,; t = 0 \end{aligned} \tag{27.4}$$

where the second equation of (27.4) expresses the assumption that the porous medium is spatially periodic and the last provides an initial condition. Results of predictions using this model based on volume averaging will be discussed in the next section.

27.2.3 Empirical Results for the Diffusion Coefficient as a Function of Structure and Comparisons with the Volume Averaged Theory

A summary of experimental and theoretical results is presented in Figure 27.1. The theoretical results of Ryan et al. (1981) are the result of predictions based on the closure theory summarized in the previous section. The calculations are for two-dimensional arrays of squares covering a range of porosities and arranged in both in-line and staggered arrays. Results are compared with experiments involving packed beds of spheres (Kim et al., 1987) and a variety of unconsolidated porous systems (Hoogschagen, 1955; Currie, 1960). In addition theories developed by Maxwell (1881), Weissberg (1963), and Wakao and Smith (1962) are included in the comparison. As can be seen in the figure, the volume averaged theory is in good agreement with the available experimental data in the range appropriate for unconsolidated porous systems including porosities from 0.2 to 0.50. Less data is available for consolidated porous systems for which both much larger and much smaller porosities are possible.

In the soils literature the tortuosity coefficient is generally utilized resulting in a model for the effective diffusion coefficient given by

$$D_{\text{eff}} = \tau \cdot \varepsilon D_g \tag{27.5}$$

For porosities typically in the range of 0.3 to 0.4 for unconsolidated materials, the tortuosity is assumed to range from 0.6 to 0.8. Recent measurements by

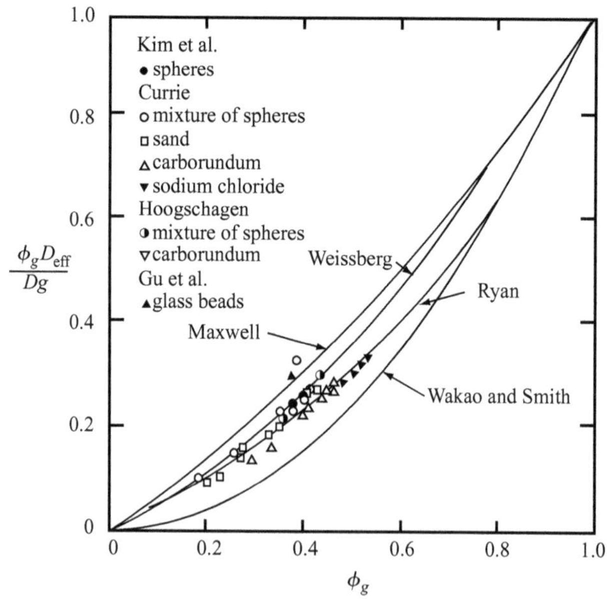

Figure 27.1. Experimental measurements and theoretical predictions for the diffusion coefficient in porous meda (adapted form Plumb and Whitaker, 1990, with permission)

Gu et al. (1998), for a packed bed of 450 micron glass spheres having a porosity of 0.37, resulted in a tortuosity coefficient of 0.78. This model is superimposed on the data of Figure 27.1. With the tortuosity taken to be 0.78 the data point falls above the other data in the figure as well as the Ryan result. However, if the tortuosity is taken to be 0.6 then the theory and previous experimental results are all in reasonable agreement.

27.2.4 The Effect of the Presence of the Liquid Phase on the Diffusion Process

During the process of drying from the saturated condition where 100% of the pore space is occupied by liquid, gas phase transport is typically not an important issue until the liquid phase saturation reaches the pendular state beyond which it is assumed to be immobile. Thus, the remaining liquid must be removed via a process involving evaporation at the liquid-gas interfaces and then diffusion to the drying surface. This, of course, assumes that the transport of liquid as bound or adsorbed liquid as in the case of hygroscopic or hydrophilic materials does not contribute significantly to the process.

In the case of immobile water the first approximation might lead to adjustments to the porosity and the tortuosity to account for the presence of the liquid islands. However, this has been shown to lead to erroneous results (Webb and Ho, 1998). Note that the Soret effect will normally exist during drying processes because of the presence of temperature gradients. Unfortunately the Soret effect leads to diffusion from regions of high temperature to regions of low temperature which is not the desired result for most practical drying processes. In addition, for very small pores, the Kelvin effect originating from the effect of the radius of curvature of the liquid-gas interface on the vapor pressure must be considered. For most processes of practical interest Soret diffusion and the Kelvin effect are second order. The Soret effect takes on significance when temperature gradients are large compared to concentration gradients and the Kelvin effect when pores are very small.

It has been found that, in the presence of the liquid phase, the diffusion of vapor is enhanced (Philip and deVries, 1957; Jackson, 1964; Cary, 1965; Jury and Letey, 1979; Cass et al., 1984; Gu et al., 1998). Thus, the effective diffusion coefficient as estimated through correcting the porosity (set equal to the gas phase volume fraction) and the tortuosity for the presence of the immobile liquid phase does not lead to results that agree with experiment. Since enhanced vapor diffusion is discussed in detail by Ho in Chapter 3 it will not be discussed here. It will suffice to say that the effective diffusion coefficient, estimated as discussed above, must be corrected by a factor of as high as 4.0 based on the experimental evidence that is available (Webb and Ho, 1998).

27.2.5 Permeability for a Dry Porous Medium

As outlined in the previous section, in order to predict gas phase transport by diffusion during the drying process one needs to first determine the effective diffusion coefficient in order to apply the diffusion equation. In order to predict gas phase transport due to convection in a dry porous medium one needs the permeability in order to apply Darcy's law or Darcy's law corrected to include the inertial effects using the so called Forchheimer extension to Darcy's law. In the next section a brief summary of theoretical and experimental results is presented.

27.2.6 Theory and Empirical Results for a Dry Porous Medium

A possible starting point that parallels the development presented for the effective diffusion coefficient in Section 27.2.2 for the theoretical development for the prediction of the permeability is volume averaging the Navier-Stokes equations. This is more challenging than that for the diffusion equation.

As the measurement of permeability is straightforward there is considerable data available for a number of porous systems. However, a general predictive theory based on the geometric properties of the material is not available. The most commonly used empirical result is the Carman-Kozeny relationship

$$K = \frac{d_p^3 \varepsilon^3}{180(1-\varepsilon)^2} \tag{27.6}$$

This relationship appears to work well for unconsolidated media with particles that are approximately spherical and of a relatively uniform size distribution. An extensive review of the relationship between permeability and geometry can be found in Dullien (1979).

27.2.7 The Effect of the Presence of the Liquid Phase on Convection, Relative Permeability

For a dry porous medium the permeability is a function of geometry alone, but for drying we expect small amounts of immobile water to be present until all of the liquid is removed. In the presence of water, the permeability is certainly, in the first approximation, dependent on the amount of water present. It might also be expected to be a function of the viscosity of the immobile liquid as well as the pressure gradient.

The simplest models for gas phase permeability in the presence of liquid or "relative permeability" assume that it depends only on the liquid saturation. Empirical models have been developed by Fatt and Klikoff (1959), van Genuchten (1980), and Verma et al. (1985) among others. Available models are reviewed in Kaviany (1991). A commonly used model is that of Fatt and Klikoff (1959)

$$k_{rg} = (1 - S^*)^3, \quad k_{rl} = S^{*3} \tag{27.7}$$

where

$$S^* = \frac{S - S_{ir}}{1 - S_{ir}} \tag{27.8}$$

We note that for liquid saturation below the irreducible saturation this model predicts a gas phase relative permeability of unity. The irreducible saturation typically falls in the range of from 10 to 15% for unconsolidated systems having a narrow particle size distribution.

27.2.8 The Effective Thermal Conductivity

The processes discussed in this chapter all include heat transfer, thus, the determination of the effective thermal conductivity for a porous medium is also of interest. Determination of the effective thermal conductivity can be approached using volume averaging as discussed above for the effective diffusion coefficient. This has been done by Nozad et al. (1985), obtaining results in good agreement with the available experimental data.

A detailed review of the various models for predicting the effective thermal conductivity for porous materials can be found in Cheng and Hsu (1999). Many investigators have used a simple parallel thermal resistance model resulting in

$$\lambda_{\text{eff}} = \lambda_f \phi + \lambda_s (1 - \phi) \tag{27.9}$$

Cheng and Hsu illustrate that for unconsolidated porous media having a porosity in the range of from 0.3 to 0.4 this model will most likely overpredict the effective thermal conductivity if the thermal conductivity of the solid is greater than the thermal conductivity of the fluid and underpredict the effective thermal conductivity in the unlikely case that the thermal conductivity of the solid is less than the thermal conductivity of the fluid.

The greatest challenge in making theoretical predictions or developing models for unconsolidated systems is estimating the contact resistance at the particle-to-particle contacts. The various assumptions used in estimating contact resistance or effective contact area are discussed by Cheng and Hsu.

Cheng and Hsu illustrate that lumped parameter models that are based on assuming a spatially periodic system and then a geometry for a unit cell yield reasonable results for dry porous media. In developing such models the unit cell is divided into layers parallel to the direction of interest. The effective thermal conductivity is then constructed by determining the effective thermal resistance of the layers in parallel with each layer analyzed as a series of fluid and solid phases. They illustrate that for various two dimensional geometries their results are in good agreement with the Nozad et al., theory and the experimental data when the ratio of contact area to total solid cross sectional area is taken to be 0.01. It is interesting to note that the Nozad et al., theory also assumes the same area ratio to be 0.01 in modeling the effects of the particle to particle contact.

Available experimental results including those discussed by Somerton et al. (1974), and Plumb (1991) for unconsolidated porous systems partially saturated with liquid indicate that small amounts of liquid have a large effect on the effective thermal conductivity. This is presumably a result of the contact resistance increasing markedly as a result of the presence of pendular liquid at the contact points. The result given by Somerton is written

$$k_{\text{eff}} = k^0_{\text{eff}} + S^{*1/2}(k_l - k^0_{\text{eff}}) \tag{27.10}$$

where k^0_{eff} is the effective thermal conductivity for the dry porous medium.

27.3 HEAT AND MASS TRANSFER IN PACKED BEDS

27.3.1 The Local Thermal Equilibrium Assumption

In the previous discussion of drying it has been assumed that local thermal equilibrium can be assumed. This implies that the process is such that the solid and gas phases are locally at the same temperature. The thermal equilibrium assumption is often not appropriate when analyzing heat and mass transfer in packed beds used routinely in a number of industrial processes. These include drying of particulate materials, drying of gas streams by passing through a desiccant bed, desiccant bed based dehumidification and air conditioning systems, absorbent beds for gas stream clean-up, and regenerative heat exchangers and heat storage devices. In addition, recent interest in micro heat exchangers has resulted in the application of heat transfer models developed for packed beds because of the obvious analogy. The use of porous surfaces to enhance heat transfer is also an application where the local thermal equilibrium assumption may not be appropriate (Huang and Vafai, 1994).

The general equations for heat transfer for a two-component system can be written (Nield and Bejan, 1992)

$$(1-\phi)(\rho c_p)_s \frac{\partial T_s}{\partial t} = k_{\text{eff}_s} \nabla^2 T_s - h(T_s - T_g) \tag{27.11}$$

$$\phi(\rho c_p)_g \frac{\partial T_g}{\partial t} + (\rho c_p)_g \vec{v} \cdot \nabla T_g = k_{\text{eff}_g} \nabla^2 T_g - h(T_g - T_s) \tag{27.12}$$

where k_{eff_s} and k_{eff_g} are effective thermal conductivities for the solid and gas phases, respectively. They are commonly approximated as the volumetrically weighted values, $k_{\text{eff}_s} = (1-\phi)k_s$ and $k_{\text{eff}_g} = \phi k_g$. The new parameter that remains to be determined is h, the heat transfer coefficient between the solid and gas phases. The above is commonly referred to as the two-equation model. Significant simplification results if local thermal equilibrium can be assumed in which case the solid and fluid phases can be treated as being at the same (local volume averaged) temperature.

In the case of heat transfer the first question that must be addressed is whether or not the bed can be assumed to be in local thermal equilibrium. That is, locally, can the gas phase and the solid phase be assumed to be at the same temperature? This question has been dealt with in detail by Whitaker (1986, 1991) and Quintard and Whitaker (1995), who demonstrate that when the convective time scale is long and the particles are small compared to the length scale associated with changes in the volume averaged temperature then it is safe to make the local thermal equilibrium assumption.

In dimensionless terms this implies, at the minimum, large Fourier number for the particles and a small Peclet number based on the pore length scale. Further analyses of the local thermal equilibrium assumption have been completed by Amiri and Vafai (1994), Nield and Kuznetsov (1999), and Lee and Vafai (1999). More recently Kim and Jang (2002), have proposed that the condition for local thermal equilibrium can be written more conveniently for steady-state engineering applications as

$$\mathrm{Pr}_{\mathrm{eff},g}\ \mathrm{Re}_{d_p} Da^{1/2} \frac{\varepsilon}{\mathrm{Nu}} \leq 1 \qquad (27.13)$$

where $\mathrm{Pr}_{\mathrm{eff},g}$, Re_{d_p}, Da, and Nu are the effective fluid Prandtl number, pore or particle Reynolds number, Darcy number, and Nusselt number, respectively.

For cases for which the constraints associated with thermal equilibrium are satisfied Equations (27.11) and (27.12) reduce to a single equation for the local temperature. This is commonly referred to as the Schuman model (Schuman, 1929) and numerous solutions have been presented for various initial and boundary conditions.

27.3.2 Non-Equilibrium System Models and the Internal Heat Transfer Coefficient

For those cases where local thermal (or mass) equilibrium cannot be assumed then it is necessary to solve the two-equation model given by Equations (27.11) and (27.12). In this case the internal heat (or mass) transfer coefficient between the solid and gas phases must be specified. There are several correlations that have been used by different authors in developing numerical solutions. These are summarized by Kaviany (1991). The most commonly used correlation is that of Wakao et al. (1979)

$$\mathrm{Nu} = \frac{hd_p}{k_g} = 2 + 1.1\mathrm{Pr}^{1/3}\mathrm{Re}_{d_p}^{0.6} \qquad (27.14)$$

where Nu, Pr, and Re_{d_p} are the particle Nusselt number, the Prandtl number and the particle or pore Reynolds number, respectively.

27.3.3 Dispersion

The final effect that must be considered in predicting heat (and/or mass) transfer in packed beds is dispersion. If the experimental data is examined [see, e.g., Plumb and Whitaker, 1990] it is observed that for Peclet number based on the particle or pore size greater than 1 the effective diffusion coefficient for a packed bed begins to deviate from the molecular diffusion coefficient. This is the result of fluctuations in velocity in the gas phase. Such fluctuations, at low Peclet number are the result of the tortuous path taken by the fluid and at higher Peclet number the result of turbulence. This problem has been examined theoretically by Eidsath et al. (1983), using the volume averaging approach. Their calculations resulted in

$$\frac{D_x^*}{D_g} \propto Pe_p^{1.7} \qquad (27.15)$$

for the longitudinal dispersion coefficient for a two-dimensional cylindrical array. The experimental data of Gunn and Pryce (1969), for a randomly packed bed of spheres results in

$$\frac{D_x^*}{D_g} \propto Pe_p^{1.2} \tag{27.16}$$

for the longitudinal dispersion coefficient.

In studies of heat transfer the lateral dispersion coefficient is often of more interest. The Eidsath et al. theoretical study under predicts the available experimental data for lateral dispersion. It was speculated that this was the result of heterogeneity in randomly packed beds leading to a theoretical study (Plumb and Whitaker, 1988) incorporating the effect of heterogeneity. In this case the theory over predicted the experimental results but provided a reasonable dependence on the Peclet number

$$\frac{D_y^*}{D_g} \propto Pe_p^{1.22} \tag{27.17}$$

Hsu and Cheng (1990) also developed a theory for lateral thermal dispersion. Their results lead to a linear relationship between the dispersion coefficient and Peclet number at high pore Reynolds number and a quadratic relationship at low pore Reynolds number. The linear relationship results in reasonable agreement with the available experimental data.

One of the greatest challenges in dealing with gas phase transport during the drying of real systems is the effect of heterogeneity if we consider some of the materials that are commonly dried including food products, wood, concrete (during the curing process), and soil at the earth's surface. In addition, many materials commonly dried are particulate materials (wood particles or fibers, pharmaceuticals, grain, etc.) that could potentially be dried in a packed bed, a good example of a non-homogeneous system.

27.4 SUMMARY

In this chapter the modeling tools commonly utilized in modeling processes that involve gas transport in porous media with heat and mass transfer are summarized. Processes considered include drying of porous materials at low moisture content and heat transfer in packed beds.

In the case of drying the focus is on the prediction of an effective diffusion coefficient that depends on porosity, tortuosity, and the presence of liquid. Relative permeability and the effective thermal conductivity are also considered.

In analyzing packed beds it is important to consider the validity of the local thermal equilibrium assumption. For those situations where the local thermal equilibrium assumption is not appropriate a method for estimating the internal heat transfer coefficient is necessary. Dispersion, both lateral and longitudinal can also play a role when the Peclet number based on the particle size is significantly greater than unity.

REFERENCES

Amiri, A. and Vafai, K., Analysis of dispersion effects and non-thermal equilibrium, non-Darcian, variable porosity incompressible flow through porous media, *Int. J. Heat Mass Transfer,* 30, 939–954, 1994.

Cary, J.W., Water flux in moist soil: thermal versus suction gradients, *Soil Science*, 100, 168–175, 1965.

Cass, A., Campbell, G.S., and Jones, T.L., Enhancement of thermal water vapor diffusion in soil, *Soil Sci. Soc. Am. J.,* 48, 25–32, 1984.

Cheng, P. and Hsu, C.T., The effective stagnant thermal conductivity of porous media with periodic structures, *J. Porous Media*, 2, 19–38, 1999.

Currie, J.A., Gaseous diffusion in porous media I: A non-steady state method, *Br. J. Appl. Phys.,* II, 314–324, 1960.

Crapiste, G.H., Rotstein, E., and Whitaker, S., A general closure scheme for the method of volume averaging, *Chem. Eng. Sci.*, 41, 227–235, 1986.

Dullien, F.A.L., *Porous Media: Fluid Transport and Pore Structure*, Academic, NY, 1979.

Eidsath, A., Carbonell, R.G., Whitaker, S., and Hermann, L.R., Dispersion in pulsed systems – III. Comparison between theory and experiments for packed beds, *Chem. Eng. Sci.,* 38, 1803–1816, 1983.

Gu, L., Ho, C.K., Plumb, O.A., and Webb, S.W., Diffusion with condensation and evaporation in porous media, *Proc. Of the 7th AIAA/ASME Thermophysics and Heat Transfer Conference,* ASME HTD-Vol. 357–2, 213–220, 1998.

Gunn, D.J. and Pryce, C., Dispersion in packed beds, *Trans. Inst. Chem. Eng.,* 47, 341–350, 1969.

Hoogschagen, J., Diffusion in porous catalysts and absorbents, *Ind. Eng. Chem.*, 47, 906–916, 1955.

Hsu, C.T. and Cheng, P., Thermal dispersion in porous medium, *Int. J. Heat Mass Transfer,* 33, 1587–1597, 1990.

Huang, P.C. and Vafai, K. Analysis of forced convection enhancement in a channel using porous blocks, *J. Thermophys. Heat Transfer*, 8, 563–573, 1994.

Jackson, R.D., Water vapor diffusion in relatively dry soil: I. Theoretical considerations and sorption experiments, *Soil Sci. Soc. Proc.,* 28, 172–176. 1964.

Jury, W.A. and Letey, J. Jr., Water vapor movement in soil: Reconciliation of theory and experiment, *Soil Sci. Soc. Am. J.,* 43, 823–827, 1979.

Kaviany, M., *Principles of Heat Transfer in Porous Media,* Springer-Verlag, NY, 1991

Kim, J.H., Ochoa, J.A., and Whitaker, S., Diffusion in anisotropic porous media, *Transport in Porous Media*, 2, 327–356, 1987.

Kim, S.J. and Jang, P.J., Effects of the Darcy number, the Prandtl number, and the Reynolds number on local thermal non equilibrium, *Int. J. Heat Mass Transfer,* 45, 3885–3896, 2002.

Lee, D.Y. and Vafai, K., Analytical characterization and conceptual assessment of solid and fluid temperature differentials in porous media, *Int. J. Heat Mass Transfer,* 42, 423–435, 1999.

Maxwell, J.C., *Treatise on Electricity and Magnetism*, Vol. I, Clarendon Press, Oxford, 1881.

Nield, D.A. and Kuznetsov, A.V., Local thermal nonequilibrium effects in forced convection in a porous medium channel: a conjugate problem, *Int. J. Heat Mass Transfer,* 42, 3245–3252, 1999.

Nield, D.A. and Bejan, A., *Convection in Porous Media*, Springer-Verlag, NY, 1992.

Nozad, S., Carbonell, R.G., and Whitaker, S., Heat conduction in multiphase systems. I: Theory and experiments for two-phase systems, *Chem. Eng. Sci.,* 40, 843–855, 1985.

Philip, J.R. and deVries, D.A., Moisture movement in porous materials under temperature gradients, *Amer. Geophys. Union Trans.,* 38, 222–232, 1957.

Plumb, O.A., Heat transfer during unsaturated flow in porous media, in *Convective Heat and Mass Transfer in Porous Media*, ed. Kakac, S., Kilkis, B., Kulacki, F.A., and Arinc, F., NATO ASI Series E, 196, Dordrecht: Kluwer, 435–464, 1991.

Plumb, O.A. and Whitaker, S., Dispersion in heterogeneous porous media II: Predictions for stratified and two-dimensional spatially periodic systems, *Water Resour. Res.,* 24, 927–938, 1988.

Plumb, O.A. and Whitaker, S., Diffusion, adsorption and dispersion in porous media: Small-scale averaging and local volume averaging, in *Dynamics of Fluids in Hierarchical Porous Media*, ed. J.H. Cushman, Academic Press, San Diego, 97–148, 1990.

Quintard, M. and Whitaker, S., Local thermal equilibrium for transient heat conduction: theory and comparison with numerical experiments, *Int. J. Heat Mass Transfer,* 38, 2779–2796, 1995.

Ryan, D., Carbonell, R.G. and Whitaker, S., *A Theory of Diffusion and Reaction in Porous Media*, American Institute of Chemical Engineers Symposium Series, No. 202, 71, 46–62, 1981.

Schuman, T.E., Heat transfer: a liquid flowing through a porous piston, *J. Franklin Institute*, 28, 405–416, 1929.

Somerton, W.H., Keese, J.A., and Chu, S.C., Thermal behaviour of unconsolidated oil sands, *SPE J.*, 14, 513–521, 1974.

Wakao, N., Kaguei, S., and Funazkri, T., Effect of fluid dispersion coefficients on particle-to-fluid heat transfer coefficients in packed beds, *Chem. Eng. Sci.*, 34, 325–336, 1979.

Webb, S.W. and Ho, C.K., *Enhanced Vapor-Diffusion in Porous Media*, Sandia Report: SAND98–2722, Sandia National Laboratories, Albuquerque, NM, 1998.

Weissberg, H.L., Effective diffusion coefficients in porous media, *J. Appl. Phys.*, 34, 2636–2639, 1963.

Whitaker, S., Simultaneous heat, mass and momentum transfer in porous media: a theory of drying. *Adv. Heat Transfer,* 13,119–200, 1977.

Whitaker, S., Local thermal equilibrium: An application to packed bed catalytic reactor design, *Chem. Eng. Sci.,* 41, 2029–2039, 1986.

Whitaker, S., Improved constraints for the principle of local thermal equilibrium, *Ind. Eng. Chem. Res.,* 30, 983–997, 1991.

INDEX

Theory and Applications of Transport in Porous Media

Series Editor:
Jacob Bear, *Technion — Israel Institute of Technology, Haifa, Israel*

---

1. H.I. Ene and D. Polisševski: *Thermal Flow in Porous Media*. 1987 ISBN 90-277-2225-0
2. J. Bear and A. Verruijt: *Modeling Groundwater Flow and Pollution*. With Computer Programs for Sample Cases. 1987 ISBN 1-55608-014-X; Pb 1-55608-015-8
3. G.I. Barenblatt, V.M. Entov and V.M. Ryzhik: *Theory of Fluid Flows Through Natural Rocks*. 1990 ISBN 0-7923-0167-6
4. J. Bear and Y. Bachmat: *Introduction to Modeling of Transport Phenomena in Porous Media*. 1990 ISBN 0-7923-0557-4; Pb (1991) 0-7923-1106-X
5. J. Bear and J-M. Buchlin (eds.): *Modelling and Applications of Transport Phenomena in Porous Media*. 1991 ISBN 0-7923-1443-3
6. Ne-Zheng Sun: *Inverse Problems in Groundwater Modeling*. 1994 ISBN 0-7923-2987-2
7. A. Verruijt: *Computational Geomechanics*. 1995 ISBN 0-7923-3407-8
8. V.N. Nikolaevskiy: *Geomechanics and Fluidodynamics*. With Applications to Reservoir Engineering. 1996 ISBN 0-7923-3793-X
9. V.I. Selyakov and V.V. Kadet: *Percolation Models for Transport in Porous Media*. With Applications to Reservoir Engineering. 1996 ISBN 0-7923-4322-0
10. J.H. Cushman: *The Physics of Fluids in Hierarchical Porous Media: Angstroms to Miles*. 1997 ISBN 0-7923-4742-0
11. J.M. Crolet and M. El Hatri (eds.): *Recent Advances in Problems of Flow and Transport in Porous Media*. 1998 ISBN 0-7923-4938-5
12. K.C. Khilar and H.S. Fogler: *Migration of Fines in Porous Media*. 1998 ISBN 0-7923-5284-X
13. S. Whitaker: *The Method of Volume Averaging*. 1999 ISBN 0-7923-5486-9
14. J. Bear, A.H.-D. Cheng, S. Sorek, D. Ouazar and I. Herrera (eds.): *Seawater Intrusion in Coastal Aquifers. Concepts, Methods and Practices*. 1999 ISBN 0-7923-5573-3
15. P.M. Adler and J.-F. Thovert: *Fractures and Fracture Networks*. 1999 ISBN 0-7923-5647-0
16. M. Panfilov: *Macroscale Models of Flow Through Highly Heterogeneous Porous Media*. 2000 ISBN 0-7923-6176-8
17. J.M. Crolet (ed.): *Computational Methods for Flow and Transport in Porous Media*. 2000 ISBN 0-7923-6263-2

18. R. de Boer: *Trends in Continuum Mechanics of Porous Media.* 2005
ISBN 1-4020-3143-2

19. R.J. Schotting, H.(C.J.) van Duijn and A. Verruijt (eds.): *Soil Mechanics and Transport in Porous Media.* Selected Works of G. de Josselin de Jong. 2006
ISBN 1-4020-3536-5

20. C. Ho and S. Webb (eds.): *Gas Transport in Porous Media.* 2006
ISBN 1-4020-3961-1

MIX
Papier aus verantwortungsvollen Quellen
Paper from responsible sources
FSC® C105338

If you have any concerns about our products,
you can contact us on
ProductSafety@springernature.com

In case Publisher is established outside the EU,
the EU authorized representative is:
Springer Nature Customer Service Center GmbH
Europaplatz 3, 69115 Heidelberg, Germany

Printed by Libri Plureos GmbH
in Hamburg, Germany